BIOASTRONOMY 2007: MOLECULES, MICROBES, AND EXTRATERRESTRIAL LIFE

COVER ILLUSTRATION:

A montage of images representing how bioastronomy integrates astronomy and biology to understand the origin and evolution of life in our solar system, and of living systems in the universe.

Montage courtesy of the Institute for Astronomy, University of Hawaii. HST image of cometary knots in the Helix nebula by Robert O'Dell and Kerry Handron (Rice University, NASA). Fish photograph courtesy Mark S. Mohlmann.

ASTRONOMICAL SOCIETY OF THE PACIFIC
CONFERENCE SERIES

A SERIES OF BOOKS ON RECENT DEVELOPMENTS IN ASTRONOMY AND ASTROPHYSICS

Volume 420

MS 179, Utah Valley University, 800 W. University Parkway, Orem, Utah 84058-5999
Phone: 801-863-8804 E-mail: aspcs@aspbooks.org
E-book site: http://www.aspbooks.org

ASPCS volumes may be found online with color images at http://www.aspbooks.org.
ASP Monographs may be found online at http://www.aspmonographs.org.

For a complete list of ASPCS Volumes, ASP Monographs, and other ASP publications see http://www.astrosociety.org/pubs.html.

All book order and subscription inquiries should be directed to the ASP at 800-335-2626 (toll-free within the USA) or 415-337-2126, or email service@astrosociety.org

ASTRONOMICAL SOCIETY OF THE PACIFIC
CONFERENCE SERIES

Volume 420

BIOASTRONOMY 2007: MOLECULES, MICROBES, AND EXTRATERRESTRIAL LIFE

Proceedings of a workshop held at
San Juan, Puerto Rico
16–20 July 2007

Edited by

Karen J. Meech
Institute for Astronomy, University of Hawaii, Hawaii, USA and NASA Astrobiology Institute

Jacqueline V. Keane
Institute for Astronomy, University of Hawaii, Hawaii, USA and NASA Astrobiology Institute

Michael J. Mumma
NASA Goddard Space Flight Center, Maryland, USA and NASA Astrobiology Institute

Janet L. Siefert
Rice University, Houston, Texas, USA and NASA Astrobiology Institute

Dan J. Werthimer
University of California Berkeley, Berkeley, California, USA

SAN FRANCISCO

ASTRONOMICAL SOCIETY OF THE PACIFIC

390 Ashton Avenue
San Francisco, California, 94112-1722, USA

Phone: 415-337-1100
Fax: 415-337-5205
E-mail: service@astrosociety.org
Web site: www.astrosociety.org
E-books: www.aspbooks.org

First Edition

ASP Conference Series

ISBN: 978-1-58381-720-9
e-book ISBN: 978-1-58381-721-6

Library of Congress (LOC) Cataloging in Publication (CIP) Data:
Main entry under title
Library of Congress Control Number (LCCN): 2009941225

Printed in the United States of America by Sheridan Books, Ann Arbor, Michigan.
This book is printed on acid-free paper.

Contents

Part 7. Origins of Viruses and Cellular Function

Part 8. Life in Extreme Environments

By David Morrison

Preface

It is profoundly human to be curious about our origins and our future. Through history, people have speculated about these questions and often sought answers in divine revelation. But in the past few decades we have begun to subject such questions to the rigors of scientific investigation. And with the realization that our planet is one among many, we can also begin to investigate whether we are alone. These fundamental themes concerning the origin of life, its distribution in the universe, and its ultimate fate, define the still-new discipline of astrobiology. In the mid-twentieth century, scientific interest the question of lifes origin was inspired by the laboratory synthesis of complex organic molecules by Miller and Urey, stimulating a new discipline of prebiotic chemistry. Soon astronomers were also detecting a remarkable array of complex molecules in interstellar space. The first deep space probes focused attention on questions of planetary habitability, culminating in the Viking missions to Mars, which searched directly for evidence of microbial life in the martian soil. At the same time a few visionary individuals recognized that the tools of radio astronomy offered the possibility to leapfrog the study of prebiotic chemistry and microbial ecosystems to search directly for transmissions by intelligent life on distant, unseen worlds. These complementary approaches have been pursued in parallel by different scientific communities: chemists researching the conditions under which life began on Earth, geologists tracing lifes history back to the first billon years of Earth history, geneticists mapping the evolutionary relationships among the life-forms on our planet, planetary scientists investigating the habitability of other worlds, and a few of us searching directly for evidence of intelligent life beyond our solar system. To these we can now add a new capability to study extrasolar planetary systems and plan the search for biosignatures on exoplanets.

Weaving these threads together to enable interdisciplinary understanding, and eventually collaboration, is a tremendous challenge. One of the most successful efforts to achieve this synthesis has been the meetings sponsored by the Bioastronomy Commission of the International Astronomical Union. From its initial narrow focus on SETI, the leaders of IAU Commission 51 have broadened their perspective and sought common ground with the biologists who study the origin of life and with the geoscientists who trace its evolution on Earth. This meeting in Puerto Rico has brought together an impressive breadth of astrobiologists. We are meeting in a beautiful environment that links a rich subtropical ecosystem with the technology, represented by the Arecibo telescope, that might reveal the presence of intelligence on a distant world. It combines the practical with the visionary, the study of our past with a hope for our future. This is what astrobiology should be a field that explores not only some of the most fundamental questions we can ask, but also pioneers ways to bring together different scientific disciplines to address our common goals.

Participants

SAM ABBAS, Chemistry Department, El Paso Community College, ⟨ dr.abbaschem@yahoo.com ⟩

JOSE ALONSO, Arecibo Observatory, ⟨ jalonso@naic.edu ⟩

DANIEL ALTSCHULER, Arecibo Observatory, ⟨ daniel@naic.edu ⟩

RICARDO AMILS, Centro de Astrobiologa, Madrid, ⟨ ramils@cbm.uam.es ⟩

DAVID ANDERSON, University of Keele, ⟨ dra@astro.keele.ac.uk ⟩

MARINA RESENDES DE SOUSA, Indiana University, ⟨ Antoniomantonio@indiana.edu ⟩

LEIGH ARINO DE LA RUBIA, Tennessee State University, ⟨ leigh.arinodelarubia@gmail.com ⟩

DIMITRA ATRI, University of Kansas, ⟨ dimitra@ku.edu ⟩

JOHN AYALA, University of Puerto Rico, ⟨ jalonso@naic.edu ⟩

HECTOR AYALA-DEL-RIO, University of Puerto Rico, ⟨ hlayala@hpcf.upr.edu ⟩

THOMAS BALONEK, Colgate University, ⟨ tbalonek@mail.colgate.edu ⟩

RORY BARNES, University of Arizona, ⟨ rory@lpl.arizona.edu ⟩

JOHN BAROSS, University of Washington, ⟨ jbaross@u.washington.edu ⟩

PAUL BARTUS, University of Puerto Rico, ⟨ jalonso@naic.edu ⟩

LUTHER BEEGLE, Jet Propulsion Laboratory, ⟨ luther.beegle@jpl.nasa.gov ⟩

JEAN-PHILIPPE BEAULIEU, Institut d'Astrophysique de Paris, ⟨ beaulieu@iap.fr ⟩

JEAN-PIERRE BIBRING, Institut d'Astrophysique Spatiale, ⟨ bibring@ias.fr ⟩

JENNIFER BIDDLE, University of North Carolina at Chapel Hill, ⟨ jfbiddle@gmail.com ⟩

LINDA BILLINGS, University of Virginia, ⟨ lbillings@seti.org ⟩

ROSALBA BONACCORSI, NASA Ames Research Center, ⟨ rbonaccorsi@mail.arc.nasa.gov ⟩

JADE BOND, University of Arizona, ⟨ cathi@as.arizona.edu ⟩

BONCHO BONEV, NASA Goddard Center for Astrobiology, ⟨ bbonev@ssedmail.gsfc.nasa.gov ⟩

ALAN BOSS, Carnegie Institute, Washington, ⟨ boss@dtm.ciw.edu ⟩

FAIK BOUHRIK, University of North Carolina, ⟨ fb5184@uncw.edu ⟩

MARK BROWN, University of New South Wales, Sydney, ⟨ markbrown@unsw.edu.au ⟩

Jorge Bueno Prieto, Universidad Nacional de Columbia, ⟨jebuenop@unal.edu.co⟩

Judy Butler, Tennessee State University, ⟨judybutler@gmail.com⟩

Richard Carrigan, Fermi National Accelerator Laboratory, ⟨carrigan@fnal.gov⟩

Cecilia Ceccarelli, Laboratoire d'Astrophysique de Grenoble, ⟨Cecilia.Ceccarelli@obs.ujf-grenoble.fr⟩

Steven Charnley, NASA Ames Research Center, ⟨Steven.B.Charnley@nasa.gov⟩

Lysa Chizmadia, UPRM, Mayagez, Puerto Rico , ⟨lysa@uprm.edu⟩

Svatopluk Civis, Czech Academy of Sciences, ⟨civis@jh-inst.cas.cz⟩

Mark Claire, University of Washington, ⟨mclaire@astro.washington.edu⟩

Maria Colin-Garcia, Instituto de Ciencias Nucleares Universidad Nacional Autonoma de Mexico, ⟨mcolin@nucleares.unam.mx⟩

Catharine A. Conley, NASA Ames Research Center, ⟨cassie.conley@nasa.gov⟩

Guy Consolmagno, University of Arizona, ⟨dps07@mac.com⟩

Curtis Cooper, NASA Astrobiology Institute, University of Arizona, ⟨curtis@lpl.arizona.edu⟩

Candy Cordwell, West Virginia University, ⟨Cordwell@nasa.wvu.edu⟩

Dale P. Cruikshank, NASA Ames Research Center, ⟨dcruikshank@mail.arc.nasa.gov⟩

Manfred Cuntz, University of Texas, ⟨cuntz@uta.edu⟩

Jane Curnutt, California State University, ⟨jcurnutt@r2labs.org⟩

Andrew Czaja, University of California, ⟨aczaja@ess.ucla.edu⟩

Eduardo De la Fuente Acosta, Universidad de Guadalajara, ⟨edfuente@inaoep.mx⟩

Prosper De Roos, Zeppers Film, ⟨prosperderoos@cooblae.nl⟩

Julia DeMarines, Astrobiology Club, University of Colorado, ⟨Julia.DeMarines@colorado.edu⟩

Kathryn Denning, York University, ⟨kdenning@yorku.ca⟩

Jackie Denson, University of Arkansas, ⟨jdenson@uark.edu⟩

David Des Marais, NASA Ames Research Center, ⟨David.J.DesMarais@nasa.gov⟩

Christelle Desnues, San Diego State University, ⟨cdesnues@projects.sdsu.edu⟩

Didier Despois, Observatoire de Bordeaux, OASU/LAB, ⟨despois@obs.u-bordeaux1.fr⟩

Edna DeVore, SETI Institute, ⟨edevore@seti.org⟩

Helen DeWitt, University of Colorado, ⟨dewitt@colorado.edu⟩

Steven Dick, NASA HQ, ⟨ steven.j.dick@nasa.gov ⟩

Michael DiSanti, NASA Goddard Space Flight Center, ⟨ disanti@ssedmail.gsfc.nasa.gov ⟩

Ulrich Dopatka, University Bern, ⟨ dopatka@ub.unibe.ch ⟩

Frank Drake, SETI Institute, ⟨ drake@seti.org ⟩

Stephane Dumas, Universite Laval, ⟨ stephane_dumas@sympatico.ca ⟩

Robert Dunn, North Carolina State University, ⟨ rob_dunn@ncsu.edu ⟩

Adam Edson, Pennsylvania State University, ⟨ are131@psu.edu ⟩

Pascale Ehrenfreund, Leiden Institute of Chemistry, ⟨ p.ehrenfreund@chem.leidenuniv.nl ⟩

Boniface Ekpenyong, Northeastern Illinois University, ⟨ bekpenyong@ccc.edu ⟩

Ernesto Esteban, University of Puerto Rico, ⟨ ep_esteban@webmail.uprh.edu ⟩

Yan Fernandez, University of Central Florida, ⟨ yan@physics.ucf.edu ⟩

David Fernandez Remolar, Universidad Rey Juan Carlos, ⟨ fernandezrd@inta.es ⟩

Bernard Foing, ESA, ⟨ bernard.foing@esa.int ⟩

Niyonkuru Francine, , ⟨ nahimana_u@yahoo.fr ⟩

Siegfried Franck, Potsdam Institute for Climate Impact Research, ⟨ franck@pik-potsdam.de ⟩

Friedemann Freund, SETI Institute, ⟨ ffreund@mail.arc.nasa.gov ⟩

Todd Gary, Tennessee State University, ⟨ tgary@coe.tsuniv.edu ⟩

Jose Gerena, SETI Institute, ⟨ edevore@seti.org ⟩

Daniel Glavin, Goddard Center for Astrobiology, ⟨ daniel.p.glavin@nasa.gov ⟩

Damhnait Gleeson, University of Colorado, ⟨ rose.hoag@lasp.colorado.edu ⟩

Javier Goicoechea, Centro de Astrobiologia (CSIC-INTA), Madrid, ⟨ jr.goicoechea@gmail.com ⟩

Stanley Goldberg, University of New Orleans, ⟨ sgoldber@uno.edu ⟩

Colin Goldblatt, NASA Ames Research Center, ⟨ colin.goldblatt@nasa.gov ⟩

Shawn Goldman, Penn State Astrobiology Research Center, ⟨ shawn.goldman@gmail.com ⟩

David Gomez-Ortiz, Universidad Rey Juan Carlos, ⟨ david.gomez@urjc.es ⟩

Natalia Gontareva, Laboratory of Exobiology, ⟨ ngontar@mail.cytspb.rssi.ru ⟩

Janneth Gonzalaz, Pontificia Universidad Javeriana, ⟨ l.lareo@javeriana.edu.co ⟩

Wilson Gonzalez-Espada, Arkansas Tech University, ⟨ wgonzalezespada@atu.edu ⟩

Jose Eduardo, gonzalezpje@inta.es, Centro de Astrobiologia ⟨ Gonzalez-Pastor ⟩

Edward Goolish, NASA Ames Research Center, ⟨ edward.m.goolish@nasa.gov ⟩

Andrew Gould, Ohio State University, ⟨ gould@astronomy.ohio-state.edu ⟩

Rose Grymes, NASA Ames Research Center, ⟨ rose.grymes@nasa.gov ⟩

Nader Haghighipour, NASA Astrobiology Institute, University of Hawaii, ⟨ nader@ifa.hawaii.edu ⟩

Anneliese Haika, Vienna Astronomy Association (WAA), ⟨ a.haika@gmx.at ⟩

DeWayne Halfen, University of Arizona, ⟨ halfendt@as.arizona.edu ⟩

Kevin Hand, Stanford University/Princeton University, ⟨ khand@stanford.edu ⟩

Pamela Harman, SETI Institute, ⟨ pharman@seti.org ⟩

David Harrington, University of Hawaii, ⟨ dmh@ifa.hawaii.edu ⟩

Christa Hasenkopf, University of Colorado, ⟨ christa.hasenkopf@colorado.edu ⟩

Diana Hernandez, SETI Institute, ⟨ edevore@seti.org ⟩

Marisol Herrera, Arecibo Observatory, ⟨ mherrera@naic.edu ⟩

Martin Hilchenbach, Max Planck Institute, ⟨ hilchenbach@mps.mpg.de ⟩

Rob Hodselmans, Zeppers Film & TV, ⟨ prosperderoos@cooblae.nl ⟩

Andrew Howard, Harvard University, ⟨ andrew@alum.mit.edu ⟩

Hui-Chun Huang, National Taiwan Normal University, ⟨ hspring@sgrb2.geos.ntnu.edu.tw ⟩

Reggie Hudson, Goddard Center for Astrobiology, ⟨ reggie.hudson@nasa.gov ⟩

Roberto Irizarry, University of Puerto Rico, ⟨ jalonso@naic.edu ⟩

William Irvine, Goddard Center forAstrobiology, ⟨ irvine@fcrao1.astro.umass.edu ⟩

Gloria Isidro, University of Puerto Rico, ⟨ jalonso@naic.edu ⟩

Hannah Jang-Condell, Carnegie Institution, ⟨ hannah@dtm.ciw.edu ⟩

Benny Jansen, Zeppers Film & TV, ⟨ prosperderoos@cooblae.nl ⟩

Libor Juha, Czech Academy of Sciences, ⟨ juha@fzu.cz ⟩

Mary Ann Kadooka, Institute for Astronomy, University of Hawaii, ⟨ kadooka@ifa.hawaii.edu ⟩

Nathan Kaib, University of Washington, ⟨ kaib@astro.washington.edu ⟩

Michael Kaufmann, Witten/Herdecke University, ⟨ mika@uni-wh.de ⟩

Jacqueline V Keane, NASA Astrobiology Institute, University of Hawaii, ⟨ keane@ifa.hawaii.edu ⟩

Hors Keller, Max-Planck-Institut, ⟨ keller@mps.mpg.de ⟩

Bishun Khare, NASA Ames Research Center, ⟨ bkhare@mail.arc.nasa.gov ⟩

Nancy Kiang, NASA Goddard Institute for Space Studies, ⟨ nkiang@giss.nasa.gov ⟩

Melissa Kirven-Brooks, NASA Ames Research Center, ⟨ mkirven-brooks@mail.arc.nasa.gov ⟩

Jan Kleyna, NASA Astrobiology Institute, University of Hawaii, ⟨ kleyna@ifa.hawaii.edu ⟩

Suzanne Knight, Caltex Australia Petroleum Pty Ltd, ⟨ sknight@caltex.com.au ⟩

Kensei Kobayashi, Yokohaam National University, ⟨ kkensei@ynu.ac.jp ⟩

Vera Kolb, University of Wisconsin, ⟨ kolb@uwp.edu ⟩

Timothy Kral, University of Arkansas, ⟨ tkral@uark.edu ⟩

Yi-Jehng Kuan, National Taiwan Normal University/Academia Sinica, ⟨ kuan@alioth.geos.ntnu.edu.tw ⟩

Anil Kumar, NMR Research Center, Indian Institute of Science, ⟨ anilcbt@yahoo.com ⟩

Robert Lamontagne, University of Montreal, ⟨ lamont@astro.umontreal.ca ⟩

Leonardo Lareo, Pontificia Universidad Javeriana, ⟨ l.lareo@javeriana.edu.co ⟩

David Latham, Harvard-Smithsonian Center for Astrophysics, ⟨ dlatham@cfa.harvard.edu ⟩

Brandon Lawton, Space Telescope Science Institute, ⟨ lawton@stsci.edu ⟩

Ester Lazaro, Centro de Astrobiologia, ⟨ lazarolm@inta.es ⟩

Antonio Lazcano, Universidad Nacional Autnoma de Mxico, ⟨ alar@correo.unam.mx ⟩

Guillermo A. Lemarchand, Instituto de Radioastronomia, ⟨ lemar@correo.uba.ar ⟩

Susan Leschine, University of Massachusetts, ⟨ suel@microbio.umass.edu ⟩

Anny-Chantal Levasseur-Regourd, Universit Pierre et Marie Curie, ⟨ aclr@aerov.jussieu.fr ⟩

Charles Lineweaver, Australian National University, ⟨ charley@mso.anu.edu.au ⟩

Ramon Lopez-Aleman, University of Puerto Rico, ⟨ elgranmoncho5@msn.com ⟩

Claudio Maccone, International Academy of Astronautics, ⟨ clmaccon@libero.it ⟩

Avi Mandell, Goddard Center for Astrobiology, ⟨ mandell@astro.psu.edu ⟩

Heidi Manning, Concordia College, ⟨ manning@cord.edu ⟩

Sergio Martin, Harvard-Smithsonian Center for Astrophysics, ⟨ smartin@cfa.harvard.edu ⟩

Eva Mateo-Marti, Centro de Astrobiologia, ⟨ mateome@inta.es ⟩

Graciela Matrajt, University of Washington, ⟨ matrajt@astro.washington.edu ⟩

Isamu Matsuyama, Carnegie Institution, ⟨ matsuyama@dtm.ciw.edu ⟩

Karen S. McBride, University of California Los Angeles, ⟨ kmcbride@nasa.gov ⟩

Curtis Mead, Harvard University, ⟨ cmead@fas.harvard.edu ⟩

Victoria Meadows, Spitzer Science Center, Caltech, ⟨ vsm@ipac.caltech.edu ⟩

Karen Meech, Institute for Astronomy, University of Hawaii, ⟨ meech@ifa.hawaii.edu ⟩

Adrian Melott, University of Kansas, ⟨ melott@ku.edu ⟩

Abel Mendez, University of Puerto Rico, ⟨ amendez@upra.edu ⟩

Robert Minard, Penn State University, ⟨ rminard@psu.edu ⟩

Hatungimana Mohamed, , ⟨ nahimana_u@yahoo.fr ⟩

Andres Moreno, Unitesidad Nacional de Colombia, ⟨ andmorenom@unal.edu.co ⟩

David Morrison, NASA Ames Research Center, ⟨ dmorrison@arc.nasa.gov ⟩

Cherilynn A. Morrow, SETI Institute, ⟨ cmorrow@seti.org ⟩

Amanda Morrow, Pennsylvania State University, ⟨ morrow@astro.psu.edu ⟩

Uwe Motschmann, University of Braunschweig, ⟨ u.motschmann@tu-bs.de ⟩

Dermott Mullan, University of Delaware, ⟨ mullan@udel.edu ⟩

Michael Mumma, NASA Goddard Space Flight Center, ⟨ mmumma@ssedmail.gsfc.nasa.gov ⟩

Paolo Musso, University of Insubria, ⟨ musso.pl@libero.it ⟩

Catherine Neish, University of Arizona, ⟨ cdneish@lpl.arizona.edu ⟩

Guillermo Nery, University of Puerto Rico, ⟨ ganeryg@aol.com ⟩

Wayne Nicholson, University of Florida, Kennedy Space Center, ⟨ WLN@ufl.edu ⟩

Delphine Nna Mvondo, Centro de Astrobiologia, ⟨ nnamvondod@inta.es ⟩

Erika Offerdahl, LAPLACE, University of Arizona, ⟨ eofferda@email.arizona.edu ⟩

Hiroshi Ohmoto, Pennsylvania State University, ⟨ hqo@psu.edu ⟩

Nilsa I. Ortiz, SETI Institute, ⟨ edevore@seti.org ⟩

Fabio Pagan, International Center for Theoretical Physics, ⟨ pagan@ictp.trieste.it ⟩

Carmen Pantoja, University of Puerto Rico, ⟨ cpantoja@prw.net ⟩

Aaron Parsons, University of California, Berkeley, ⟨ aparsons@astron.berkeley.edu ⟩

Matthew Pasek, University of Arizona, ⟨ mpasek@lpl.arizona.edu ⟩

Yvonne Pendleton, NASA Ames Research Center, ⟨ ypendleton@mail.arc.nasa.gov ⟩

Amy Perkins, University of Florida, ⟨ amye.perkins@gmail.com ⟩

Gayle Philip, NASA Astrobiology Institute, University of Hawaii, ⟨ gayle.philip@gmail.com ⟩

Alan Pickwick, Manchester GrammarSchool/EAAE, ⟨ Alan_C_Pickwick@btinternet.com ⟩

Carl Pilcher, NASA, ⟨ carl.b.pilcher@nasa.gov ⟩

Jana Pittichova, Institute for Astronomy, University of Hawaii, ⟨ jana@ifa.hawaii.edu ⟩

Andrew Pohorille, NASA Ames Research Center, ⟨ pohorill@max.arc.nasa.gov ⟩

Gustavo Porto de Mello, Universidade Federal do Rio de Janeiro/Obs, ⟨ gustavo@ov.ufrj.br ⟩

Edward Prather, University of Arizona, ⟨ eprather@as.arizona.edu ⟩

Aaron Price, Tufts University, ⟨ aaronp@aavso.org ⟩

Andreas Quirrenbach, ZAH, ⟨ A.Quirrenbach@lsw.uni-heidelberg.de ⟩

Francisco Ramos Stierle, University of California, Berkeley, ⟨ ramospancho@gmail.com ⟩

Francois Raulin, LISA -Universits Paris, ⟨ raulin@lisa.univ-paris12.fr ⟩

Jose Rivera, University of Puerto Rico, ⟨ jrivera@cnnet.upr.edu ⟩

Jose Robles, The Australian National University, ⟨ josan@mso.anu.edu.au ⟩

Steven Rodgers, NASA Ames Research Center, ⟨ rodgers@dusty.arc.nasa.gov ⟩

Celia Rogero, Centro de Astrobiologia, ⟨ rogerobc@inta.es ⟩

Patricio Rojo, Cornell University, ⟨ pato@das.uchile.cl ⟩

Michael Rubin, University of Puerto Rico at Cayey, ⟨ mrubin@cayey.upr.edu ⟩

Oscar N. Ruiz, University of Puerto Rico, ⟨ oruiz@bc.inter.edu ⟩

Marta Ruiz Bermejo, Centro de Astrobiologia, Madrid, ⟨ ruizbm@inta.es ⟩

John Rummel, NASA HQ, ⟨ john.d.rummel@nasa.gov ⟩

Edgardo Sanchez, SETI Institute, ⟨ edevore@seti.org ⟩

Scott Sandford, NASA Ames Research Center, ⟨ ssandford@mail.arc.nasa.gov ⟩

Cara Santelli, ,

Nitza E. Santiago, University of Puerto Rico at Humacao,

Jan Sapp, York University, ⟨ jsapp@yorku.ca ⟩

Luis Sarmiento, Arecibo Observatory, ⟨ lsarmiento@arecibo.inter.edu ⟩

Susana Irene Schneider, Pennsylvania State University, ⟨ sschneid@geosc.psu.edu ⟩

J. William Schopf, University of California Los Angeles, ⟨ Schopf@ess.ucla.edu ⟩

Matthew Schrenk, Carnegie Institution, ⟨ m.schrenk@gl.ciw.edu ⟩

David Schwartzman, Howard University, ⟨ dschwartzman@gmail.com ⟩

Owoyemi Segun, , ⟨ trans2ng@hotmail.com ⟩

Antigona Segura, Universidad Nacional Autnoma de Mxico, ⟨ antigona@nucleares.unam.mx ⟩

Young Min Seo, Seoul National University, ⟨ seo3919@gmail.com ⟩

Todd Sformo, University of Alaska, Fairbanks, ⟨ rfts@uaf.edu ⟩

Seth Shostak, SETI Institute, ⟨ seth@seti.org ⟩

Janet Siefert, Rice University, ⟨ siefert@rice.edu ⟩

Michael Simakov, Russian Academy of Sciences, ⟨ exobio@mail.ru ⟩

David Smith, Space Studies Board, National Research Council, ⟨ dhsmith@nas.edu ⟩

Pablo Sobron, University of Valladolid-CAB, ⟨ psobron@iq.uva.es ⟩

Michael F. Sterzik, ESO, ⟨ msterzik@eso.org ⟩

Karen Stockstill, NASA Astrobiology Institute, University of Hawaii, ⟨ kstockst@higp.hawaii.edu ⟩

Jessica Sunshine, University of Maryland, ⟨ jess@astro.umd.edu ⟩

Jun-ichi Takahashi, NTT, ⟨ jitaka@ba2.so-net.ne.jp ⟩

Jill Tarter, SETI Institute, ⟨ cneller@seti.org ⟩

Inge Ten Kate, NASA Goddard Space Flight Center, ⟨ tenkate@umbc.edu ⟩

Richard Terrile, Caltech, ⟨ rich.terrile@jpl.nasa.gov ⟩

Giovanna Tinetti, Institut d'Astrophysique de Paris, ⟨ tinetti@iap.fr ⟩

Karla Torres, Instituto Nacional de Pesquisas Espaciais (INPE), ⟨ karlchen79@gmail.com ⟩

Hrant Tovmasyan, Instituto Nacional Astrofisica, Optica y Electronica, ⟨ hrant@inaoep.mx ⟩

Melissa Trainer, NASA Goddard Space Flight Center, ⟨ melissa.trainer@nasa.gov ⟩

Catherine Tsairides, NASA Ames Research Center, ⟨ ctsairides@mail.arc.nasa.gov ⟩

Alexandre Tsapin, Jet Propulsion Laboratory, Caltech, ⟨ tsapin@jpl.nasa.gov ⟩

Rebecca Turk, University of Colorado, ⟨ turk@colorado.edu ⟩

Edwin Turner, Princeton University, ⟨ elt@astro.princeton.edu ⟩

Douglas Vakoch, SETI Institute, ⟨ vakoch@seti.org ⟩

Jaap Van Hoewijk, Zepper Films & TV, ⟨ prosperderoos@cooblae.nl ⟩

Geronimo Villanueva, NASA Goddard Space Flight Center, ⟨ villanueva@ssedmail.gsfc.nasa.gov ⟩

Ted Von Hippel, University of Texas, Austin, ⟨ ted@astro.as.utexas.edu ⟩

Gabe Waggoner, NASW, ⟨ gwaggoner@tamu.edu ⟩

Lori Walton, Tigerstar Geoscience, ⟨ lawalton@telus.net ⟩

Jillian Ward, University of Hawaii, ⟨ jmward@hawaii.edu ⟩

Martin Weiss, New York Hall of Science, ⟨ mweiss@nyscience.org ⟩

Dan Werthimer, University of California, Berkeley, ⟨ danw@ssl.berkeley.edu ⟩

Laura Woodney, California State University, San Bernardino, ⟨ woodney@csusb.edu ⟩

Nick Woolf, University of Arizona, ⟨ nwoolf@as.arizona.edu ⟩

Katherine Wright, University of Colorado, ⟨ katherine.wright@colorado.edu ⟩

Weijun Zheng, NASA Astrobiology Institute, University of Hawaii, ⟨ zhengw@hawaii.edu ⟩

H.J. Ziock, Los Alamos National Laboratory, ⟨ ziock@lanl.gov ⟩

Lucy Ziurys, University of Arizona, ⟨ lziurys@as.arizona.edu ⟩

Session I

Introduction

Bioastronomy 2007: Molecules, Microbes, and Extraterrestrial Life
ASP Conference Series, Vol. 420, 2009
K. J. Meech, J. V. Keane, M. J. Mumma, J. L. Siefert, and D. J. Werthimer, eds.

Overview of the Bioastronomy 2007 Meeting

K. J. Meech[1] and W. M. Irvine[2]

[1]*Institute for Astronomy, 2680 Woodlawn Dr., Honolulu, HI 96822 USA; NASA Astrobiology Institute*

[2]*Univ. of Massachusetts, Dept. of Astronomy, LGRT-B 619E, 710 N. Pleasant Street, Amherst MA 01003-9305, USA*

Abstract.
The 9th Bioastronomy conference, "Molecules, Microbes and Extraterrestrial Life" was held in San Juan Puerto Rico from July 16–20, 2007. Originally established as a triennial meeting of the Commission on Bioastronomy of the International Astronomical Union, the Bioastronomy meetings have grown in size and scope, exemplifying the growth and breadth of this interdisciplinary field. The 2007 meeting had a record attendance of 234 participants, and hosted a rich variety of ancillary workshops, meetings and Education and Public Outreach activities. This paper presents an overview of the conference.

1. Introduction

The International Astronomical Union's Commission 51 (Bioastronomy) was established in 1982 with objectives that included the search for planets around other stars; the search for radio transmissions, intentional or unintentional, of extraterrestrial origin; the search for biologically relevant interstellar molecules and the study of their formation processes; detection methods for potential spectroscopic evidence of biological activity; and the coordination of efforts in all these areas at the international level. In fact, the Commission's principal activity over the years has been the organization of a series of Bioastronomy meetings which have played an important role in integrating the broader interests and techniques of both astronomy and biology to understand the origin of our solar system and the origin and evolution of living systems, and to generate a context for exploration in our solar system and in extrasolar planetary systems. These conferences provide an opportunity for astronomers, biologists, geologists, planetary scientists, and those from other disciplines to meet and discuss research of mutual interest for addressing the question of the origin and evolution of life on this planet and elsewhere in the universe, and more generally topics that have come to be included in what is now often called astrobiology. The first International Bioastronomy Conferences were visionary in tackling problems that many scientists felt were premature to attack, given our knowledge at the time (Lemarchand 2000). However, these conferences increasingly reflect a world that realizes these seemingly impenetrable questions are becoming amenable to at least provisional answers.

Historically, there were two main groups dealing with the investigation of extraterrestrial life and habitable worlds that engaged scientists. The first is IAU

Commission 51, composed of astronomers, physicists and engineers who focus on the search for extrasolar planets, formation and evolution of planetary systems, and the astronomical search for intelligent signals. The second group, the International Society for the Study of the Origin of Life (ISSOL), is composed largely of biologists and chemists focusing research on the biogenesis and evolution of life on Earth and in the solar system. More recently the NASA Astrobiology Institute (NAI) has hosted Institute meetings every 2 years, and in alternate years the NASA Astrobiology program has hosted a large interdisciplinary meeting called AbSciCon for the whole scientific and educational community. Held in years without an IAU General Assembly, originally the Bioastronomy series of meetings was out of phase with meetings hosted by ISSOL. However, in 1992, ISSOL changed the dates of their triennial meetings to avoid the Olympics and they began to hold their meetings in the same year as the Bioastronomy meetings, usually close in time to the Bioastronomy meetings, but not typically close geographically. In order to break this conflict, Commission 51 decided to hold the Bioastronomy meeting a year early in 2004.

Table 1. Previous Bioastronomy Meetings

Year	IAU Symposium?	Location
1984	IAU Symposium 112	Boston, USA
1987	IAU Colloquium 99	Balaton, Hungary
1990		Val Cenis, France
1993		Santa Cruz, USA
1996‡		Capri, Italy
1999‡		Hawai'i, USA
2002	IAU Symposium 213	Hamilton Island, Australia
2004		Reykjavik, Iceland
2007		San Juan, Puerto Rico

Note: ‡These two meetings overlapped with the years of the ISSOL meetings.

Figure 1. A variety of past Bioastronomy meeting venues.

The Federation of Astrobiology Organizations (FAO), which fosters collaboration between an international set of astrobiology networks, associations, institutes and societies, challenged these organizations (IAU C51, ISSOL, NAI, AbSciCon) to develop a 5-year plan of conference dates to avoid duplications and conflicts. On September 2, 2004, representatives from ISSOL (A. Lazcano, President; W. Schopf, ISSOL member), IAU C51 (K. Meech, President), NAI

(R. Grymes, Deputy Director, also representing EANA, the European Astrobiology Network Association) and AbSciCon (L. Rothschild; organizer) met at the Center for the Study of the Evolution and Origin of Life at the Univ. of CA, Los Angeles. The result was a visionary plan reflecting the worldwide astrobiology community's desire for one self-sustaining, triennial, international meeting addressing the interdisciplinary breadth and focus of our field. The first step in this plan was an agreement between ISSOL and IAU Comm. 51 to undertake to hold a joint (overlapping) meeting at a site to be selected by both groups. Unfortunately, there was insufficient time to plan a joint meeting for 2008, so IAU C51 proposed an independent meeting for 2007.

Table 2. Previous ISSOL Meetings

Year	**Location**	**Year**	**Location**
1957	Moscow, Russia	1989	Prague, Czech Republic
1963	Walkulla Spring, USA	1993	Barcelona, Spain
1970	Pont-a-Mousson, France	1996	Orleons, France
1973	Barcelona, Spain	1999	San Diego, CA
1977	Kyoto, Japan	2001	Oaxaca, Mexico
1980	Jerusalem, Israel	2005	Beijing, China
1983	Mainz, Germany	2008	Florence, Italy
1986	Berkeley, USA	2011	Montpellier, France

Figure 2. The Arecibo Observatory, one of the astrobiologically relevant sites visited during the 2007 Bioastronomy meeting, as seen from an Earth orbital satellite. Image provided by Geoeye and satellite imaging corporation www.satimagingcorp.com <http://www.satimagingcorp.com/>.

During the past several years, Commission 51 had attempted to select meeting venues that had unique environments relevant to astrobiology. The 2004 meeting held in Reykjavik Iceland, was especially attuned to the conference theme of Habitability. Iceland boasts a unique sub-arctic ecosystem, at the separating boundary of the Eurasian and North American continental plates. Subglacial Icelandic volcanic eruptions compared to volcanic deposits in Antarctica help us to understand the ice/volcano interaction both on earth and other astrobiologically important environments in the solar system. One of the spectacular successes of the Bioastronomy 2004 meeting was the unique synergy of the meeting theme and the post-conference scientific tours which allowed attendees to explore the environments we had just discussed. This concept was a driver in the selection of venues for the 2007 meeting.

During March 2006, proposals for the Bioastronomy 2007 meeting were quickly solicited, and a sub-committee from IAU Comm 51, ISSOL, and the NAI met at AbSciCon in Washington DC to discuss site options. San Juan, Puerto Rico was selected as the site that would provide the most affordable astrobiology access. One obvious scientific theme for Puerto Rico centers on the Arecibo observatory, and the role it has played both in the search for extraterrestrial life programs, and in the radio identification of molecules in space, relevant to solar system and biological origins.

Table 3. Local Organizing Committee

Member	Institutition
José Alonso	University of Puerto Rico
Daniel Altschuler	Arecibo Observatory
Hector Ayala-del-Rio	University of Puerto Rico, Mayagüez
Frank Drake	SETI Institute
Edna DeVore	SETI Institute
Sixto Gonzales	Arecibo Observatory
Pamela Harmon	SETI Institute
Karen Meech	University of Hawai'i
Carlos Rogriguez	University of Puerto Rico, Mayagüez
Michele Sonoda	University of Hawai'i
Melita Thorpe	MWT Associates

2. Conference Arrangements

The meeting was held at the Condado Plaza Hotel, which was well located, a 10 minute drive from the international airport and within within walking distance of Old San Juan. The Local Organizing Committee (LOC) was chaired by K. Meech (see Table 3) with a lot of help from local scientists, in particular José Alonso. All of the outreach activities, in particular, would not have been possible without the generous local support, and the post-conference scientific tour was enabled through generous support from the University of Puerto Rico and other sponsors (see Table 6).

3. Meeting Science Content

The Scientific Organizing Committee (SOC), chaired by W. Irvine (see Table 4), was fortunate in that the broad themes, "Molecules, Microbes, and Extraterrestrial Life" had already been selected by the then President and Vice-President of Commission 51, Karen Meech and Alan Boss. Since the Commission is, after all, a part of the International Astronomical Union, it made sense to divide the proposed program so that the first half considered astronomical topics both within the Milky Way and other galaxies (day 1) and more locally within the Solar System (day 2). This was followed by a day focused on more biologically related subjects, with a final day devoted to extrasolar planets and the search for life elsewhere in the universe.

The Scientific Organizing Committee (Table 4) was composed of members of the bioastronomy community drawn from IAU Commission 51, ISSOL and the NASA Astrobiology Institute community.

Table 4. Scientific Organizing Committee

Member	Institute	Country	C51	NAI	ISSOL
Peter Backus	SETI Institute	USA		⋆	
John Baross	Univ. Washington	USA		⋆	⋆
Alan Boss	Carnegie Institution	USA	⋆	⋆	⋆
N. Haghighipour	Univ. Hawai'i	USA		⋆	
William Irvine	Univ. Massachusetts	USA	⋆	⋆	⋆
Karen Meech	Univ. Hawai'i	USA	⋆	⋆	⋆
Michael Mumma	NASA Goddard	USA		⋆	
Carol Oliver	Austral. Ctr Astrobio.	Australia	⋆	⋆	
John Rummel	NASA Headquarters	USA			
Janet Siefert	Rice University	USA			⋆
Th. Thorsteinsson	Orkustofnun	Iceland		⋆	
Dan Werthimer	UC Berkeley	USA	⋆	⋆	

The 12 member SOC had two general telecons to plan for the conference, with most of the details being subsequently worked out through email. Because of the extremely broad nature of the subject matter, the SOC Chair was particularly aided by John Rummel, who took responsibility for organizing the sessions on day 3 ("Microbes"), and Peter Backus, who was the corresponding "sub-chair" for day 4 ("Extraterrestrial Life"), which in the present case included all topics associated with extrasolar planets. Scientific topics on the first two days went from distant objects to those closer to home, beginning with the molecular complexity of material in the interstellar medium of external galaxies, passing to organic chemistry and isotopic fractionation within the Milky Way, and on to theoretical and ground and spacecraft studies of solar system objects possibly relevant to bioastronomy. Each of the four days of scientific presentations was divided into two morning plus two afternoon sessions, with each session normally beginning with one or two invited reviews, followed by several contributed talks (see Table 5). The SOC made an effort to distribute several presentations devoted to education and public outreach throughout the scientific sessions, rather than grouping them together in a separate (and often ill-attended) special session. Time was also set aside for viewing contributed posters, and we were fortunate to have an evening public lecture by the noted astrobiologist, and International Society for the Study of the Origin of Life (IS-SOL) President, Antonio Lazcano, on "The Emergence of Life on Earth: Old Problems and New Perspectives."

The SOC is particularly grateful to both the invited speakers and to those giving contributed talks for their response to our request to avoid the jargon of their particular discipline and to make their presentations accessible to the diverse audience present. The real success of the conference is due in large part to the speakers' success in this regard. We are also extremely grateful to all those of our colleagues who served as session chairs and who kept the sessions and their discussions lively and interesting. In total there were 19 invited talks, 57 contributed talks, and 113 poster presentations for a total of 189 papers.

Table 5. Conference Paper Distribution

Day	Inv†	Oral†	Post†	Session Topic
7/16	1	4	3	How Early Could Life Originate in the Universe?
	1	4	8	Nature of Interstellar Organics
	1	4	8	From Interstellar to Protoplanetary Organics
	2		7	Organic Matter in Comets, Meteorites & IDPs
7/17	1	3	3	Life in the Outer Solar System
	2	2	11	Mars as a Setting for Life
	2	2	2	Comet Space Missions: Stardust & Deep Impact
		6		Prebiotic Chemistry
7/19	1	4	4	Origin of Viruses and Cellular Function
	1	4	10	Life in Extreme Environments
	2	4	1	CoEvolution of Life & Early Earth Environment
	1	2	2	Life Beyond or Without Earth
7/20	1	4	7	Extrasolar Planets / Planet Formation
	1	4	19	Habitable Planets and Their Stars
	1	4	8	Biosignatures
	1	6	16	SETI

Notes: †Invited, Oral and Poster paper counts.

4. Education and Public Outreach

4.1. AbGradCon

The Astrobiology Graduate Conference AbGradCon, which had been held in 2004 in Tucson Arizona and 2005 in San Diego, is being transformed into an on-going conference series. The goal of AbGradCon is to foster communication within the broad astrobiology-related graduate student community, and to give early graduate students a chance to practice speaking in a collegial atmosphere. The intent is to hold AbGradCon in association with major astrobiology conferences whenever possible to benefit from the conference infrastructure for logistics. Each conference will be organized by a graduate student committee, but input and assistance from the larger graduate community is critical for the continued success of the series.

AbGradCon 2007 was held as 1.5-day mini-conference held July 14 and 15 2007, in conjunction with the Bioastronomy 2007 meeting. The organizers were Avi Mandell (Goddard Spaceflight Center) and Shawn Goldman (Penn State Univ.). For AbGradCon 2007, the logistics were handled by the LOC in conjunction with the main meeting, but the science organization was independent. The program included approximately 25 short talks available to all graduate students, with four keynote talks given by late-stage graduate students or young postdocs at the beginning of each of the sessions.

4.2. AAS Teaching Workshop

A program entitled, "Advanced Strategies for Creating a Learner-Centered Introductory College Astronomy Course: An Introductory Workshop", was developed by Timothy Slater (the AAS Education officer), E. Prather and G. Brissenden (Univ. AZ), and sponsored by the Chautauqua summer program, was held in conjunction with the Bioastronomy meeting on July 13-15, at the meeting venue. The workshop targeted scientific participants of the Bioastronomy meeting, with the goal of familiarizing them with learner-centered teaching and assessment materials, and their implementation in their college astronomy courses. Participants learned how to create productive learning environments by reviewing research on the nature of teaching and learning; setting course goals and objectives; and using interactive lectures, peer instruction, engaging demonstrations, collaborative groups, tutorials, and ranking tasks.

4.3. Mid-Conference Teacher Workshop

In collaboration with the Office for the Public Understanding of Science (OPUS) of the Arecibo Observatory, the Puerto Rico NASA Space Grant Consortium, the SETI Institute, the ALACiMA project of the University of Puerto Rico, the University of Hawai'i and the NASA Astrobiology Institute we held a special 1 day astrobiology teacher workshop aimed at science teachers of Puerto Rico. The public school system in Puerto Rico is organized under a centralized unit through the Puerto Rico Department of Education (DoE). The DoE serves 580,000 students in 1520 schools in the K-12 levels. There are 43,000 teachers who are now required to be certified in their teaching subject, of which $\sim$4000 are middle and high school science teachers.

The enrollment was limited to 90 teachers, recruited through the Arecibo Observatory and the Puerto Rico Department of Education. The teachers worked in smaller groups in 3 topical programs: "The Universe at your Fingertips", "Habitable Worlds", and "Voyages Through Time". The day also afforded opportunities for the teachers to interact with selected scientists, and materials were provided to the teachers for classroom use. The workshop instructors included the following scientists and educators:

- Daniel Altschuler – a senior research associate at the National Astronomy and Ionosphere Center's Arecibo Observatory. Although his research focuses on the study of active galactic nuclei, he has been very engaged in public outreach, and has been the director of the Arecibo Observatory office for the public understanding of science (OPUS).

- Jose Alonso – the Education Officer at Arecibo Observatory, whose research is in the area of young stellar objects. As head of the observatory's Angel Ramos Foundation Visitor Center, he has helped to develop its outreach programs.

- Edna DeVore – Director of EPO at the SETI Institute, Edna is a science and astronomy educator. She's been CoI on several NASA missions, and has co-developed many integrated high school science curricula.

- Pamela Harman – EPO Manager at the SETI Institute, Pamela is an established educator, and a CoI for the NASA Sofia and Kepler Mission outreach programs.

- Mary Kadooka – In her capacity as the UH Institute for Astronomy and UH NAI Education Outreach coordinator, Mary has been actively developing outreach and training programs for students and teachers in astrobiology.

- Carol Oliver – In her position as Director of outreach at the Australian Center of Astrobiology, she has been working to bring NASA learning technology tools to allow students to experience a virtual field trip to the Pilbara of Western Australia.

- Erika Offerdahl – Graduating in Fall 2007 from the Dept. of Biochemistry at the Univ. of AZ, Erika has been researching learning-centered teaching strategies.

A highlight of the astrobiology teacher workshop was a public evening lecture on Wed. July 18, entitled "The Emergence of Life on Earth: Old Problems, New Perspectives" delivered by Antonio Lazcano in spanish. Antonio Lazcano is a Mexican profesor of biochemistry of the Science Faculty of National Autonomous University of Mexico (UNAM) in Mexico City. He has studied the origin and early evolution of life for more than 30 years. He has been professor-in-residence or visiting scientist in France, Spain, Cuba, Switzerland, Russia, and the United States. He has written several books in Spanish, including The Origin of Life (1984) which became a best-seller with more than 600,000 sold copies. In addition, he has been a member of several advisory and review boards of scientific organizations, such as NASA He is the current president of the International Society for the Study of the Origin of Life, and also the first Latin American scientist to occupy this position. Lazcano has devoted considerable efforts to promote scientific journalism and teaching.

4.4. Public Event – Contact

The year 2007 was the 10th anniversary of the release of the movie *Contact*, which was filmed in part in Puerto Rico at the Arecibo Observatory. Because of the anniversary, we used the opportunity to highlight astronomical and astrobiological

Figure 3. Scenes from the public "Contact Event". [Top] Seth Shostak (SETI Institute) was the moderator for the panel [Bottom Left] consisting of Jill Tarter (L), Br. Guy Consolmagno (center), and John Rummel (R). [Bottom Right] An attentive audience in a full room!

topics for the public through a dynamic panel forum which discussed the possibilities for other habitable worlds, the science that goes into the understanding of habitable zones elsewhere, where we are in searches for extraterrestrial life, and the philosophical implications of discovery of life elsewhere. Panel members included Jill Tarter (SETI Institute) for her pioneering work in the field of SETI giving the perspective of the science of how to make contact with other civilizations and the protocols to be followed when potential signals are discovered. Brother Guy Consolmagno (planetary astronomer, Vatican Observatory) gave his perspective on the philosophical implications related to the detection of extraterrestrial life, and John Rummel, the NASA Planetary Protection officer talked about the precautions we should take with physical exploration. The lively debate was moderated by Seth Shostak (SETI Institute) and was followed by a showing of the movie *Contact*. It is a testament to the event that on the last day of the meeting, Friday evening, the meeting room was packed.

Table 6. Conference Sponsors

Agency	Purpose	Amount
NASA	Travel Grants	$ 62,151
NASA Astrobiology Institute	AbGradCon, EPO	$ 44,777
National Science Foundation	Travel Grants	$ 30,000
Puerto Rico Space Grant Consortium	Teacher Workshop	$ 10,000
Alacima	Teacher Workshop	$ 7,500
NAIC/Arecibo Observatory	Teacher Workshop	$ 5,000
American Astronomical Society	Teaching Workshop	$ 2,378
Australian Center for Astrobiolgy	Poster printing	$ 1,000
Institute for Astronomy, Univ. HI	In Kind staffing	
Geoeye Satellite Imaging Corporation	Photo use rights	
	Total Contributions	$162,806

5. Meeting Finances

Although meeting planning started late by most meeting standards, we were aggressive about obtaining outside funding to support all of the ancillary meeting activities, as well as to get support for travel grants (see Table 6). The registration fee was set to cover basic meeting expenses, including continental breakfasts and lunches, as well as the conference proceedings. The early registration (prior to 1/15/07) was set at $465 and the late registration was set at $595.

The travel grants were administered with an online system for applications, and reviewed by the SOC. In total, 55 travel grants were given out, 27 to faculty and researchers, and 28 to postdocs and students (not including the AbGradCon funding which was separate). Grants were only given out to people whose papers had been accepted for presentation (either oral or poster). The average award was for $1375, and the range was from $465-$2000. Table 7 itemizes the income and meeting expenses. Thanks to some generous discounts for A/V use from the hotel, we were able to break even on the meeting.

Table 7. Program Budget

Expenses		**Income**	
Category	**Cost**	**Category**	**Cost**
Auditorium and A/V	$ 24,127	Registration fees	$109,855
Food & Beverage	$ 68,103	Grants / Support	$162,806
Staffing & Fees	$ 24,223		
Printing	$ 19,256		
Miscellaneous	$ 1,878		
AAS Workshop	$ 2,378		
AbGradCon	$ 21,618		
EPO Workshop	$ 35,423		
Travel Grants	$ 75,423		
Total Expenses	$272,661	**Total Income**	$272,661

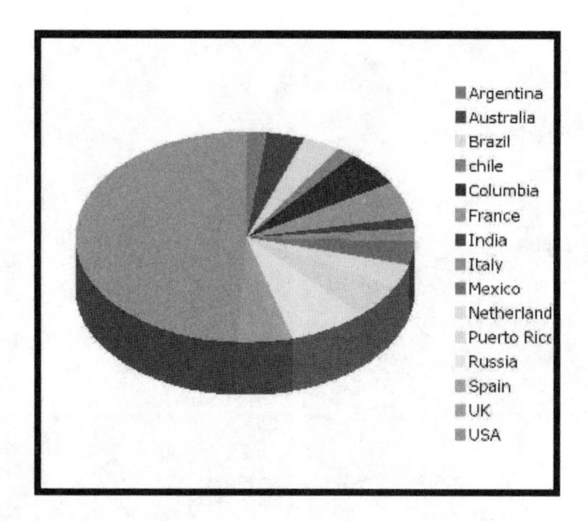

Figure 4. Distribution of travel grants by country. Approximately 49% of the grants went to senior scientists, and of this 1/3 were from the USA.

6. Ancillary Activities and Tours

It has been traditional at these conferences to organize excursions to locations with bioastronomical relevance, in the present instance to the Arecibo National Observatory and nearby Camuy caves. Because of the distances involved, this excursion occupied the entire Wednesday of the five day conference. The Arecibo observatory, built in 1963, is a national facility operated by Cornell University which pursues research in radio astronomy, planetary radar astronomy and atmospheric science. The observatory is the site of the world's largest single-dish telescope. Notably for bioastronomy, this facility has been a centerpiece for projects devoted to SETI.

The conference participants split into two groups to keep the tours a manageable size, and while one group got to explore the Arecibo facilities, the second group was visiting the nearby Camuy caverns. At Arecibo, NAIC Director, R. L. Brown and radar astronomer M. Nolan gave presentations on the observatory science before the group went out to see the antenna, and explore the facilities.

The Camuy Cave Park hosts one of the world's largest cave complexes and was carved over a million years ago by a tropical river which still runs through it. Located in NW Puerto Rico in karst topography, it is the result of dissolution of the middle Tertiary limestone. The area is underlain by a thick sequence of platform carbonates. The caverns represent one of the largest systems of caverns in the Western hemisphere, and represents some unique extreme ecosystems.

A fascinating scientific post-conference tour was also organized to the salterns at Cabo Rojo, Puerto Rico, home to examples of extremophilic organisms (in this case halophiles, living in a concentrated salt solution) and to a neighboring a bay with bioluminescent organisms. Cabo Rojo represented some of the New world's oldest salt mines. The salterns are composed of an estuary surrounded by natural mats feeding a series of artificial salt ponds with seawater. After a 4 hour bus trip from San Juan, the group was lead by Dr. Rafael

Figure 5. Images from Arecibo observatory tour on 18 July 2007. Top Left: 305m diameter primary reflector; Top Right: Image of the Gregorian dome, a state of the art feed system that uses two mirrors to focus the radio-wave energy, suspended 137m above the main reflector. Bottom: Images from the control room and computer room at the observatory.

Montalvo of the University of Puerto Rico to explore the salterns and the extremophiles inhabiting them.

On the return trip, the group took a short boat tour to look at bioluminscent organisms in the La Parguera Bay. The bioluminescence is generated by the dinoflagellates of the species Pryodinium bahemense.

Acknowledgments. This material is based upon work supported by the National Aeronautics and Space Administration through the NASA Astrobiology Institute under Cooperative Agreement No. NNA04CC08A issued through the Office of Space Science.

References

Lemarchand, G. 2000, in ASP. Conv. Ser. Vol. 213: A New Era in Bioastronomy, ed. G. Lemarchand & K. Meech, (San Francisco, ASP), 7

Figure 6. Camuy Caverns. [Top] Entrance and exit from the 170 foot high Cueva Clara. [Middle] The Rio Camuy, which has carved the caverns; dense jungle outside the caverns. [Bottom] View of the rich tropical vegetation just outside the caverns.

Figure 7. [Top] Pre-tour orientation; View of the salterns at Cabo Rojo, SW Puerto Rico. The pink hue to the water is caused by the halophiles inhabiting the salt ponds. [Row 2] Salterns Dr. Rafael Montalvo showing participants the various layers (green, pink, cream and black) in the microbial mats in the waters near the edge of the salterns. [Row 3] Algal mat. The cyanobacterial mats contain an abundant population of sulfate-reducing bacteria. Close up of the evaporating surface of shallow pond areas. [Bottom] Cabo Rojo coastline; participants on night bioluminescent tour.

Session II

How Early Could Life Originate in the Universe?

Bioastronomy 2007: Molecules, Microbes, and Extraterrestrial Life
ASP Conference Series, Vol. 420, 2009
K. J. Meech, J. V. Keane, M. J. Mumma, J. L. Siefert, and D. J. Werthimer, eds.

Observing the Molecular Composition of Galaxies

Sergio Martín

Harvard-Smithsonian Center for Astrophysics

Abstract. The recent availability of wideband receivers and high sensitivity instruments in the mm and submm wavelengths has opened the possibility of studying in detail the chemistry of the interstellar medium in extragalactic objects. Within the central few hundred parsec of galaxies, we find enormous amounts of molecular material fueling a wide variety of highly energetic events observed in starbursts (galaxies undergoing an intense burst of star formation) and active galactic nuclei (AGN, where activity is driven by the accretion of material onto the nuclear black hole).

This paper presents a brief summary of both the history and the latest results in observational chemistry in distant galaxies. It will be shown how the molecular emission is a powerful tool to explore the physics of the dust-enshrouded, buried nuclei of distant ultraluminous galaxies, which are heavily obscured at other wavelengths. Special attention will be given to the possibilities offered by next generation instruments such as ALMA (Atacama Large Millimeter Array), expected to have a vast impact on the field of Extragalactic Chemistry. Molecular studies in the early Universe will become available at unprecedented sensitivity and resolution.

1. Introduction: A Brief Look Back

It was back in the mid 1930's when the discovery of a series of diffuse unidentified interstellar lines by Merrill (1934) aroused the interest in interstellar molecules in space. Several species such as CO_2, Na_2 and NaK were invoked to explain these lines. It was shortly after, when Swings & Rosenfeld (1937) identified the spectral feature observed by Dunham (1937) at $\lambda = 4300.3$ Å as due to the CH molecule. Being the first accurate identification of a molecular species in space, this detection became the starting point of the fruitful field of astronomical molecular spectroscopy. This field has experienced an enormous progress during the last four decades thanks to the development of sensitive instruments operating at radio frequencies, where most of the rotational molecular spectrum lies. Almost three decades after the identification of CH, Weinreb et al. (1963) carried out the first molecular radio observation, namely OH in absorption towards the supernova remnant Cassiopeia A. It was followed by a series of pivotal detections such as that of NH_3 (Cheung et al. 1968) and CO (Wilson et al. 1970), marking the starting line of astrochemistry. A few years later, the chemistry in external galaxies would be opened with the CO detection towards M 82 and NGC 253 by Rickard et al. (1975).

Table 1. Complete census of extragalactic molecular species detected to date. Isotopical substitutions are also included.

2 atoms	**3 atoms**	**4 atoms**	**5 atoms**	**6 atoms**	**7 atoms**
OH	H_2O	H_2CO	c-C_3H_2	CH_3OH	CH_3C_2H
CO {^{13}CO, $C^{18}O$, $C^{17}O$}	HCN {$H^{13}CN$, $HC^{15}N$, DCN}	NH_3	HC_3N	CH_3CN	
H_2, HD	HCO^+ {$H^{13}CO^+$, $HC^{18}O^+$, DCO^+}	HNCO	CH_2NH		
CH	C_2H	H_2CS	NH_2CN		
CS {^{13}CS, $C^{34}S$}	HNC {$HN^{13}C$, DNC}	$HOCO^+$			
CH^+	N_2H^+, N_2D^+	C_3H			
CN	OCS	H_3O^+			
SO, ^{34}SO	HCO				
SiO	H_2S				
CO^+	SO_2				
NO	HOC^+				
NS	C_2S				
LiH	H_3^+				

Forty years after the first molecular radio observation, a total of 151 molecular species have been detected in space [1]. The vast majority of them have been detected towards the extremely rich chemical environments of hot molecular cores associated with massive star formation and towards the envelopes of evolved stars. Outside the Milky Way, on the other hand, beam dilution effects together with much broader spectral line features prevent us from such a prolific detection. As shown in Table 1, with the recent addition of the detection of H_3O^+ towards M 82 and Arp 220 (Van der Tak et al. 2007) the number of molecular species detected outside the Milky Way has risen to 40, plus 16 isotopical substitutions of these species. We notice how only some of the simplest complex organic molecules have been detected in the extragalactic interstellar medium (ISM). Several key species in the organic chemistry such as ethanol (C_2H_5OH) and formic acid (HCOOH) have been elusive to several detection attempts.

2. Extragalactic Interstellar Medium

Most of the molecular emission observed in galaxies stems from giant molecular clouds complexes (GMCs) of dense ($> 10^3 cm^{-3}$) and cold (T $\sim 10 - 50$ K) gas. These GMCs with masses of $10^4 - 10^6\ M_\odot$ and sizes of $30 - 50$ parsecs represent the future birthplaces of stars. The study of the molecular emission allows us to extract not only the distribution but also the kinematics and physical conditions (density and temperature) of the molecular gas component in galaxies (e.g. the M 31 mapping of the CO emission by Nieten et al. 2006).

Molecular gas is the fuel powering the energetic events taking place in the nuclear regions of galaxies such as starbursts (SBs), where enhanced star forma-

[1] As compiled in astrochemistry.net (Oct. 2007)

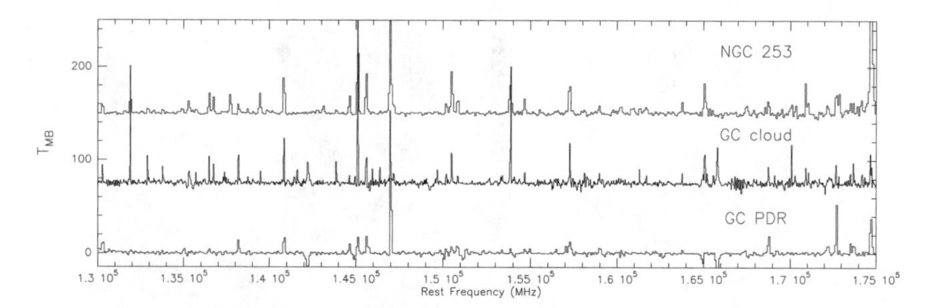

Figure 1. *Upper spectrum*: Composite emission spectra of NGC 253 in the 2 mm atmospheric window (Martín et al. 2006b). *Middle and Lower spectra*: Comparison of the same 2 mm window towards a position in the Sgr B2 molecular envelope, as example of a Galactic Center (GC) giant molecular cloud, and a position in the Sgr A* circumnuclear disk (CND), as an example of a Galactic photodissociation region (Martín et al. in prep.). The spectrum of NGC 253 clearly resembles that in GC molecular clouds, suggesting a similar origin for their chemistries.

tion is observed, or active galactic nucleus (AGN), as a result of the accretion onto a supermassive black hole at the center of the galaxy. The main heating mechanisms, closely related to the nuclear energetic events, affecting the molecular material are: the pervasive and photodissociating UV radiation of young OB associations; the presence of shock waves due to cloud-cloud collisions, expanding envelopes driven by stellar winds of evolved stars or surpernovae explosions; X-ray radiation at the vicinity of the nuclear black hole; and cosmic rays. These heating mechanisms determine the chemistry and therefore the molecular composition of the ISM in galaxies. While most of the molecular material is in the form of molecular hydrogen ($\gg 99\%$), the emission of the different molecular species other than H_2, and thus the study of the chemical composition, provides a key information on the dominating heating mechanisms and physical processes on each galaxy nucleus.

3. The Molecular Composition in Starbursts

Due to the brightness of their lines and the prolific molecular detection towards starburst galaxies, these sources have turned into the target of the vast majority of molecular studies outside the Milky Way. Among the two most extensively studied examples of starbursts we find NGC 253 and M 82.

NGC 253 was recently the target of the first unbiased millimeter molecular line survey in the 2 mm atmospheric window with the IRAM 30m telescope (Martín et al. 2006b). This survey (shown in the upper spectrum of Fig. 1) not only lead to the detection of new extragalactic species such as SO_2, NO and NS (Martín et al. 2003), out of the more than a hundred of identified transitions, but it also represented the first step in the spectral classification of the molecular gas in the nuclei of galaxies. Such complete studies allowed the description, for the first time, of the chemistry of sulfur-bearing molecules in NGC 253 as well as the direct comparison with chemical templates within the galaxy such as that shown in Fig. 1. NGC 253 shows a molecular emission significantly comparable

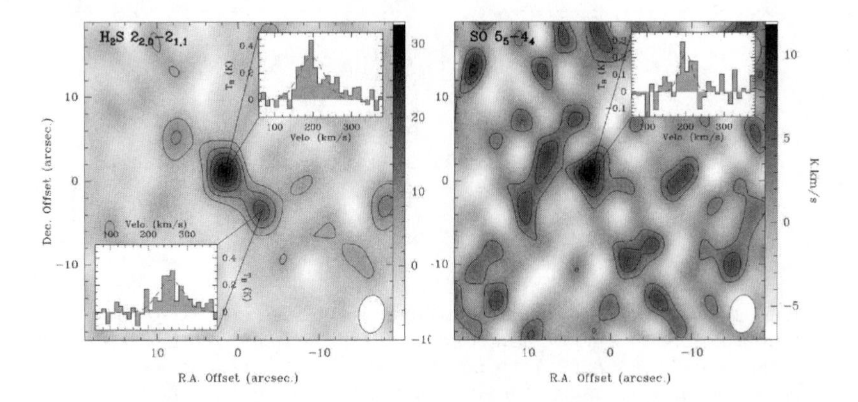

Figure 2. High-resolution observations of H_2S and SO towards the starburst galaxy NGC 253 with the Submillimeter Array (SMA). The emission from these molecules trace the warm dense molecular material associated with the ongoing star formation within the nuclear region of NGC 253. Figure extracted from Minh et al. (2007).

to that observed towards the GMCs we find in the Galactic Center region. This comparison suggests that similar to what we observe towards GMCs, the molecular gas in NGC 253 is mainly heated by low-velocity shock waves, likely due to cloud-cloud collisions in a barred potential orbital motions. High resolution imaging with instruments such as the SMA, allows us to trace the warm and dense molecular cloud complexes within the nuclear region of NGC 253, which represent the fuel of the intense observed star formation in this galaxy. Fig 2 shows an example of the observation of two sulfur-bearing species such as H_2S or SO (Minh et al. 2007).

M 82, on the other hand shows a particularly interesting chemistry. It has long been known that *M 82* has surprisingly relative low abundances derived by the non detection of several species such as SiO, HNCO and CH_3OH (Henkel et al. 1987; Mauersberger & Henkel 1991; Nguyen-Q-Rieu et al. 1991) as well as the non detection of a hot NH_3 component that was observed in other starbursts (Weiß et al. 2001; Takano et al. 2002; Mauersberger et al. 2003). The recent detection of widespread emission of species commonly used as photodissociation tracers, namely HCO, CO^+, HOC^+, and C_2H (García-Burillo et al. 2002; Fuente et al. 2006), turned the nuclear region of M 82 into the prototype of extragalactic photodissociation region (PDR). Nevertheless, the nuclear regions of galaxies are heterogeneous composites of cloud complexes affected by different environmental conditions. It is therefore difficult to find pure chemical prototypes in external galaxies. This is particularly suggested by the detection of methanol in M 82 with abundances well above that expected by pure gas-phase chemistry, which is an evidence of the presence of a significant amount of molecular material shielded from the pervading UV radiation and similar to that found in other starbursts galaxies such as NGC 253 or Maffei 2 (Martín et al. 2006a).

NGC 4945 has also been the target of one of the most ambitious multi-line studies of the molecular content of the ISM in a starburst galaxy. A total of

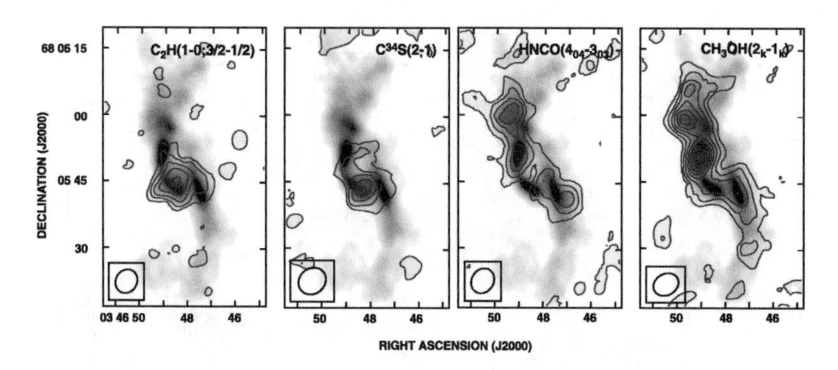

Figure 3. Integrated intensity of C_2H, $C^{34}S$, HNCO, and CH_3OH towards the nuclear region of IC 342 with the ^{12}CO (1-0) in grey scale for comparison. Figure extracted from Meier & Turner (2005).

80 transitions of 19 species were observed with the SEST 15 m telescope in the range between 82 and 354 GHz towards this galaxy (Wang et al. 2004).

The overall comparison of the fractional abundances observed towards these prototype galaxies has allowed us to build up a picture of the chemical evolutionary sequence of their nuclear starbursts. Within this sequence, NGC 253 and M 82 would represent the extreme evolutionary states. Thus NGC 253 appears to be in an early stage with huge amounts of dense molecular gas feeding the ongoing burst of star formation where the dominant chemical fingerprints we observe are those of the shocks affecting this material. On the other side, M 82, with a similar star forming rate, has consumed most of its molecular gas reservoir and its ISM shows the traces of the photodissociation produced by the young hot stars formed during star burst period. Galaxies such as NGC 4945 would represent a prototype of intermediate stage between those of NGC 253 and M 82 (Wang et al. 2004; Martín et al. 2006b).

The understanding of the chemical complexity in galaxies altogether with the high spatial resolution provided by current interferometers enables us to spatially disentangle the different physical processes taking place in the nuclear region of galaxies. That is the case of the high angular resolution molecular line observations towards IC 342 carried out by Meier & Turner (2005). The mapping of selected molecular tracers (Fig. 3) shows how the molecular gas is highly pervaded by the photodissociation radiation around the nuclear star clusters (traced by C_2H, and $C^{34}S$,) while the denser and warmer material is observed towards the circumnuclear spiral arms where the gas is mostly affected by cloud-cloud collisions (traced by HNCO and CH_3OH), in agreement with the SiO observations by Usero et al. (2006).

4. Beyond Starbursts

During the last few years, an increasing interest is coming up on unveiling the molecular gas content of galaxies at early times as this is likely to be directly related to the star formation history of the Universe. As shown in Fig. 4, molecular gas has been detected up to a redshift of z=6.42 (Walter et al. 2003), when

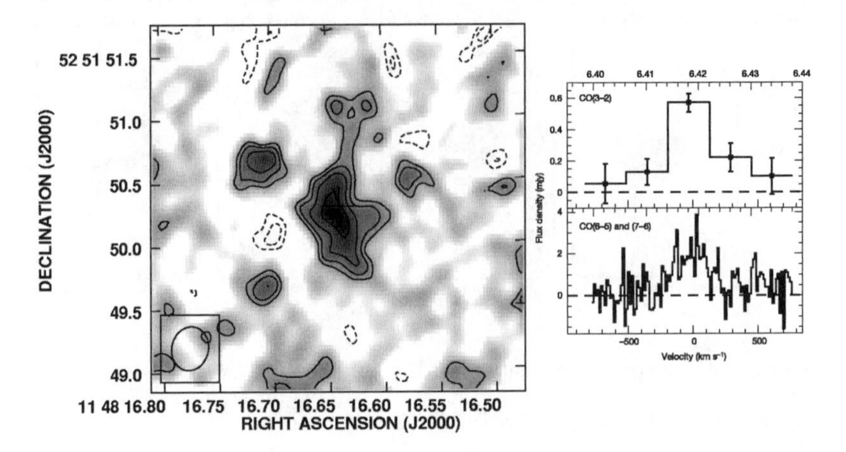

Figure 4. CO observations towards J1148+5251, the highest redshift quasar currently known, at a redshift z=6.42 which corresponds to an age of the Universe younger than 1 Myr. *Left:* CO (3-2) obtained with the VLA. A total of ~ 60 hours of integration time were necessary to get the resolved CO structure of this source *Right:* CO transitions detected with the VLA (*upper*) and PdBI (*lower*). Figures extracted from Walter et al. (2003, 2004).

the Universe had an age below 1 million years. Unfortunately, we are still far from being able to study the chemical composition of these sources in such detail as we reach in nearby galaxies. The limited sensitivity of current instruments prevent us from going further in the search for molecules more complex than CO as evidenced by the recent non detection of HCN towards J1148+5251 (Riechers et al. 2007). This is the reason of the increased effort to study the chemistry of Ultra Luminous Infrared Galaxies (ULIRGs) which, with enormous star formation rates, are thought to be reasonable templates of the ISM conditions at high redshift. Among these studies, Arp 220 (at a z$\sim$ 0.02) has been the target of recent sensitive multiline studies (Greve et al. 2009).

At not so high redshifts, when the Universe was $\sim 2-4$ Myr old, detections have been obtained of HCN and HCO^+. These observations, close to the limit of current instrumentation capabilities, require many hours of observing time in order to reach a significant detection (e.g. Wagg et al. 2005; Riechers et al. 2006).

Nevertheless, the observation of the molecular material at high redshift provides us with some interesting advantages, such as the possibility of detecting the high-J CO transitions as shown by the $J = 11-10$ observations towards a z=3.9 quasar by Weiß et al. (2007). The rest frequency of this transition, at 1.3 GHz, is almost impossible to be observed from ground telescopes, but it is easily detected with the available mm telescopes at its redshifted frequency of $\sim$ 1 mm.

5. A Bright Future with ALMA

The Atacama Large Millimeter Array (ALMA) is creating great expectations among the astronomical comunity. In particular, ALMA will be a breakthrough in the field of extragalactic chemistry because of two main reasons: First, the combination of resolution and sensitivity. With baselines ranging from 150 m to 18 km, ALMA will reach resolutions between $0.7''$ and $0.01''$ at 1 mm with a sensitivity of more than an order of magnitude with respect of current available mm and sub-mm instruments. It will be possible to observe the nearby galaxies with the resolution we have currently in the Galactic Center with single dish telescopes and we will be able to fully resolve the molecular emission of galaxies at the early ages of the Universe. Similar to what we observe today in the Galaxy, we will be able to study in detail the chemical complexity towards the star forming region in nearby galaxies; Secondly, the instantaneous wideband of at least 8 GHz will allow the mapping of the emission of many molecular lines at a time. With an unprecedented observing speed, as compared to the hundreds of hours which would require nowadays, ALMA will provide us with similar line surveys as those in Fig. 1 with the additional spatial distribution of each of the molecular species.

ALMA (the Spanish word for *SOUL*) will be the starting point of the study of the complex organic molecules in external galaxies. This will open the possibility of observing the emission of complex bio-molecules from nearest to the highest redshifted galaxies and therefore we will get a hint on the evolution of the building blocks of life through the history of the Universe.

References

Cheung, A. C., Rank, D. M., Townes, C. H., Thornton, D. D., & Welch, W. J. 1968, Physical Review Letters, 21, 1701
Dunham, T., Jr. 1937, PASP, 49, 26
Fuente, A., García-Burillo, S., Gerin, M. et al. 2006, ApJ, 641, L105
García-Burillo, S., Martín-Pintado, J., Fuente, A., & Usero, A. 2002, ApJ, 575, L55
Greve, T. R., Papadopoulos, P. P., Gao, Y., & Radford, S. J. E. 2009, ApJ, 692, 1432
Henkel, C., Jacq, T., Mauersberger, R., Menten, K. M., & Steppe, H. 1987, A&A, 188, L1
Martín, S., Mauersberger, R., Martín-Pintado, J., García-Burillo, S., & Henkel, C. 2003, A&A, 411, L465
Martín, S., Martín-Pintado, J., & Mauersberger, R. 2006a, A&A, 450, L13
Martín, S., Mauersberger, R., Martín-Pintado, J., Henkel, C., & García-Burillo, S. 2006b, ApJS, 164, 450
Mauersberger, R., & Henkel, C. 1991, A&A, 245, 457
Mauersberger, R., Henkel, C., Weiß, A., Peck, A. B., & Hagiwara, Y. 2003, A&A, 403, 561
Meier, D. S., & Turner, J. L. 2005, ApJ, 618, 259
Merrill, P. W. 1934, PASP, 46, 206
Minh, Y. C., Muller, S., Liu, S.-Y., & Yoon, T. S. 2007, ApJ, 661, L135
Nguyen-Q-Rieu, Henkel, C., Jackson, J. M., & Mauersberger, R. 1991, A&A, 241, L33
Nieten, C., Neininger, N., Guélin, M., Ungerechts, H., Lucas, R., Berkhuijsen, E. M., Beck, R., & Wielebinski, R. 2006, A&A, 453, 459
Rickard, L. J., Palmer, P., Morris, M., Turner, B. E., & Zuckerman, B. 1975, ApJ, 199, L75

Riechers, D. A., Walter, F., Carilli, C. L., Weiss, A., Bertoldi, F., Menten, K. M., Knudsen, K. K., & Cox, P. 2006, ApJ, 645, L13
Riechers, D. A., Walter, F., Carilli, C. L., & Bertoldi, F. 2007, ArXiv e-prints, 710, arXiv:0710.4525
Swings, P., & Rosenfeld, L. 1937, ApJ, 86, 483
Takano, S., Nakai, N., & Kawaguchi, K. 2002, PASJ, 54, 195
Usero, A., García-Burillo, S., Martín-Pintado, J., Fuente, A., & Neri, R. 2006, A&A, 448, 457
Van der Tak, F., Aalto, S., & Meijerink, A&A, submitted (2007).
Wagg, J., Wilner, D. J., Neri, R., Downes, D., & Wiklind, T. 2005, ApJ, 634, L13
Walter, F., et al. 2003, Nat, 424, 406
Walter, F., Carilli, C., Bertoldi, F., Menten, K., Cox, P., Lo, K. Y., Fan, X., & Strauss, M. A. 2004, ApJ, 615, L17
Wang, M., Henkel, C., Chin, Y.-N., Whiteoak, J. B., Hunt Cunningham, M., Mauersberger, R., & Muders, D. 2004, A&A, 422, 883
Weinreb, S., Barrett, A. H., Meeks, M. L., & Henry, J. C. 1963, Nature, 200, 829
Weiß, A., Neininger, N., Henkel, C., Stutzki, J., & Klein, U. 2001, ApJ, 554, L143
Weiß, A., Downes, D., Neri, R., Walter, F., Henkel, C., Wilner, D. J., Wagg, J., & Wiklind, T. 2007, A&A, 467, 955
Wilson, R. W., Jefferts, K. B., & Penzias, A. A. 1970, ApJ, 161, L43

Bill Irvine

Steven Charnley

Bioastronomy 2007: Molecules, Microbes, and Extraterrestrial Life
ASP Conference Series, Vol. 420, 2009
K. J. Meech, J. V. Keane, M. J. Mumma, J. L. Siefert, and D. J. Werthimer, eds.

Theoretical Models of Complex Molecule Formation on Dust

S. B. Charnley,[1] and S. D. Rodgers,[2]

[1]*Astrochemistry Laboratory & Center for Astrobiology, Solar System Exploration Division, Code 691, NASA Goddard Space Flight Center, Greenbelt, MD 20771, USA*

[2]*Space Science and Astrobiology Division, NASA Ames Research Center and Carl Sagan Center, SETI Institute, USA*

Abstract. Catalytic reactions on interstellar grain surfaces can lead to a large variety of organic molecules. The basic theoretical ideas necessary to model these processes correctly are presented.

1. Introduction

Chemical reactions on cold interstellar dust grains are potentially a rich source of organic molecules throughout the Galaxy (Ehrenfreund & Charnley 2000). As some cometary volatiles, at least, appear to originate from dense cloud chemistry (Bockelee-Morvan et al. 2004; Ehrenfreund et al. 2004; Charnley & Rodgers 2008), and both comets and asteroids may have served to provide the first organic material to be utilized in prebiotic chemistry (Botta & Bada 2002; Chyba & Hand 2005), it is of great interest to understand the details of catalysis on interstellar grains from an astrobiological perspective.

The high sensitivity of the Spitzer Space Telescope is now allowing grain mantle characteristics to be determined in fainter luminosity sources, such as low-mass protostars (Boogert et al. 2004). Apart from the compositional information, Infrared Space Observatory (ISO), Spitzer and ground-based observations at large telescopes (Keck, VLT) have allowed the molecular ice *structures* to be investigated in protostellar (YSO) environments and toward field stars (Ehrenfreund et al. 1998; van Dishoeck 2004; Boogert et al. 2004; Bergin et al. 2005; Knez et al. 2005; Whittet et al. 2007). These include the polarity of the mixture in which a molecule (CO) resides, methanol-CO_2 complexes seen towards some YSOs, and CO_2 in both polar and apolar environments in quiescent media.

In dense molecular clouds, the first icy mantle constituents form via surface reactions on cold grains. These reactions occur in two steps: atoms and molecules accrete from the gas phase on to the surface where they diffuse (by tunneling or hopping) and react, essentially in a Langmuir-Hinchelwood kinetics. Interstellar organic molecules may be produced by cold grain-surface reactions (Charnley 1997) or, once a mantle has formed, by secondary energetic processing by UV photons or cosmic rays (Bernstein et al. 1995; Moore & Hudson 1998; Bennett et al. 2005). In the first case, the scheme of Figure 1 (Charnley

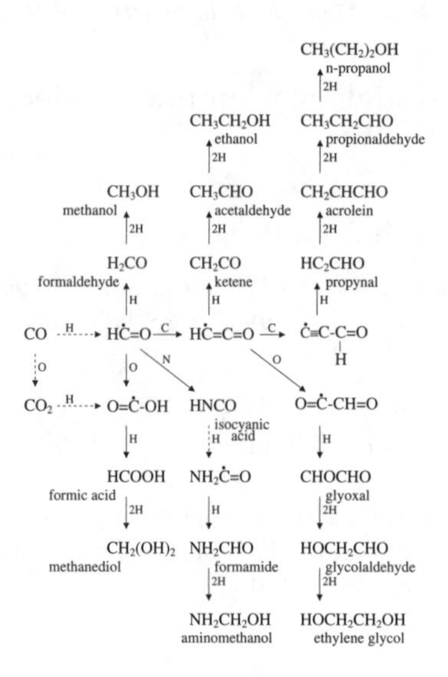

Figure 1. Grain surface reaction scheme forming interstellar organics via atom addition reactions (see Charnley & Rodgers 2006). Broken arrows indicate reactions with activation energy barriers and solid arrows denote barrierless reactions.

2001a; Charnley & Rodgers 2006) indicates that a defining feature of interstellar grain catalysis is the growth of linear chains by single atom additions and their subsequent saturation. In this contribution we outline how the composition and structure resulting from such theoretical surface schemes can be computed and compared with observations of interstellar ices.

2. Stochastic Surface Kinetics

A stochastic treatment of both the gas (Charnley 1998, paper I) and the grain-surface chemical kinetics is necessary (Charnley 2001b, paper II). For N chemical species connected by M chemical reactions, the occurrence of a particular reaction μ (including accretion and desorption) changes the surface populations. If X_i denotes the population of molecule i, the chemical composition of the surface is defined by a set of N independently-distributed random variables $\mathbf{x} = \{X_i\}$ and the probability that the system is in state $\{X_i\}$ at time t is given by the joint probability distribution function (PDF), $P(\mathbf{x};t)$. The statistical formulation of chemical reaction kinetics self-consistently takes account of the intrinsically random nature of the gas-grain interaction and involves the solution of the chemical master equation for the surface evolution of $P(\mathbf{x};t)$.

$$\frac{\partial}{\partial t}P(\mathbf{x};t) \;=\; \sum_{\mu}\Big[W_\mu(\mathbf{x}-\mathbf{s}_\mu)P(\mathbf{x}-\mathbf{s}_\mu;t) \;-\; W_\mu(\mathbf{x})P(\mathbf{x};t)\Big] \qquad (1)$$

where $\mathbf{s}_\mu$ is a vector of integer quantities accounting for the stoichiometry of reaction μ and $W_\mu(\mathbf{x})$ is the probability per unit time of reaction μ causing chemical change in the infinitesimal time interval $[t,\ t+dt]$, given that it is in state $\mathbf{x}$ at time t.

Solution of the master equation by a Monte Carlo technique (Gillespie 1976) allows a straightforward computation of the surface populations. The transition rates $W_\mu(\mathbf{x})$ for each surface process are defined as follows. For a surface reaction between two distinct species a and b it is $W_\mu(\mathbf{x}) = \gamma_\mu q_\mu(\mathbf{x}) p_\mu X_a X_b$. Here γ_μ is the stochastic surface rate coefficient and is given by τ^{-1}, the inverse of the characteristic time-scale for a particle to move to an adjacent surface site (i.e. by quantum tunneling or thermal hopping). The combinatorial factor $q_\mu(\mathbf{x})$ accounts for the number of distinct ways[1] two species, having surface populations X_a and X_b, can react on a surface with a finite number of binding sites $\mathcal{N}$. The probability factor p_μ allows reactions with activation-energy barriers to be considered. For exothermic activationless reactions $p_\mu = 1$. For a reaction possessing an activation-energy barrier (idealized as a square potential of height E_μ and width L), the probability of quantum-mechanical penetration is

$$p_\mu = \exp\left[-13\left(\frac{L}{1\text{Å}}\right)\left(\frac{E_\mu}{1000\ \text{K}}\right)^{1/2}\right] \tag{2}$$

whereas thermal hopping over the reaction barrier has a probability of

$$p_\mu = \exp\left[-100\left(\frac{E_\mu}{1000\ \text{K}}\right)\left(\frac{kT_{\text{dust}}}{10\ \text{K}}\right)^{-1}\right] \tag{3}$$

As there is a natural hierarchy of time-scales on grain surfaces (collision and physisorption, tunneling, hopping, reaction), we need only know the populations and the rates of the different surface processes to compute the surface kinetics. Hence, the model, as formulated, effectively bypasses the need to consider the actual 2D problem of surface catalysis. The great utility of this approach is that the physics assumptions made in the models, such as whether H atoms migrate by tunneling or hopping, can be easily modified. However, it is important to recognize that chemical reactions are restricted to a *surface monolayer* and thus it is necessary to consider separately the surface and bulk components of each particle's population (X_i^s and X_i^b). Only X_i^s is active in the the simulation of the above master equation. Not making this distinction can lead to gross errors since, usually, $X_i^b \gg X_i^s$. Algorithms designed to account for the covering and reaction (or uncovering) of surface particles are used to keep track of the surface population; X_i^b is monitored since it is the quantity that can be compared to observations (Charnley 2001b).

3. Stochastic Simulation of Gas-Grain Chemistry

Stochastic simulation of both the gas and grain chemistries in tandem is necessary because the chemical composition (e.g. the atom/molecule ratios : C/CO,

[1]Equation (8) of Paper II contains a typographical error and should read (for surface populations X_a and X_b) $q_\mu(\mathbf{x}) = \frac{\binom{X_a}{1}\binom{X_b}{1}}{\binom{\mathcal{N}}{2}} = \frac{X_a X_b}{\mathcal{N}(\mathcal{N}-1)}$

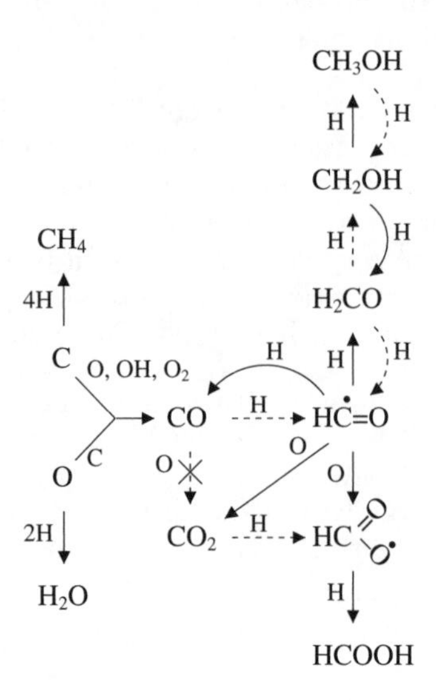

Figure 2. Grain-surface chemistry constrained by the observed composition of ice mantles. Broken arrows indicate reactions with activation energy barriers Rodgers & Charnley 2008). The reaction of atomic oxygen to CO is not supported by current laboratory experiments and is crossed-out.

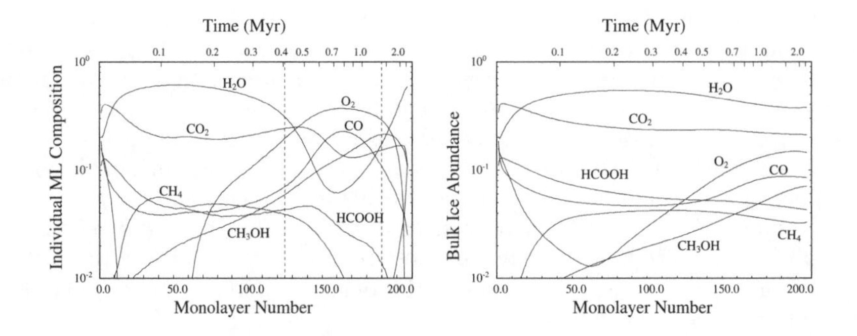

Figure 3. Ice compositions predicted by our model. Left panel: abundances in individual monolayers (ML) vs. depth and time. Right panel: Total bulk abundances. The dashed lines in the left panel show three chemically-distinct layers.

O/CO, N/N_2, O/O_2) of the gas changes in time. We have therefore recently completed a unified gas-grain chemistry code (Charnley & Rodgers 2006; Rodgers & Charnley 2009) which at present is the state-of-the art in interstellar surface kinetics. However, the theoretical model outlined above assumes reactions on perfect *regular* lattices and that the energy barrier heights and widths (E_μ & L) are well known. Neither of these assumptions may in fact apply. For example, interstellar grains are likely to have a fractal morphology and a random walk on such a structure involves longer migration times than on a regular lattice (Charnley 2005). Hence, we treat the $W_\mu(\mathbf{x})$ as parameters to be adjusted so that the composition and structure of the grain mantles can be reproduced. We also include reverse reactions for H additions (Figure 2). As we can track the molecular composition of each monolayer we can calculate the chemical abundance as a function of depth into the mantle. Figure 3 shows an example of the growth and chemical evolution grain mantles in a simulation of the surface chemistry of Figure 1 (Rodgers & Charnley 2008). One important finding from this study was that O atom addition to CO on grains cannot be efficient, otherwise it is not possible to have observable amounts of CO in the ices. Instead, the CO_2 kinetics is governed by O addition to HCO, as proposed by Ruffle and Herbst (2001). We always find that H_2O, CO(polar), CO_2, HCOOH layers form first. As the atomic O/H ratio in the gas falls, an apolar layer of CO, CO_2 and O_2 forms where the O_2 present is accreted directly from the gas. The precise composition of this layer is quite sensitive to the elemental C/O ratio in the gas. In this model, the highest abundances of CH_3OH occur later in the evolution, when reduction dominates oxidation. The methanol abundance grows slowly due to the influence of reverse reactions (Figure 2), reaching its highest abundance in the outermost layers where it is mixed with water. All of these characteristics are consistent with those determined for astronomical ices (van Dishoeck 2004; Boogert et al. 2004; Bergin et al. 2005; Knez et al. 2005; Whittet et al. 2007) and this is the first model to reproduce them.

4. Conclusions

We have outlined what are likely to be the dominant pathways to organic molecules on cold interstellar dust grains. Computation of the predicted mantle compositions requires a stochastic treatment of both the gas and grain-surface chemistries. The fact that such models can be adjusted to yield, for the dominant ice molecules, both their observed composition and how the molecules are mixed together, leads us to expect that future simulations will allow the abundances of larger organic molecules to be accurately predicted.

Acknowledgments. This work was supported by NASA's Exobiology Program and by the NASA Goddard Center for Astrobiology.

References

Bennett, C.J. et al. 2005, 634, 698
Bergin, E.A., et al. 2005, ApJ, 627, L33
Bernstein, M.P. et al. 1995, ApJ, 454, 327

Bockelée-Morvan D., et al. 2004, in *COMETS II*, Eds. M. Festou, H.U. Keller, H.A. Weaver, University of Arizona Press, 391

Boogert, A.C.A. et al. 2004, ApJS, 154, 359

Botta, O., Bada, J.L. 2002, Surv. Geophys. 23, 411

Charnley, S.B. 1997, in Astronomical and Biochemical Origins and the Search for Life in the Universe, eds. C.B. Cosmovici et al. (Editrice Compositori), p. 89

Charnley S.B., 1998, ApJ, 509, L121 (Paper I)

Charnley S.B., 2001a, in The Bridge Between the Big Bang and Biology, Ed. F. Giovannelli, Consiglio Nazionale delle Ricerche President Bureau, (Rome:Italy), Special Volume, p. 139

Charnley S.B., 2001b, ApJ, 562, L99 (Paper II)

Charnley S.B., 2005, Adv. Space Res., 36,, 132

Charnley, S.B. & Rodgers, S.D. 2006, in Astrochemistry Throughout the Universe: Recent Successes and Current Challenges , D. C. Lis, et al. (Eds.), (CUP), 237

Charnley S.B., Rodgers S.D., 2008, Space Sci. Rev., in press

Chyba, C.F. & Hand, K.P. 2005, ARA&A, 43, 31

Ehrenfreund, P. et al. 1998, A&A, 339, L17

Ehrenfreund, P, Charnley S.B. 2000. ARA&A, 38, 427

Ehrenfreund P., Charnley S.B., Wooden D.H., 2004, in *COMETS II*, Eds. M. Festou, H.U. Keller, H.A. Weaver, University of Arizona Press, p. 115

Gillespie, D.T. 1976, J. Comp. Phys. , 22, 403

Knez, C. et al. 2005, ApJ, 635, L145

Moore M.H., Hudson R.L., 1998, Icarus, 135, 518

Rodgers, S.D., Charnley, S.B. 2009, in preparation

Ruffle, D.P., Herbst, E. 2001, MNRAS, 324, 1054

van Dishoeck, E.F. 2004, Annu. Rev. A&A, 42, 119

Whittet, D.CB., et al. 2007, Ap. J. , 655, 332

Bioastronomy 2007: Molecules, Microbes, and Extraterrestrial Life
ASP Conference Series, Vol. 420, 2009
K. J. Meech, J. V. Keane, M. J. Mumma, J. L. Siefert, and D. J. Werthimer, eds.

Searching for the Precursors of Life in External Galaxies

B. Lawton,[1,2] C. W. Churchill,[1] B. A. York,[3] S. L. Ellison,[3]
T. P. Snow,[4] R. A. Johnson,[5] S. G. Ryan,[6] and C. R. Benn[7]

[1]*Department of Astronomy, MSC 4500, New Mexico State University, P.O. Box 30001, Las Cruces, NM 88003*

[2] *Space Telescope Science Institute (STScI), 3700 San Martin Drive Baltimore, MD 21218*

[3]*Department of Physics and Astronomy, University of Victoria, P.O. Box 3055, Station CSC, Victoria, BC V8W 3P6, Canada*

[4]*Center for Astrophysics and Space Astronomy, University of Colorado at Boulder, 389 UCB, Boulder, CO 80309*

[5]*Oxford Astrophysics, Denys Wilkinson Building, Keble Road, Oxford OX1 3RH, UK; raj@astro.ox.ac.uk*

[6]*Centre for Astrophysics Research, University of Hertfordshire, College Lane, Hatfield AL10 9AB, UK*

[7]*Isaac Newton Group, Apartado 321, E-38700 Santa Cruz de La Palma, Spain*

Abstract. Are the organic molecules crucial for life on Earth abundant in early-epoch galaxies? To address this, we searched for organic molecules in extragalactic sources via their absorption features, known as diffuse interstellar bands (DIBs). There is strong evidence that DIBs are associated with polycyclic aromatic hydrocarbons (PAHs) and carbon chains. Galaxies with a preponderance of DIBs may be the most likely places in which to expect life.

We use the method of quasar absorption lines to probe intervening early-epoch galaxies for the DIBs. We present the equivalent width measurements of DIBs in one neutral hydrogen (H I) abundant galaxy and limits for five DIB bands in six other H I-rich galaxies (damped Lyman-α systems–DLAs). Our results reveal that H I-rich galaxies are dust poor and have significantly lower reddening than known DIB-rich Milky Way environments. We find that DIBs in H I-rich galaxies do not show the same correlation with hydrogen abundance as observed in the Milky Way; the extragalactic DIBs are underabundant by as much as 10 times. The lower limit gas-to-dust ratios of four of the H I-rich early epoch galaxies are much higher than the gas-to-dust ratios found in the Milky Way. Our results suggest that the organic molecules responsible for the DIBs are underabundant in H I-rich early epoch galaxies relative to the Milky Way.

1. Introduction

Since their discovery in 1921 (Heger 1922), the diffuse intersteller bands (DIBs) have remained the longest known interstellar absorption features without a positive identification. There have been several hundred DIBs discovered to date (Jenniskens et al. 1994; Tuairisg et al. 2000; Weselak et al. 2000). The DIBs span the visible spectrum between 4000 and 13000 Å. Despite no positive identifications, several likely organic molecular candidates have emerged as the sources of the DIBs, including polycyclic aromatic hydrocarbons (PAHs), fullerenes, long carbon chains, and polycyclic aromatic nitrogen heterocycles (PANHs) (Herbig 1995; Snow 2001; Cox & Spaans 2006a; Hudgins et al. 2005). The organic-molecular origin of the DIBs may give them an importance to astrobiology; they are now considered an important early constituent to the inventory of organic compounds on Earth (Bada & Lazcano 2002).

There are several environmental factors that are known to enhance or inhibit DIB strengths. Thus, measuring DIB strengths can give clues to the environments of galaxies like those with high neutral hydrogen (H I) content, known as damped Lyman-α systems (DLAs). Measuring DIB strengths allow researchers to explore the quantities of gas, metallicity, dust, and radiation in galaxies and their influence on the strengths of the bands. Furthermore, observing DIBs in galaxies at a higher redshift, z, (distance or lookback time related to the expansion of the Universe) allows us to test the evolution of these organics in cosmic time.

Two important environmental factors that are often probed in galaxies are the H I content and the reddening. The H I content in DLAs is typically measured as a column density via their Lyman-α line in absorption, as observed using a bright background source such as a quasar. The column density is the number of atoms of a certain element as seen along a line-of-sight projected to a unit area at the observer (typically in units of atoms cm^{-2} and denoted as N(H I) for neutral hydrogen). The reddening is a measure of the dust content in the galactic environment. Reddening is expressed as $E(B-V)$, or the magnitude of blue light minus visible light observed relative to what is expected from typical stars or galaxies. Because dust preferentially obscures blue light, a high reddening is a signature for a high dust content. N(H I) is a measure of the gas phase and $E(B-V)$ is a measure of the dust phase of the ISM in galaxies.

Due to their relatively weak absorption strengths, the DIBs have been difficult to detect in extragalactic sources. Aside from the hundreds of detections within the Milky Way (Jenniskens et al. 1994; Tuairisg et al. 2000; Weselak et al. 2000), DIBs have been detected in the Magellanic Clouds (Welty et al. 2006; Cox et al. 2006b, 2007), seven starburst galaxies (Heckman & Lehnert 2000), the active galaxy Centaurus A via supernova 1986A (Rich 1987), spiral galaxy NGC 1448 via Supernovae 2001el and 2003hn (Sollerman et al. 2005), one DLA galaxy at $z = 0.524$ toward the quasar AO 0235+164 (Junkkarinen et al. 2004; York et al. 2006), and one galaxy selected by singly ionized calcium (Ca II), J0013–0024, at $z = 0.157$ from the Sloan Digital Sky Survey (Ellison et al. 2008).

We further the knowledge of DIB lines in extragalactic environments by cataloguing the strengths of the λ4428, λ5780, λ5797, λ6284, and λ6613 DIBs relative to the $E(B-V)$ and N(H I) content of each of the seven DLAs in our sample. Observations were obtained, with seven facilities, of seven DLAs toward six QSO sightlines. The facilities and instruments used for this project are the VLT/FORS2, VLT/UVES, APO/DIS, Keck/HIRES, WHT/ISIS, and Gemini/GMOS-S.

2. Analysis & Results

There are two detections included in this work, the λ5705 and λ5780 DIBs first reported by York et al. (2006), in the $z = 0.524$ DLA toward AO 0235+164. For all other DLAs in our sample, we report upper limits on the λ4428, λ5780, λ5797, λ6284, and λ6613 DIB equivalent widths. We measured the equivalent width limits using a generalized method of the Schneider et al. (1993) technique for finding lines and limits. We compare our measured limits to the expected DIB equivalent widths from the known Milky Way DIB–$E(B-V)$ and DIB–N(H I) relations (Welty et al. 2006). The N(H I) quantities are known for the DLAs; however, Junkkarinen et al. (2004) published the only reddening known for the DLA galaxies in our sample, AO 0235+164, with a measured $E(B-V) = 0.23$. We estimate the upper limit to the reddening using our equivalent width limits and the Milky Way DIB–$E(B-V)$ correlation. Our equivalent width limits are robust enough to constrain the upper reddening limits near the $E(B-V) < 0.04$ limit found by Ellison et al. (2005) for the highest redshift DLA galaxies.

The results from the N(H I) model suggests that the organics that give rise to the DIBs in DLAs are underabundant relative to Milky Way sightlines of the same hydrogen column density. Fig. 1 shows this by plotting the measured equivalent widths and upper equivalent width limits for the DLAs in our sample. The line is the best-fit to the Milky Way data from Welty et al. (2006). The Milky Way points are observed to lie within the dotted region while the Large Magellanic Cloud sightlines are observed to lie within the dashed region. The Small Magellanic Cloud sightlines are all within the dot-dashed region. The λ6284 DIB gives the best constraints and shows that this DIB is at least 4-10 times weaker in four of our DLAs compared to what is expected in the Milky Way. As is the case for the Magellanic Clouds, the Milky Way DIB–N(H I) relation does not apply to DIBs in DLAs. Many environmental factors can potentially work to inhibit the organics responsible for the DIBs so this alone can not be used as evidence that the environments of DLAs are similar to the Large or Small Magellanic Cloud. For example, the dust content, ionizing radiation, and metallicity may be quite different in DLAs relative to the Magellanic Clouds.

From our equivalent width limits we can estimate the upper limit to the reddening assuming the Milky Way DIB–$E(B-V)$ relation holds for DLAs. There are little data on DIBs in DLAs; however, Ellison et al. (2008) create a fit to all known extragalactic points for the λ5780 DIB. The λ5780 DIB–$E(B-V)$ relation appears to remain valid when the extragalactic equivalent width measurements are included, although the slope is slightly steeper. Our upper limits for $E(B-V)$ yield lower limits to the gas-to-dust ratios for our DLAs; these results are shown in Fig. 2. $E(B-V)_{lim}$ are the upper limits for the reddening determined by our best equivalent width limits. The best-fit lines for the Milky Way and the Large Magellanic Cloud are given from the literature. Our limits are robust enough to constrain the gas-to-dust ratio for the DLAs as being much higher than the Milky Way for four of the DLAs. The discrepency between the limit and the measurement of the DLA toward AO 0235+164 may be due to problematic measurements. Our best gas-to-dust lower limit comes from the reddening constraints for the λ6284 DIB; however, there is a large atmospheric absorption band at the DIB's expected location. It could also be that the λ6284 DIB–$E(B-V)$ relation is different for DLAs that is measured for

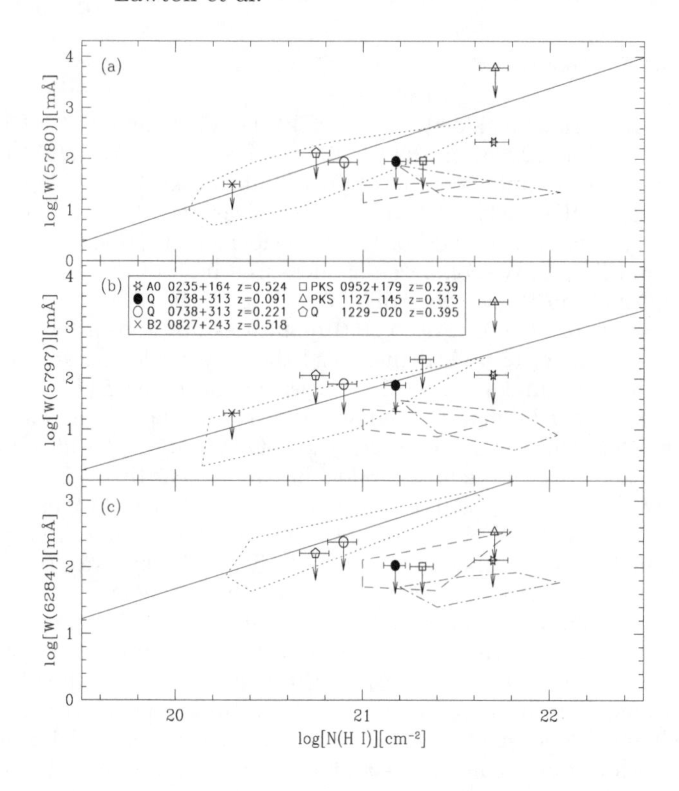

Figure 1. The DIB equivalent width–N(H I) relations (Welty et al. 2006) with our DLAs added. —(a) λ5780 DIB. —(b) λ5797 DIB. —(c) λ6284 DIB. The solid lines are the best-fit weighted Milky Way lines. The region enclosed by the dotted lines contain the Milky Way data. The regions enclosed by the dashed lines contain the LMC data. The regions enclosed by the dot-dash lines contain the SMC data. Error bars are 1 σ, and upper limits are marked with arrows. The vertical error bars for AO 0235+164 in panel (a) are smaller than the point size and all values for this DLA are from York et al. (2006).

the Milky Way. The reddening limits allow us to extend the results by Ellison et al. (2005) to lower redshift DLAs. The $E(B-V) < 0.04$ from Ellison et al. (2005) appears to also be a robust result for moderate redshift DLAs.

3. Implications for Astrobiology

Despite having comparable H I abundances to the Milky Way, DLA galaxies have DIB strengths that are weaker than expected. The weakness of DIBs in DLA selected galaxies suggests that the environments of these early epoch H I-rich galaxies are much less suitable to create and/or sustain organic molecules. DLA galaxies do have considerably lower reddening than the Milky Way regions where DIBs are observed. Our results imply that the low reddening in DLAs will inhibit the DIBs if they are present. Reddening is dependent on dust grain sizes and abundances. It is conceivable that the organics require dust grains

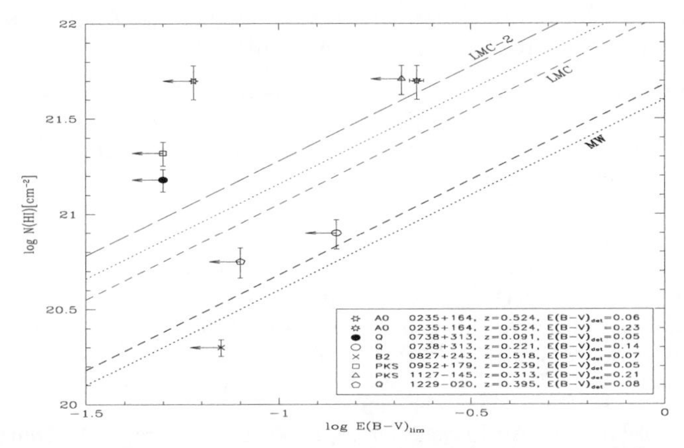

Figure 2. The gas-to-dust ratios of the DLAs in our sample relative to measured values in the Milky Way (MW) and the Large Magellanic Cloud (LMC). The figure is modified from Cox et al. (2006b). The plot measures the log column density [cm^{-2}] versus the upper limit to the log reddening for each DLA determined. The A0 0235+164 reddening measurement of 0.23±0.01 is from Junkkarinen et al. (2004). The top three lines represent the LMC while the bottom two lines represent the MW. The long-dashed LMC line gives the gas-to-dust ratio of 19.2 × 10^{21} cm^{-2} mag^{-1} from the LMC-2 data of Gordon et al. (2003). The dotted LMC line is a linear fit to the LMC data in Cox et al. (2006b) and gives a gas-to-dust ratio of 14.3 × 10^{21} cm^{-2} mag^{-1}. The short-dashed line is the average LMC regions from (Gordon et al. 2003) and has a gas-to-dust ratio of 11.1 × 10^{21} cm^{-2} mag^{-1}. The dashed MW line gives a gas-to-dust ratio of 4.8 × 10^{21} cm^{-2} mag^{-1} (Bohlin et al. 1978), and the dotted MW line is the fit to the Milky Way data from Cox et al. (2006b) which yields a ratio of 4.03 × 10^{21} cm^{-2} mag^{-1}. Several of the DLAs in our sample are consistent with having higher gas-to-dust ratios than the MW and the LMC.

for their formation or are created from the same sources, such as carbon stars (Herbig 1995). A large carbon abundance may be required for the DIB–$E(B-V)$ relations to hold true. Therefore, it is feasible that early epoch H I galaxies are not conducive to the formation of the organics because they lack the necessary carbon. However, Herbig (1995) points out that the scatter in the Milky Way DIB–$E(B-V)$ relation is real, *i.e.* not due to noise or systematic errors. Thus, other environmental factors such as the local ionizing radiation field and the metallicity may be important (Cox et al. 2007). Little is published about the radiation in the DLA galaxies in our sample. Their metallicities are known to be low, on the order of ~0.1 Solar metallicity, with the exception of AO 0235+164, which is metal abundant (Junkkarinen et al. 2004). Based on published work, it is not surprising that the DLA galaxy in our sample with the highest reddening and metallicity has the only detected DIBs. Thus, the lack of dust and metals plays a large role in inhibiting the organics in our sample of early epoch H I-rich galaxies whereas, the abundance of H I is not a strong determinent in their strengths. Our DLA selection method is not representative of all early epoch galaxies, it is biased toward galaxies with low reddening. Another selection method for early epoch galaxies may be more fruitful.

Acknowledgments. BL acknowledges the support of NASA via the Graduate Student Researchers Program (GSRP). BL also thanks the NASA Astrobiology Institute and the National Science Foundation for travel support arranged through the conference organizers.

References

Bada, J. L. & Lazcano, A. 2002, Science, 296, 1982
Bohlin, R. C., Savage, B. D., & Drake, J. F. 1978, ApJ, 224, 132
Cox, N. L. J., & Spaans, M. 2006a, A&A, 451, 973
Cox, N. L. J., Cordiner, M. A., Cami, J., Foing, B. H., Sarre, P. J., Kaper, L., & Ehrenfreund, P. 2006b, A&A, 447, 991
Cox, N. L. J., Cordiner, M. A., Ehrenfreund, P., Kaper, L., Sarre, P. J., Foing, B. H., Spaans, M., Cami, J., Sofia, U. J., Clayton, G. C., Gordon, K. D., & Salama, F. 2007, A&A, 470, 941
Ellison, S. L., Hall, P. B., & Lira, P. 2005, AJ, 130, 1345
Ellison, S. L., York, B. A., Murphy, M. T., Zych, B. J., Smith, A. M., & Sarre, P. J. 2008, MNRAS 383, L30
Gordon, K. D., Clayton, G. C., Misselt, K. A., Landolt, A. U., & Wolff, M. J. 2003, ApJ, 594, 279
Heckman, T. M. & Lehnert, M. D. 2000, ApJ, 537, 690
Heger, M. L. 1922, Lick Observatory Bull. 10, 337, 146
Herbig, G. H. 1995, ARA&A, 33, 19
Hudgins, D. M., Bauschlicher, C. W., Jr., & Allamandola, L. J. 2005, ApJ, 632, 316
Jenniskens, P., & Desert, F. -X. 1994, A&A, 106, 39
Junkkarinen, V. T., Cohen, R. D., Beaver, E. A., Burbidge, E. M., & Lyons, R. W. 2004, ApJ, 614, 658
Rich, R. M. 1987, AJ, 94, 651
Schneider, D. P., Hartig, G. F., Jannuzi, B. T., et al. 1993, ApJS, 87, 45
Snow, T. P. 2001, Spectrochimica Acta Part A, 57, 615
Sollerman, J., Cox, N., Mattila, S., Ehrenfreund, P., Kaper, L., Leibundgut, B., & Lundqvist, P. 2005, A&A, 429, 559
Tuairisg, S. O., Cami, J., Foing, B. H., Sonnentrucker, P., & Ehrenfreund, P. 2000, A&A, 142, 225
Welty, D. E., Federman, S. R., Gredel, R., Thorburn, J. A., & Lambert, D. L. 2006, ApJS, 165, 138
Weselak, T., Schmidt, M., & Krelowski, J. 2000, A&A, 142, 239
York, B. A., Ellison, S. L., Lawton, B., Churchill, C. W., Snow, T. P., Johnson, R. A., & Ryan, S. G. 2006, ApJ, 647, L29

Ceccelia Ceccarelli

Javier Goicoechea

Bioastronomy 2007: Molecules, Microbes, and Extraterrestrial Life
ASP Conference Series, Vol. 420, 2009
K. J. Meech, J. V. Keane, M. J. Mumma, J. L. Siefert, and D. J. Werthimer, eds.

Simple Organic Chemistry in the Horsehead Nebula

J. R. Goicoechea,[1] J. Pety,[1,2] M. Gerin,[1]
P. Hily-Blant,[2] D. Teyssier,[3] and E. Roueff[4].

[1]*Centro de Astrobiologia (CSIC-INTA), Madrid, Spain*
[2]*IRAM, Grenoble, France.*
[3]*European Space Astronomy Centre, Villafranca del Castillo, Spain*
[4]*LUTH, UMR 8102, CNRS, Observatoire de Paris, France.*

Abstract. We present our latest results on carbon chemistry in the Horsehead nebula, one of the most famous objects in the sky and a unique laboratory to understand the chemistry of interstellar clouds illuminated by UV radiation from nearby stars. Photodissociation regions (PDRs) are interesting intermediate media between diffuse and dense dark clouds, thus enabling astrochemists to probe a large variety of physical and chemical processes. In particular, our high resolution astronomical observations show that the Horsehead edge is a realistic template to determine the molecular inventory in PDRs and to investigate the photostability of simple organic molecules. In this contribution we show that simple carbon chains and rings (CCH, c-C_3H_2 and C_4H) are tightly spatially correlated with each other and with the emission of polycyclic aromatic hydrocarbons (PAHs). We show how molecules such as HCO^+ start to be enriched in deuterium ($DCO^+/HCO^+ > 0.02$) as the gas cools down in the densest and UV protected prestellar condensations. We also determine the gas phase sulfur abundance in the UV irradiated gas from CS and HCS^+ observations and chemical modeling. We finally present the first results of our search of gas phase species with a probable dust grain surface origin (e.g., H_2S). We stress the need of well conceived astronomical observations together with models that treat consistently both the photochemistry of simple organic species and the radiative transport of their emission lines.

1. Introduction

Far–UV (FUV) radiation ($h\nu < 13.6$ eV, i.e., photons with energies below the hydrogen ionization threshold) strongly affects the physical and chemical state of dusty molecular clouds in many evolutionary stages: from star forming regions and protoplanetary disks, to circumstellar envelopes around evolved stars and supernova remnants. Astrochemists generally speak about *photodissociation regions* (PDRs) when they refer to regions where the *physics* (e.g., gas heating and cooling) and the *chemistry* (e.g., main chemical reaction routes) are dominated by the presence of such a FUV radiation field (e.g., Goicoechea & Le Bourlot 2007). This radiation ionizes key elements such as carbon, sulfur or silicon, and ionizes and dissociates most of gas phase molecules. Because of this broad definition of PDRs, its study includes many different astronomical environments, all sharing common properties. Complex theoretical and computational models have been developed in the last 2 decades to simulate the main properties and physical conditions in PDRs (e.g., Röllig et al. 2007). These models basically have to solve: (*i*) the FUV radiation field attenuation through the cloud (caused mainly by the dust extinction but also by the absorption of hundreds of discrete electronic lines from H, H_2 and CO), (*ii*) the cloud thermal balance; by including the dominant gas heating mechanisms (e.g., photoelectric effect on dust grains) and gas cooling mechanisms (e.g. atomic and molecular line emission) and (*iii*) the chemistry (by following thousands of chemical reactions and calculating the abundance of a few hundreds of molecular species). Assuming that clouds are homogeneous, these models predict the classical *layered* structure of a PDR with the transition from atomic to molecular hydrogen H/H_2 near the cloud surface (at visual extinctions $A_V \leq 1$) and the $C^+/C/CO$ conversion deeper inside the cloud (at $A_V \geq 2$-3). Larger carbon macromolecules such as the Polycyclic Aromatic Hydrocarbons (PAHs) are particularly resistant species to the dissociating radiation. Therefore, they are easily observed at the FUV-irradiated surfaces of PDRs. In summary, astronomers observe PDRs because these regions are the best laboratories to understand the chemical complexity and spatial segregation in the FUV-illuminated interstellar medium (ISM).

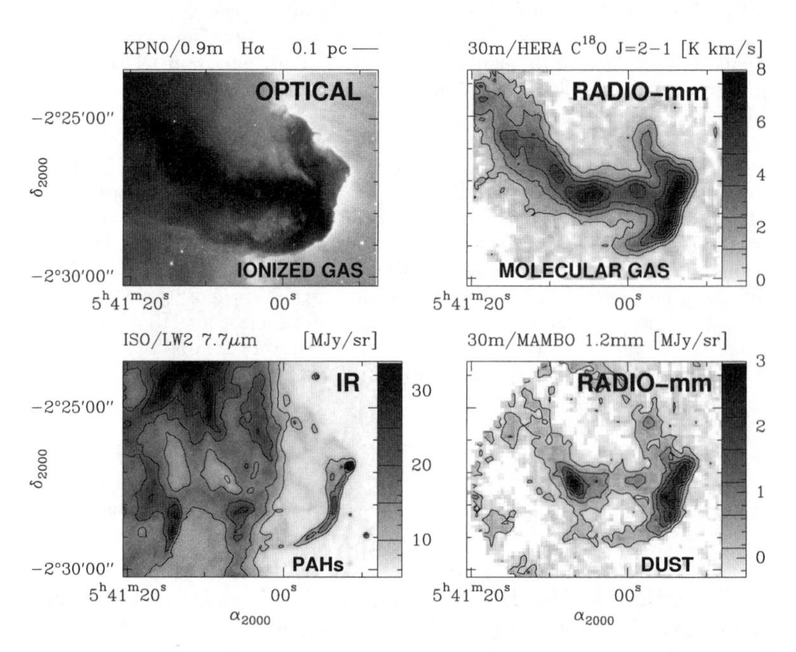

Figure 1. The Horsehead nebula observed at different wavelengths. *Top left*: Hα observations in the visible revealing the ionized gas (*black* means *no signal* only in this box). Due to the dust extinction, the molecular cloud is veiled in this range. *Botton left*: Mid–IR emission due to PAHs showing the FUV–photoactive layers of the cloud. *Top right*: Radio observations penetrate deeper inside the cloud and allow us to trace the molecular cloud where most of the chemical activity takes place. *Botton right*: The thermal emission of dust grains can also be observed as radio continuum providing information of the column density of material.

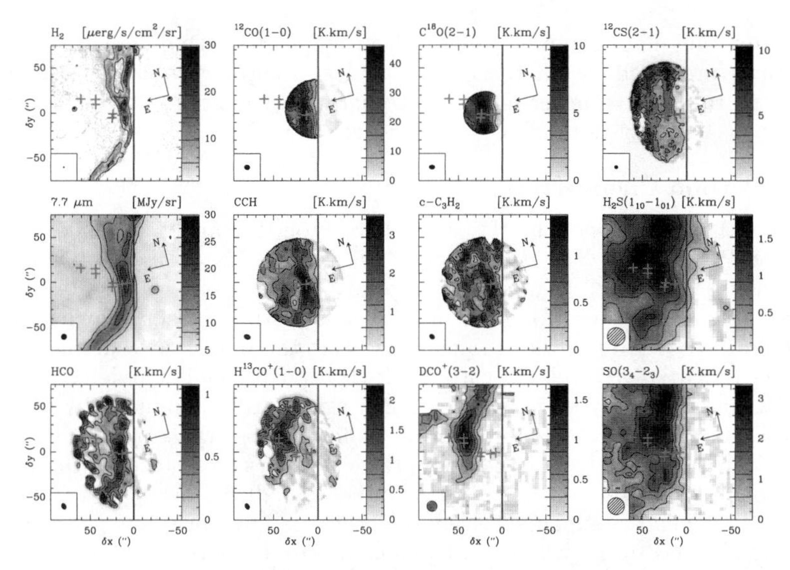

Figure 2. Emission maps obtained with the IRAM Plateau de Bure Interferometer or 30m single-dish radiotelescope, except for the H_2 v=1-0 S(1) emission observed with the NTT/SOFI and the PAH mid–IR emission observed with the Infrared Space Observatory. The maps have been rotated by 14° counter–clockwise around the image center to bring the exciting star direction in the horizontal direction as this eases the comparison of the PDR tracer stratifications. The nebula is illuminated by the σOri star located at the west of the cloud (right side of the boxes). Either the synthesized beam or the single dish beam (i.e., the angular resolution of the observations) is plotted in the bottom left corner. The emission of all the lines is integrated between 10.1 and 11.1 $\mathrm{km\,s^{-1}}$. Values of contour level are shown on each image lookup table (contours of the H_2 image have been computed on an image smoothed to 5″ resolution). The green crosses display the position where we typically derive column densities and abundances.

2. The Horsehead Edge as a Photochemical Laboratory

The Horsehead nebula, appears as a dark patch of ∼5′ diameter against the bright H II region IC434. Emission from gas and dust associated with this globule has been detected from IR to millimeter wavelengths, although the first astronomical plates were taken ∼120 yr ago. In particular, the Horsehead western edge is a PDR viewed nearly edge-on and illuminated by the O9.5V star σOri at a projected distance of ∼3.5 pc. The intensity of the incident FUV radiation field is $\chi \simeq 60$ times the mean interstellar radiation field (ISRF) in Draine's units (Abergel et al. 2003). Radio observations reveal that the Horsehead is not only one of the most famous and beautiful objects in the sky. It is also a fantastic PDR laboratory to investigate the spatial abundance gradients and chemistry of simple organics including O, C, N and S elements and ^{18}C, ^{34}S and D isotopes.

Since 2001, we started to study the Horsehead PDR with the IRAM Plateau de Bure interferometer (France) and the IRAM-30m radiotelescope (Spain) reaching angular resolutions down to 3 to 11″. Figure 2 displays all the observed, high resolution maps acquired so far. Those maps trace the different segregated structures predicted by photochemical models according to chemical reaction networks, radiative transfer effects and excitation conditions (through the different physical conditions of the emitting regions).

2.1. Carbon Chemistry:

Through the modeling of the H_2 and CO emission, Habart et al. (2005) showed that the PDR has a very steep density gradient, rising to $n_H \sim 10^5\,\mathrm{cm}^{-3}$ in less than $10''$ (0.02 pc), at a roughly constant pressure of $P \sim 4 \times 10^6\,\mathrm{K\,cm}^{-3}$. These observations were followed by a chemical study of small hydrocarbons (CCH, c-C_3H_2, C_4H). Observations show that the emission from these small organics is spatially correlated with the PAH aromatic emission detected in the mid–IR, and thus they arise from the same regions still illuminated by the FUV radiation field. Nevertheless, Pety et al. (2005) showed that the abundances of the hydrocarbons are higher than the predictions based on state–of–the–art gas phase photochemical models (e.g., Le Petit et al. 2006). We have argued that these results could be explained either if (*i*) the small hydrocarbons are a product of the PAHs and small carbon grains photoerosion (Joblin et al. 2007) or (*ii*) if turbulent mixing in the cloud transports molecular species formed deeper inside the cloud to the illuminated surface of the PDR (Gerin et al. 2007).

2.2. Sulfur Chemistry:

Goicoechea et al. (2006) linked the chemical abundances predicted by the PDR model with detailed non-LTE radiative transfer models adapted to the Horsehead geometry and able to predict the line profiles and intensities observed in radiotelescopes. They analyzed the CS and HCS^+ photochemistry, excitation and radiative transfer to obtain their abundances and the physical conditions prevailing in the PDR. Since the CS abundance scales to that of sulfur, they determined the gas phase sulfur abundance in the PDR itself, S/H=$(3.5\pm1.5)\times10^{-6}$, an interesting intermediate medium between translucent clouds (where sulfur remains in the gas phase) and dark clouds (where large depletions onto dust grains have been invoked). They showed that the gas sulfur depletion invoked to account for CS and HCS^+ abundances in the PDR is orders of magnitude lower than in previous studies of the sulfur chemistry. Therefore, the abundance of species such as CS may indicate that something important is lacking from chemical models or that an abundant sulfur–bearing carrier has been missed. Hence, the abundances of sulfur species remain interesting puzzles for interstellar chemistry. In order to analyze the sulfur chemistry in full detail, we have started a line survey of several S-bearing molecules in the Horsehead. Preliminary results indicate that abundant species such as SO or H_2S show widespread emission regardless their position in the nebula (either in the PDR surface or in regions deeper inside the cloud and shielded from the FUV radiation).

2.3. Deuterium Chemistry:

Molecules are enriched in deuterium over the elemental D/H abundance ratio of $1 - 2 \times 10^{-5}$ in many different astrophysical environments (see Ceccarelli et al. in this volume). These include cold and dense cores, the mid–planes of protoplanetary disks, hot molecular cores, and even PDRs. In Pety et al.

Species	Environment	$(\delta x,\delta y)$ [arcsec]	Col. Dens. N(X) [cm^{-2}]	Abundance n(X)$/n_H$
H_2	PDR	$(+7.4, -1.0)$	$(3.6 \pm 1.7)10^{21}$	0.5
$C^{18}O$			$(1.0 \pm 0.3)10^{15}$	$1.4\,10^{-7}$
CCH			$(1.1 \pm 0.3)10^{14}$	$1.5\,10^{-8}$
c-C_3H_2			$(9.5 \pm 5.0)10^{12}$	$1.3\,10^{-9}$
C_4H			$(3.7 \pm 1.0)10^{13}$	$5.2\,10^{-9}$
$H^{13}CO^+$	cold	$(+44.0, +15.0)$	$(0.5 - 1)10^{13}$	(0.5-1) 10^{-10}
DCO^+	condensation		$(0.5 - 1)10^{13}$	(0.5-1) 10^{-10}
CS		$(+44.0, +9.5)$	$(1.2 \pm 1.0)10^{14}$	$3.9\,10^{-9}$
$C^{34}S$			$(5.3 \pm 0.5)10^{12}$	$1.8\,10^{-10}$
HCS^+			$(6.8 \pm 0.5)10^{11}$	$2.31\,0^{-11}$

Table 1. Column densities and abundances of several chemical species at the FUV–illuminated PDR (upper part) and the shielded cold region (lower part) of the Horsehead. Coordinate offsets are given in the coordinate system adapted to the source geometry and used in Figure 2. Abundances are computed with respect to the density of protons $n_H{=}n(H) + 2n(H_2)$.

(2007), we have studied the deuterium fractionation in the Horsehead edge from observations of several $H^{13}CO^+$ and DCO^+ lines. A large $[DCO^+]/[HCO^+]$ abundance ratio (≥ 0.02) is inferred at the DCO^+ emission peak, a condensation shielded from the illuminating FUV radiation field where the gas must be cold (10–20 K) and dense ($n_H \geq 4 \times 10^5\,\mathrm{cm}^{-3}$). This prestellar core is likely confined by the pressure from the PDR/H II region.

To our knowledge, this is the brightest DCO^+ emission detected in an interstellar cloud close to a bright H_2/PAH emitting region. This opens the interesting possibility to probe at high resolution the chemical transition from FUV photo–dominated gas to "dark cloud" shielded gas in a small field of view. DCO^+ is not detected in the warmer PDR front (the cloud surface), implying a lower $[DCO^+]/[HCO^+]$ ratio ($< 10^{-3}$).

2.4. The Horsehead: Towards an Observational Benchmark

We have shown that the Horsehead is a good source to serve as reference for photochemical models and to investigate the main chemical formation routes in FUV–irradiated gas. A situation that applies for many astronomical scenarios, from the diffuse ISM to star forming regions and protoplanetary disks. An ideal observational benchmark would provide to astrochemists a set of molecular abundances (with their associated uncertainties) as a function of the distance to the illuminating star (or extinction through the cloud). This goal is difficult to achieve for several reasons: (*i*) the geometry of the source is never as simple as one may wish, (*ii*) the spectra produced by the instruments must be inverted to obtain abundances and (*iii*) the spectra are often measured at different angular resolutions, implying that different cloud components with different physical conditions are heavily diluted or overlapped within their beams. For several years now, we have started to systematically study the western edge of the Horsehead nebula because its geometry is not only well understood but also quite simple (almost 1D and viewed edge-on). The density profile across the PDR is well constrained and there are several current efforts to constrain the thermal profile. The combination of low distance to Earth (400 pc), low FUV illumination ($\chi \sim 60$) and high density ($n_H \sim 10^5\,\mathrm{cm}^{-3}$) implies that all the interesting physical and chemical processes can be probed in a small field-of-view (less than 50″). Table 1 shows a summary of the molecular abundances derived so far at two very different environments of the Horsehead, the illuminated PDR and the cold and shielded condensation (i.e., where DCO^+ peaks)

The next generation of observatories such as the Herschel Space Observatory (launched in 2009) or ALMA (~2012) will provide a huge step forward in our understanding of the ISM chemical complexity. It will then be possible to investigate the role of key diagnostics (such as water) and chemically related species (such as light hydrides) in the ISM (Herschel), and to map the abundance gradients with angular resolutions down to 1″ (ALMA).

Where the telescope ends, the microscope begins. Who can say which has the grander view?. Victor Hugo, ***Les Misérables****, 1862.*

Acknowledgments. Javier R. Goicoechea was supported by a *Marie Curie Intra-European Individual Fellowship* within the 6th European Community Framework Programme, contract MEIF-CT-2005-515340.

References

Abergel, A., Teyssier, D., et al. A&A, 410, 577

Gerin, M., Lessafre, P. Goicoechea, J.R. et al. 2007, in *Molecules in Space & Laboratory*, Paris,. Eds. J.L. Lemaire & F. Combes.

Goicoechea, J. R., Pety, J., Gerin, et al. 2006, A&A, 456, 565

Goicoechea, J. R., & Le Bourlot, J. 2007, A&A, 467, 1

Habart, E.,Abergel, A., Walmsley, C. M., Teyssier, D., & Pety, J. 2005, A&A, 437, 177

Joblin, C. et al. 2007, in *Molecules in Space & Laboratory*, Paris,. Eds. J.L. Lemaire & F. Combes.

Le Petit, F., Roueff, E., & Le Bourlot, J. 2002, A&A, 390, 369

Pety, J., Teyssier, D., Fossé, et al. 2005, A&A, 435, 885

Pety, J., Goicoechea, J. R., Hily-Blant, et al., 2007, A&A, 464, L41

Röllig, M., Abel, N. P., Bell, T., et al. 2007, A&A, 467, 187

Bioastronomy 2007: Molecules, Microbes, and Extraterrestrial Life
ASP Conference Series, Vol. 420, 2009
K. J. Meech, J. V. Keane, M. J. Mumma, J. L. Siefert, and D. J. Werthimer, eds.

Organic Molecules in the Orion KL Hot Molecular Core

K. S. Wang,[1] Y. J. Kuan,[1,2] S. Y Liu,[1]
H. C. Huang,[2] and S. B. Charnley[3]

[1] *Academia Sinica Institute of Astronomy and Astrophysics, Taipei 106, Taiwan, ROC*

[2] *National Taiwan Normal University, Taipei 116, Taiwan, ROC*

[3] *Astrochemistry Laboratory & Center for Astrobiology, Solar System Exploration Division, Code 691, NASA Goddard Space Flight Center, Greenbelt, MD 20771, USA*

Abstract. We report Submillimeter Array (SMA) observations of the Orion KL region in lines of CH_3CN, C_2H_3CN, C_2H_5CN, CH_3OH, $HCOOCH_3$ and $(CH_3)_2O$. High spatial-resolution mapping shows remarkable chemical differentiation on small scales. These observations provide new insights into the organic chemistry associated with young protostars.

1. Introduction

The dense, warm gas surrounding both massive and low-mass protostars exhibits strong molecular emission at radio wavelengths and many organic molecules are detected in these environments (van Dishoeck & Blake 1998; Charnley et al. 2001; Cazaux et al. 2003; Kuan et al. 2004). As such, they are prime candidates for understanding the nature of interstellar organic chemistry and its astrobiological implications (Ehrenfreund & Charnley 2000). The Orion KL region contains two subsources that have been of particular interest from this perspective: the Hot Core and the Compact Ridge. Both subsources are rich in organic molecules and early studies suggested that the Hot Core is selectively richer in N-bearing organics, whereas O-bearing organics appear to be favored in the Compact Ridge (Blake et al. 1987). Such chemical differentiation appears to be common in star-forming regions (e.g. Fontani et al. 2007) and various chemical explanations have been proposed, including hot gas-phase chemistry and grain-surface reactions (e.g. Charnley et al. 1992; Caselli et al. 1993; Rodgers & Charnley 2001; Charnley & Rodgers 2006). We have made an extensive high-resolution study of this chemical dichotomy with the Submillimeter Array (Wang et al. in preparation), here we present some preliminary results.

2. Observations

Submillimeter observations of Orion KL with the SMA in the compact-north configuration were made at 330.6-332.6 GHz (LSB) and 340.6-342.6 GHz (USB). The uniformly-weighted synthesized beam is of the size $\sim 1.6'' \times 0.9''$; a spectral

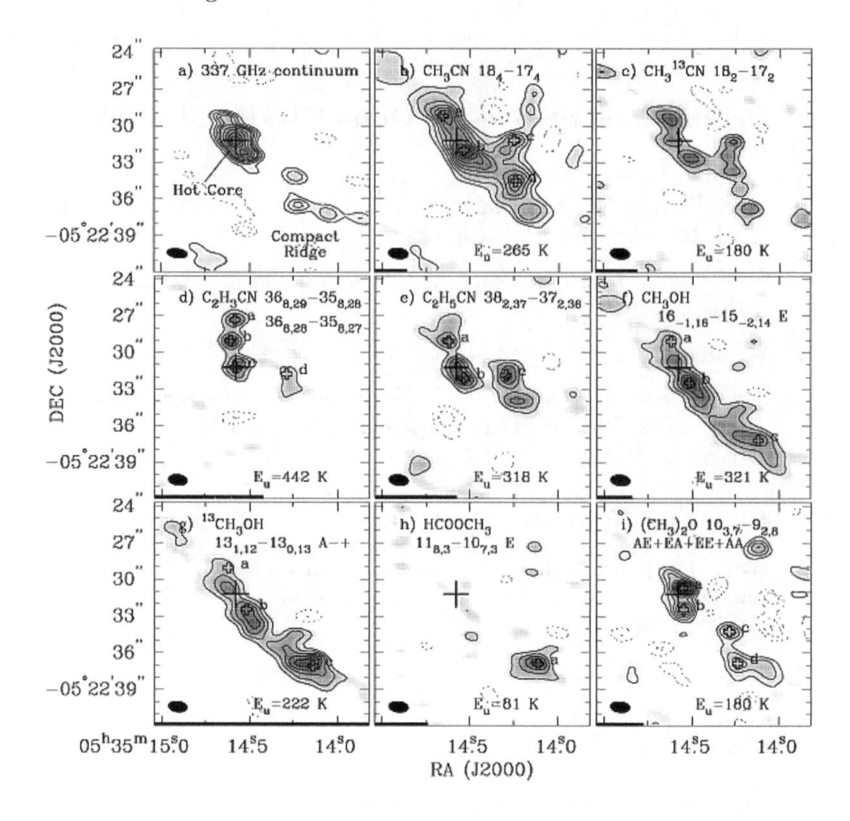

Figure 1. Orion KL spectral images of emission of various nitrogen-bearing and oxygen-bearing molecules. Also shown in each panel are molecular transitions (top) and the upper energy levels (bottom). The big cross marks the continuum peak position; open crosses indicate the spectral emission peaks, which are labeled alphabetically from north to south, of each corresponding molecule.

resolution of 0.356 km s^{-1} ch^{-1} was employed. The nominal LSR velocity of 8.0 km s^{-1} was adopted for the observations. The spectral emission of multiple transitions of the N-bearing organic species CH_3CN, C_2H_3CN and C_2H_5CN, and the O-bearing molecules CH_3OH, $HCOOCH_3$ and $(CH_3)_2O$ was imaged. USB and LSB averaged continuum at 337 GHz indicates much stronger thermal dust emission at the Hot Core than toward the Compact Ridge (Figure 1(a)).

3. Spatial Distribution of N-Bearing and O-Bearing Molecules

The integrated intensity maps of CH_3CN, $CH_3{}^{13}CN$, C_2H_3CN and C_2H_5CN are shown in Figures 1.(a)-(e). CH_3CN emission can be seen peaking toward the Hot Core, the Compact Ridge and the Western Cloud (Fig. 1.(b)). Nevertheless, C_2H_3CN and C_2H_5CN emission is found mainly toward the Hot Core (Fig.

1(d) and (e)). Spectral images of oxygen-bearing molecules CH_3OH, $^{13}CH_3OH$, $HCOOCH_3$ and $(CH_3)_2O$ are presented in Figures 1(f)-(i). CH_3OH emission is observed toward the Hot Core and Compact Ridge (Fig. 1(f)). $HCOOCH_3$, on the other hand, can only be seen toward the Compact Ridge (Fig. 1(h)). In contrast, the double-rotor $(CH_3)_2O$ is detected in the Hot Core, the Compact Ridge and the Western Cloud (Fig. 1(i)). Observing at arcsecond resolution, we found the clumpy nature of the Hot Core and the Compact Ridge, and it can be readily seen that they are both composed of multiple gas clumps (labeled alphabetically from north to south in Figure 1). Distinct velocity components are also uncovered toward these gas clumps, e.g. at the C_2H_5CN emission peaks shown in Figure 1(e).

The usual rotation-diagram analysis was performed to derive the column densities and excitation temperatures of C_2H_3CN, C_2H_5CN and $HCOOCH_3$. The population-diagram analysis described in Goldsmith and Langer (1999) was carried out on CH_3CN transitions in order to account for their large optical depth (Wang et al. in preparation). Due to a limited number of transitions observed, the column densities of CH_3OH, $^{13}CH_3OH$ and $(CH_3)_2O$ in optically thin case were calculated with the assumed excitation temperatures of 300 K and 200 K, respectively, for the Hot Core and the Compact Ridge. By adopting H_2 column densities of 8.85×10^{23} cm^{-2} for the Hot Core and of 1.90×10^{23} cm^{-2} for the Compact Ridge (Wright et al. 1996), we converted the individual column densities of all observed molecular species to fractional abundances. Figure 2 gives the schematic view of the inferred fractional abundances in Orion KL.

4. Chemical Differentiation in Orion KL

Our arcsecond-resolution SMA observations reveal a significant chemical diversity between the Hot Core and the Compact Ridge (Fig. 2). It is most noticeable that N-bearing molecules are concentrated mainly toward the Hot Core. A high fractional abundance of CH_3OH is reported toward the Hot Core, yet an even higher one is inferred toward the Compact Ridge presumably due to recent evaporation of icy mantles. High abundances of $HCOOCH_3$ and $(CH_3)_2O$ are also found in the Compact Ridge.

Charnley et al. (1992) suggested that such chemical diversity could be a result of different mantle compositions, with the Hot Core rich in NH_3 and the Compact Ridge in CH_3OH, followed by different gas-phase chemistry after mantle evaporation. The distinct mantle compositions between the Hot Core and the Compact Ridge may be simply a reflection of a different thermal history of the two (Charnley et al. 1992; Caselli et al. 1993). Nonetheless, Rodgers et al. (2001) showed that different cloud-core temperatures and the presence of NH_3 in mantles could be crucial in determining N/O differentiation. Indeed, from the analysis of CH_3CN K-ladder transitions, we find the kinetic temperature in the Hot Core can be as high as 400 K (Wang et al. in preparation). Although the true cause of chemical differentiation observed in Orion KL is still unclear, our SMA results will be very useful in testing these different hypotheses explaining the chemical diversity, and will provide new insight to organic chemistry in massive star-forming region.

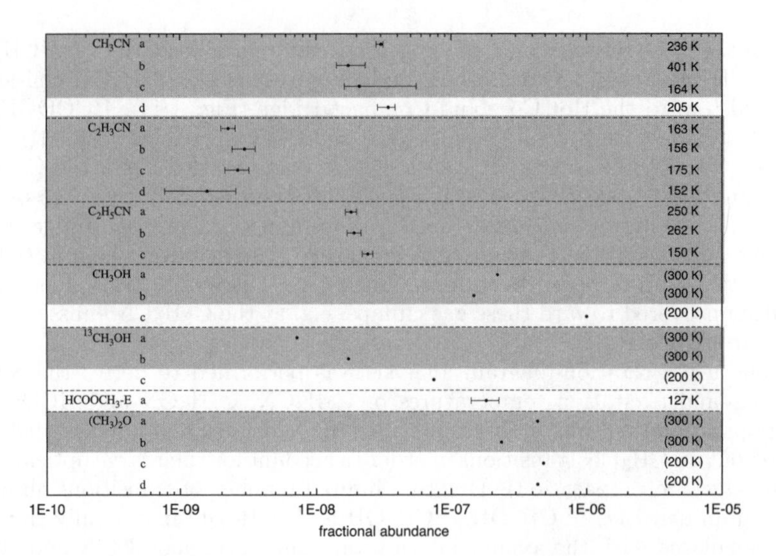

Figure 2. Fractional abundances of N-bearing and O-bearing molecules in Orion KL. The numbers shown in the right are rotational temperatures derived from excitation analysis; temperatures in parentheses are not derived but assumed. The rows shown in grey shading indicate the gas-clump positions in the Hot Core, while those without shading are at the positions toward the Compact Ridge.

Acknowledgments. The research of Y.-J. K. was supported by NSC 95-2112-M-003-016 grant. This work was supported by NASA's Exobiology Program and by the NASA Goddard Center for Astrobiology.

References

Blake, G.A., Sutton, E.C., Masson, C.R. and Phillips, T.G., 1987, ApJ, 315, 621

Caselli, P., Hasegawa, T.I. and Herbst, E. 1993, ApJ, 408, 548

Cazaux, S., Tielens, A. G. G. M., Ceccarelli, C., Castets, A., Wakelam, V., Caux, E., Parise, B., & Teyssier, D. 2003, ApJ, 593, L51

Charnley S.B., Tielens A.G.G.M., Millar T.J., 1992, ApJ, 399, L71

Charnley, S.B., Ehrenfreund, P. and Kuan Y- J. 2001. Spectrochimica Acta 57, 685

Charnley S.B. & Rodgers, S.D. 2006 , in Astrochemistry Throughout the Universe: Recent Successes and Current Challenges, Proceedings of IAU Symposium No. 231, Eds. D. C. Lis, G. A. Blake & E.Herbst, CUP, 237

Ehrenfreund, P., Charnley S.B., 2000, Annu. Rev. A & A, 38, 427

Fontani, F. et al. 2007, A&A, 470, 639

Goldsmith, P.F. & Langer, W.D., 1999, ApJ, 517, 209

Kuan, Y.-J., et al., 2004, ApJ, 616, L27

Rodgers S.D., Charnley S.B. 2001, ApJ, 546, 324

van Dishoeck E.F., Blake G.A., 1998, Annu. Rev. A & A, 36, 317

Wang, K.-S. et al. in preparation
Wright, M.C.H., Plambeck, R.L. and Wilner, D.J., ApJ, 1996, 469, 216

Boncho Bonev

Yvonne Pendleton

Session III

The Nature of Interstellar Organics

Pascale Enhrenfreund

Jacqueline Keane

Bioastronomy 2007: Molecules, Microbes, and Extraterrestrial Life
ASP Conference Series, Vol. 420, 2009
K. J. Meech, J. V. Keane, M. J. Mumma, J. L. Siefert, and D. J. Werthimer, eds.

Establishing the Synthetic Contingencies for Life: Following the Carbon from AGB Stars to Planetary Surfaces

L. M. Ziurys, D. T. Halfen, and N. J. Woolf

LAPLACE Center for Astrobiology, Steward Observatory, University of Arizona, 933 N. Cherry Avenue, Tucson, AZ 85721

Abstract. The biochemistry of living systems is remarkably selective, utilizing certain chemical compounds and not other, closely related molecules. This selectivity might be a result of interstellar chemistry. Organic synthesis in space is remarkably active. It begins with the carbon-rich chemistry of circumstellar envelopes, which appears to survive in part as these objects evolve into planetary nebula and then into diffuse clouds. Diffuse clouds subsequently collapse into dense clouds, carrying along the chemical imprint of previous syntheses. This cycling of molecular gas provides dense clouds with C-rich starting material, which accelerates organic chemistry in these objects. The low temperature environment of interstellar space restricts available chemical pathways, resulting in molecule formation that is non-combinatorial. Comets, meteorites, and perhaps interplanetary dust particles then bring these select organic compounds to planetary surfaces. Thus, the "synthetic contingencies" that led to life on Earth may have been established by interstellar chemistry.

1. Introduction

Many paradoxes currently exist in the biochemistry of living systems. For example, why are there only twenty amino acids used in proteins on Earth? Experiments have shown that other such acids can be readily incorporated into proteins (e.g. Hohsaka & Masahiko 2002). Why does Terran genetics only use the pentose sugar ribose in RNA? Why not a hexose or a tetrose? Some of these sugars have been found to work equally well in RNA analogs (Eschenmoser 1999). There are other examples as well, suggesting that life evolved relatively quickly on Earth and did not have time to sample all the chemical possibilities (e.g. Benner, Ricardo, & Carrigan 2004). Therefore, modern biochemistry reflects what starting materials happened to be present on Earth when life began, i.e. a matter of "synthetic contingency."

These starting materials had to be composed of largely organic compounds. Where did the organic material that set the stage for life on Earth originate? Studies of the history of early Earth suggest that the planet lost most of its original carbon with its first atmosphere (Harper & Jacobsen 1996). It is unclear how much primordial carbon survived. The organic material that was incorporated into living systems thus had to come from somewhere other than Earth.

The Earth has been bombarded by meteorites, comets and interplanetary dust particles since its formation, with particular heavy bombardment 3.8-4.0 billion years ago (Gomes et al. 2005). Shortly thereafter, life is known to have de-

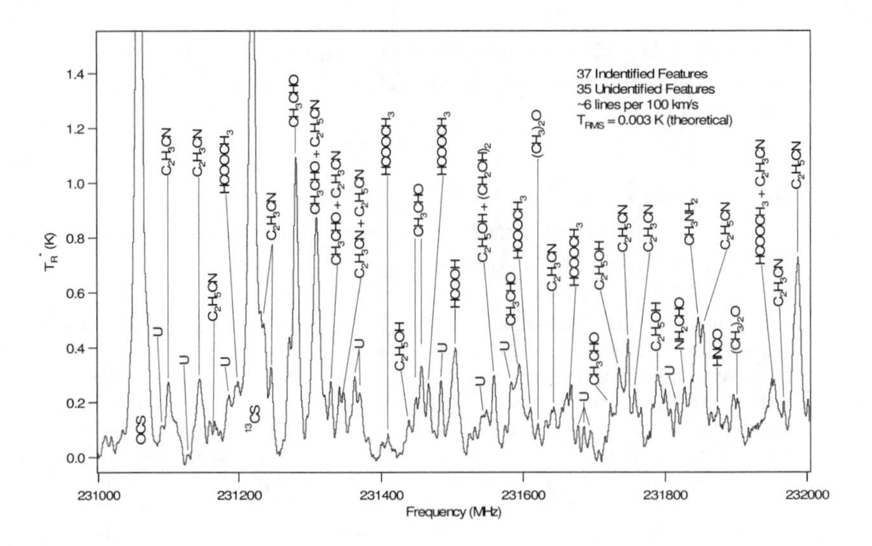

Figure 1. A spectrum of the dense molecular cloud in the Galactic Center, Sgr B2(N), observed at a frequency of 231.5 GHz with the Sub-mm Telescope (SMT) of the Arizona Radio Observatory (ARO). The spectrum shows the variety of organic compounds found in this cloud, as well as unidentified features, marked "U" (from Halfen et al. 2009b).

veloped (Rosing 1999; Furnes et al. 2004). Meteorites, comets and dust particles may have brought back the carbon, but from where? The only other region of the universe where organic molecules are found in large quantities is in interstellar space. As shown in Figure 1, organic compounds such as HCOOH, CH_3NH_2, CH_3CH_2CN, CH_3CHO, CH_3OH, and $HCOOCH_3$ are common constituents of giant molecular gas clouds found throughout our Galaxy; the two-carbon sugar $HCOCH_2OH$ and the simple protein analog CH_3CONH_2 have been identified in the dense cloud core Sgr B2(N) located in the Galactic center (Halfen et al. 2006, Hollis et al. 2006, Halfen et al. 2009a). Could there be a connection between this organic material and living systems on Earth?

2. Origin of Interstellar Carbon: Massive Stars and Their Envelopes

Perhaps this question can be best addressed by following the pathways of carbon from its origin in nucleosynthesis, its ejection into the interstellar medium via circumstellar gas, and then its arrival on planets. Carbon is formed from helium by nuclear reactions in the interiors of more massive stars (M > 1.5 solar masses: Karakas et al. 2002). The carbon makes its way from stellar interiors and into the interstellar medium via supernovae explosions and mass loss from giant and supergiant stars (Carigi et al. 2005). The net result of this processing is an average Galactic C/O ratio of about 0.7, with a gradient with distance from the Galactic Center, with a maximum value of 1 (Carigi et al. 2005). Given such ratio values, all the carbon should be contained in the very stable CO molecule. (CO is the second most abundant interstellar molecule after H_2). Therefore, it seems unlikely that much carbon is available to create the wide variety of organic species observed in molecular clouds.

It should be noted that the chemical compounds claimed to be present in such clouds are, in most cases, secure identifications. These species have been detected primarily using techniques of radio astronomy, which utilize heterodyne receivers with multiplexing spectrometers. Spectral resolution is typically 1 part in 10^7-10^8. Combined with precise gas-phase laboratory measurements of the candidate molecules, radio astronomical observations are able to identify the unique "fingerprint" of a given species created by its rotational motion. Therefore, the organic chemistry observed in the gas phase in interstellar space is based on concrete measurements.

In order to form compounds with multiple carbon atoms, a carbon-rich environment would be desirable. There are some regions in interstellar space where the C/O ratio is significantly greater than 1: the circumstellar envelopes of certain types of stars, so-called asymptotic giant branch (AGB) stars. These stars typically have masses between 1.5-8 solar masses, and have reached an advanced stage in their evolution where they have exhausted both their core hydrogen and helium supplies. They thus have a carbon core surrounded by helium and hydrogen-burning shells (Herwig 2006). By this stage, these stars have begun to shed their outer atmosphere via stellar winds and continue to lose mass at a high rate throughout the AGB phase. Thus, these objects become surrounded by a dense, thick circumstellar shell. This material cools and forms molecules as it flows from the stellar photosphere. In addition, AGB stars can undergo so-called "third dredge-up", which is triggered by the sporadic He-shell burning. It results in mixing of interior nucleosynthetic products to the stellar surface and then into the circumstellar shell. Carbon is one of the main products that undergo efficient dredge-up; as a consequence, the C/O ratio becomes larger than 1, and carbon-rich AGB envelopes are created. Their composition is reflected in the molecular content of the shell, which is abundant in carbon-containing species composed of multiple C-C bonds (see Table 1). For example, the well-studied C-rich envelope of IRC+10216 contains over 50 carbon-bearing species, including the chain compounds HC_5N, HC_7N, HC_9N, $HC_{11}N$, C_6H, C_7H, and C_8H (Ziurys 2006). It also appears to have an active phosphorus chemistry, demonstrated by the species PN, CP, HCP, and, most recently, CCP (Halfen et al. 2008).

A large fraction of circumstellar envelopes remain oxygen-rich, however, with $C < O$. Even in these environments, the chemistry of carbon is not confined to CO. For example, in the oxygen-rich shell of the supergiant star VY Canis Majoris, six carbon-containing species have been identified, including HCO^+, CS, HCN, and HNC (Ziurys et al. 2007). HCN has also been found in several other O-rich shells (e.g. Nercessian et al. 1989), such as TX Cam, NML Tau, and NML Cyg (see Figure 2). The active chemistry of carbon in these objects is thought to be a result of photospheric shock waves that alter abundances from their equilibrium values (Cherchneff 2006).

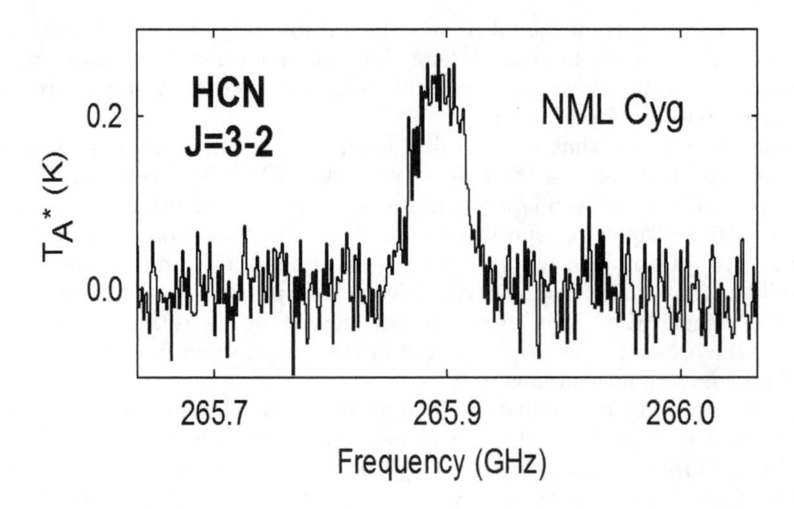

Figure 2. Spectrum of the J=3-2 transition of HCN observed towards the O-rich shell of the star NML Cyg using the ARO SMT.

3. From Planetary Nebulae and into Diffuse Clouds

AGB stars evolve into planetary nebulae (PN), the final stage in their history, which lasts about 10,000 years (Balick & Frank 2002). In this phase, the remainder of the stellar atmosphere is lost and contributes to the residual shell. The exposed star becomes a hot white dwarf, a strong emitter of ultraviolet (UV) radiation. This harsh radiation field ionizes the surrounding remnant envelope, which by the end of the PN phase, could potentially destroy all remaining circumstellar molecules. Yet, these species appear to at least partly survive, despite the harsh environment. Studies of the Helix Nebula, a very evolved PN, have shown the presence of a large, extended shell in CO lying almost coincident with ionized, atomic material (Young et al. 1999). Additional molecules have been found in the Helix and other, younger planetary nebula as well, including HCN, HNC, HCO^+, and CN (see Table 2). Recently, CCH, C_3H_2, and H_2CO have also been detected in the Helix Nebula (Tenenbaum et al. 2009). Sample spectra of H_2CO from this source are shown in Figure 3. Therefore, not only are simple

Table 1. Known Circumstellar Carbon Compounds

CO	CCH	HCP	CCS	CH_3CCH	H_2C_4	HC_9N
CS	C_3H	CCP	SiC	H_2CO	H_4C_4	C_4H^-
C_2	C_4H	MgCN	c-SiCC	CH_3CN	HC_6H	C_6H^-
C_3	C_5H	MgNC	c-SiC_3	CH_3NC	H_2C_6	C_8H^-
C_5	C_6H	NaCN	SiC_4	HCCNC	HCCN	H_4C_2
CN	C_7H	AlNC	SiCN	HCCH	HC_3N	HC_4N
HCN	C_8H	C_3N	C_3O	c-C_3H	HC_5N	CH_4
HNC	CP	C_5N	C_3S	c-C_3H_2	HC_7N	

Table 2. Molecules Observed in Planetary Nebulae

CO	HCO^+	H_2CO	CH^+	CH
H_2	N_2H^+	OH	CO^+	CN
C_3	c-C_3H_2	HNC	HCN	CCH

diatomic species surviving in evolved planetary nebulae, but also more complex compounds containing multiple carbon-carbon bonds.

The molecules may be protected from the UV radiation because they exist in dense, self-shielding dusty clumps. Infrared images of dust emission in the Helix, taken by the Spitzer Space Telescope, suggest the presence of such clumps (Hora et al. 2006). They have also been predicted by theoretical calculations of Howe, Hartquist, & Williams (1994), whose work suggests that they are created by Parker instabilities in the outflow from the central star.

The material from planetary nebulae eventually becomes part of the diffuse interstellar medium, where particle densities are on the order of 1-100 particles cm^{-3}. What becomes of the molecules from the remnant shell? Recent observations by H. Lizst and collaborators (e.g. Liszt, Lucas & Pety 2006) have demonstrated that diffuse clouds have an unexpectedly rich chemical content.

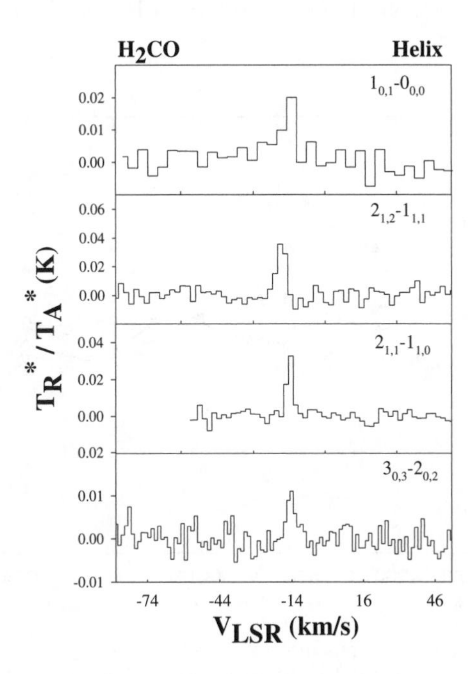

Figure 3. Spectra of H_2CO observed toward the Helix Nebula, a highly evolved PN with excessive UV radiation. The spectra were measured with the ARO 12 m and SMT by Tenenbaum et al. (2009).

Table 3. Molecules Observed in Diffuse Clouds

CO	HCO^+	H_2CO	HNC	H_2S	CH^+
CN	NH_3	OH	c-C_3H_2	SO	CO^+
H_2	CCH	HCN	CS	CH	C_3

As shown in Table 3, a surprising number of polyatomic molecules appear to be present in diffuse gas, including such species as CCH, C_3H_2, and H_2CO. Recent observations by Maier et al. (2001) have shown that C_3 exists in diffuse clouds as well. In fact, most of the species that have been identified in planetary nebulae also exist in diffuse gas. This remarkable coincidence suggests that the molecular content in diffuse clouds arises from remnant molecules from planetary nebulae. It is difficult to explain the diffuse cloud abundances otherwise; the densities in these objects are very low such that polyatomic molecules cannot easily be created. Chemical models of diffuse clouds (e.g. van Dishoeck & Black 1986) predict that in general only diatomic species form, given the physical conditions in these sources. A comparison can be made between molecular abundances in young and evolved planetary nebulae and diffuse clouds (Figure 4). As this figure illustrates, the abundances in diffuse clouds are about a factor of 10-100 lower than in the older planetary nebulae, consistent with the scenario that the molecules from these objects are the survivors from PN outflows.

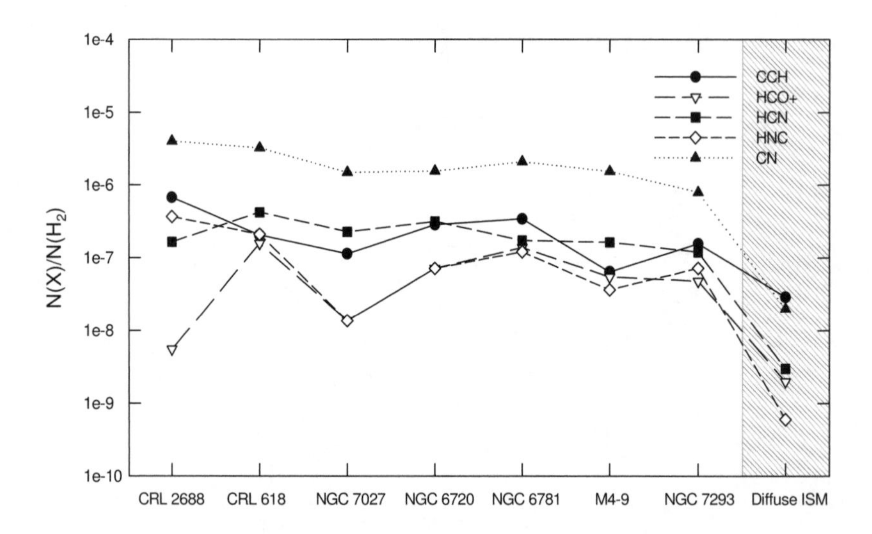

Figure 4. Abundances of various molecules, relative to H_2, plotted vs. nebula age, and then extrapolated to diffuse clouds. The youngest planetary nebula are plotted on the far left, with increasing age to the right. The abundances suggest that planetary nebulae could seed the molecular content of diffuse clouds.

4. From Diffuse Gas To Dense Molecular Clouds

Dense molecular clouds are thought to be created from the gravitational collapse of diffuse clouds (e.g. Elmegreen 2007). Therefore, it is possible that the molecular content from diffuse clouds seeds the chemistry of the denser objects. As postulated, the chemical content of diffuse gas can be traced back to planetary nebulae and then to circumstellar shells, in particular the carbon-rich variety. If this scheme is correct, then dense cloud organic chemistry may get a "jump-start" from the presence of multiple carbon-carbon bonds that are leftovers of the former, C-rich environment of AGB stars. Certainly a wide range of organic functional groups are present in molecules in dense clouds such as Sgr B2(N) and Orion-KL, as shown in Figure 5. Alcohols, organic acids, aldehydes, ketones, ethers, esters, and even a simple sugar are part of the dense cloud inventory. Multiple carbon-carbon bonds appear in many of these species. It seems difficult to explain their relatively high abundance in material where, on average, C < O. Furthermore, the limit of chemical complexity has not yet been ascertained in these objects. Organic species with four or more carbon atoms are likely to be present.

Although typical organic functional groups are present in interstellar molecules, interstellar chemistry is not combinatorial. Certain species are present (e.g. acetone), but other closely-related ones are not (hydroxyacetone: Apponi et al. 2006). This selectivity results from the low temperatures commonly present in dense clouds (T = 10-50 K). At such temperatures, only a fraction of chemical pathways are available in comparison to room temperature. Interstellar synthesis is thus kinetically as opposed to thermodynamically-controlled.

5. Onto Planetary Surfaces to Create Synthetic Contingencies?

A rich organic chemistry is present in dense interstellar gas. It could be passed on to planets through several avenues, such as comets. These objects are thought to contain pristine interstellar material that originated in the molecular cloud that formed the solar system (e.g. Bockelée-Morvan & Crovisier 2002). Studies of comet volatiles by spectroscopic methods have shown that the chemical

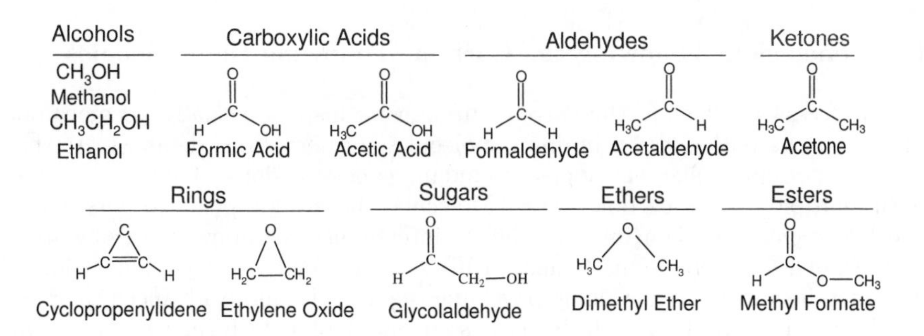

Figure 5. Organic functional groups and representative molecules identified in dense molecular clouds.

inventory in such objects is a subset of the species found in molecular clouds. Compounds such as CH_3OH, CH_3CN, H_2CO, HCOOH, NH_2CHO, and even $HOCH_2CH_2OH$ have been observed in these objects (Crovisier et al. 2004), indicating their extensive organic composition. The volatile materials sublimate as the orbits of comets bring them near the sun. Water, the main "glue" holding the refractory parts of the nucleus together, is vaporized as well, causing comet fragmentation. These fragments can fall into planetary atmospheres, thus bringing organic compounds to planetary surfaces. There is some thought that a large fraction of the water found on Earth came from comets, as well (e.g. Hunten et al. 1989).

Other possible mechanisms of transport include meteorites and interplanetary dust particles. Certain carbonaceous chondrite meteorites such as Tagish Lake and Murchison have been found to contain a wide variety of organic compounds, such as carboxylic acids, amino acids, aliphatic hydrocarbons, and dicarboximides (Pizzarello, Cooper, & Flynn 2004). Based on the D/H and $^{12}C/^{13}C$ ratios measured in this material, some of these compounds are highly fractionated, indicating an interstellar origin in cold gas (T = 10-20 K). The organic matter extracted from meteorites is naturally biased towards heavier, less volatile molecules, as opposed to the simpler compounds such as CH_3OH. These findings suggest that more complex species exist in interstellar gas, and that they can be successfully conveyed to planetary surfaces.

The availability of unprocessed interstellar material relates back to the question of synthetic contingency. As discussed, the biochemistry in living systems evolved along particular pathways because of the starting materials present. Chemistry in interstellar space can and does provide chemical selectivity. Hence, it could be responsible for the current paradoxes in biological systems. An example is ribose. The prebiotic source of this compound is thought to be the Formose reaction, where under basic, aqueous conditions, H_2CO is converted to a wide range of sugars (Müller et al. 1990). How did ribose alone end up in RNA (deoxyribose in DNA)? Laboratory experiments have shown that ribose is the kinetically-controlled product in the Formose reaction; hence, it would be the favored product at low temperatures. If a gas-phase Formose reaction was producing sugars in interstellar gas, as theoretical calculations suggest (Jalbout et al. 2007), ribose would be the primary result.

6. Conclusion: Following the Carbon...Following the Molecules

The process of following the carbon through its history actually becomes one of tracing the paths of C-containing molecules. Observations demonstrate that in C-rich circumstellar envelopes, abundant concentrations of molecules exist with multiple carbon-carbon bonds. Molecular material appears to persist into the PN stage, and then into the diffuse interstellar medium, preserving some of the previous carbon-enrichment. Diffuse clouds collapse into dense clouds, carrying the molecular gas along, providing the seeds for more advanced chemical synthesis. Finally, this material is passed on to planets from the clouds via comets, meteorites, and interplanetary dust particles. This sequence of events (Figure 6) means that the some molecular material is preserved as one type of molecular source evolves to the next. Therefore, synthesis in dense clouds

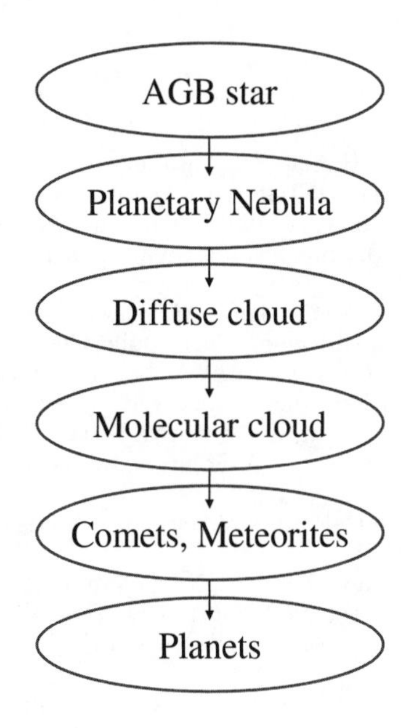

Figure 6. "Following the carbon" from AGB stars to planets: the possible life-cycle of carbon-carbon bonds, originating in the C-rich environment of AGB envelopes, passed on to planetary nebulae, diffuse clouds, dense clouds, and then to planets via comets and meteorites.

starts with a molecular base, not just with atoms. As a consequence, a higher degree of chemical complexity can be achieved than chemical models predict (e.g. Horn et al. 2004). This scenario, starting with C-rich AGB shells, also offers an explanation for the flourishing interstellar organic chemistry in dense gas, despite an O-rich environment.

The remarkable fact is that gas-phase interstellar organic chemistry is robust and widespread, despite the extreme physical conditions of space. The large scale presence of interstellar organic material suggests a link to prebiotic chemistry on planets. Interstellar chemistry may have provided the starting materials for life. If so, it could explain a number of biochemical paradoxes, such as, why ribose? Consequently, it is important to accurately evaluate the chemical content of molecular clouds. Finally, the origin of organic chemistry and life itself may ultimately be traced back to the carbon-rich circumstellar shells of AGB stars.

Acknowledgments. This research is based on work supported by the National Aeronautics and Space Administration through the NASA Astrobiology Institute under Cooperative Agreement No. CAN-02-OSS-02 issued through the Office of Space Science. This research was also supported in part by NSF grants AST-06-07803 and AST-06-02282.

References

Apponi, A. J., Hoy, J. J., Halfen, D. T., Ziurys, L. M., & Brewster, M. A. 2006, Hydroxyacetone (CH_3COCH_2OH): A Combined Microwave and Millimeter-wave Laboratory Study and Associated Astronomical Search. ApJ, 652, 1787

Balick, B., & Frank, A. 2002, Shapes and Shaping of Planetary Nebula. ARA&A, 40, 439

Benner, S. A., Ricardo, A., & Carrigan, M. A. 2004, Is There a Common Chemical Model for Life in the Universe? Curr. Opin. Chem. Biol., 8, 672

Bockelée-Morvan, D., & Crovisier, J. 2002, Lessons of Comet Hale-Bopp for Coma Chemistry: Observations and Theory. EM&P, 89, 53

Carigi, L., Peimbert, M., Esteban, C., & Garcia-Rojas, J. 2005, Carbon, Nitrogen, and Oxygen Galactic Gradients: A Solution to the Carbon Enrichment Problem. ApJ, 623, 213

Cherchneff, I. 2006, A Chemical Study of the Inner Winds of Asymptotic Giant Branch Stars. A&A, 456, 1001

Crovisier, J., Bockelée-Morvan, D., Biver, N., Colom, P., Despois, D., & Lis, D. C. 2004, Ethylene glycol in comet C/1995 O1 (Hale-Bopp). A&A, 418, L35

Elmegreen, B. G. 2007, On the Rapid Collapse and Evolution of Molecular Clouds, ApJ, 668, 1064

Eschenmoser, A. 1999, Chemical Etiology of Nucleic Acid Structure. Science, 284, 2118

Furnes, H., Banerjee, N. R., Muehlenbachs, K., Staudigel, H., & de Wit, M. 2004, Early Life Recorded in Archean Pillow Lavas. Science, 304, 578

Gomes, R., Levison, H. F., Tsiganis, K., & Morbidelli, A. 2005, Origin of the cataclysmic Late Heavy Bombardment period of the terrestrial planets. Nature, 435, 466

Halfen, D. T., Apponi, A. J., Polt, R., Woolf, N. J., & Ziurys, L. M. 2006, A Systematic Study of Glycolaldehyde in Sgr B2(N) at 2 and 3 Millimeters: Criteria for Detecting Large Interstellar Molecules. ApJ, 639, 237

Halfen, D. T., Apponi, A. J., Woolf, N. J., & Ziurys, L. M. 2008a, A Comprehensive Study of Acetamide in Sgr B2(N). in prep.

Halfen, D. T., Apponi, A. J., Adande, G., Woolf, N. J., & Ziurys, L.M. 2008b, A Confusion-Limited Survey of Sgr B2(N) at 1, 2, and 3 mm. in prep.

Halfen, D. T., Sun, M., Clouthier, D. J., & Ziurys, L.M. 2008c, Detection of the CCP ($X^2\Pi_r$) Radical in IRC+10216: A New Interstellar Phosphorus-Containing Species. ApJ, 677, L101

Harper, C. L., Jr., & Jacobsen, S. B. 1996, Noble Gases and Earth's Accretion. Science, 273, 1814

Herwig, F. 2006, AGB Stars Evolution and Nucleosynthesis, Proceedings of the International Symposium on Nuclear Astrophysics IX. CERN., p.206.

Hohsaka, T., & Masahiko, S. M. 2002, Incorporation of Non-Natural Amino Acids into Proteins. Curr. Opin. Chem. Biol., 6, 809

Hollis, J. M., Lovas, F. J., Remijan, A. J., Jewell, P. R., Ilyushin, V. V., & Kleiner, I. 2006, Detection of Acetamide (CH_3CONH_2): The Largest Interstellar Molecule with a Peptide Bond. ApJ, 643, L25

Hora, J. L., Latter, W. B., Smith, H. A., & Marengo, M. 2006, Infrared Observations of the Helix Planetary Nebula, ApJ, 652, 426

Horn, A., Mollendal, H., Sekiguchi, O., Uggerud, E., Roberts, H., Herbst, E., Viggiano, A. A., & Fridgen, T. D. 2004, The Gas-Phase Formation of Methyl Formate in Hot Molecular Cores. ApJ, 611, 605

Howe, D. A., Hartquist, T. W., & Williams, D. A. 1994, Molecules in Dense Globules in Planetary Nebula, MNRAS, 271, 811

Hunten, D. M., Donahue, T. M., Walker, J. C. G., & Kasting, J. F. 1989, Escape of Atmospheres and Loss of Water, in Origin and Evolution of Planetary and Satellite Atmospheres (Tucson, University of Arizona Press), 386

Jalbout, A. F., Abrell, L., Adamowicz, L., Polt, R., Apponi, A. J., & Ziurys, L. M. 2007, Sugar Synthesis From a Gas-phase Formose Reaction, Astrobio., 7, 433

Karakas, A. I., Lattanzio, J. C., & Pols, O. R. 2002, Parameterising the Third Dredge-up in Asymptotic Giant Branch Stars. Publ. Astron. Soc. Aust. 19, 515

Liszt, H., Lucas, R., & Pety, J. 2006, Comparative Chemistry of Diffuse Clouds V. Ammonia and Formaldehyde, A&A, 448, 253

Maier, J. P., Lakin, N. M., Walker, G. A. H., Bohlender, D. A., 2001, Detection of C_3 in Diffuse Interstellar Clouds, ApJ, 553, 267

Müller, D., Pitsch, S., Kittaka, A., Wagner, E., Wintner, C. E., Eschenmoser, A., & Ohlofjgewidmet, G. 1990, Chemistry of α-Aminonitriles. Aldomerisation of Glycolaldehyde Phosphate to rac-Hexose 2,4,6-Triphosphates and (in Presence of Formaldehyde) rac-Pentose 2,4-Diphosphates: rac-Allose 2,4,6-Triphosphate and rac-Ribose 2,4-Diphosphate Are the Main Reaction Products. Helv. Chim. Acta, 73, 1410

Nercessian, E., Omont, A., & Benayoun, J. J., & Guilloteau, S., 1989, HCN Emission and Nitrogen-Bearing Molecules in Oxygen-Rich Circumstellar Envelopes, A&A, 210, 225

Pizzarello, S., Cooper, G., & Flynn, G. 2006, The Nature and Distribution of the Organic Material in Carbonaceous Chondrites and Interplanetary Dust Particles. in Meteorites and the Early Solar System II, ed. D. S. Lauretta & H. Y. McSween, Jr. (Tucson, University of Arizona Press), 625

Rosing, M. T. 1999, ^{13}C Carbon Microparticles in >3700-Ma Sea-Floor Sedimentary Rocks from West Greenland. Science, 283, 674

Tenenbaum, E. D., Milam, S. N., Woolf, N. J., & Ziurys L. M. 2009 Molecular Survival in Evolved Planetary Nebulae: Detection of H_2 CO, c-C_3 H_2 , and CCH in the Helix, Ap.J.//,/ /in press

van Dishoeck, E. F., & Black, J. H. 1986, Comprehensive models of diffuse interstellar clouds - Physical conditions and molecular abundances. ApJS, 62, 109

Young, K., Cox, P., Huggins, P. J., Forveille, T. & Bachiller, R. 1999, The Molecular Envelope of the Helix Nebula. ApJ, 522, 387

Ziurys, L.M. 2006, The Chemistry in Circumstellar Envelopes in Evolved Stars: Following the Origin of the Elements to the Origin of Life, PNAS, 103, 12274

Ziurys, L. M., Milam, S. N., Apponi, A. J., & Woolf, N. J. 2007, Chemical Complexity in the Winds of the Oxygen-Rich Supergiant Star VY Canis Majoris. Nature, 447, 1094

Ted von Hipple

Bioastronomy 2007: Molecules, Microbes, and Extraterrestrial Life
ASP Conference Series, Vol. 420, 2009
K. J. Meech, J. V. Keane, M. J. Mumma, J. L. Siefert, and D. J. Werthimer, eds.

White Dwarf Debris Disks: The Fate of Planetary Systems ?

T. von Hippel,[1] W. T. Reach,[2] F. Mullally,[3] M. Kilic,[4] and M. J. Kuchner[5]

[1]*Department of Astronomy, University of Texas at Austin, 1 University Station C1400, Austin, TX 78712-0259; ted@astro.as.utexas.edu*
[2]*Spitzer Science Center, MS 220-6, California Institute of Technology, Pasadena, CA 91125*
[3]*Department of Astrophysical Sciences, Princeton University, Peyton Hall - Ivy Lane, Princeton, NJ 08544-1001*
[4]*Columbus Fellow, Department of Astronomy, Ohio State University, 140 West 18th Avenue, Columbus, OH, 43210*
[5]*NASA Goddard Space Flight Center, Greenbelt, MD 20771*

Abstract. White dwarfs may offer new insights into how common extrasolar minor bodies such as asteroids are. One quarter of all white dwarfs appear to have heavy elements such as Ca, Mg, Si, and Fe in their atmospheres, despite the fact that the extreme gravity of these stars causes heavy elements to sink through their atmospheres on a timescale of weeks to years. These white dwarfs are currently accreting from their circumstellar environments, and we argue that they are all accreting from debris disks left over from tidally disrupted asteroids, comets, or other minor rocky bodies. This interpretation leads to the tentative conclusion that $\geq 25\%$ of all stars have planetary systems composed of at least minor, rocky bodies.

1. How Are White Dwarfs Relevant to Astrobiology?

White dwarf (WD) stars, the compact remnant end state of the vast majority of stars, do not, at first glance, seem relevant to astrobiology. White dwarfs range in surface temperature from 100,000 to maybe 4000 K, with a constantly declining energy production as they cool. Their precursor stars went through a red giant phase that scoured out the inner planetary system to 1–2 AU, and the subsequent mass loss that led to the white dwarf causes further orbital expansion. No one has yet argued that a white dwarf is a good place to start looking for life. Yet, white dwarfs do have certain advantages for the study of conditions for life, particularly for the study of planetary systems and their remnants. The low luminosity of white dwarfs, typically 10^{-3} as bright as their precursor star, will decrease the contrast ratio between white dwarfs and any of their planets massive enough to be self-luminous. In addition, we are now able to detect small quantities of dust around white dwarfs, and this dust may be derived from shredded asteroids or other rocky planetary material. By looking at the fate of the majority of stars, we may have a unique window into the composition, ubiquity, and even dynamics of planetary systems. We may also learn something about the future of our Earth.

2. Heavy Metal White Dwarfs: Where do the Metals Come From?

The presence of Ca, Mg, Fe, or other heavy elements in the atmospheres of hydrogen-dominated (DA) white dwarfs poses a long-standing problem. The high surface gravities of WDs cause their atmospheres to become highly stratified and theory predicts that these heavy elements will settle on timescales of 10^{-2} to 10^6 years (Dupuis et al. 1992; Koester, & Wilken 2006), depending on the mass and surface temperature of the white dwarf. Yet most WDs are old, and almost every known DAZ, as these stars are now called (Sion et al. 1983), has been cooling as a WD for at least 10^8 years, and usually 10^9 years or more. These surface metals cannot be inherited from the precursor stars and whatever their source, the surface metals require on-going accretion from the WDs' surroundings. (Metals can also be present in the rarer He-dominated WDs, and the story for these stars is probably similar, though less well developed. This class of WDs will not be discussed further here.)

The first convincing demonstration of metals in a H-atmosphere WD was over 20 years ago (Lacombe et al. 1983), with the second and third DAZs (Koester, Provencal, & Shipman 1997; Holberg, Barstow, & Green 1997) coming substantially later. The DAZ sample grew with the survey of Zuckerman & Reid (1998), which yielded seven to nine new DAZs. Subsequent modern surveys (e.g., Zuckerman et al. 2003) have shown that ~25% of all single H-dominated WDs possess metals in their atmospheres. On the theoretical side, early work focused on understanding how WDs might accrete sufficient material from the interstellar medium (ISM) to maintain their atmospheric metals. Some researchers (Dupuis, Fontaine, & Wesemael 1993; Koester, & Wilken 2006) concluded that ISM accretion is the source of the atmospheric metals, while other researchers (Aannestad et al. 1993; Zuckerman & Reid 1998; Zuckerman et al. 2003; Kilic & Redfield 2007) concluded that ISM accretion is insufficient.

3. G29-38's Infrared Excess: Circumstellar Dust in a Debris Disk

The white dwarf known as Giclas 29-38 (G29-38) was discovered by Zuckerman & Becklin (1987) to radiate substantially more light than expected at wavelengths longer than 2 microns. Initially, it was unclear whether the IR excess indicated a brown dwarf (Zuckerman & Becklin 1987) or particulate debris. Eventually, it became clear that G29-38 had a debris disk, based on its infrared energy distribution (Tokunaga, Becklin, & Zuckerman 1990; Chary, Zuckerman, & Becklin 1999), as well as time-resolved photometry that constrained the dust geometry (Graham et al. 1990; Patterson et al. 1991), and limits on the presence of companions from pulsation-timing studies (Kleinman et al. 1994) and speckle imaging (Kuchner et al. 1998). Recently, *Spitzer* 7–14 μm spectroscopy (Reach et al. 2005b) showed a strong 10 μm emission feature that can only be caused by small silicate dust particles. Might G29-38 be a prototype for understanding all white dwarfs with metals in their atmospheres, or is this object just a single example of some rare evolutionary path? Are there other debris disk white dwarfs, and if so, are their properties consistent with a picture where debris disks are the sources for the metals being continuously accreted by these WDs?

4. Other Debris Disk White Dwarfs

The second debris disk white dwarf, GD 362, was reported by Becklin et al. (2005) and Kilic et al. (2005). Subsequent work has found a hand full of other white dwarfs with debris disks (Kilic et al. 2006b; Jura, Farihi, & Zuckerman 2007; Mullally et al. 2007, von Hippel et al. 2007). A related object, SDSS 1228+1040, was found by Gänsicke et al. (2006) to have a gaseous disk of Ca and other metals. The WD in this system is substantially hotter, with a surface temperature of ~22,000 K compared to 9000 to 14,000 K for the WDs hosting dust debris. It is thus expected that the orbiting material will be heated past the sublimation temperature of dust (~1200 K) and be found in the gas phase.

When these WDs are considered as a group, they share two properties: (1) all are DAZs, and in fact, all have higher metal abundances than most other DAZ WDs and (2) they are at a temperature appropriate for warming their circumstellar material to the point where it is detectable by current generation instruments. Colder WDs would not warm dust debris sufficiently to be detected by ground-based telescopes or *Spitzer*. And WDs with lower metal abundances, and therefore presumably lower accretion rates, would have more anemic disks, which would escape detection. The fact that 25% or more of all WDs are DAZs, yet only a fraction of DAZs have detectable debris disks, is consistent with every DAZ hosting a debris disk simply because debris disks are hard to detect and those that are the easiest to detect are the ones we find to have debris disks. Thus, the absence of debris disks detected around all DAZ WDs could be a simple observational selection effect.

5. Planetary System Destruction or Late Stages of Stellar Evolution?

What could cause 25% or more of all stars to host debris disks during their final evolutionary state? Whatever the scenario, it must result in at least 10^3 km^3 of dust orbiting within ~1 $R_\odot$ of the WD (Reach et al. 2005). The two most compelling current scenarios, with neither yet developed into a theory with quantitative predictive power, are planetary sources or stellar material left over after mass loss.

In the former scenario, a rocky body such as an asteroid, moon, or volatile-poor comet need only wander within about 1.5 $R_\odot$ of a WD to be tidally disrupted (Jura 2003) and create a near instantaneous debris disk. There are sufficient such bodies in the Solar System, for instance. To feed G29-38 over its lifetime (i.e. assuming this disk is not transient) would only require about 1/4 the mass of the Moon. Yet, the difficulty for this scenario is that 25% of all WDs are DAZs, making it necessary for a large fraction of evolved stars to somehow dynamically move material from outside the giant star-scoured region, at 200–500 $R_\odot$, to within 1 $R_\odot$. While Debes & Siguardsson (2002) showed that it is dynamically possible to move minor bodies from outer to inner orbits after the central star loses mass, this area of research has been insufficiently explored and there is currently no identified mechanism that could perform the required dynamics sufficiently often (H. Levison, 2007, private communication).

In the latter scenario, a small fraction of the material ejected during the stellar mass loss phase does not achieve escape velocity. This material settles into orbit about the WD and evolves into the debris disk we see today. This scenario is analogous to creating pulsar planets (Wolszczan & Frail 1992) via a fallback disk (see, e.g., Lin, Woosley, & Bodenheimer 1991). The chief difficulty with this scenario is that the WD disks seen to date show no H nor He, and have low C abundances compared to typical values for stars. A mass losing giant star typically gives off copious amounts of He, and because most WD precursors produce C and O, there should be an excess of C relative to Fe, Mg, Ca, etc., counter to what is observed.

The abundance argument causes us to currently favor a planetary origin for the WD debris disks. Under this hypothesis, the prevalence of DAZ WDs indicate that at least 25% of all stars host rocky bodies in orbit and probably also planets to destabilize those orbits after the central star evolves and loses mass. This is consistent with the results of radial velocity surveys (e.g., Udry & Santos 2007) that report $\sim$5% of solar-type stars having Jupiter mass planets, since the WD debris disks have substantially lower mass. Indeed, it may be that the WDs are guiding us as we extrapolate the radial velocity work to lower masses and anticipate what we will eventually find for the fraction of extrasolar Earth-mass planets.

Acknowledgments. This work is based in part on observations made with the Spitzer Space Telescope, which is operated by the Jet Propulsion Laboratory, California Institute of Technology under NASA contract 1407. Support for this work was provided by NASA through award project NBR 1269551 issued by JPL/Caltech to the University of Texas.

References

Aannestad, P. A., Kenyon, S. J., Hammond, G. L., & Sion, E. M. 1993, AJ, 105, 1033

Becklin, E. E., Farihi, J., Jura, M., Song, I., Weinberger, A. J., & Zuckerman, B. 2005, ApJ, 632, L119

Chary, R., Zuckerman, B., & Becklin, E. E. 1999, in The Universe as Seen by ISO, ed. P. Cox & M. F. Kessler (ESA-SP 427; Garching: ESA), 289

Debes, J. & Siguardsson, S. 2002, ApJ, 572, 556

Dupuis, J., Fontaine, G., Pelletier, C., & Wesemael, F. 1992, ApJS, 82, 505

Dupuis, J., Fontaine, G., & Wesemael, F. 1993, ApJS, 87, 345

Gänsicke, B.T., Marsh, T.R., Southworth, J., & Rebassa-Mansergas, A. 2006, Science, 304, 1908

Graham, J. R., Matthews, K., Neugebauer, G., & Soifer, B. T. 1990, ApJ, 357, 216

Holberg, J. B., Barstow, M. A., & Green, E. M 1997, ApJ, 474, L127

Jura, M. 2003 ApJ, 584, L91

Jura, M., Farihi, J., & Zuckerman, B. 2007 ApJ, 663, 1285

Koester, D., Provencal, J., & Shipman, H. L. 1997, A&A, 320, L57

Kilic, M., & Redfield, S. 2007, ApJ, 660, 641

Kilic, M., von Hippel, T., Leggett, S. K., & Winget, D. E. 2005a, ApJ, 632, L115

Kilic, M., von Hippel, T., Leggett, S. K., & Winget, D. E. 2005b, ApJ, 646, 474

Kleinman, S. J. et al. 1994, ApJ, 436, 875

Koester, D., & Wilken, D. 2006, A&A, 453, 1051

Kuchner, M. J., Koresko, C. D. & Brown, M. E. 1998, ApJ, 508, L81

Lacombe, P., Wesemael, F., Fontaine, G., & Liebert, J. 1983, ApJ, 272, 660

Lin, D. N. C., Woosley, S. E., & Bodenheimer, P. H. 1991, Nat, 353, 827

Mullally, F., Kilic, M., Reach, W. T., Kuchner, M. J., von Hippel, T., Burrows, A., & Winget, D. E. 2007, ApJS, 171, 206

Patterson, J., Zuckerman, B., Becklin, E. E., Tholen, D. J., & Hawarden, T. 1991, ApJ, 374, 330

Reach, W. T., Kuchner, M. J., von Hippel, T., Burrows, A., Mullally, F., Kilic, M., & Winget, D. E. 2005b, ApJ, 635, L161

Sion, E. M., Greenstein, J. L., Landstreet, J. D., Liebert, J., Shipman, H. L., & Wegner, G. A. 1983, ApJ, 269, 253

Tokunaga, A. T., Becklin, E. E., & Zuckerman, B. 1990, ApJ, 358, L21

von Hippel, T., Kuchner, M. J., Kilic, M., Mullally, F., & Reach, W. T. 2007, ApJ, 662, 544

Udry, S., & Santos, N. C. 2007, ARA&A, 45, 397

Wolszczan, A., & Frail, D. A. 1992, Nat, 355, 145

Zuckerman, B. & Becklin, E. E. 1987, Nat, 330, 138

Zuckerman, B., Koester, D., Reid, I. N., Hünsch, M. 2003, ApJ, 596, 477

Zuckerman, B., & Reid, I. N. 1998, ApJ, 505, L143

Kathryn Denning

Bioastronomy 2007: Molecules, Microbes, and Extraterrestrial Life
ASP Conference Series, Vol. 420, 2009
K. J. Meech, J. V. Keane, M. J. Mumma, J. L. Siefert, and D. J. Werthimer, eds.

The Near-Star Environment: Spectropolarimetry of Herbig Ae/Be Stars

D. M. Harrington and J. R. Kuhn

University of Hawaii, Institute for Astronomy, 2680 Woodlawn Drive, Honolulu HI 96822

Abstract. The near-star environment around young stars is very dynamic with winds, disks, and outflows. These processes are involved in star and planet formation, and influence the formation and habitability of planets around host stars. Even for the closest young stars, this will not be imaged even after the completion of the next generation of telescopes decades from now and other proxies must be used. The polarization of light across individual spectral lines is such a proxy that contains information about the geometry and density of circumstellar material on these small spatial scales. We have recently built a high-resolution spectropolarimeter (R 13000 to 50000) for the HiVIS spectrograph on the 3.67m AEOS telescope. We used this instrument to monitor several young intermediate-mass stars over many nights. These observations show clear spectropolarimetric signatures typically centered on absorptive components of the spectral lines, with some signatures variable in time. The survey also confirms the large spectroscopic variability in these stars on timescales of minutes to months, and shows the dynamic bullets and streamers in the stellar winds. These observations were largely inconsistent with the traditional scattering models and inspired the development of a new explanation of their polarization, based on optical-pumping, that has the potential to provide direct measurements of the circumstellar gas properties.

1. Introduction

The formation of planets around stars is a dynamic process. Over the course of a few million years, the circumstellar gas and dust will either accrete onto the star, turn in to planets, or dissipate in the form of winds and jets. The star also evolves from an active young star showing emission lines (TTauri or Herbig Ae/Be stars) to a stable main-sequence star. There is evidence of accretion, outflows, ionized disk structures, and strong stellar winds in young stars, many times happening simultaneously. All of these processes influence the environment of the planets around these stars. The star's physical properties influence the habitable zone and its planets' atmospheres through planet-star interactions. By measuring properties of the circumstellar material in young stars, we can put direct constraints on these processes.

There are only a few techniques that can put meaningful constraints on the environment around a star. Spectropolarimetry is one, but it is still considered a difficult specialty field. In the stellar astronomical community, only a handful of spectropolarimeters exist. There are only two high-resolution instruments, ESPaDOnS and HiVIS on large telescopes (over 3m). However, this technique is a powerful probe of small spatial scales, being sensitive to the geometry and density of the material very close the central star. Typically, material in the region a few to several stellar radii away from the star produces the polarization. Even for the closest young stars (150pc), these spatial scales are smaller than 0.1 milliarcseconds across and will not be imaged directly, even by 100m telescopes. Since the circumstellar material is involved in accretion, outflows, winds and disks, with many of these phenomena happening simultaneously, spectropolarimetry can put unique constraints on the types of densities and geometries of the material involved in these processes.

Spectropolarimetry in it's most simple form is just spectroscopy through a polarizer. Exact instrumental techniques vary, but they all measure intensity *and* polarization as a function of wavelength. With a high-resolution instrument ($\frac{\lambda}{\delta\lambda} \geq 15000$) clear polarization changes across individual emission and absorption lines are seen. Things like circumstellar disks, rotationally distorted winds, magnetic fields, or asymmetric radiation fields (optical pumping) can cause these signatures. The polarization comes directly from the circumstellar material and can be used as a proxy for the physical properties of the circumstellar material, just like spectroscopy. Typical spectropolarimetric signals are small, often a few tenths of a percent change in polarization. Measuring these signals requires very high signal to noise observations and careful control of systematics errors. We have recently built and calibrated a polarimetry package for the HiVIS spectrograph on the 3.67m AEOS telescope on Haleakala, HI to address these important issues (Harrington et al. 2006, Harrington & Kuhn 2008).

There are many models of spectropolarimetric effects. Early analytical studies showed the possibility of spectropolarimetric effects from scattering very close to the central star (McLean 1979, Wood et al. 1993 & 1994). Recent modeling of scattering by circumstellar materials has shown a wealth of possible polarimetric line-effects from disks, winds, and envelopes (Harries 2000, Ignace 2004, Vink et al. 2005a). Unpolarized line emission that forms over broad stellar envelopes can produce a depolarization in the line core relative to the stellar continuum. Small clumps in a stellar wind that scatter and polarize significant amounts of light can enhance the polarization at that clump's specific velocity and orientation. Magnetic fields can, through the Zeeman effect, shift the wavelength of polarized atomic emission, giving rise to a field-dependent polarization. Optically pumped gas can produce polarization through absorption (Kuhn et al. 2007).

2. Spectropolarimetry

To date, only a few detections of spectropolarimetric signals in young stars have been reported, and the variability of these signatures has not yet been studied in detail (Vink et al. 2002, 2005b Mottram et al. 2007, Harrington & Kuhn 2007). Our study, as well as most others, have focused on the H_α line, a strong line of Hydrogen. We initially chose bright, well-known stars for close study but have since expanded our target list to include many young stars. The H_α line in these stars is typically a very strong emission line with additional absorptive components. The line can be up to 20 times continuum, typically with evidence for winds or disks: P-Cygni absorption profiles or central absorption components respectively. Our observations of windy-stars showed strong variability of the absorption component, often over 10-minute timescales. In disky stars, the H_α

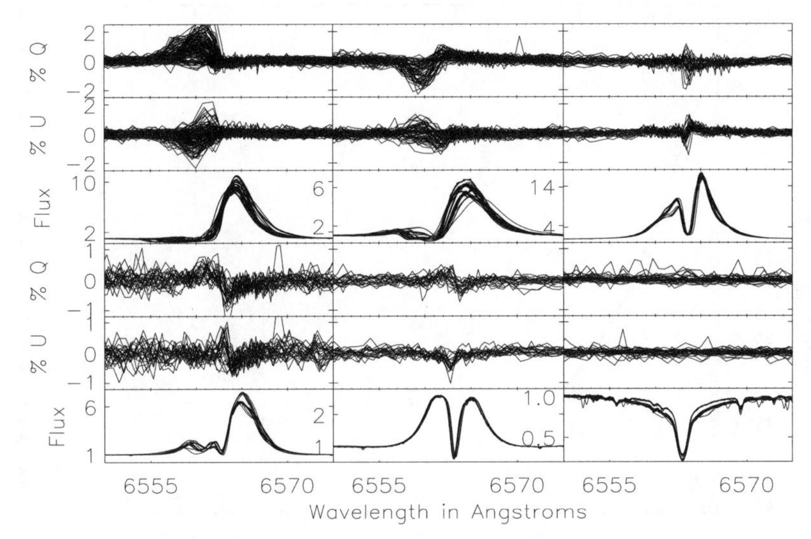

Figure 1. The spectroscopy and spectropolarimetry of 5 Herbig AeBe's and unpolarized stars in the lower right for comparison. Each star has three boxes. The left column is AB Aurigae above and MWC120 below. The middle column is MWC480 and HD58647. The right column, is MWC158, and unpolarized calibrators. Stokes q, the vertical (+) and horizontal (-) polarization in percent in the top box. Stokes u, the $\pm 45°$ polarization in percent is in the middle box. The average nightly spectra are in the bottom box. See Harrington & Kuhn 2007 for details.

line showed strong central absorptions but were much less variable. Modeling the spectra and variability alone can provide information about the near-star environment (cf. Beskrovnaya et al. 1995, Bouret & Catala 1998, Catala et al. 1999).

Preliminary studies at medium resolution showed many different signatures with amplitudes up to 2% change (Vink et al. 2002, 2005b). Our observations of the H_α line for our first five stars, shown in figure 1, show similar amplitudes. Instrumental effects cause a lot of the apparent variability in this figure, but there is intrinsic variability as well (see Harrington & Kuhn 2008). The signatures are roughly 0.3% to 1.5% centered on the absorption component, not on line center. We have since observed polarization signatures that span the entire H_α line and have several non-detections, but the polarization-in-absorption is a common feature for intermediate-mass young stars.

3. A New Model - Optical Pumping

The simple scattering polarization and depolarization models all predicted signatures centered on the line core (McLean 1979, Wood et al. 1993, Vink et al. 2005a).

In most of our stars, the change in polarization occurred in and around the absorptive component, whether central or blue-shifted. The polarization

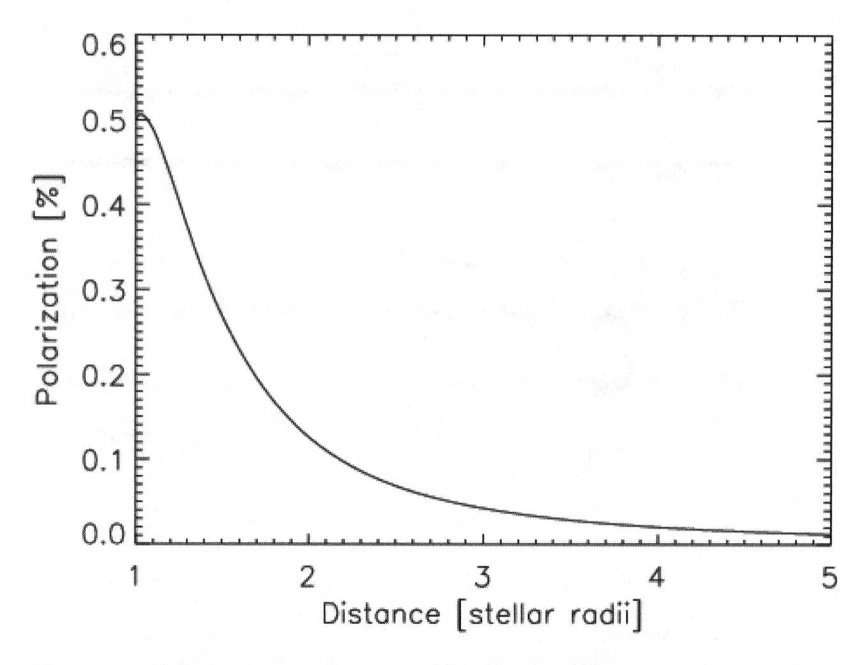

Figure 2. Polarization as a function of distance between a gas cloud and the star assuming a T=10,000K star with limb darkening. The intervening cloud is obscuring the edge of the stellar disk. See Kuhn et al. 2007 for details.

of the emission peak was nearly identical to the continuum polarization. This problem led us to explore alternative explanations that would require the absorbing material to also be the polarizing material. We developed a new model where the stellar radiation causes the absorbing material to polarize the transmitted light (Kuhn et al. 2007). The circumstellar gas is optically pumped by the strong anisotropic radiation from the star causing the absorbing gas to absorb different incident polarizations by different amounts. The main difference between this optical pumping model and the scattering model is that only the absorbing material is responsible for the changing polarization, whereas the scattering models integrate scattered light from the entire circumstellar region with each part contributing to the polarization change. An example of an optical pumping calculation is shown in figure 2. This new model is simple and deterministic. With some further development, this model can be used to put a direct constraint on the circumstellar gas conditions. This in turn will clarify our understanding of the circumstellar environment and interactions.

Acknowledgments. This program was partially supported by the NSF AST-0123390 grant, the University of Hawaii and the AirForce Research Labs (AFRL). This program also made use of the Simbad data base operated by CDS, Strasbourg, France.

References

Beskrovnaya et al. 1995, A&A, 295, 585
Bouret J.-C. & Catala C., 1998, A&A, 340, 163
Catala C. et al., 1999, A&A, 345, 884
Harries T.J., 2000, MNRAS, 315, 722
Harrington D.M., Kuhn. J.R., & Whitman K., 2006, PASP, 118, 845
Harrington D.M. & Kuhn. J.R., 2007, ApJL, 667, L89
Harrington D.M. & Kuhn. J.R., 2008, PASP, Submitted
Ignace R., Nordsieck K.H., & Cassinelli J.P., 2004, ApJ, 609, 1018
Kuhn J.R. et al., 2007, ApJL, 668, L63.
McLean I.S., 1979, MNRAS, 186, 265
Mottram J.C., Vink J.S., Oudmaijer R.D., & Patel M., 2007, MNRAS, 377, 1363
Vink J.S., Drew, J.E., Harries T.J., and Oudmaijer R.D., 2002, MNRAS, 337, 356
Vink, J.S., Harries T.J., and Drew J.E., 2005a, A&A 430, 213
Vink J.S. et al., 2005b, MNRAS, 359, 1049
Wood K., Brown J.C., and Fox G.K., 1993, A&A, 271, 492
Wood K. and Brown J.C., 1994, A&A, 291, 202

Charles Lineweaver

Bioastronomy 2007: Molecules, Microbes, and Extraterrestrial Life
ASP Conference Series, Vol. 420, 2009
K. J. Meech, J. V. Keane, M. J. Mumma, J. L. Siefert, and D. J. Werthimer, eds.

Comparing the Chemical Compositions of the Sun and Earth

C. H. Lineweaver and J. A. Robles

Planetary Science Institute, Research School of Astronomy & Astrophysics and Research School of Earth Sciences, The Australian National University, Canberra Australia

Abstract. We compare the solar and terrestrial bulk elemental compositions. We show that the Sun and Earth share the same refractory elemental abundances, and that, to first order, the Earth is a devolatilized piece of the Sun left over from the Sun's formation. We discuss the advantages of an aluminium normalization for this comparison over the traditional silicon normalization.

1. Introduction

If we did not know which star (from a group of stars) was our own, a comparison of the refractory (not volatile) elemental abundances of each star and the Earth's, would identify our Sun from the stellar lineup. Conversely, we can infer the elemental compositions of extrasolar earths from spectroscopic measurements of the elemental compositions of other stars. An important part of this procedure is to quantify as accurately as possible the chemical fractionation that occurred in our Solar System 4.5 billion years ago. We do this by comparing the bulk elemental abundances of the Sun and Earth.

2. The Sun's Bulk Composition

The mass fractions of H, He and other elements in the Sun and elsewhere in the universe, are referred to as X, Y and Z respectively, where $X + Y + Z =$ 100%. After big bang nucleosynthesis (BBN) the universe can be described by $(X, Y, Z) = (75.2, 24.8, 0)\%$ (Spergel et al. 2007). After $\sim$ 13.8 billion years of burning H into He and Z, we have a current cosmic baryonic average of approximately $(X, Y, Z) = (74.2, 25.3, 0.5)$. The 0.5% increase in Y after BBN is from Figure 1b of Fukugita & Kawasaki (2006). This 0.5% increase in Z is a crude estimate based on the idea that the net result of stellar fusion produces about as much He as it does everything else. This is also plausible because the Sun may have about twice as much Z as the cosmic baryonic average, since the material in the Sun has probably been more processed in stellar interiors than has most of the baryonic material in the universe. Also, the Sun is more metal-rich than $65 \pm 2\%$ of near-by stars of a variety of ages (Robles et al. 2008).

The difference between the proto-Sun and the current Sun is that as the Sun burns H to He, X has gone down by $\sim 4.6\%$ and Y has gone up by the same amount. The equivalent mass loss of the Sun due the Sun's luminosity over the 4.5 Gyr lifetime of the Sun ($\delta m = L_{\odot} t / c^2$ where $L_{\odot}$ is the solar luminosity and

Table 1. Cosmic and Solar Mass Fraction Estimates [%]

X	Y	Z	Reference
75.2	24.8	0	cosmic values after big bang nucleosynthesis ((Spergel et al. 2007)
74.2	25.3	0.5	current cosmic baryonic average ((Fukugita & Kawasaki 2006)
71.2	27.4	1.4	proto-solar ((Lodders 2003, Grevesse et al. 2005)
66.6	32.0	1.4	current bulk Sun
73.9	24.9	1.2	solar photosphere ((Grevesse et al. 2005, Asplund et al. 2005)

t is the age of the Sun) is negligible, since $\delta m \approx 0.03\%$ of the Sun's mass. These numbers are summarized in Table 1.

There is an on-going controversy over the precise value of the solar Z. Helioseismological estimates of ~ 1.8 (Antia & Basu 2006, Chaplain et al. 2007) are hard to reconcile with recent photospheric estimates of ~ 1.2 (Asplund et al. 2005, Grevesse et al. 2005). However, the investigation of the systematic errors of these techniques is still in its infancy.

The $\sim 1.5\%$ of the mass of the Sun that is not H and not He, consists of oxygen (43.0%), carbon (17.3%), iron (9.7%), neon(8.2%), silicon(5.7%), magnesium(5.1%), nitrogen(5%), sulfur(2.9%), (Lodders 2003, Grevesse et al. 2005). These elements are unsurprisingly, just the right ingredients to form life (= C, H, O, N, S) and terrestrial planets (= Fe, O, Si, Mg).

The solar Z value and the abundances of the individual elements which scale with Z, depend to some degree on whether we are talking about the proto-solar nebula (= current bulk = proto bulk Sun) or the Solar photospheric values. To avoid this ambiguity in future work, when we compare the Sun to other stars, we use the solar photospheric values, because our observations of other stars are limited to their photospheres. When we compare the elemental abundances of the Sun to those of the Earth, we use the proto-solar values which are estimates of the bulk composition of the Sun.

In Figure 1 we compare the bulk elemental composition of the Sun to that of the Earth. The solar abundances plotted are the protosolar abundances from Lodders (2003, Table 2, p. 1224), with the exception of oxygen, nitrogen, and neon which are photospheric abundances from Asplund et al. (2005, Table 1) using the Lodders (2003) protosolar correction factor = 0.074 dex (see p. 1235).

3. Normalization

The relative elemental abundances of the Sun and Earth can be compared only after normalization. For example, it has been traditional to normalize to the element silicon. The normalization can be described as follows: take an average piece of the Sun that contains 10^6 atoms of silicon and then count how many atoms of the other elements are contained in that piece. Then take an average piece of the Earth that also contains 10^6 atoms of silicon and then count how many atoms of the other elements are contained in that piece. When plotted, this provides a comparison of the Sun and Earth, normalized to silicon.

We normalize to aluminium because it is a more refractory element than silicon. Silicon is about 10 times more abundant in the Earth than is aluminium,

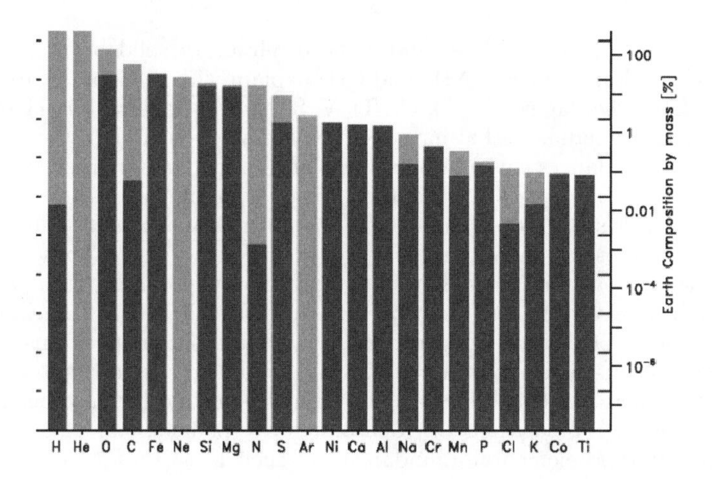

Figure 1. Comparison between the elemental abundances of the Sun (Lodders 2003, Asplund et al. 2005) and bulk Earth (Morgan & Anders 1980, McDonough 2003) (non-weighted averages from these two references) normalized to aluminium. To first order, the Earth is a devolatilized piece of the protosun. The x-axis shows the elements arranged in decreasing abundance in the Sun by mass.

so in the past, when measurement accuracy was low, Si was the preferred normalization. However, as precision has increased to parts-per-billion levels, the rational to use Si has disappeared. For example, with a silicon normalization, the Earth is enriched relative to the Sun in all the elements which are more refractory than Si (e.g. Fe, Mg, Ni, Ca, Al, Cr, Ti, V, Sr Zr). However, when normalized to Al we see that the Earth is not enriched in anything, but depleted in the volatile elements and has maintained solar abundances of the refractory elements.

Our preferred normalization is an element that has not been depleted in either the Sun or the Earth. Volatile loss is not an issue for the Sun because the Sun's gravitational field is strong enough to hold all elements. This is not true for the Earth. Thus, we normalize the Earth-Sun comparison to the most refractory elements — elements that have not been lost during Earth's formation. Thus, Figure 1 has been normalized to the same number of aluminium atoms and then converted to percent mass. Thus, for Al, the light grey Sun histogram bar and dark grey Earth histogram bar are of equal height).

Notice that the blue bars of the histogram are always lower or equal to the yellow bars. The difference between the bars is the degree to which the Earth has lost its volatiles. Any element that can only appear in a gaseous state will be severely depleted on Earth (e.g. He, Ne, Ar). Other elements that are largely, but not exclusively found in gaseous state (H, O, C, N) have been partially depleted. Gaseous states that have led to their depletion include: H_2, O_2, CO, CH_4 and N_2. In the protoplanetary disk that formed our Solar System, H_2O was a volatile within about 4 AU (the snowline) and a more refractory solid beyond 4 AU. Similar consideration of the ices NH_3 and CH_4 explain why H, C and N are only partially depleted. Fe, Mg Ni,Ca, Al, Cr, Ti, V, Sr and Zr are refractory elements and have relative abundances that have not changed since the Earth's formation. These relative abundances closely track the solar abundances for these elements and would be the abundances used in the thought experiment described above to identify the host star of the Earth. Of these refractory elements, aluminium seems to be the most abundant refractory element, so we use it to normalize the Sun-Earth comparison. This allows us to meaningfully quantify the pattern of volatiles that have been lost from the Earth. We will extend this work to quantify the range of devolatilization that led to the bulk compositions of the rocky planets of our Solar System. This will yield our best estimate of the devolatilization that has led to formation of rocky planets orbiting other stars. Thus, the chemical compositions of extrasolar rocky planets, can be extracted from the spectra (and elemental abundances) of their host stars.

Acknowledgments. We would like to thank Mount Stromlo Observatory at The Australian National University, the Bioastronomy 2007 and ABGradCon local organising committee for travel support. Our special thanks to K. Meech.

References

Antia, H.M. & Basu, S. 2006 Determining Solar Abundances Using Helioseismology, Apj 644, 1292-1298

Asplund, M., Grevesse, N. and Sauval A. J., 2005. The Solar Chemical Composition in "Cosmic Abundances as Records of Stellar Evolution and Nucleosynthesis" ed xxx, ASP Conf. Series 336, 25

Chaplain, W.J. et al. 2007 Solar Heavy-element abundance: constraints from frequency separation ratios of low-degree p-modes. ApJ, 670, 872-884

Fukugita, M. Kawasaki, M, 2006 Primordial Helium Abundance: A Reanalysis of the Izotov-Thuan Spectroscopic Sample ApJ 646, 2, 691-695

Grevesse, N., Asplund, M. and Sauval A. J., 2005. The New Solar Composition. in "Element Stratification in Stars, 40 years of Atomic Diffusion", ed. Alecian, G., Richard, O. and Vauclair S., EAS Publications Series

Lodders, K. 2003. Solar System Abundances and Condensation Temperatures of the Elements. ApJ, 591:1220–1247,.

Lodders, K. & Fegley, B. 1997. An Oxygen Isotope Model for the Composition of Mars. Icarus, 126:373–394.

McDonough, W. 2003. Compositional Model for the Earth's Core, volume 2. Elsevier-Pergamon, Oxford.

Morgan, J. W. & Anders, E. 1980. Chemical composition of earth, Venus, and Mercury. In National Academy of Sciences, Proceedings, vol. 77, Dec. 1980, p. 6973-6977., pages 6973–6977.

Robles, J. A., Lineweaver, C. H., Grether, D., *et al.* 2008, A comprehensive comparison of the Sun to other stars: searching for self-selection effects, ApJ 689, 1457

Spergel, D. et al. 2007, Three-year Wilkinson Microwave Anisotropy Probe (WMAP) Observations: Implications for Cosmology ApJSS, 170, 377-408

Bioastronomy 2007: Molecules, Microbes, and Extraterrestrial Life
ASP Conference Series, Vol. 420, 2009
K. J. Meech, J. V. Keane, M. J. Mumma, J. L. Siefert, and D. J. Werthimer, eds.

Towards Quantifying the Prevalence of Primitive Membranes in the Galaxy: The Millimeter-Wave Rotational Spectrum of Pyruvic Acid

Z. Kisiel,[1] L. Pszczółkowski,[1] E. Białkowska-Jaworska,[1] and S. B. Charnley[2]

[1]*Institute of Physics, Polish Academy of Sciences Warszawa, Poland*

[2]*Astrochemistry Laboratory & Center for Astrobiology, Solar System Exploration Division, Code 691, NASA Goddard Space Flight Center, Greenbelt, MD 20771, USA*

Abstract. If present in the interstellar medium, pyruvic acid is a molecule that could initiate the formation of primitive, vesicle-like structures throughout the Galaxy. We outline how it could be formed in interstellar chemistry and report recent laboratory work to measure its rotational spectrum. The results from this work will enable a search for interstellar pyruvic acid to be undertaken.

1. Introduction

Understanding the chemistry of interstellar molecules which could act as precursors or building blocks of biologically-important molecules is an active area of research in astrobiology (see reviews by Ehrenfreund & Charnley 2000 and by Ehrenfreund et al. 2002). Many interstellar organic molecules have also been detected in the comae of comets (Bockelée-Morvan et al. 2004) and so, as cometary and asteroidal impacts may have been important for delivering organics to the early Earth (Chyba & Hand 2005), there is a putative connection between interstellar molecules, cometary volatiles and prebiotic chemistry (Ehrenfreund et al. 2004).

It is likely that the first living cellular structures resulting from prebiotic chemistry on the early Earth used the ingredients available and did not resemble those known today (e.g. Ehrenfreund et al. 2006). In the primitive minimal cell (e.g. Ganti 1997), three basic autocatalytic processes occur that determine: 1) metabolism, 2) information transfer between generations and 3) the development and retention of a membrane-enclosed vesicle that contains these biochemicals and mediates the interaction with the environment, taking in nutrients and expelling waste. Most discussion has centred on extraterrestrial chemicals that could contribute to 1 and 2. However, it has been argued that the formation of a container structure should be the crucial first step to forming primitive cells (Walde 2006). The formation of membranes in an aqueous environment requires the presence of amphiphilic molecules, i.e. molecules that possess both hydrophilic and hydrophobic functional groups (e.g. long-chain organic acids). Intriguingly, it has been demonstrated experimentally that organic material from the Murchison carbonaceous meteorite can spontaneously

glyoxal, glycolaldehyde, methyl formate, glyoxylic acid, oxalic acid

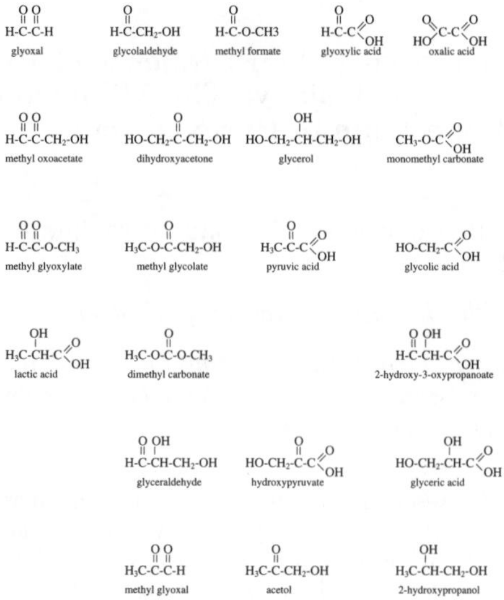

methyl oxoacetate, dihydroxyacetone, glycerol, monomethyl carbonate

methyl glyoxylate, methyl glycolate, pyruvic acid, glycolic acid

lactic acid, dimethyl carbonate, 2-hydroxy-3-oxypropanoate

glyceraldehyde, hydroxypyruvate, glyceric acid

methyl glyoxal, acetol, 2-hydroxypropanol

Figure 1. Possible reaction products from reactions between mobile organic radicals on dust grains.

self-assemble into cell-sized membraneous vesicles in aqueous solution (Deamer & Pashley 1989). Furthermore, the organic residue remaining after ultraviolet radiation processing of interstellar ice analogs can also form vesicles (Dworkin et al. 2001). Although the precise chemical composition and structure of the membranes formed have not yet been determined, it seems certain that the original meteoritic and interstellar analog materials contained amphiphilic molecules or their precursors.

Very recently, Hazen & Deamer (2007) have shown that, at the high temperatures and pressures of hydrothermal systems, pyruvic acid ($CH_3COCOOH$) can undergo a series of reactions including polymerization and cycloaddition. The remarkable result from these experiments is that a water-soluble organic residue is created which also forms cell-sized vesicles which bear a striking similarity to those produced from the Murchison extract. The volatile fraction of carbonaceous meteorites contain organic hydroxy and carboxylic acids (Botta & Bada 2002). Several keto acids have been extracted from the Murchison meteorite and pyruvic acid is expected to also be present (Cooper et al. 2005). In interstellar grain mantles, formyl (HCO), methoxy (CH_3O), hydroxymethyl (CH_2OH) and COOH radicals are all expected to be present, resulting from the hydrogenation and oxidation of CO ice (e.g. Charnley & Rodgers 2006, 2008). If the ices are warmed then radical migration and reaction can produce many large organic molecules (Fig. 1), including lactic acid, glycolic acid and pyruvic acid. To facilitate a radioastronomical search for these compounds we have undertaken a program to measure their rotational spectrum in the laboratory. In this contribution we summarise our results to date.

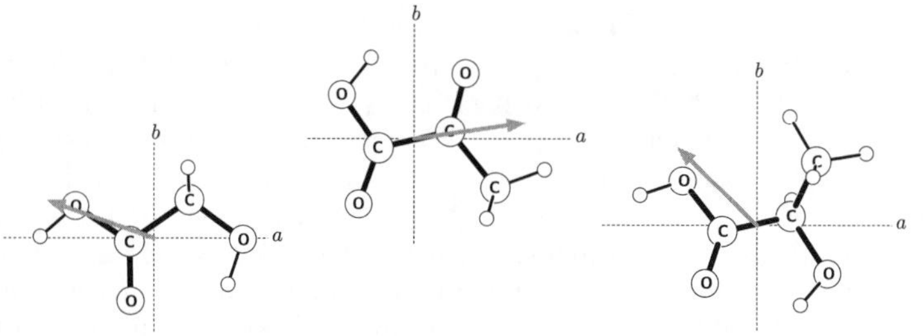

Figure 2. Molecular geometries, inertial axes and dipole moment vectors for the 9-atom glycolic acid (left), 10-atom pyruvic acid (center), and 12-atom lactic acid (right).

2. Rotational Spectroscopy

Rotational spectra were recorded in the mm-wave region with the broadband scanning spectrometer in Warsaw (Kisiel et al. 2003) operating at 120-350 GHz. Additional cm-wave measurements were carried out at conditions of supersonic expansion with the Fourier transform microwave spectrometer, also in Warsaw (Kisiel et al., 1997). Investigation of the pyruvic acid spectrum is part of a systematic series of studies of molecules of biological relevance performed with the aim of understanding their rotational spectrum at a level of detail sufficient for astrophysical searches. Other molecules recently studied with the same techniques have been the remaining two acids in Fig. 2, glycolic acid (Kisiel et al. 2007) and lactic acid (Pszczółkowski et al. 2005), and the important small heterocycles: pyrimidine (Kisiel et al. 1999), and quinoline and isoquinoline (Kisiel et al. 2003).

The rotational spectrum of pyruvic acid, as well as those of the other two acids in Fig. 2, reflect the presence of multiple low lying vibrational states associated with low frequency torsional motions. The main skeletal torsional mode in these molecules is characterised by a wavenumber of around 100 cm^{-1}, and in pyruvic acid there is an added complication from the low-frequency methyl torsion (118 cm^{-1}), which induces an appreciable splitting between transitions in A and E internal rotation substates. Interestingly, in lactic acid internal rotation splitting could not be observed. Finally all molecules have nonzero μ_a and μ_b dipole moment components, which makes the spectra even more complex. There is, however, a bonus in that quadruple degeneracies between the most intense rotational transitions are possible leading to useful candidate lines for detection.

As a result of this work the data set for the ground state of pyruvic acid has been extended to over 1500 lines in A and E substates, transitions have been measured up to J=70 and K_a=43, and measured lines span the complete range of values of K_a at the mid values of J. Measured transition frequencies have been fitted to within the experimental accuracy by means of three different codes for dealing with internal rotation, and the A-E energy difference was determined to be 0.72355(43) GHz. Extensive results were also obtained for three vibrationally excited states, each consisting of A and E internal rotation substates. Such a detailed analysis was made possible by the use of a newly developed AABS package for *Assignment and Analysis of Broadband Spectra* (Kisiel et al. 2005a, 2007a). The package allows work on multi-million point spectra and can keep track of assignments made in many different states and/or isotopic species, associated with hundreds of thousands of predicted lines. A key feature of the package is the possibility for graphical assignment of spectra by means of Loomis-Wood type diagrams. Very efficient mechanisms are provided for measuring lines in the spectrum and for setting up data sets for several different fitting programs. The package can be used for any type of broadband, highly resolved spectra and is freely available at the PROSPE website (http://info.ifpan.edu.pl/~kisiel/prospe.htm).

3. Conclusions

We have measured the mm-wave rotational spectra for several large organic molecules believed to be present in the interstellar medium and which may be important for prebiotic chemistry. The extensive results for pyruvic acid have now been published (Kisiel et al. 2007b). A discussion on how to read and compare the derived spectroscopic constants for the case of strong torsional coupling was also made. Of the studied molecules, pyruvic acid may be the most astrobiologically important since, under hydrothermal vent conditions, pyruvic acid can initiate the production of amphiphilic molecules and the formation of cell-sized, membrane-bounded, structures. Radio astronomical observations of emission from rotational transitions of pyruvic acid molecules, evaporated from ice mantles in *hot molecular cores* (e.g. Charnley et al. 2001), could allow the potential for primitive vesicle formation throughout the Galaxy to be evaluated.

Acknowledgments. command. This work was supported by NASA's Exobiology Program and by the NASA Goddard Center for Astrobiology. Support from the Polish Ministry of Science and Higher Education grant no. N202-0541-33 is also acknowledged.

References

Bockelée-Morvan, D., Crovisier, J., Mumma, M. J., & Weaver, H. A. 2004, in Comets II, eds. M. Festou, H.U. Keller & H.A. Weaver, Univ. Arizona Press, p. 391

Botta, O., Bada, J.L. 2002, Surv. Geophys. 23, 411-467.

Charnley, S.B., Ehrenfreund, P. and Kuan Y- J. 2001. Spectrochimica Acta, Part A, 57, 685

Charnley, S.B., Rodgers, S.D. 2006, in: D.C. Lis, G.A. Blake, E. Herbst (Eds.) Astrochemistry: Recent Successes and Current Challenges, CUP, p. 237

Charnley, S.B., Rodgers, S.D. 2008, these proceedings,

Chyba, C.F., Hand, K.P. 2005, Annu. Rev. A & A, 43, 31

Cooper, G. et al. 2005, in: XXXVI LPSC, 2005, abstract no. 2381

Deamer, D.W., Pashley, R.M. 1989, OLEB, 19, 21

Deamer, D.W. et al. 2002, Astrobiology, 2, 371

Dworkin, J.P. et al. 2001, PNAS, 98, 815

Ehrenfreund, P., Charnley S.B., 2000, Annu. Rev. A & A, 38, 427

Ehrenfreund, P. et al. 2002, Rep. Prog. Phys., 65, 1427

Ehrenfreund, P., Charnley, S. B., & Wooden, D. 2004, in Comets II, eds. M. C. Festou, H. U. Keller, & H. A. Weaver, University of Arizona Press, Tucson, p. 115

Ehrenfreund, P. et al. 2006, Astrobiology, 6, 490

Ganti, T. 1997, J. theor. Biol., 187, 583

Hazen, R.M., Deamer, D.W. 2007, OLEB, 37, 143

Kisiel, Z. Kosarzewski, J., Pszczółkowski, L., 1997, Acta Phys. Pol. A, 92, 507

Kisiel, Z., Pszczółkowski, L., , Lopez J.C., Alopnso J.L., Maris A., Caminati W., 1999, J. Mol. Spectrosc. 195, 332

Kisiel, Z. Desyatnyk, O., Pszczółkowski, L., Charnley, S.B., Ehrenfreund P.2003, J. Mol. Spectrosc. , 217, 115

Kisiel, Pszczółkowski, L., Medvedev, I.R., Winnewisser, M., De Lucia, F.C., Herbst, E., 2005a, J. Mol. Spectrosc., 233, 231

Kisiel, Z., Pszczółkowski, L., Białkowska-Jaworska, E, Charnley, S.B. 2007, 62nd OSU Symposium on Molecular Spectroscopy, communication WG10.

Kisiel, Z., Dorosh O., Winnewisser, M., Behnke, M., Medvedev I.R., De Lucia, F.C., 2007a, J. Mol. Spectrosc., 246, 39

Kisiel, Z. Pszczółkowski, L., Białkowska-Jaworska, E, Charnley, S.B. 2007b, J. Mol. Spectrosc. , 241, 220

Pszczółkowski, L., Białkowska-Jaworska, E, Kisiel, Z. 2005, J. Mol. Spectrosc. , 234, 106

Walde, P. 2006, OLEB 36, 109

Yi-Jehng Kuan

Session IV

Organic Material in Comets, Meteorites, and IDPs

Br. Guy Consolmagno

Vera Kolb

Bioastronomy 2007: Molecules, Microbes, and Extraterrestrial Life
ASP Conference Series, Vol. 420, 2009
K. J. Meech, J. V. Keane, M. J. Mumma, J. L. Siefert, and D. J. Werthimer, eds.

Application of the Organic Synthetic Designs to Astrobiology

V. M. Kolb

Department of Chemistry, University of Wisconsin-Parkside, Kenosha, WI 53141

Abstract. In this paper we propose a synthesis of the heterocyclic compounds and the insoluble materials on the meteorites. Our synthetic scheme involves the reaction of sugars and amino acids, the so-called Maillard reaction. We have developed this scheme based on the combined analysis of the regular and retrosynthetic organic synthetic principles. The merits of these synthetic methods for the prebiotic design are addressed.

1. Introduction

The prebiotic syntheses of the molecules that are now part of the chemistry of life, such the nucleic acids, probably required multiple steps. Laboratory simulation of these steps typically involves synthetic design in the direction from the prebiotically feasible starting materials towards the desired target molecules. The examples of the prebiotically feasible starting materials are the interstellar molecules, the compounds that have been found on meteorites, comets and interstellar dust particles, and the compounds that can be synthesized easily in the experiments which utilize prebiotic experimental conditions. Good examples of the latter compounds are the amino acids that are synthesized in the Miller's spark discharge experiments (Miller and Orgel, 1974). The examples of the desired target molecules are various bio-organic materials or their precursors, such as RNA (ribonucleic acid), ATP (adenosine tri-phosphate), lipids and others. The desired products may be more complicated and more fragile organic molecules than those that have been isolated from the meteorites.

2. Design Methodolgy and Discussion

The synthetic design which progresses from the starting materials to the target molecules is presented in the Fig 1.

This scheme works well for the organic synthesis in general, such as the industrial syntheses of desired molecules. Typically, chemists do not just mix the starting materials and reagents all at once, and let them react on their own. Instead, they add reagents in a particular sequence. They protect fragile functional groups, and de-protect them later. They isolate and purify the intermediate compounds, and remove the undesired products. These interventions by the chemists assure that the final desired product is obtained in a reasonable yield and a reasonably pure state. The situation differs in the simulated prebiotic syntheses. For the reaction conditions to be considered prebiotic, the experimentalist cannot do any interventions in the reaction, after the starting materials are mixed. As a consequence, the final desired product, when formed, is typically obtained in a poor yield and is mixed with many other compounds. Isolation of the desired product from the mixture may be a challenge. The only input we have in the synthetic design is the choice of the starting materials.

The second synthetic strategy is called the "retro-synthesis", (Warren, 1982). It requires scientists to think "backwards",, and to mentally break up the target molecule into the smaller sub-targets until the starting materials are reached. The actual synthesis would be the reversal of the process. The synthetic design involves the hypothetical bond breaking to various reasonable fragments, as dictated by the rules of the organic chemistry. These subtargets are fragmented further, until we arrive to the starting materials. This method produces dead ends when some subtargets cannot be fragmented further in a chemically meaningful way. Figure 2 summarizes this method.

Both synthetic methods suffer from the teleological approach, since we label various compounds as "desirable", "undesirable", or "dead-ends", and we focus only on the "desirable" compounds. However, the "undesirable" or "dead-end" compounds may be compounds of interest for understanding chemical evolution and chemistry of meteorites.

In this paper we focus on the prebiotic chemistry on asteroids and meteors, as reflected in the organic composition of the meteorites, which are our best prebiotic laboratory (Sephton, 2002). We propose a primordial synthesis of the insoluble organic material (IOM) on meteorites, based on the combined analysis of the regular and retrosynthetic principles. The ISO is what we would superficially consider an undesirable material. It is too complex, essentially unreactive, and it does not contain any biologically interesting materials. So, why should we be concerned with it? We shall show that this material, when looked at in a retrosynthetic way as "desired", can provide us with some important subtargets and meaningful "dead ends" and may shed light also on the chemical evolution on meteorites.

Our study was motivated by the desire to elucidate the following mysteries in the meteoritic chemistry (Pizzarello, 2004, Sephton, 2002):

1. Sugar compounds are found in the meteorites in very small amounts, and only as sugar alcohols or acids, although the formose reaction, by which they can be obtained, is prebiotically feasible.

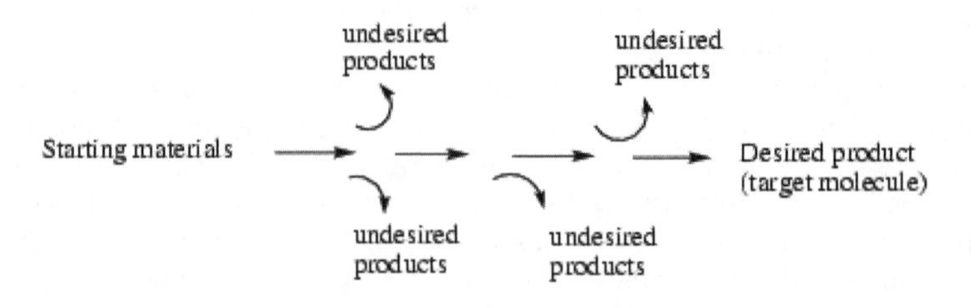

Figure 1. Scheme 1: Regular organic synthetic design.

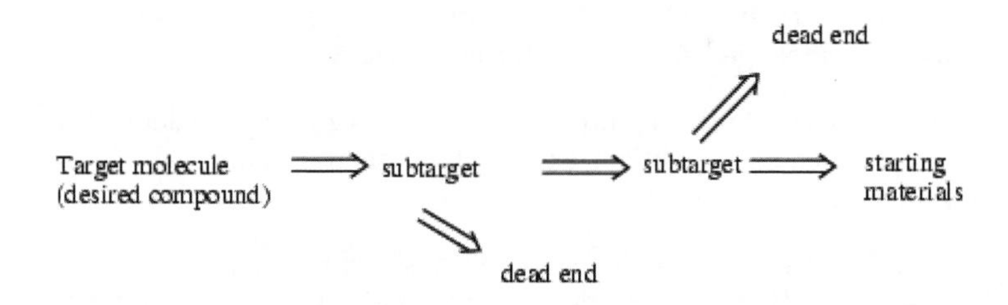

Figure 2. Scheme 2: Retrosynthetic design.

2. Various heterocyclic compounds have been found on the meteorites, but their origin has not been fully elucidated. There are no feasible synthetic pathways proposed for the formation of many of these compounds.

3. Most of the organic material on the meteorites is found in the form of insoluble organic material (IOM), which is characterized as a condensation polymer, having carbonyl, amino, carboxyl, aromatic, C=C and other functional groups. The origin of this material has not been fully elucidated.

We first hypothesized that if sugars can be easily formed but are missing (the point 1 above) they were probably consumed in a reaction with another material on the meteorites. The best candidates for the synthetic partners for the sugars, we reasoned, are the amino acids. These are found on all the meteorites that contain organic material (such as Murchison, Tagish Lake meteorite, and others). Moreover, the reaction of sugars and amino acids, so-called Maillard reaction, is well known and is described in the literature (Fayle and Gerrard, 2002). The reaction is robust, it produces various heterocyclic compounds (c.f. the point 2 above) and it also gives an insoluble organic material (c.f. the point 3 above). The Maillard reaction is extremely complex, but it was studied in great detail since it has important applications in the food industry. Many volatile by-products of the Maillard reaction are flavors and aromas associated with cooking or baking. We have shown that the Maillard reaction occurs also in the solid state, with both regular (Kolb et al., 2005) and meteoritic amino acids (Kolb et al., 2006). The retrosynthetic pathway from the insoluble Maillard product leads to the missing sugars. The synthetic pathways of the Maillard reaction are summarized in Scheme 3 below.

Scheme 3. Steps of the reaction between the sugar aldoses (ketoses react similarly) and the amino acids (the Maillard reaction) (Fayle and Gerrard, 2002).

1. The aldehyde group of a sugar and the amino group of an amino acid react together, to give first an addition intermediate. The latter loses water to give a Shiff base, which is an imine.

2. The Shiff base undergoes Amadori rearrangement, to give a keto-intermediate.

3. The Amadori intermediate undergoes an irreversible enolisation and elimination of the amino acid to give reductones, which are alpha-di-keto com-

pounds, and their enols, the dehydroreductones. These are also referred to as the rearranged Amadori intermediates.

4. The rearranged Amadori intermediates undergo a retro-aldolization reaction, in which they split off carbonyl fragments. Different alpha-di-keto products are also obtained.

5. The alpha-di-keto products react with the amino acids in the so-called Strecker oxidative degradation of amino acids. In this very important step of the Maillard reaction various aldehydes and alpha-amino ketones are obtained, which are involved in the formation of heterocyclic compounds.

6. In the late stages of the Maillard reaction various heterocyclic compounds and other reactive intermediates and products react among themselves in many different ways, but the condensation is an important step. Eventually, the dark brown of black polymeric material is obtained, which is not water soluble. This material is called a melanoidin. It is not well characterized, other than in terms of the main functional groups, such as carbonyl, carboxyl, amino, and C=C functions, among others.

Our hypothesis that the IOM on meteorites are related to the Maillard melanoidins has been supported experimentally. The analysis of the insoluble organic materials which we have obtained from the Maillard reaction utilizing both regular and meteoritic amino acids showed that this material is remarkably similar to the IOM from Murchison meteorite, by the solid state C-13 NMR (nuclear magnetic resonance) analysis (Kolb et al., 2005, 2006) and by the IR (infra-red) analysis (Kolb and Bajagic, 2006).

3. Conclusion

In conclusion, the retrosynthetic analysis, in combination with the regular synthetic analysis, can lead to the reasonable explanation for the low concentration of sugars on the meteorites, and for the origins of the heterocyclic compounds and the insoluble organic material on the meteorites.

Acknowledgments. Thanks are expressed to the Wisconsin Space Grant Consortium/NASA for the Higher Education Award, and to the University of Wisconsin System, for the Wisconsin Teaching Scholar appointment, which made this work possible.

References

Fayle, S. E. and Gerrard, J. A. 2002, The Maillard Reaction, Royal Society of Chemistry, Cambridge, England.

Kolb, V. M., Bajagic,M., Zhu, W., and Cody, G. D. 2005, Prebiotic Significance of the Maillard Reaction, in "Astrobiology and Planetary Missions", R. B. Hoover, G. V. Levin, A. Y. Rozanov, and G. R. Gladstone, Editors, Proc. of SPIE Vol. 5906, pp. 59060T (1-11).

Kolb, V. M., Bajagic, M., Liesch, P. J., Philip, A. and Cody, G. D. 2006, On the Maillard reaction of meteoritic amino acids, in "Instruments, Methods, and Missions for

Astrobiology IX" R. B. Hoover, G. Y. Levin, and A. Y. Rozanov, Eds., Proc. of SPIE, Vol. 6309, pp. 63090B (1-13).

Kolb, V. M. and Bajagic, M. 2006, Prebiotic Significance of the Maillard Reaction: An Infra-red Study of the Maillard Melanoidins, in "Continuing the Voyage of Discovery", R. A. Yingst, S. D. Brandt, J. Borg, S. Dutch, M. Gustafson, M. Rudd, and A. Roethel, Eds., Proceedings of the 15th Annual Wisconsin Space Conference, Wisconsin Space Grant Consortium, Green Bay, WI, Part Six, Chemistry.

Miller, S. and Orgel, L. 1974, "The Origins of Life on Earth", Prentice-Hall, Englewood Cliffs, New Jersey, pp. 83-95.

Pizzarello, S., 2004, Chemical evolution and meteorites: An update, Origins of Life and Evolution of the Biosphere, 34, 25-34.

Sephton, M. A., 2002, Organic compounds in carbonaceous meteorites, Natural Products Reports, 19, 292-311.

Warren, S., 1982, Organic Synthesis, The Disconnection Approach, Wiley, New York.

Dale Cruikshank

Bioastronomy 2007: Molecules, Microbes, and Extraterrestrial Life
ASP Conference Series, Vol. 420, 2009
K. J. Meech, J. V. Keane, M. J. Mumma, J. L. Siefert, and D. J. Werthimer, eds.

Imaging of Comet 21P/Giacobini-Zinner

J. Pittichová,[1,2] M. S. Kelley,[3] C. E. Woodward,[4] K. J. Meech[5]

[1]*Institute for Astronomy, 2680 Woodlawn Drive, Honolulu, HI 96822*

[2]*Astronomical Institute of SAV, Bratislava, Slovakia*

[3]*Department of Physics, University of Central Florida, Orlando, FL*

[4]*Department of Astronomy, University of Minnesota, Minneapolis MN*

[5]*NASA Astrobiology Institute, Institute for Astronomy, HI*

Abstract. We report on ground-based optical and space infrared observations of comet 21P/Giacobini-Zinner. Our data indicate that the comet is quite active with fairly uniform coma emission out to $r \sim 3.8$ AU, absent large numbers of jets or other coma features. Dust production measurements combined with no visibly detected extended coma emission suggest that comet activity turned off ~ 3.8 AU post-perihelion. The color of the coma is slightly redder than the Sun, and no variations in the light curve of 21P were evident. The 24 micron Spitzer image shows a strong dust coma and tail, but no evidence for a dust trail, even though this comet is the parent body of the Draconid meteor stream.

1. Introduction

The study of the physical properties of cometary nuclei and coma are both equally important to our understanding of the outer solar system environment during the era of icy planetesimal formation. A basic goal of modern astrophysics is to understand conditions extant in early protoplanetary disks during the epoch of planetesimal formation. In our own solar system, comets are frozen archives of this early epoch. We present here ten sets of optical and IR imaging observations of comet 21P/Giacobini-Zinner.

Comet 21P/Giacobini-Zinner was discovered by the French astronomer Michel Giacobini in 1900 and rediscovered two apparitions later, in 1913, by the German astronomer Ernst Zinner. 21P, the parent body of the Draconids (also know as the Giacobinids) meteor shower, is a Jupiter-family (i.e., short-period) comet with an orbital period of 6.61 years, and an aphelion distance just exterior to Jupiter's.

Visible imaging with the UH 2.2-m telescope conducted between June 2004 and March 2006 enabled us to obtain precise optical photometry and morphology of the comet's near-nucleus structures, including its jets and coma. Images have allowed us to map how the activity of comet is developing with time and to observe the morphological changes of surrounding dust. During the time of the observations, the comet passed through perihelion on July 2, 2005, and the heliocentric distances, r, were from 3.80 AU pre-perihelion (June 2004) to 1.78 AU (October 2005), and 2.91 AU (March 2006) post-perihelion. Comet activity rapidly changed from non-active to very strongly out-gassing over this time period.

IR imaging of comet 21P was performed with the Spitzer Space Telescope IRAC camera at 4.5 and 7.9 microns, and the MIPS 24 micron imager. The 4.5 micron image has a different morphology than the 7.9 micron image, likely due to CO_2 (4.3 micron) and CO (4.7 micron) gaseous emission from distributed sources in the coma contributing to the 4.5 micron image surface brightness distribution.

2. Observations

2.1. Ground-based and Spitzer Infrared Observations

During eight runs in June 2004, October and December 2005, and March 2006, we observed comet 21P/Giacobini-Zinner with the University of Hawaiʻi 2.2-m telescope on Mauna Kea, using the Tek2048 CCD camera with a $7.5' \times 7.5'$ FOV, read-noise 6.0 e^-, and gain 1.74 e^-/ADU. The specifics of the observations, the orbital geometry for the comet, and the photometric measurements as well as the computed values of a quantity known as Afρ are shown in Table 1. Depending on the comet's brightness during each run, several of the Kron-Cousins filters were used (V: λ_o = 5450 Å, $\Delta\lambda$ = 836 Å; R: λ_o = 6460 Å, $\Delta\lambda$ = 1245 Å; I: λ_o = 8260 Å, $\Delta\lambda$ = 1888 Å). We used non-sidereal guiding at the cometary motion rates for all of the images.

The Spitzer Space Telescope consists of a 0.85-m telescope and three cryogenically cooled science instruments (Werner et al. 2004). We used the IRAC camera to obtain images at 4.5 and 7.9 μm on December 31, 2005, and MIPS 24 (Multi-band Imaging Photometer) on December 2, 2005, for photometric information at 24 μm.

3. Data Reduction

The optical CCD frames were reduced with standard routines in IRAF (the Image Reduction and Analysis Facility). All the images were reduced using flat-field images taken during the evening and morning twilight sky and cleaned of bad pixels and cosmic rays. The frames were calibrated with the standard stars of Landolt (1992), which were observed on each photometric night. Observations of typically 20 standard stars were obtained over a range of air-masses, and with a wide dispersion of color to fit for both extinction and color terms. We measured the magnitudes of as many field stars as possible (usually 20–30) with equal or greater brightness to the comet on each frame in order to do relative photometry. After correcting the measured magnitudes for extinction, we used the deviations of the field star magnitudes in each frame from their nightly average values to correct for frame-to-frame extinction in the comet's measured signal.

The IRAC and MIPS observations were calibrated with *Spitzer* pipelines S13.2.0 and S13.0.1, respectively. The images were mosaicked in the rest frame of the comet with the MOPEX software (Makovoz & Khan, 2005) at the native IRAC and MIPS pixel scales ($1.22''$ pixel^{-1} for IRAC, $2.5''$ pixel^{-1} for MIPS 24). The MIPS background observation was mosaicked with the same size, scale,

Table 1. Observing Circumstances and Photometric Results

UT Date	Inst.	r [AU]	Δ [AU]	α [deg]	No.	Fil.	m(1,1,0) [mag]	Afρ [cm]
Jun/21/04[†]	Tek	3.80	3.21	13.61	2	R	17.19	4.80
Jun/22/04[†]	Tek	3.79	3.21	13.76	3	R	17.24	4.55
Oct/20/05[§]	Tek	1.76	1.74	33.02	11	R	12.82	130.66
Oct/22/05[§]	Tek	1.78	1.74	32.80	7	V,R,I	12.89	124.47
Dec/02/05[§]	MIPS	2.14	1.94	28.19	14	24 μm		
Dec/21/05[§]	Tek	2.32	1.72	22.56	15	R,I	13.74	83.00
Dec/22/05[§]	Tek	2.32	1.72	22.32	24	V,R,I	13.76	82.07
Dec/31/05[§]	IRAC	2.40	1.89	23.51	60	4.5, 7.9 μm		
Mar/05/06[§]	Tek	2.91	2.24	16.37	6	R,I	14.80	39.06
Mar/06/06[§]	Tek	2.92	2.25	16.45	6	V,R,I	14.61	46.50

Notes: All photometry results are for 3.0″ aperture radii. [†] denotes UT date pre-perihelion and [§] post-perihelion observations; r and Δ are the heliocentric and geocentric distances; α is the phase angle; No. is the number of images per night; Fil. is the filter; m(1,1,0) is the average reduced comet magnitude for the R filter, Afρ is the quantity of dust production.

and orientation as the primary observation, then subtracted from the primary. Before mosaicking, all images were masked to remove cosmic rays and bad pixels.

4. Analysis and Discussion

4.1. Photometry

The comet fluxes were extracted using the circular aperture photometry routine "PHOT". This routine automatically finds the centroid of the image within the user-specified photometry aperture, with the sky background determined in an annulus lying immediately outside the photometry aperture (for stellar images), or selected from an average of many sky positions outside the extent of any coma. For our optical data, the the optimum aperture radii was 3.0″. Figure 1 shows appearance of the comet for selected observing runs.

For the non-photometric data from the nights of June 21 and 22, 2004, and March 5, 2006, the comet fields were re-imaged on May 18, 2005 and June 3, 2006, and as many field stars as possible were measured on both the calibration images and the non-photometric images. Differential photometry was used to calculate the calibrated brightness of the comet. This technique works well for up to ~0.5 mag of extinction.

4.2. Heliocentric Light Curve

The R-band photometry from December 2005 run is shown in Figure 2a,b. Only these two nights have enough data points (15 and 24) to obtain the light curve for rotation. Figure 3a shows the reduced magnitude m(1,1,0) to unit heliocentric distance r, geocentric distance Δ, and zero phase α as a function of

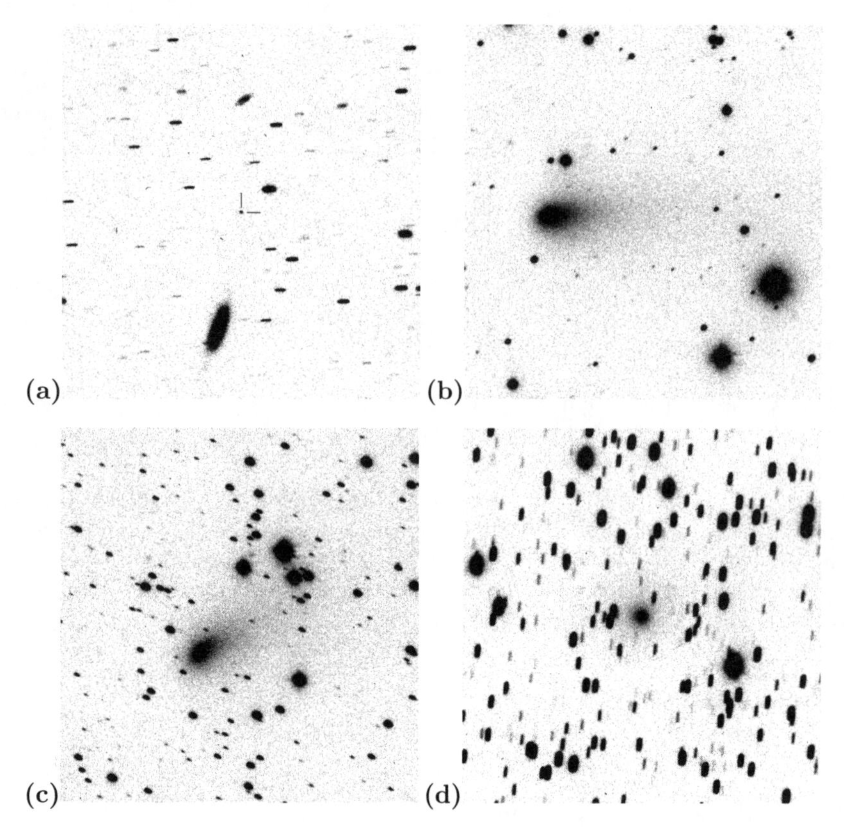

Figure 1. Selected images of 21P/Giacobini-Zinner. The fields of view are $180'' \times 180''$. North is up, and East is to the left in all images. **(a)** Jun 21, 2004; **(b)** October 22, 2005; **(c)** December 21, 2005; **(d)** March 5, 2006.

heliocentric distance r. The horizontal lines correspond to the supposed of the flux contribution from a bare nucleus. Based on our current observations, the reduced magnitude value at ~ 3.8 AU is consistent with that of a bare nucleus and little, if any, coma contribution. However, additional observations of 21P at $r > 4.0$ AU are necessary to confirm this assertion. From the Figure 3a,b, it is clear that activity probably ceased near heliocentric distance $r = 3.8$ AU, and all our data except June 2004 were taken during the active stage of the comet.

The reduced magnitude for an active comet is computed from the following equation:

$$m(1,1,0) = m_R - 5\ \log(r) - 2.5n\log(\Delta) - \alpha\phi. \qquad (1)$$

where n is constant, usually ~ 1.

4.3. Other Indicators of Coma and Activity Level

The first and the most direct indicator of nucleus activity is the presence of a coma around the comet nucleus. However, in individual images taken at a large

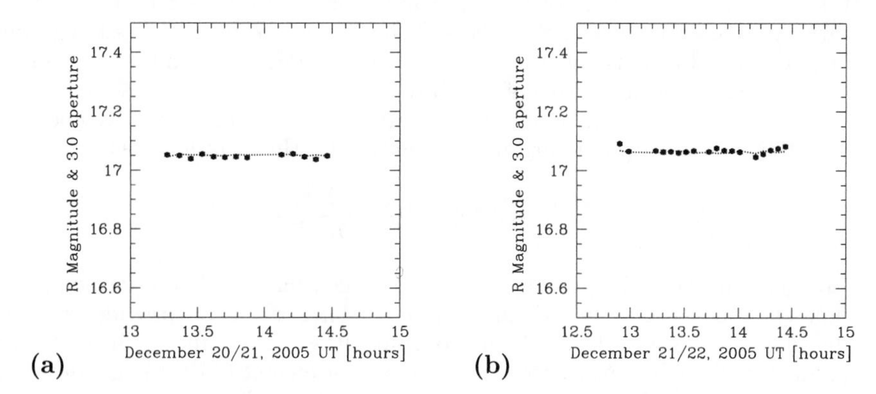

Figure 2. R-band photometry of comet 21P/Giacobini-Zinner through a 3.0″ aperture. The dashed line is the average of normalized field stars. **(a)** December 20/21, 2005, **(b)** December 21/22, 2005.

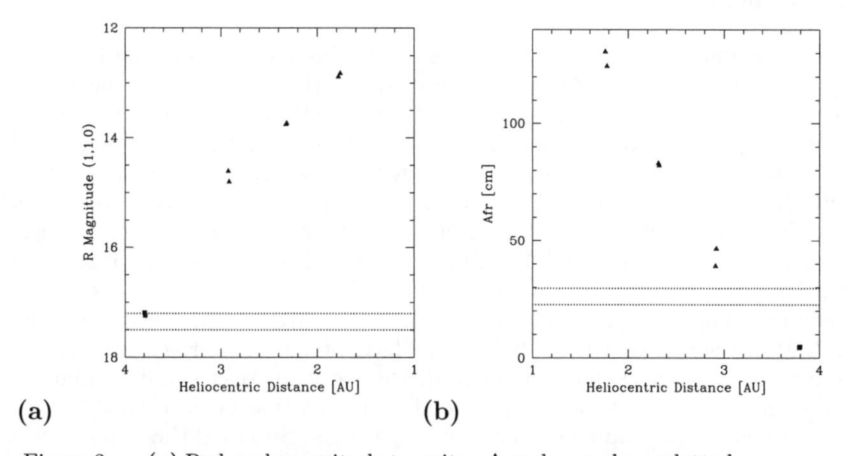

Figure 3. **(a)** Reduced magnitude to unit r, Δ and zero phase plotted versus r. The square symbols are for pre-perihelion data and the triangular symbols are for post-perihelion data. The dotted lines represent the likely brightness range for the bare nucleus. **(b)** Afρ values plotted against heliocentric distance for a period between June 2004 and March 2006. The square symbols are for pre-perihelion data and the triangular symbols are for post-perihelion data. The dotted lines is an estimate borderline value for an active comet (A'Hearn et al. 1984).

heliocentric distance, as is the case with our June 2004 image (Fig. 1a, taken at 3.80 AU), a coma was not visible. Unfortunately we have only two to three images from each night, which is not sufficient to create a composite image with adequate signal-to-noise off the nucleus pixels to detect extended, low surface brightness emission indicative of a diffuse coma.

The second indicator of nucleus activity is the value of a quantity known as Afρ, often used as a proxy for dust production (A'Hearn et al. 1984), given by

$$Af\rho = 2.467 \times 10^{19} \frac{r^2\Delta}{\theta} \frac{F_{obs}}{F_{\odot}}. \quad (2)$$

The $Af\rho$ parameter is typically expressed in centimeters and is formulated to be independent of photometric aperture size. This parameter provides a metric quantifying a comet's level of dust output (large Afρ values indicate higher activity), and enables us to ascertain the level or change of dust contribution in a set of photometric measurements made of a cometary body over a given period of time. Figure 3b shows R-band Afρ values depend on a heliocentric distance. As can be seen from the figure, the values for 3.80 AU are 4.8 cm, which is much smaller then 25 cm, which is the borderline value for an active comet (A'Hearn et al. 1984). This low $Af\rho$ value is consistent with no visible activity (i.e., no extended coma) in the June 2004 images.

5. Conclusions

21P/Giacobini-Zinner CCD data obtained during eight visible and two infrared nights reveals that the comet is very dusty, with activity out to near $r = 3.8$ AU. However, the emission is fairly uniform, without the presence of a large number of jets or other coma features. When the nucleus is surrounded by a bright coma, nucleus observations can be difficult. A previous unfortunate interpretation (Delsemme 1982), based on experience with the development of activity in predominantly water-ice nuclei in other short-period comets, was that the coma contamination will be minimal for $r > 3$ AU for comets inbound to the Sun, but can present a considerable problem for outbound comets at comparable distances. For comet 21P/Giacobini-Zinner, the latter argument is not germaine as 21P was active past $r = 3.0$ AU. The level of activity derived from our study is consistent with that expected from the work of Meech and Svoreň (2004) who derive new gas production rates for sublimation of pure volatiles. Their work suggests, depending on the dust-to-gas mass ratio and the grain sizes, that significant observable dust comae at larger heliocentric distances are possible, consistent with that observed this last apparition of 21P/Giacobini-Zinner.

- The comet was observed at 4 heliocentric distances over a period from June 2004 through March 2006. The brightening of the comet as a function of heliocentric distance r suggests that the activity began near $r = 3.3$ AU and ceased at a somewhat greater distance.

- The color of the coma is slightly redder than the Sun, $V-R = 0.524 \pm 0.003$, $R-I = 0.487 \pm 0.004$.

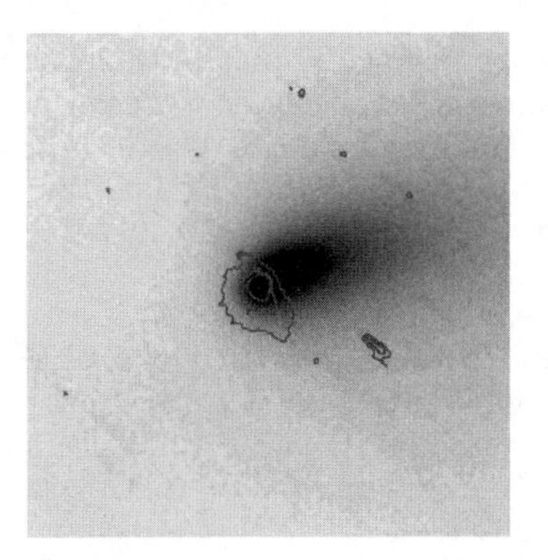

(a)

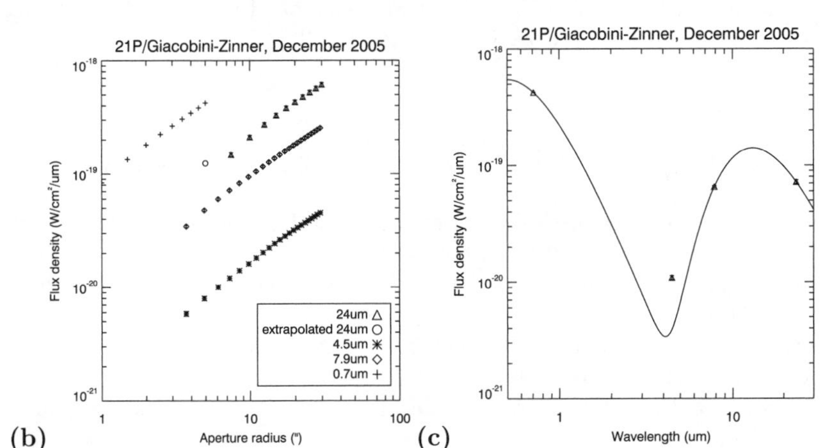

(b) (c)

Figure 4. **(a)** Grey scale image of 21P/Giacobini-Zinner in the IRAC 7.9 μm and gas emission in IRAC 4.5 μm images (CO + CO_2) - black contours. **(b)** Aperture profiles for the optical 0.7 μm, IRAC 4.5 and 7.9 μm and MIPS 24.0 μm data. **(c)** Spectral Energy Distribution (SED) for 5.0″ aperture. Two thermal black body model fitted for the R-band point and to fit the 8 and 24 μm data.

- No variations in the light curve of comet 21P/Giacobini-Zinner (Fig. 2a,b) were observed; however, our data-set did not span a sufficient length of time to determine a possible (long duration) rotation period.

- The slope of the channel 2 (ch2) 4.5μm images - gas and channel 4 (ch4) 7.9 μm images - dust at 4 –16 pixels (6.6$\times$ 10^3 to 2.6$\times$ 10^4 km) from the nucleus was determined by subtracting a scaled version of the ch4 image (dust) from the ch2 image (dust + gas). The gaseous emission is a long-lived species at 2.4 AU. The lifetimes of CO and CO_2 at 2.4 AU are τ_{CO}=89 and τ_{CO_2} = 34 days (Huebner et al. 1994). For an outflow velocity of 1.0 km/s, the characteristic length scales are 7.7$\times$ 10^6 km (CO) and 2.9$\times$ 10^6 km (CO_2). There appears to be emission out to 60 pixels from the nucleus (1.0$\times$ 10^5 km), but background stars make it difficult to measure the logarithmic slope beyond 20 pix. Both species are long lived at 4 –16 pixels., even if the outflow velocity is ~ 0.1 km/s. From our analysis, both CO and CO_2 are equally viable candidates for the ch2 excess (Figure 3). For comparison, in comet C/1995 O1 (Hale-Bopp; Crovisier et al. 1999), the CO_2 was the dominant species at 2.9 AU (21P was observed at 2.4 AU).

Acknowledgments. This work is based on observations made with the 2.2-m telescope of University of Hawaii and with the Spitzer Space Telescope, which is operated by the Jet Propulsion Laboratory, California Institute of Technology under a contract with NASA. Support for this work was provided by NASA through contracts 1278383 issued by JPL/Caltech to the University of Hawaii and 125606, 123741, 1275835 to the University of Minnesota. Partial support was also provided by the SAA Grant Vega 2/7040/27.

References

A'Hearn, M.F., Schleicher, D.G., Feldman, P.D., Millis, R.L., & Thompson, D.T. 1984, AJ, 89, 579

Allen, C.W. 1973, Astrophysical Quantities. (3rd ed; London: Athlone Press), 162

Crovisier, J., et al. 1999, LPI Contribution 969, 7

Delsemme, A. 1982, in Comets, ed. L.L. Wilkenning (Tucson: Univ. Arizona Press), 85

Fernie, J.D. 1983, PASP, 95, 782

Huebner, W.F. ,& Boice, D.C. 1994, Origins Life Evol. Biosphere, 24, 2 - 4, 91

Landolt, A.U. 1992, AJ, 104, 340

Makovoz, D., & Khan, I. 2005, Astronomical Data Analysis Software and Systems XIV, 347, 81

Meech, K.J., & Svoreň, J. 2004, in Comets II, ed. M. Festou et al. (Tucson: Univ. Arizona Press), 317

Werner, M. W., et al. 2004, ApJS, 154, 1

Cherrilyn Morrow

Scott Sandford

Bioastronomy 2007: Molecules, Microbes, and Extraterrestrial Life
ASP Conference Series, Vol. 420, 2009
K. J. Meech, J. V. Keane, M. J. Mumma, J. L. Siefert, and D. J. Werthimer, eds.

Organics in the Samples Returned by the Stardust Spacecraft from Comet 81P/Wild 2

S. A. Sandford

NASA-Ames Research Center, Astrophysics Branch, Moffett Field, CA 94035

Abstract. Cometary organics are of great interest because these materials represent a reservoir of the original carbon-containing materials from which everything else in our Solar System was made and that may have played key roles in the origin of life on Earth. These organics are products of a series of universal chemical processes expected to operate in all galaxies, so they also provide insights into the abundance of Life elsewhere in the universe. Our understanding of cometary organics has made a quantum leap forward due to the recent availability of samples collected from Comet P81/Wild 2 and returned to the Earth by the Stardust mission.

1. Introduction

Most of the material in the protosolar disk from which our Solar System was made was incorporated into the Sun, was ejected from the system, or ended up in planets. However, minor amounts of material were incorporated into planetesimals that survived as asteroids and comets. The material in asteroids and comets have undergone considerably less parent body processing than planetary materials and therefore likely contain more pristine examples of the raw starting materials of the Solar System. However, the study of meteorites has demonstrated that they are not completely pristine samples, but show evidence of varying degrees of thermal processing and aqueous alteration (Kerridge & Matthews 1988; Lauretta & McSween 2006). Also, most meteorites are likely from asteroids that formed in the inner Solar System and did not incorporate a full share of the more volatile components in the original protosolar disk. In contrast, comets are thought to have formed in the outer Solar System, probably contain a larger portion of the volatile components of the original disk, and have likely undergone less parent body processing since formation. Thus, cometary materials may represent the best samples of pristine early Solar System materials available for study and they may provide powerful insights into the formation of the entire Solar System, not just comets. The nature of cometary volatiles and organics are also of great astrobiological importance since comets may have delivered key volatiles and organics to the early Earth (Oró et al. 1980; Thomas et al. 1996). Finally, since comets are thought to be the end result of a series of universal processes involving stellar, interstellar, and star-forming environments, their compositions should be generally representative of the compositions of these bodies in other planetary systems. Thus, insofar as comets have played a role in the formation of life on Earth, they would be expected to be available to play a similar role in other planetary systems containing appropriate conditions for the origin of life (whatever those are!). Understanding comets in this context may therefore provide insights into the frequency of life elsewhere in the universe.

Our current understanding of comets has made a significant leap forward due to the study of cometary samples recently returned to the Earth from Comet 81P/Wild 2 by the Stardust spacecraft (Brownlee et al. 2006). One of the central scientific goals of the Stardust Discovery Mission was to establish whether comets contained complex organic materials and to establish the abundance, chemical, and isotopic nature of any organics present (Brownlee et al. 2003). The study of samples from Wild 2 represents a significant analytical challenge. The Wild 2 cometary particles that hit the Stardust collector were typically smaller than 25 μm in diameter and contained less than a nanogram of material (Brownlee et al. 2006). To further complicate matters, most of these particles consisted of heterogeneous aggregates of subgrains in the μm and sub-μm size range that broke upon impact (Brownlee et al. 2006; Hőrz et al. 2006; Matrajt et al. 2008). Finally, only a portion of the material in the samples is organic. Despite these challenges, a surprising amount has been learned from these samples and more will be learned in the future.

2. Organics from Comet 81P/Wild 2: The Stardust Comet Sample Return Mission

2.1. The Collection of Material from Comet 81P/Wild

Stardust was the first mission in history to return solid samples from beyond the Earth-Moon system. The mission retrieved samples from Comet 81P/Wild 2, an ~4.5 km diameter body currently in an orbit between Jupiter and Mars. Wild 2 entered this orbit on 10 September 1974 after a close encounter with Jupiter. During Stardust's flyby of Wild 2 on 2 January 2004, the comet showed evidence of at least 20 dust jets coming from the nucleus (Brownlee et al. 2004; Sekanina et al. 2004). Stardust approached to within 234 km of the comet's surface and cometary particles were collected as they impacted at 6.12 km/sec into silica aerogel. Particles ejected from the comet were exposed to space for only a few hours before collection, but solar heating probably vaporized most ices during transit from Wild 2 to the Stardust spacecraft. The aerogel collector medium consists of a porous glass having bulk densities that varied from < 0.01 g/cm^3 at the impact surface to 0.05 g/cm^3 at 3 cm depth. Stardust aerogel tiles collected over a thousand 5–300 μm (and many more smaller) comet particles. Onboard impact sensors indicate that the collected particles were largely associated with two specific dust jets (Tuzzolino et al. 2004).

Particle impacts into aerogel produced tracks whose shapes depended on the nature of the impacting particle. Non-fragmenting particles produced carrot-shaped tracks, but many tracks show bulbous upper regions and sometimes multiple 'roots'. These tracks were produced by weakly bound aggregate particles that broke apart upon impact with the aerogel (Brownlee et al. 2006; Hőrz et al. 2006). In addition to aerogel, ~15% of the Stardust collector surface was the aluminum frame and aluminum foils used to hold the aerogel. Impacts on the frame produced bowl-shaped craters lined with melted, and in some cases unmelted, projectile residues (Hőrz et al. 2006).

2.2. Measurement of Wild 2 Organics in the Returned Stardust Samples

Approximately 200 investigators around the world participated in the preliminary examination (PE) of the samples returned by Stardust and their findings appeared in a special issue of Science (Brownlee et al. 2006; Flynn et al. 2006; Hőrz et al. 2006; Keller et al. 2006; McKeegan et al. 2006; Sandford et al. 2006; Zolensky et al. 2006). The analysis of the samples was challenging because the samples were complex, nanogram-sized aggregates that broke up into smaller particles and were distributed along the entire lengths of aerogel tracks. For organics analyses, addition attention had to be paid to the possibility of contaminants associated with the collector, spacecraft flight, and return and recovery of the sample return capsule (SRC). Fortunately, contaminants are found to be of low enough abundance or were sufficiently well characterized that they can usually be distinguished from the cometary organics (Sandford et al. 2006). Most problematic was the aerogel collector medium itself, which contained up to a few wt% C. This C is largely in the form of Si–CH_3 groups easily distinguishable from the cometary organics described below. There is evidence that at least some organic compounds were generated or altered by impact heating of the aerogel itself (Sandford et al. 2006; Sandford & Brownlee 2007; Spencer & Zare 2007). Despite these difficulties, a great deal has been learned about cometary organics and the origin of the Solar System from these samples.

Analytical techniques used during the preliminary examination (PE) of organics in the Wild 2 samples include two-step laser desorption laser ionization mass spectrometry (L^2MS), Liquid Chromatography with UV Fluorescence Detection and Time of Flight Mass Spectrometry (LC-FD/TOF-MS), Scanning Transmission X-ray Microscopy (STXM), X-ray Absorption Near Edge Spectroscopy (XANES), infrared and Raman spectroscopy, Ion Chromatography with conductivity detection (IC), Secondary Ion Mass Spectrometry (SIMS), and Time-of-Flight Secondary Ion Mass Spectrometry (TOF-SIMS). A brief summary of the combined findings of these analyses is provided below. More complete discussions of the Organics PE results can be found in Sandford et al. (2006), its associated Supporting Online Material, and in an upcoming special issue of Meteoritics and Planetary Science.

Multiple experimental techniques demonstrate that the samples contain polycyclic aromatic hydrocarbons (PAHs). L^2MS mass spectra obtained from individual particles and along impact tracks show PAHs and their alkylated derivatives. Two distinct types of PAH distributions are seen (Sandford et al. 2006; Clemett et al. 2009; Spencer et al. 2009). In some cases, PAH populations dominated by benzene and naphthalene (1–2 ring PAHs), including alkylation out to several –CH_3 additions, are observed in the absence of larger PAHs along track surfaces. Many of these lower mass PAHs may originate from impact processing of C original to the aerogel (Sandford & Brownlee 2007; Spencer & Zare 2007). The second population of PAHs shows complex distributions that resemble those seen in some meteorites and IDPs. The more complex comet mass spectra also include additional mass peaks not observed in meteorite mass spectra, but seen in some IDP spectra, that suggest the presence O- and N-substituted aromatic species having hetero-functionality external to the aromatic structure.

The presence of PAHs is confirmed by TOF-SIMS analyses of terminal particles extracted from aerogel tracks and residues found in a crater in aluminum foil. PAH abundances decrease sharply with increasing number of C atoms (Sandford et al. 2006; Stephan et al. 2008a,b). Raman spectra also confirm the presence of aromatic materials (Sandford et al. 2006; Rotundi et al. 2008). All the Raman spectra are dominated by broad bands centered at $\sim$1360 Δcm^{-1} and $\sim$1580 Δcm^{-1} that are characteristic of graphitic sp^2-bonded C in condensed rings. The profiles and positions of these bands indicate that at least some organics were captured with relatively little alteration. Many Raman spectra of Wild 2 particles also contain very high backgrounds that suggest they may be rich in heteroatoms such as N. In a few cases, aromatic materials were also identified by detection of an aromatic CH stretching mode band using infrared spectroscopy (Sandford et al. 2006; Rotundi et al. 2008; Bajt et al. 2009).

Individual particles extracted from Stardust aerogel were also microtomed into slices thin enough to be amenable to STXM and C,N,O-XANES analyses. C-XANES spectra of many of these thin sections confirmed the presence of aromatics. Full C,N,O-XANES spectra show variable abundances of aromatic, keto/aldehydic, and carboxyl moieties, as well as amides and nitriles (Sandford et al. 2006; Cody et al. 2007). While XANES data confirm aromatics are present, they also show the particles contain abundant non-aromatic C. XANES analyses also show that Wild 2 organics are considerably richer in O and N relative to both meteoritic organic matter and the average composition of Comet Halley particles measured by Giotto, but have O/C and N/C ratios similar to the average of IDPs (Sandford et al. 2006; Cody et al. 2007). Both the O and N exist in a wide variety of bonding states; particles rich in N exhibit abundant amide C in their XANES spectra.

To test whether Stardust may have returned a 'diffuse' sample of molecules that struck the aerogel directly from the cometary coma or that diffused away from grains after impact, a search was made for excess primary amines in flight aerogel using LC-FD/TOF-MS. Methylamine (MA), ethylamine (EA), and glycine were detected above background and control sample levels, suggesting they may have a cometary origin (Sandford et al. 2006; Glavin et al. 2008). MA and EA concentrations were similar in aerogels near and not near impact tracks, suggesting that these amines, if cometary, originate from sub-μm particles or gas that directly impacted the collector. No MA, EA, or glycine was detected in non acid-hydrolyzed aerogel extracts, suggesting that they are present in an acid soluble bound form, rather than as a free primary amine.

IR spectra taken from tracks and individual extracted particles show both aromatic and non-aromatic chemical functional groups are present (Keller et al. 2006; Sandford et al. 2006; Rotundi et al. 2008; Bajt et al. 2009). Infrared spectra of particles and tracks often contain absorption features at 3322 cm^{-1} (–OH), 3065 cm^{-1} (aromatic CH), 2968 cm^{-1} ($-CH_3$), 2923 cm^{-1} ($-CH_2-$), 2855 cm^{-1} ($-CH_3$ and $-CH_2-$), and 1706 cm^{-1} (C=O). One particle also showed a weak 2232 cm^{-1} band consistent with $-C\equiv N$ stretching vibrations. Combined, the IR data indicate the presence of aromatic, aliphatic, carboxylic, and N-containing functional groups consistent with the results of other analytical techniques. The observed $-CH_2-/-CH_3$ band depth ratios in the samples is typically $\sim$2.5, a value similar to that seen from anhydrous IDPs, but considerably larger than seen in carbonaceous chondrites ($\sim$1.1) and the diffuse ISM (1.1–1.25). This indicates that the aliphatics in Wild-2 samples are longer or less branched than those in meteorites and the diffuse ISM. The ratio of aromatic to aliphatic C-H is quite variable in the IR spectra, consistent with the variations seen in the XANES data. IR spectral maps of entire impact tracks and their surrounding aerogel show that the organics that produce the –OH, $-CH_3$, $-CH_2-$, and C=O IR absorption bands sometimes extend well beyond the visible track edge (Sandford et al. 2006). This implies that some particles contained organics that volatilized and diffused into the surrounding aerogel during impact. Since similar length tracks are also seen in the same aerogels that show no excess IR-detectable organics, this material is unlikely to be due to impact-altered aerogel carbon.

SIMS ion imaging elemental maps of Wild 2 particle sections show that N and S are associated with organic molecules and that the N is very heterogeneously distributed with N/C ratios ranging from 0.005 to almost 1 (Sandford et al. 2006; Matrajt et al. 2008). Some particles exhibit the entire range of values, while others fall more uniformly at the high N/C end of the range. Sulfur is often associated with C and N. SIMS H/D isotopic measurements of Wild 2 particles show D enrichments up to about three times the terrestrial value within about half of the particles (McKeegan et al. 2006; Matrajt et al. 2008). The D enrichments are heterogeneously distributed and associated with C, indicating the carrier is probably organic. Isotopic anomalies were also observed in N (McKeegan et al. 2006). As with D, the anomalies are often heterogeneously distributed within the particles. These D and ^{15}N enrichments are likely due to materials with an interstellar/protostellar chemical heritage (Messenger 2000; Sandford et al. 2001; Aléon & Robert 2004) and provide clear evidence of a cometary origin for the organics. A number of interstellar chemical processes and environments can fractionate H and N isotopes. These processes are expected to leave characteristic isotopic 'fingerprints' in their products (Sandford et al. 2001) that can provide important clues about the carriers' natal environments. One would expect many organic molecules having interstellar origins to show enrichments in D and ^{15}N. Isotopic analyses of the Wild 2 samples appear to be qualitatively consistent with these expectations. The heterogeneous distribution of D and ^{15}N excesses in these samples suggests that materials having an interstellar heritage were mixed with solar nebular materials during Solar System formation, but that they were not exposed to significant processing after the component subgrains were assembled, either within the protosolar nebula or in the cometary parent body.

3. Summary and Conclusions

Overall, the organics in the Comet Wild 2 samples show many similarities with those in IDPs, and to a lesser extend, with those in primitive meteorites, but there are some very distinct differences. Like meteoritic organics, Wild 2 organics contain both aromatic and non-aromatic fractions. However, the Stardust samples exhibit a greater range of compositions (higher O and N concentrations), include an abundant organic component that is poor in aromatics, and a more labile fraction (possibly the same material). The non-aromatic fraction appears to be far more abundant relative to aromatics than in meteorites. Some of the cometary material represents a new class of organics not previously seen in other extraterrestrial samples, making these samples unique among currently available extraterrestrial samples.

The distribution of organics (overall abundance, functionality, and relative elemental abundances of C, N, and O) is remarkably heterogeneous both between particles and within individual particles. Cometary organics clearly represent a highly unequilibrated reservoir of materials. This indicates that these materials did not undergo much parent body processing. In general terms, the organics in Stardust samples appear to be even more "primitive" than those in meteorites and IDPs, at least in terms of being highly heterogeneous and unequilibrated. The presence of organics with high O and N contents and high ratios of $-CH_2-/-CH_3$ indicate that the Stardust organics are not identical to the organics seen in the diffuse ISM. This implies that cometary organics are not the direct result of stellar ejecta or diffuse ISM processes, but rather the ultimate result of dense cloud and/or protosolar nebular processes.

Since the processes that led to life on Earth are poorly understood, it is difficult to assess the importance these cometary organics may have had for the origin of life. However, it is clear that comets contain a complex population of organic materials. This population appears to include a wider range of organics than is found in primitive meteorites, which themselves contain many species of astrobiological interest. Clearly comets contain unique organics that have much to tell us about the nature of interstellar and protostellar chemistry and the roles they may play in the origin of life.

Acknowledgments. This work was supported by NASA grants from the Origins of Solar System, Exobiology, Cosmochemistry, and Discovery Mission Programs. The author is grateful to the many, many people who made the Stardust Mission both a reality and a joy to work on.

References

Aléon, J., & Robert, F. 2004, Icarus, 167, 424

Bajt, S., Sandford, S. A., Flynn, G. J., Matrajt, G., Snead, C. J., Westphal, A. J., & Bradley, J. P. 2009, Infrared Spectroscopy of Wild 2 Particle Hypervelocity Tracks in Stardust Aerogel: Evidence for the presence of Volatile Organics in Comet Dust. Meteoritics and Planetary Science 44, 471-484.

Brownlee, D. E., et al. 2003, J. Geophys. Res., 108, #E10, 8111, p. 1-1

Brownlee, D. E., et al. 2004, Science, 304, 764

Brownlee, D., et al. 2006, Science, 314, 1711

Clemett, S. J., Spencer, M. K., Sandford, S. A., McKay, D. S., & Zare, R. N. 2009, MAPS, in press

Cody, G. D., et al. 2008, MAPS 43, 1, 353-365

Flynn, G. J., et al. 2006, Science, 314, 1731

Glavin, D. P., Dworkin, J. P., & Sandford, S. A. 2008, MAPS 43, 1, 399-413

Hőrz, F., et al. 2006, Science, 314, 1716

Keller, L. P., et al. 2006, Science, 314, 1728

Kerridge, J. F., & Matthews, M. S., eds. 1988, Meteorites and the Early Solar System. (Tucson, AZ: Univ. Arizona Press)

Lauretta, D. S., & McSween, H. Y. Jr., eds. 2006, Meteorites and the Early Solar System II. (Tucson, AZ: Univ. Arizona Press)

Matrajt, G., et al. 2008, MAPS 43, 1, 315-334

McKeegan, K. D., et al. 2006, Science, 314, 1724

Messenger, S. 2000, Nature, 404, 968

Oró, J., Holzer, G., & Lazcano-Araujo, A. 1980, in Vol. 18 – Proceedings of the Open Meeting of the Working Group on Space Biology (Oxford: Pergamon Press), 67

Rotundi, A., Baratta, G. A., Borg, J., Brucato, J. R., Busemann, H., Colangeli, L., d'Hendecourt, L., Djouadi, Z., Ferrini, G., Franchi, I. A., Fries, M., Grossemy, F., Keller, L. P., Mennella, V., Nakamura, K., Nittler, L. R., Palumbo, M. E., Sandford, S. A., Steele, A., & Wopenka, B. 2008, Combined Micro-Raman, Micro-Infrared, and Field Emission Scanning Electron Microscope Analyses of Comet 81P/Wild 2 Particles Collected by Stardust. Meteoritics and Planetary Science 43, 367-397.

Sandford, S. A., Bernstein, M. P., & Dworkin, J. P. 2001, MAPS, 36, 1117

Sandford, S. A., et al. 2006, Science, 314, 1720

Sandford, S. A., & Brownlee, D. E. 2007, Science 317, 1680

Sekanina, Z., Brownlee, D. E., Economou, T. E., Tuzzolino, A. J., & Green, S. F. 2004, Science 304, 769

Spencer, M. K., & Zare, R. N. 2007, Science 317, 5845, 1680

Spencer, M. K., Clemett, S. J., Sandford, S. A., McKay, D. S., & Zare, R. N. 2009, Organic Compound Alteration during Hypervelocity Collection of Carbonaceous Materials in Aerogel. Meteoritics and Planetary Science 44, 15-24.

Stephan, T., et al. 2008a, MAPS 43, 1, 233-246

Stephan, T., Flynn, G. J., Sandford, S. A., & Zolensky, M. E. 2008b, MAPS 43, 1, 285-298

Thomas, P. J., Chyba, C. F., & McKay, C. P., eds. 1996, Comets and the Origin and Evolution of Life (New York: Springer)

Tuzzolino, A. J., et al. 2004, Science, 304, 1776

Zolensky, M. E., et al. 2006, Science, 314, 1735

Rose Grymes

Session V

Mars as a Setting for Life

Bioastronomy 2007: Molecules, Microbes, and Extraterrestrial Life
ASP Conference Series, Vol. 420, 2009
K. J. Meech, J. V. Keane, M. J. Mumma, J. L. Siefert, and D. J. Werthimer, eds.

CO_2 Clathrates on Mars and the CH_4 Question: A Laboratory Investigation

M. G. Trainer,[1,2] C. P. McKay,[3] M. A. Tolbert,[4] and O. B. Toon[1,5]

[1]*Laboratory for Atmospheric and Space Physics, University of Colorado, UCB 392, Boulder, CO 80309*

[2]*NASA Goddard Space Flight Center, Code 699, Greenbelt, MD 20771*

[3]*Space Sciences Division, NASA Ames Research Center, Moffett Field, CA 94035*

[4]*CIRES and Department of Chemistry and Biochemistry, University of Colorado, UCB 216, Boulder, CO 80309*

[5]*Department for Atmospheric and Oceanic Sciences, University of Colorado, UCB 392, Boulder, CO 80309*

Abstract. Recent detection of methane (CH_4) on Mars has generated interest in evaluating the source of this trace species and its role in the current atmosphere. While the photochemistry of CH_4 is well understood, little is known about the heterogeneous (gas-surface) reactions that may take place on Mars. Understanding the complete processing of CH_4 will help determine whether this trace species may be an indicator of past or present life. The existence of currently unknown sources or sinks of CH_4 may help explain the variability observed in CH_4 concentration, despite its long photochemical lifetime. It has been suggested that the Martian polar deposits may contain carbon dioxide (CO_2) clathrate hydrates. These clathrates may store large amounts of CO_2, and may also trap CH_4 molecules from the atmosphere. We have begun a series of laboratory studies to explore the possibility that CO_2 clathrates may serve as a sink for CH_4 gas on Mars. Here we report the first results on the formation and characteristics of CO_2 clathrates at low temperatures and the ability of these structures to serve as a sink for CH_4 on Mars.

1. Introduction

Recent detection of CH_4 in the atmosphere of Mars at ppb levels has generated a great deal of interest as a possible biomarker. The photochemical lifetime of CH_4 in the atmosphere is on the order of hundreds of years, and it is expected to be well distributed globally. However, reported observations to date suggest both spatial and temporal variations in CH_4 concentration, from 0 to 200 ppbv (Formisano et al. 2004; Krasnopolsky, Maillard, & Owen 2004; Mumma et al. 2005). The unexpected variability indicates that current modeling does not fully account for all of the processing of CH_4 in the Martian atmosphere. Little is known about possible heterogeneous (gas-surface) reactions that may exist as local sources or sinks of CH_4. While it has been suggested that the presence of CH_4 may be an indicator for past or current life, such a determination cannot be made without a complete inventory of the abiogenic reactions of CH_4. Thus, an exploration of the heterogeneous chemistry of CH_4 on Mars is warranted.

One surface of interest for potential heterogeneous chemistry of CH_4 on Mars is CO_2 clathrate hydrates. A clathrate hydrate is a crystalline solid in which water ice forms an open lattice of cages that are occupied and stabilized by trapped gas molecules, which would likely be CO_2 on Mars. It has been suggested that CO_2 clathrates may be a means of sequestering CO_2 from the Martian atmosphere (Jakosky et al. 1995). In addition, CH_4 molecules may be inserted as guests in the open cage structures of the CO_2 clathrate. It has been speculated that CO_2 clathrates may serve as reservoirs for trace volatiles by trapping these species as secondary guest molecules (Musselwhite & Lunine 1995). Thus, the inclusion of CH_4 molecules into any existing CO_2 clathrates in the Martian polar caps may be a temporary sink for this trace gas species. However, there have been limited laboratory studies exploring the formation of CO_2 clathrates from low temperature vapor deposition (Fleyfel & Devlin 1991), and none meant to simulate depositional formation in the Mars atmosphere. Also, some laboratory work has shown that CH_4 molecules can be incorporated into CO_2 clathrates (Adisasmito, Frank, & Sloan 1991; Uchida et al. 2005), but this has not been studied at Mars pressures. Here, we present results from our study which looks at the properties of clathrates formed under low temperature and pressure conditions, and evaluates the possibility that CO_2 clathrates could be a sink for CH_4.

2. Experimental Methodology

The experimental apparatus used in this study has been described previously (Glandorf et al. 2002). For this study, a silicon wafer was placed in a vacuum chamber and cooled to temperatures from 130 - 145 K using a compressed He cryostat. A substrate of tetrahydrofuran (THF) clathrate was formed from H_2O/THF vapor deposition to promote the formation of the CO_2 clathrate (Fleyfel & Devlin 1991). Once the THF clathrate substrate was formed, CO_2 vapor was added to the vacuum chamber at mTorr pressures, and the clathrate was formed from codeposition of H_2O and CO_2. The vapor phase in the chamber was maintained with CO_2 in excess of H_2O, to simulate approximate atmospheric abundances on Mars. The surface was continually probed with FTIR transmission spectra collected every 4 sec, and these spectra were used to identify the clathrate formation in the condensed phase. Once clathrate formation was complete, the chamber was evacuated and the sample was warmed at 1 K min^{-1}. Gas molecules desorbed from the ice structure were monitored using a mass spectrometer, thus producing a Temperature Programmed Desorption (TPD) profile which is used to quantify the ratio of constituents in the clathrate (Koehler et al. 1992). For experiments studying CH_4 uptake, the above procedure was performed with the addition of CH_4 vapor to the H_2O/CO_2 mixture in the chamber. The TPD was used to measure any CH_4 incorporation into the clathrate structure.

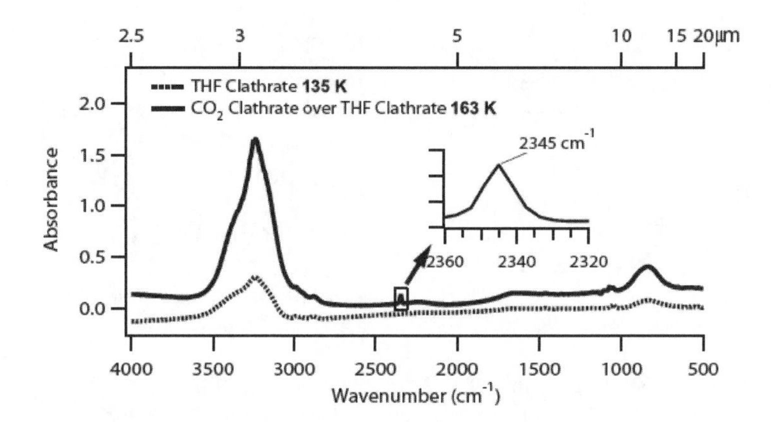

Figure 1. Infrared spectrum of CO_2 clathrate over a THF clathrate base. Clathrate was formed by vapor deposition from a mixture of 1.6 mTorr CO_2 and 0.04 mTorr H_2O. The region in the inset shows the absorption peak for enclathrated CO_2 at 2345 cm^{-1}.

3. Results and Discussion for CO_2 Clathrate Formation

Figure 1 shows the IR spectrum of the CO_2 clathrate formed over a THF clathrate at approximately 135 K, then warmed to 163 K. The primary peak used for clathrate identification is at 2345 cm^{-1}, and is shown in the inset (Blake et al. 1991). Trapped, but not enclathrated, CO_2 has an absorption peak at 2340 cm^{-1}, and has been observed in our study under different formation conditions. Features used to identify the THF clathrate are observed from 3050 - 2800 cm^{-1} and 1350 - 1000 cm^{-1}. Using a mass spectrometer to monitor gas-phase components during TPD, we have been able to quantify the H_2O/CO_2 ratio, and thus calculate a cage occupancy for the clathrate. The clathrate shown in Fig. 1 has a H_2O:CO_2 ratio of 10.6:1. This correlates to a cage occupancy of 54%, assuming a Structure I clathrate. The cage occupancy measurement indicates that approximately 50% of the available cages are empty, assuming that the entire volume of the ice film is a clathrate. This result has implications for the potential of clathrate reservoirs to store CO_2 on Mars. In addition, the prevalence of empty cages indicates that trace gases in the Mars atmosphere could potentially be stored in the clathrates, as previously proposed in theoretical studies Musselwhite & Lunine (1995).

4. Results and Discussion for CH_4 Uptake

Once the CO_2 clathrate formation procedure was established, experiments were performed to explore whether the CO_2 clathrate structure could serve as a heterogeneous sink for CH_4. Figure 2 shows the results from one of these experiments, in which CO_2 clathrate was formed in the presence of CH_4 vapor. The IR spectrum in Fig. 2a does not show any indication of CH_4 incorporation into the ice structure. The region near a pure CH_4 absorbance feature at 1300 cm^{-1}

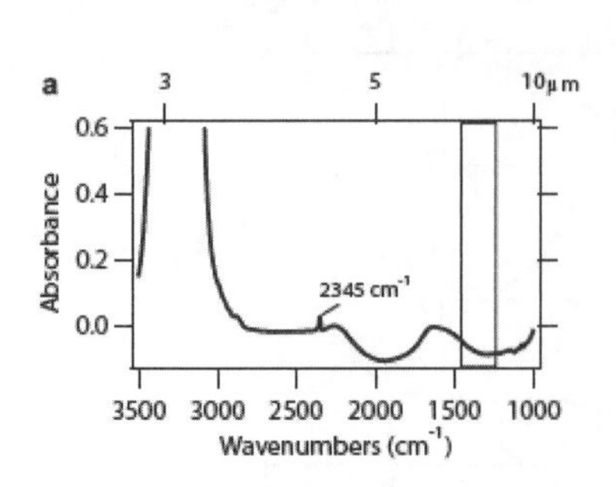

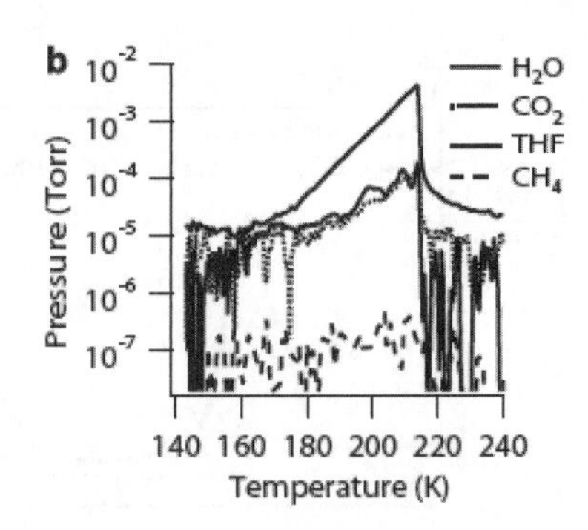

Figure 2. *(a)* Infrared spectrum of CO_2 clathrate over a THF clathrate base at 140 K. Clathrate was formed by vapor deposition from a mixture of 5.5 mTorr CO_2 and 0.18 mTorr H_2O in the presence of 3 mTorr CH_4. Box is given to indicate where CH_4 peak may appear. *(b)* TPD profile of the CO_2 clathrate shows no CH_4 was trapped in the clathrate structure.

was monitored, since there was little interference here from the CO_2 and THF clathrates. Fig. 2b shows the TPD profile from this experiment, which is more sensitive to CH_4 uptake than the IR spectrum of the ice film. The film desorption does not show any trace of CH_4 gas above the noise level, which indicates no detectable CH_4 incorporation into the clathrate structure. The upper limit on CH_4 incorporation provided from the desorption results is $CH_4/H_2O \leq 5$ x 10^{-4}. The cage occupancy for the CO_2 clathrate shown in Fig. 2 is 40%, thus confirming there were empty cages available for CH_4 trapping, but that the trapping did not take place. Further experiments, not shown here, have involved exposing a CO_2 clathrate to CH_4 gas for several hours at approximately 130 and 160 K, again with no observed uptake of CH_4. Thus, so far we have not observed CH_4 trapping into ice structures, and this suggests that the clathrates formed in the laboratory may not serve as a large heterogeneous sink for CH_4 on Mars. These results can be quantified to provide an upper-limit for the amount of CH_4 that might be expected to interact with ice surfaces on Mars.

Results presented here show no evidence of CH_4 trapping in CO_2 clathrate. CO_2 clathrates have been formed in the presence of high partial pressures of CH_4 gas relative to Mars (approximately 1000x Martian abundance). However, upon desorption the CO_2 clathrates show no CH_4 incorporation. Thus, we have not identified a significant sink for CH_4 in the clathrate structure.

Future work includes studying new substrates for CO_2 clathrate formation. All of the experiments discussed here have used a THF clathrate as a substrate, which is not a relevant surface for Mars. Future plans include testing whether CO_2 clathrate will form on an analog for Mars soil, Fe-montmorillonite. Modeling studies have shown that the structure of montmorillonite might provide an excellent lattice match for the clathrate structure, and may promote clathrate formation (Park & Sposito 2003). It is expected that the structure of clathrate formed on montmorillonite will more closely represent that of any clathrates that may exist on Mars. This may have implications for the structure of the CO_2 clathrate and the CO_2 cage occupancy (Fleyfel & Devlin 1991). Thus, it cannot be completely determined whether CO_2 clathrates will serve as a CH_4 sink until a more Mars-relevant clathrate system is studied. Once formed on the clay analog, this clathrate will also be studied for potential trapping of CH_4 gas.

Acknowledgments. This research is funded by NASA Mars Fundamental Research Program Award NNH05ZDAS001N. M.G.T. was supported by an appointment to the NASA Postdoctoral Program at the the Center for Astrobiology at the University of Colorado, administered by Oak Ridge Associated Universities through a contract with NASA. Travel Support was received from the NASA Astrobiology Institute and the National Science Foundation through the conference organizers.

References

Adisasmito, S., Frank, R.J., & Sloan, E.D. 1991, J. Chem. and Eng. Data, 36, 68
Blake, D., Allamandola, L., Sandford, S., Hudgins, D., & Freund, F. 1991, Sci, 254, 548
Fleyfel, F., & Devlin, J.P. 1991, J. Phys. Chem., 95, 3811
Formisano, V., Atreya, S., Encrenaz, T., Ignatiev, N., & Giuranna, M. 2004, Sci, 306, 1758
Glandorf, D.L., Colaprete, A., Tolbert, M.A., & Toon, O.B. 2002, Icarus, 160, 66
Jakosky, B.M., Henderson, B.G., & Mellon, M.T. 1995, JGR-Planets, 100, 1579
Koehler, B.G., Middlebrook, A.M., & Tolbert, M.A. 1992, JGR-Atm., 97, 8065
Krasnopolsky, V.A., Maillard, J.P., & Owen, T.C. 2004, Icarus, 172, 537
Mumma, M.J., DiSanti, M.A., Novak, R.E., Bonev, B.P., Dello Russo, N., Hewagama, T., & Smith, M. 2005, Astrobiology, 5, 300
Musselwhite, D., & Lunine, J.I. 1995, JGR-Planets, 100, 23301
Park, S.H., & Sposito, G. 2003, J. Phys. Chem. B, 107, 2281
Uchida, T., Ikeda, I.Y., Takeya, S., Kamata, Y., Ohmura, R., Nagao, J., Zatsepina, O. Y., & Buffett, B. A. 2005, Chemphyschem, 6, 646

Ricardo Amils

Bioastronomy 2007: Molecules, Microbes, and Extraterrestrial Life
ASP Conference Series, Vol. 420, 2009
K. J. Meech, J. V. Keane, M. J. Mumma, J. L. Siefert, and D. J. Werthimer, eds.

Production of Phyllosilicates and Sulfates on Mars Through Acidic Weathering: The Río Tínto Mars Analog Model

D. C. Fernández-Remolar,[1] O. Prieto-Ballesteros,[1] R. Amils,[1] F. Gómez,[1] L. Friedlander,[2] R. Arvidson,[2] R. V. Morris,[3] and D. Gómez[4]

[1]*Centro de Astrobiología, INTA-CSIC, Ctra Ajalvir km 4, Torrejón de Ardoz, 28850 Spain, email: fernandez@inta.es*

[2]*Department of Earth and Planetary Sciences, Washington University in St. Louis, St. Louis, MO, USA*

[3]*NASA Johnson Space Center, Houston, TX, USA*

[4]*Geología, ESCET, Universidad Rey Juan Carlos, Móstoles, 28933 Spain*

Abstract. Understanding the paleoclimate of Mars and the nature and extent of the interaction of water with crustal materials is essential to evaluate if life emerged once on Mars. Using recent results obtained by orbiters and surface rovers we propose that mildly acidic aqueous conditions produced phyllosilicate minerals during the Noachian period on Mars related to a CO_2 -rich atmosphere and an active hydrosphere. Underground neutralization of these mild acidic meteoric solutions with crustal materials would produce subsurface carbonates inside the crust after. Moreover, hydrothermalism provides mineralization to favor subsurface sulfide orebodies, but also secondary geochemical processes would also favor its formation. After cessation of the internal magnetic dynamo, the CO_2 -rich atmosphere was eroded and decreased by interactions with the solar wind and the hydrologic cycle that induced a climatic change to dry conditions where hydrological processes had subsurface dominance. Under this thin atmospheric aridic conditions photochemistry played an essential role in generating oxidizing and acidifying compounds that after entering the Mars crust promoted carbonate dissolution and sulfide oxidation. As a consequence, acid-sulfate evaporite deposits were precipitated in areas where acidic subsurface solutions emerged. Analogous processes have been observed in the underground fluids that feed the Río Tínto Mars analog. Seasonal subsurface of sulfur continued by a rapid oxidation induced by rainwaters, which induces a strong acidification through proton releasing ($FeS_2 + 3\cdot O_2 + 2\ H_2O \rightarrow 2\cdot SO_4{}^{2-} + Fe^{2+} + 4\ H^+$). This stage is followed by a neutralization and reduction of the ferric and sulfate rich acidified waters, which favors the subsurface carbonate precipitation. In addition, phyllosillicate sedimentation under the strong acidic conditions of Río Tínto, supports preservation of same minerals under early Noachian mildly acidic conditions.

1. Introduction

The discovery of phyllosilicate minerals exposed on the surfaces of ancient Noachian cratered terrains on Mars (Poulet *et al.* 2005; Bibring *et al.* 2006) is important because the formation of these minerals has been suggested to be coincident with neutral to alkaline aqueous conditions. The hydrated sulfate-rich deposits that unconformably overlie the Noachian cratered terrain in Terra Meridiani (Arvidson *et al.* 2005, 2006) and that are found as interior layered deposits in Valles Marineris (Gendrin *et al.* 2005) are equally important discoveries because the environment of deposition must have been dominated by acid-sulfate conditions (Bigham & Nordstrom 2000). In this paper we develop an overall model for the global mineralogical and geochemical evolution of Mars that accounts for these changes, and we make predictions for the accumulation and subsequent dissolution of early carbonate and sulfide minerals as part of the overall change from mildly acidic to strongly acidic hydrologic conditions. We begin with a description of the acid-sulfate system in Río Tínto (Fernández-Remolar *et al.* 2005), Spain, with its highly seasonal rainfall, as a process analog from which we draw lessons for understanding the global evolution of aqueous systems on Mars.

2. The Río Tínto Extreme Environment

Río Tínto is an acid-sulfate dominated fluvial system in which a variety of hydroxylated and hydrated sulfate minerals form annually as evaporite and efflorescent minerals during the summer dry season (Fernández-Remolar *et al.* 2005). Sulfate-bearing minerals found along the river include jarosite, copiapite, coquimbite, schwertmannite, gypsum, and epsomite (Fernández-Remolar *et al.* 2005). Springs at the head waters of the Río Tínto are located in the Peña de Hierro area and are highly acidic because of ground water percolation through the sulfide ore deposits within the Iberian Pyritic Belt (IPB) (Fig. 1) (Fernández-Remolar *et al.* 2008; Leistel *et al.* 1998). Biotic and abiotic alteration of the pyrite deposits produces acidic solutions (pH $\sim$ 0.8-3) enriched in sulfur- and iron-bearing ions. In shallow areas of the aquifer, pyrite weathering is mediated by aerobic microbes using O_2 as the electron acceptor (Sand *et al.* 2001). After O_2 depletion anaerobic microbes oxidize sulfides following the reaction S^{2-} + 8 Fe^{3+} + 4 $H_2O \rightarrow$ $(SO_4)^{2-}$ + 8 Fe^{2+} + 8 H^+, resulting in reduction of Fe^{3+} to Fe^{2+} and generation of sulfate ions (Malki *et al.* 2006). The acidic spring waters empty into the Río Tínto and give this fluvial system its acid-sulfate environment.

Several boreholes were recently drilled through the ore body of the Iberian Pyrite Belt and into surrounding country rock to explore the ground water chemistry and subsurface mineralogy (Fernández-Remolar *et al.* 2008) (Fig. 1). Borehole BH1 was drilled into shale-rich bedrock south of the main ore body whereas boreholes BH4 and BH8 were drilled into the ore body proper. BH4 is near an open-pit mine site which allowed oxygenated meteoric waters to easily reach the subsurface during the rainy season (Fig. 1). SEM-EDAX and Mssbauer analyses of rock samples obtained below 30 m depth from BH1 and SEM-EDAX analyses from BH4 (Fernández-Remolar *et al.* 2008) indicate the

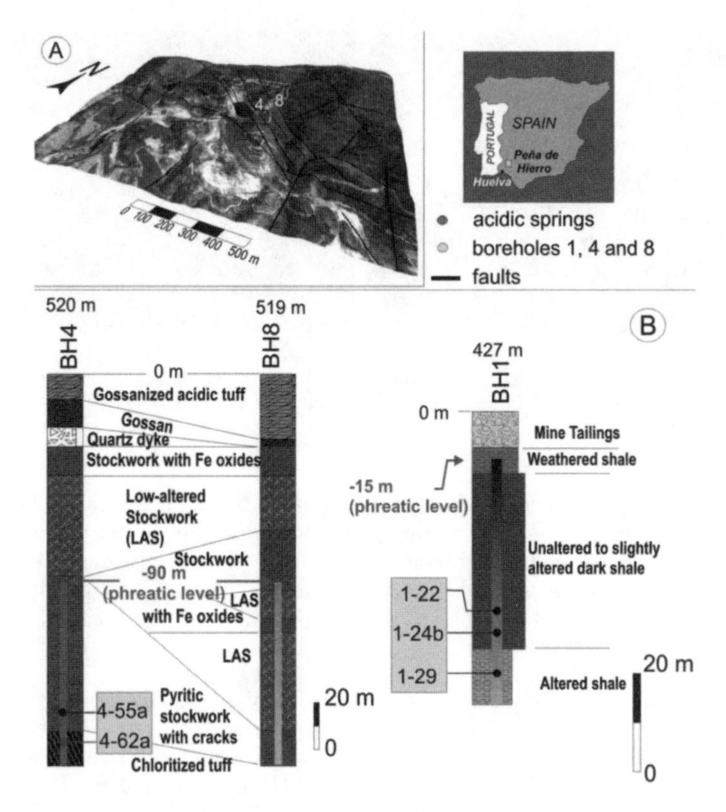

Figure 1. Geological and hydrogeochemical constraints of the Río Tínto subsurface environments. (A) Three dimensional view of the Peña de Hierro area showing the location of the sampled boreholes. (B) Lithological and hydrochemical characteristics of the three boreholes around Peña de Hierro. Carbonate containing samples are shown (4-55a, 4-62a, 1-22, 1-24b and 1-29).

presence of iron carbonate minerals. Siderite is present in BH4 (measured pH ~ 5.5) whereas dolomite and ankerite are present in BH1 (measured pH ~ 7). Water collected from BH8 indicates a saturated state for carbonate minerals, although no carbonates have yet been identified from core samples. This variation is consistent with a thermodynamically stable gradient for carbonate formation from a siderite zone in BH4 and BH8 to dolomite and ankerite zone (Buil *et al.* 2006) in BH1 (Fig. 1). Carbonates are produced when the supply of meteoric water ceases during the dry season when the ground water changes from acidic and oxidizing to neutral and reducing conditions. This induces the reduction of sulfate and ferric ions and the water becomes saturated with respect to iron carbonate minerals. The presence of carbonates in core materials shows that formation exceeds dissolution rates, leading to net carbonate accumulation.

Other issues relate to the occurrence of phyllosillicate deposits in the Río Tínto riverbanks, which are mainly transported during the rainfall seasons (early spring and fall), but also during fast flood events occurring in summer. These mineralogies are formed by simple erosion and transport from weathered and

Figure 2. Illite, chlorite and smectite mineralogies in a highly weathered shale (a) and as part of the mineral association of transported fine-grained sediments of the Río Tínto.

non-weathered hydrothermal deposits composed by phyllosillicates as primary chlorite, and illite, but also smectite (e.g. montmorillonite and corrensite) and kaolinite (Fig. 2). Moreover, during the summer dry season, high fluid saturation produces isolation of pools containing acidic brines where precipitation of amorphous silica co-occurs with sulfate efflorescences at pH below 2.

3. Model for the Evolution of Aqueous Systems on Mars

The presence of phyllosilicate minerals in Noachian cratered terrains provides important evidence for early aqueous conditions (Poulet *et al.* 2005; Bibring *et al.* 2006). Specifically, the pH for waters present during the formation of these minerals must have been between 4 (averaged stability boundary for phyllosilicate minerals) and 5.5 (lower boundary for precipitating siderite under a CO_2-rich atmosphere (Squyres and Kasting 1994; Owen 1994) with moderately-oxidizing conditions (Chevrier *et al.* 2007). An active magmatic activity provided the mechanisms of formation of primary sulfide deposits (Burns and Fisher, 1990), but also secondary can be expected by weathering and atmospheric sulfur recycling. Assuming that early Martian atmospheric conditions were similar to Archean conditions on Earth, the Mars hydrosphere equilibrated to slightly oxidizing and acidic conditions resulting from (a) the photochemical production of oxidants (i.e. $PO_2 > 10^{-5}$ PAL is needed for weathering Archean S-bearing mineralogies (Ohmoto, 2004; Ohmoto *et al.* 2006)), and (b) the combined effect of a CO_2-rich atmosphere (Squyres & Kasting 1994; Owen 1994) and a high input of ferrous cations. Specifically, weathering and hydrothermal activity produced the ferrous ions and helped provide a modestly effective acidic buffer (i.e. $Fe^{2+} + H_2O \rightarrow Fe(OH)^{+} + H^{+}$). These early conditions led to the formation of phyllosilicates at the near-surface (i.e., in contact with meteoric water) and the production of subsurface carbonates (Carr 1999) via reactions between crustal materials and groundwater. This mechanism is analogous to the neutralization of groundwater in Río Tínto that produces subsurface carbonates (Fig. 1). During this stage of Martian evolution, liquid water was stable on the surface and the planet (Gendrin *et al.* 2005) was subjected to a vigorous surface and subsurface hydrologic cycle. The system was supported by a carbon-dioxide rich atmosphere protected from the solar wind by the presence of a magnetic field (Solomon *et al.* 2005).

A significant change in climatic conditions occurred during the late Noachian to the Hesperian as the magnetic field disappeared and the carbon dioxide rich atmosphere was stripped by solar wind (Jakoski & Phillips 2001). Arid conditions prevailed and the hydrologic cycle became much less vigorous. Ground water contact with subsurface sulfide deposits begun during hydrothermal events remained for extended periods, generating acid-sulfate charged waters. In addition, the thinner atmosphere allowed enhanced photochemical production of oxidants and acidifying agents (Krasnopolski 1993; Wong *et al.* 2003) fed by the volcanic emission (Moore and Howard 2004) of H_2O, CO_2, NO_2, SO_2 and other gases (i.e. $2\ O_2 + 2\ H + h\nu \rightarrow H_2O_2 + O_2$, where O_2 and H are supplied through CO_2 photolysis via reaction such as $CO_2 + h\nu \rightarrow CO + O$, $O + O + CO_2 \rightarrow CO_2 + O_2$, and $H_2O + h\nu \rightarrow OH + H$ (Krasnopolski 1993; Wong *et al.* 2003)). Photo-oxidation of complexed Fe^{2+} increased the acidic buffering effect, leading to the formation of ferric oxyhydroxides ionic aqueous complexes (Braterman *et al.* 1983; Burns 1988). Acidic ground water dissolved carbonate minerals, which could eventually precipitate in deeper regions of the Mars crust (Fernández-Remolar *et al.* 2008). Emerging acid-sulfate ground waters in lowlands produced spring deposits and lacustrine deposits dominated by sulfates in a manner similar to what is found in the Río Tínto fluvial system. Without significant rainfall these deposits would have been preserved, except for wind erosion.

4. Implications for life

Photosynthetic microbes mediate Earth's contemporary aqueous carbonate system through removing CO_2 from the solutions (i.e. $CO_2 + H_2O \leftrightarrow HCO_3^- + H^+$). On Earth this system is a buffer maintaining an ocean pH favorable to life (van Kranendonk, 2006), which might have started by the microbial removing of CO_2 during the early periods of Earth's history through photosynthetic surface life. The apparent absence of surface carbonates on Mars could eliminate the possibility of a similar system for the development of Martian life. However, Martian life may have developed utilizing the energy transfer potential of an iron and sulfur-based system. Microbes on Mars would have favored the formation of sulfates and ferric iron minerals to the same extent that carbonate precipitation was favored by the early terrestrial microbial communities. Under these subsurface conditions, underground carbonates can be formed through redox processes involving CH_4 that is a compound detected as a component of the Mars atmosphere (Mumma *et al.* 2003).

Acknowledgments. We thank to Prof Juan Prez-Mercader for the continuous support to our research. This paper was supported by the Project ESP2006-09487 funded by the Ministry of Science and Education of Spain.

References

Arvidson, R. E., Poulet, F., Morris, R. V., Bibring, J. P., Bell III, J. F., Squyres, S. W., Christensen, P. R., Bellucci, G., Gondet, B., Ehlmann, B. L., Farrand, W.

H., Fergason, R. L., Golombek, M., Griffes, J. L., Grotzinger, J. P., Guinness, E. A., Herkenhoff, K. E., Johnson, J. R., Klingelhfer, G., Langevin, Y., Ming, D., Seelos, K., Sullivan, R. J., Ward, J. G., Wiseman, S. M. & Wolff, M. 2006, JGR 111, doi:10.1029/2006JE002728.
Bibring, J.-P., Langevin, Y., Gendrin, A., Gondet, B., Poulet, F., Berthe, M., Soufflot, A., Arvidson, R., Mangold, N., Mustard, J., Drossart, P. & the OMEGA team (2005) Science 307, 1576-1581.
Bigham, J. M. & Nordstrom, D. K. 2000, in Sulfate Minerals. Crystallography, geochemistry, and environmental significance, ed. C. N. Alpers, J. L. Jambor & D. K. Nordstrom (Washington, DC), 40, 351
Braterman, P., Cairns-Smith, A. & Sloper, R. 1983, Nature 303, 163
Buil, B., Gómez, P., Turrero, M. J., Garralón, A., Lago, M., Arranz, E. & de la Cruz, B. 2006, J Iberian Geol 32, 113
Burns, R. G. & Fisher, D. S. 1990, JGR 95, 14415
Burns, R. G. 1988, in 18th Lunar and Planetary Science Conference, 713
Chevrier, V., Poulet, F. & Bibring, J.-P. 2007, Nature 448, 60
Fernández Remolar, D. C., Prieto Ballesteros, O., Rodríguez, N., Gómez, F., Amils, R., Gómez-Elvira, J. & Stoker, C. 2008, Astrobiology 8, 1023
Fernández-Remolar, D. C., Morris, R. V., Gruener, J. E., Amils, R. & Knoll, A. H. 2005, EPSL 240, 149
Gendrin, A., Mangold, N., Bibring, J.-P., Langevin, Y., Gondet, B., Poulet, F., Bonello, G., Quantin, C., Mustard, J., Arvidson, R., LeMouelic, S., Berthe, M., Erard, S., Forni, O., Soufflot, A., Combes, M., Drossart, P., Encrenaz, T., Fouchet, T., Merchiorri, R., Belluci, G., Altieri, F., Formisano, V., Capaccioni, F., Cerroni, P., Coradini, A., Fonti, S., Kottsov, V., Ignatiev, N., Moroz, V., Titov, D., Zasova, L., Pinet, P., Doute, S., Schmitt, B., Sotin, C., Hauber, E., Hoffmann, H., Jaumann, R., Keller, U., Mustard, J., Duxbury, T. & Forget, F. 2005, Science 307, 1587
Jakosky, B. M. & Phillips, R. J. 2001, Nature 412, 237
Krasnopolsky, V. A. 1993, Icarus 101, 313
Leistel, J. M., Marcoux, E., Thiblemont, D., Quesada, C., Snchez, A., Almodvar, G. R., Pascual, E. & Sez, R. 1998, Min Dep 33, 2
Carr, M. H. 1999, JGR 104, 21897
Malki, M., Gonzalez-Toril, E., Sanz, J. L., Gomez, F., Rodriguez, N. & Amils, R. 2006, Hydrometallurgy 83, 223
Moore, J. M. & Howard, A. D. 2004 in Second Conference on Early Mars (Houston), paper 8014
Mumma, M. J., Novak, M. T., DiSanti, M. A. & Bonev, B. P. 2003, American ASB 35, 937
Ohmoto, H. (2004) in The Precambrian Earth: Tempos and Events, ed. P. G. Eriksson, W. Altermann, D. R. Nelson, W. U. Mueller & O. Catuneanu (Elsevier, Amsterdam), 12, 361
Ohmoto, H., Watanabe, Y., Ikemi, H., Poulson, S. R. & Taylor, B. E. 2006, Nature 442, 908
Owen, T. 1994, Phyl Trans Royal Soc: Phys Sci Eng 349, 209
Poulet, F., Bibring, J.-P., Mustard, J. F., Gendrin, A., Mangold, N., Langevin, Y., Arvidson, R. E., Gondet, B. & Gomez, C. 2005, Nature 438, 623
Sand, W., Gehrke, T., Jozsa, P.-G. & Schippers, A. 2001, Hydrometallurgy 59, 159
Solomon, S. C., Aharonson, O., Aurnou, J. M., Banerdt, W. B., Carr, M. H., Dombard, A. J., Frey, H. V., Golombek, M. P., Hauck II, S. A., Head III, J. W., Jakosky, B. M., Johnson, C. L., McGovern, P. J., Neumann, G. A., Phillips, R. J., Smith, D. E. & Zuber, M. T. 2005, Science 307, 1214
Squyres, S. W. & Kasting, J. F. 1994 Science 265, 744
van Kranendonk, M. J. 2006, ESR 74, 197
Wong, A.-S., Atreya, K. & Encrenaz, T. 2003, JGR 108, 10.1029/2002JE002003

William Schopf

Timothy Kral

Bioastronomy 2007: Molecules, Microbes, and Extraterrestrial Life
ASP Conference Series, Vol. 420, 2009
K. J. Meech, J. V. Keane, M. J. Mumma, J. L. Siefert, and D. J. Werthimer, eds.

Methane Production on Rock and Soil Substrates by Methanogens: Implications for Life on Mars

H. A. Kozup and T. A. Kral

[1] *West Virginia University, WV 26505*

[2] *Arkansas Center for Space and Planetary Sciences, University of Arkansas, Fayetteville 72701*

Abstract. In order to understand the methanogens as models for possible life on Mars, and some of the factors likely to be important in determining their abundance and distribution, we have measured their ability to produce methane on a few types of inorganic rock and soil substrates. Since organic materials have not been detected in measurable quantities at the surface of Mars, there is no reason to believe that they would exist in the subsurface. Samples of three methanogens (*Methanosarcina barkeri, Methanobacterium formicicum,* and *Methanothermobacter wolfeii*) were placed on four substrates (sand, gravel, basalt, and a Mars soil simulant, JSC Mars-1) and methane production measured. Glass beads were used as a control substrate. As in earlier experiments with JSC Mars-1 soil simulant, a crushed volcanic tephra, methane was produced by all three methanogens when placed on the substrates, sand and gravel. None produced methane on basalt in these experiments, a mineral common in Martian soil. While these substrates do not represent the full range of materials likely to be present on the surface of Mars, the present results suggest that while some surface materials on Mars may not support this type of organism, others might.

1. Introduction

The possibility of life on Mars has been a subject of intense interest for many decades, but only recently have we learned enough about Martian surface conditions to seriously address the issue. The Viking missions of the mid-1970s, though a spectacular engineering accomplishment, shed little light on the issue, yielding results that initially appeared to be positive, but were subsequently interpreted by most researchers as reflecting an unexpected inorganic chemistry. However, the Viking missions did confirm that conditions at the surface are rather harsh, conceivably too harsh to support any known life forms (Klein, 1978; Klein, 1979; Klein *et al.*, 1992). The possibility of life below the surface is another matter. Recent evidence for global distributions of subsurface ice (Boynton *et al.*, 2002; Feldman *et al.*, 2002; Mitrofanov *et al.*, 2002) and indications that liquids flowed on the surface in recent times (Christensen *et al.*, 2004; Herkenhoff *et al.*, 2004; Klingelhofer *et al.*, 2004; Rieder *et al.*, 2004; Squyres *et al.*, 2004a; Squyres *et al.*, 2004b) mean that the possibility of extremophiles existing below the surface cannot be entirely excluded. We have been exploring the possibility that methanogens, microorganisms in the domain Archaea, could survive on Mars in subsurface environments. In general, methanogens metabolize H_2, using CO_2 as carbon source, without a need for organic carbon (even though they are ubiquitous in environments associated with the decomposition of organic matter [Sowers, 1995]). The recent discovery of methane in the atmosphere of Mars is highly significant in connection with the possible presence of these organisms on Mars (Formisano *et al.*, 2004; Malik, 2004; Krasnopolsky *et al.*, 2004; Krasnopolsky, 2005). One factor that must be considered is whether there are reasonable soil/rock substrates – reasonable in terms of what we know about the surface of Mars – that would support methanogenesis on the planet. We have shown that three species of methanogens produce methane when incubated on the Mars soil simulant, JSC Mars-1 (Kral *et al*, 2004), a crushed volcanic tephra from Hawaii. However, as in all simulation experiments, there is an issue as to whether this substrate is relevant to Mars, and if so, how widespread such material might be. We have therefore proposed to examine the suitability of a few other inorganic substrates, in order to explore whether the identity of the substrate is important and, if so, gain some insight into the type of materials likely to support methanogens on Mars. Since organic materials have not been detected in measurable quantities at the surface of Mars (Biemann *et al.*, 1977; Biemann, 1979), there is no reason to believe that they would exist in the subsurface. Here we report attempts to measure methanogenesis on a number of substrates, sand and gravel which are typical streambed deposits, basalt which might be expected in the vicinity of the huge volcanic regions on Mars, and a Mars soil simulant, JSC Mars-1.

2. Materials and Methods

2.1. Cultures and Growth Media

The methanogens were obtained from David Boone, Portland State University, OR. Each methanogenic strain was grown in a medium that supported growth (MS medium [Boone *et al.*, 1989] for *Methanosarcina barkeri* [OCM 38] and *Methanobacterium formicicum* [OCM 55]; MM medium [Xun *et al.*, 1988] for *Methanothermobacter wolfeii* [OCM 36; formerly *Methanobacterium wolfei*]. Growth media were prepared under 90% carbon dioxide and 10% hydrogen in a Coy anaerobic environmental chamber. Media made in the described atmosphere were saturated with carbon dioxide. This resulted in a pH of 6.6, which allows for good growth of the methanogens being studied. The 10% hydrogen is only required for the oxygen-removing palladium catalysts to function properly. The catalysts facilitate the reaction of the residual molecular oxygen with the molecular hydrogen to form water. (Methanogens are strict anaerobes and will not grow or produce methane in the presence of molecular oxygen [Zinder, 1993].) The anaerobically-prepared media were added to anaerobic culture tubes and sealed as described by Boone *et al.* (1989). At least one hour prior to inoculation, a sterile 2.5% sodium sulfide solution was added to each tube (0.15 mL per 10 mL medium) to eliminate any residual molecular oxygen (Boone *et al.*, 1989).

2.2. Preparation of the Soil and Rock Samples

Four different soil and rock samples were used in these experiments. They included a Mars soil simulant [JSC Mars-1], sand, gravel, and basalt. A fifth substrate, glass beads (Pyrex brand), served as a negative control. JSC Mars-1 is a fraction from altered volcanic ash from a Hawaiian cinder cone that approximates the reflectance spectrum, composition, grain size, density, and magnetic properties for the oxidized soil of Mars (Allen *et al.*, 1998). The sand (silicate [SiO_2] mineral [Leet, 1982]) and the gravel (natural mixture of various rock fragments resulting from erosion [Leet, 1982]) were from Quickrete Companies, Inc., Atlanta GA. The basalt (pulverized) was obtained from Todd Stevens, Portland State University. It is composed of plagioclase feldspars and ferromagnesium silicates (Leet, 1982).

The sand and gravel were washed in deionized water to remove any loose debris and fine-grained dust. Because of the small grain size of the JSC Mars-1 and the basalt, washing was not feasible. Following the washing procedure, the sand, gravel and basalt were analyzed for total organic compounds at the Arkansas Water Resource Center, Water Quality Laboratory, University of Arkansas, Fayetteville, using an Inductively Coupled Argon Plasma Spectrophotometer. (JSC Mars-1 had previously been tested for total organic compounds [Kral *et al.*, 2004]). Five grams of each type of soil or rock were placed into anaerobic culture tubes. All of the tubes containing soil and rock samples were placed into a Coy anaerobic chamber, allowed to acclimate to the hydrogen:carbon dioxide atmosphere for 24 hours, then stoppered. The sealed tubes were removed from the chamber, crimped, and autoclaved at 121C for 30 minutes at 100 kPa above ambient.

2.3. Methane Production on Different Soil and Rock Substrates

Actively growing cells (approximately 0.1 O.D. at 675nm) of all three methanogens were centrifuged at 4200 rpm for 45 min, and then washed with sterile carbonate buffer (the same buffer used to make methanogenic growth medium). At least an hour earlier, sterile sodium sulfide solution was added to the buffer to eliminate residual oxygen. This washing procedure was repeated three times. Following the final washing, the cell pellets were suspended in the same buffer. Four milliliters of each organism were added to individual tubes of each of the soil and rock substrates. (We have shown previously [Kral *et al.*, 2004] that 4 mL is ideal for growth on 5 g of JSC Mars-1.) All tubes were pressurized with 180 kPa (above ambient) of 75:25 H_2:CO_2. The tubes were incubated at temperatures within the growth range for the respective methanogens (37 ^oC for *M. barkeri* and *M. formicicum*; 55 ^oC for *M. wolfeii*). Methane was measured as a percentage of the total headspace gas. Headspace gas samples (1 mL) were removed with a plastic syringe at time intervals of 0, 72, 144, 312 and 480 hours after inoculation and manually injected into the port of a Hewlett Packard model 5890 gas chromatograph with a thermal conductivity detector at an oven temperature of 40 ^oC using argon as the carrier gas.

2.4. Transfer Experiments

Washed cell cultures mixed with soil and rock samples were prepared as described in the previous section. Gas chromatographic measurements were recorded at 0, 72, and 144 hours. At 168 hours after inoculation, 4 mL of sterile methanogenic medium buffer (with sodium sulfide) were added to each culture tube in the anaerobic chamber using a sterile syringe. The tubes were inverted a few times and then allowed to settle for five minutes. The tubes were then carefully inverted, one at a time, allowing the liquid fraction to collect at the stopper end of each tube. Using a clean syringe, 4 mL were removed and transferred to a fresh, sterile tube of the exact same soil or rock substrate. All tubes were pressurized with 75:25 H_2:CO_2 at 180 kPa and incubated at their respective temperatures. Gas chromatographic measurements were again recorded at 0, 72 and 144 hours. This transfer procedure was repeated two more times. All transfer experiments were performed in triplicate.

3. 2.5. pH Measurements

One gram of each type of substrate was added to an anaerobic tube. Each tube was placed into the Coy anaerobic chamber and allowed to acclimate for three hours. They were then sealed with rubber stoppers, crimped and autoclaved. They were placed back into the anaerobic chamber, crimps and rubber stoppers removed, and then 1 mL of sterile buffer solution (with sodium sulfide) was added to each. The contents of each tube were agitated at 5-minute interval for 25 minutes. The tubes were then sealed with rubber stoppers, removed from the chamber, and the contents of each tube were immediately subjected to pH measurement.

4. Results

In the first set of experiments, where medium-grown washed cells were added directly to soil and rock substrates, increasing methane concentrations were observed in all tubes including the glass beads (data not shown). Methane production in the negative control led to the transfer experiments where, theoretically, any residual nutrients from the original growth medium would eventually be diluted out and endogenous energy reserves from the medium-grown cultures would be exhausted. All three organisms showed no methane production following two transfers on glass beads (*M. barkeri* showed no methane production after the first transfer). Figures 1, 2 and 3 show methane production for *M. barkeri, M. formicicum* and *M. wolfeii*, respectively, following the third transfer on various soil and rock substrates (the third transfer increases our confidence that residual nutrient carryover and endogenous energy stores are insignificance). All three methanogens produced methane on sand, gravel, and JSC Mars-1. *M. barkeri* (Figure 1) and *M. wolfeii* (Figure 3) showed no significant differences on the three substrates as indicated by overlapping error bars. *M. formicicum* (Figure 2) produced significantly greater methane on sand than on the other two substrates. All three organisms failed to produce methane on basalt in these experiments.

Analysis of the sand, gravel and basalt revealed no measurable organic compounds, thus the methanogenesis observed on sand and gravel was not due to metabolism of organic material. Even though JSC Mars-1 contains traces of organic matter (79 mg/kg), we have demonstrated (Kral *et al.*, 2004) that these organic compounds cannot support methane production in the absence of hydrogen, carbon dioxide or both. Results from the pH studies showed a rather

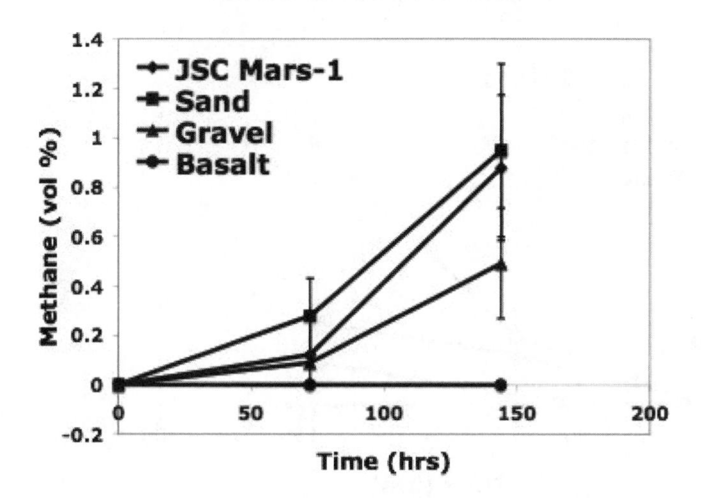

Figure 1. Methane production by Methanosarcina barkeri on different soil and rock substrates following a third transfer on the same substrates. Each point represents the average of three measurements. Error bars represent +/- one standard deviation.

narrow range of pH values from a low of 6.6 for basalt to a high of 6.8 for sand and gravel.

5. Discussion

Results from the first set of experiments, where washed cells were added directly to soil and rock substrates, showed methane production on all substrates, including the glass beads. Possible explanations for methane production on the negative control include endogenous energy reserves and residual nutrients following the washing procedure. With respect to endogenous energy reserves, most organisms store reserve material when in excess for times when there are limited or exhausted energy sources. Reserve polymers of glycogen and polyphosphate have been detected in methanogens (Zinder, 1993). Research on *Methanosarcinaceae* has shown that they contain storage granules, as described by Sprott and Beveridge (1993) that are polyphosphate-like electron dense particles. Using X-ray microanalysis, these particles were shown to contain stored macronutrients of P, Ca, Fe and sometimes Mg, S and Cl. In one species, 14% of the cell's weight was made up of the stored nutrients in these granules. With respect to the washing procedure, it may have been incorrectly assumed in the first set of experiments that washing the cells three times with a buffer would have been sufficient to dilute the nutrient pool to insignificant levels. Nonetheless, the follow-up transfer experiments were very successful in demonstrating which substrates supported methane production for each of the three methanogens.

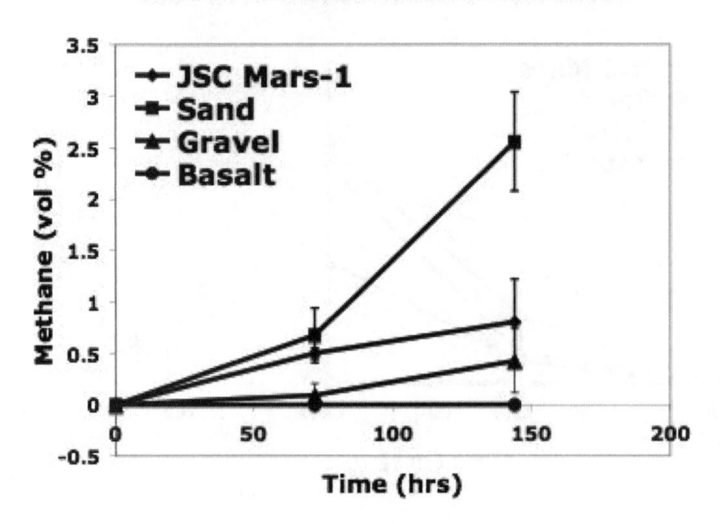

Figure 2. Methane production by Methanobacterium formicicum on different soil and rock substrates following a third transfer on the same substrates. Each point represents the average of three measurements. Error bars represent +/- one standard deviation.

In all cases, the second transfer resulted in lack of methane production on glass beads. We chose to report the results for the third transfer, two removed from the negative controls that were positive for methane. Since we were only interested in knowing which substrates supported methane production, we only have three measurements for each organism on each substrate for 144 hours of incubation. However, with three replicates for each experiment, we are confident in concluding which substrates supported methane production. Results from pH measurements would seem to eliminate pH as a factor contributing to the observed differences. The narrow range of pH values (6.6 – 6.8) falls within the ranges for optimal growth of these organisms (6.5-7.5 for *M. barkeri*, 6.6-7.8 for *M. formicicum* [Sowers, 1995], 5.8-7.7 for *M. wolfeii* [data from our laboratory]).

It should be noted that methane production is what was measured and reported here. Measurement of cell growth by gas chromatographic analysis of methane production is unique to the methanogens and is the most rapid and commonly used technique to measure growth (Sowers, 1995). It is certainly conceivable that methane production could be uncoupled from actual growth (increase in cell mass) under certain conditions. However, we have demonstrated (Kral *et al.*, 2004) that increasing methane production on JSC Mars-1 is correlated with increase in total organic carbon. In light of the insignificant differences in methane production by the three methanogens on most of the substrates (the only case of significantly greater methane production was by *M. formicicum* on sand [Figure 2]) reported here, we have no reason to believe that

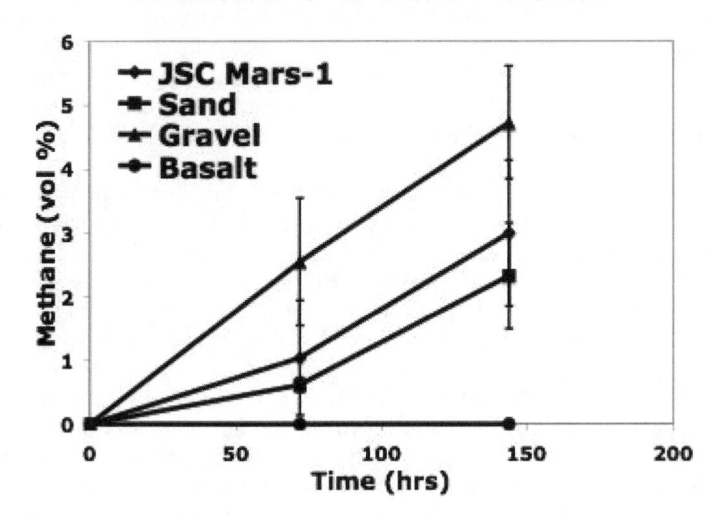

Figure 3. Methane production by Methanothermobacter wolfeii on different soil and rock substrates following a third transfer on the same substrates. Each point represents the average of three measurements. Error bars represent +/- one standard deviation.

the methanogens are not growing in these experiments. The observation that all three methanogens produced methane on sand, gravel and JSC Mars-1 means that they must be obtaining their nutrient requirements (other than H_2 and CO_2) from these substrates. (There is typically about 1% atmospheric N_2 in the headspace gas [from the anaerobic chamber where the buffer is stored]. Some methanogens including *M. barkeri* [Murray and Zinder, 1984] and *M. formicicum* [Magingo and Stumm, 1991] have been shown to fix nitrogen.) In addition to nitrogen, methanogens growing on H_2 and CO_2 typically require phosphate, sulfur, magnesium, iron, cobalt, nickel and molybdenum (Scherer *et al.*, 1982; Schonheit *et al.*, 1979) and/or tungsten (Winter *et al.*, 1984; van Bruggen *et al.*, 1986; Widdel, 1986). The sodium sulfide added to scavenge residual oxygen would supply sulfur. Sodium hydroxide pellets (J.T. Baker, 98% min) are used to make the buffer, and like any chemical reagent, contain trace impurities. Analyses show the presence of nitrogen (<3 ppm), phosphate (<1 ppm), sulfate (<5 ppm), iron (<2 ppm), and nickel (<5 ppm). Other possible sources of trace contaminants are the sodium sulfide solution (sodium sulfide, nonahydride, Sigma, trace analysis not listed), the carbon dioxide gas that is bubbled through the buffer, the deionized water, and even the stainless steel needle on the syringe used for inoculum transfer. The bottom line is that there was no measurable methane production by any of the three methanogens on glass beads. Additionally, there was no methane production on basalt which contains iron and magnesium as part of its chemical makeup (Leet, 1982). The combination of contaminant nutrients along with the H_2, CO_2 and N_2 are apparently not sufficient to support methanogenesis alone. We can also rule out some inhibitory factor in the basalt that might be preventing methanogenesis because in the first set of experiments, where washed cells were added directly to the substrates, there was methane production by all three methanogens, comparable to that seen on the other substrates (data not shown). The "additional" required nutrients must be present in sand, gravel and JSC Mars-1, but not in basalt.

Methanogens are commonly associated with environments with high carbon loads (sewage digestors, peat bogs, cattle pastures, bovine rumens), and are ubiquitous in environments associated with decomposition of organic matter (Sowers, 1995). In these rich organic environments, methanogens are the terminal members of a three-member consortium of microorganisms. The first two members, using interspecies hydrogen transfer, ultimately produce hydrogen and acetate as the major intermediates, which the methanogens then consume (Sowers, 1995; Bryant *et al.*, 1967; Bryant, 1979). Thus, the observation that all three methanogens did produce methane on inorganic substrates is not necessarily unexpected. Today, one of the most widely favored hypotheses concerning the origin of life on Earth is that the first microorganisms were chemoautotrophs (Bennett *et al*, 2003), and that methanogens may have been one of the first of these to evolve (Ehrlich, 1990). Hydrogen and iron compounds were abundant on the early Earth, so energy sources were available (Bennett *et al.*, 2003). They probably would have existed at a time when our planet was more Mars-like, before atmospheric oxygen concentrations began to rise (between 2 and 3 billion years ago) due to oxygenic photosynthesis, and before the deposition of the tremendous quantities of organic matter that we observe today (Bennett *et al.*, 2003).

6. Summary

In summary, these three methanogenic species metabolize (methanogenesis) on sand, gravel and JSC Mars-1, substrates that may have similar counterparts on Mars. The observation that the methanogens tested here did not metabolize on basalt is interesting, especially since Mars is known to contain a considerable quantity of this mineral (McSween, 1994). However, this may be a moot point since methanogens are ubiquitous on Earth (Sowers, 1995), another basalt-rich planet (Leet, 1982). While these substrates do not represent the full range of materials likely to be present on the surface of Mars, the present results suggest that while some surface materials on Mars may not support this type of organism, others might.

Acknowledgments. This work was supported by a grant from the NASA-Ames University Consortium Office, and funding from the Arkansas Center for Space and Planetary Sciences, University of Arkansas, Fayetteville. Special thanks go to Derek Sears for reviewing this manuscript.

References

Allen, C.C., Jager, K.M., Morris, R.V., Lindstrom, D.J., Lindstrom, M.M., Lockwood, J.P. 1998, EOS, 79, 405.

Bennett, J., Shostak, S., Jakosky, B. 2003, in Life in the Universe. Addison Wesley, San Francisco, p. 127.

Biemann, K. 1979, J. Mol. Evol. 14, 65.

Biemann, K., Oro, J., Toulmin, P., III, Orgel, L.E., Nier, A.O., Anderson, D.M., Simmonds, P.G., Flory, D., Diaz, A.V., Rushneck, D.R., Biller, J.E., LaFleur, A.L. 1977, J. Geophys. Res. 82, 4641.

Boone, D.R., Johnson, R.L., Liu, Y. 1989, Appl. Environ. Microbio. 55, 1735.

Boynton, W.V., Feldman, W.C., Squyres, S.W., Prettyman, T., Bruckner, J., Evans, L.G., Reedy, R.C., Starr, R., Arnold, J.R., Drake, D.M., Englert, P.A.J., Metzger, A.E., Mitrofanov, I., Trombka, J.I., d'Uston, C., Wanke, H., Gasnault, O., Hamara, D.K., Janes, D.M., Mancialis, R.L., Maurice, S., Mikheeva, I., Taylor, G.J., Tokar, R., Shinohara, C. 2002, Science Online, *http://www.sciencemag.org/cgi/content/abstract/1073722v1* *http://www.sciencemag.org/cgi/content/abstract/1073722v1.*

Bryant, M.P. 1979, J. Animal Sci. 48, 193.

Bryant, M.P., Wolin, E.A., Wolin, M.J., Wolfe, R.S. 1967, Arch. Microbiol. 59, 20.

Christensen, P.R, Wyatt, M.B., Glotch, T.D., Rogers, A.D., Anwar, S., Arvidson, R.E., Bandfield, J.L., Blaney, D.L., Budney, C., Calvin, W.M., Fallacaro, A., Fergason, R.L., Gorelick, N., Graft, T.G., Hamilton, V.E., Hayes, A.G., Johnson, J.R., Knudson, A.T., McSween, H.Y., Mehall, G.L., Mehall, L.K., Moersch, J.E., Morris, R.V., Smith, M.D., Squyres, S.W., Ruff, S.W., Wolff, M.J. 2004, Science 306, 1733.

Ehrlich, H.L. 1990, Geomicrobiology (2nd ed.) Marcel Dekker, Inc. New York.

Feldman, W.C., Boynton, W.V., Tokar, R.L., Prettyman, T.H., Gasnault, O., Squyres, S.W., Elphic, R.C., Lawrence, D.J., Lawson, S.L., Maurice, S., McKinney, G.W., Moore, K.R., Reedy, R.C. 2002, Science Online, http://www.sciencemag.org/cgi/content/abstract/1073541v1.

Formisano, V., Atreya, S., Encrenaz, T., Ignatiev, N., and Giuranna, M. 2004, *Science* 306, 1758.

Herkenhoff, K.E., Squyres, S.W., Arvidson, R., Bass, D.S., Bell, J.F., Bertelsen, P., Ehlmann, B.L., Farrand, W., Gaddis, L., Greeley, R., Grotzinger, J., Hayes, A.G., Hviid, S.F., Johnson, J.R., Jolliff, B., Kinch, K.M., Knoll, A.H., Madsen, M.B., Maki, J.N., McLennan, S.M., McSween, H.Y., Ming, D.W., Rice, J.W., Richter, L., Sims, M., Smith, P.H., Soderblom, L.A., Spanovich, N., Sullivan, R., Thompson, S., Wdowiak, T., Weitz, C., Whelley, P. 2004, Science 306, 1727.

Klein, H.P. 1978, Icarus 34, 666.

Klein, H.P. 1979, Rev. Geophys. Space Phys. 17, 1655.

Klein, H.P., Horowitz, N.H. and Biemann, K. 1992, in Mars, ed. H.H. Kieffer, B.M. Jakosky, C.W. Snyder, M.S. Matthews, University of Arizona Press, Tucson, 1221.

Klingelhofer, G., Morris, R.V., Bernhardt, B., Schroder, C., Rodonov, D.S., de Souza, P.A., Yen, A., Gellert, R., Evlanov, E.N., Zubkov, B., Foh, J., Bonnes, U., Kankeleit, E., Gutlich, P., Ming, D.W., Renz, F., Wdowiak, T., Squyres, S.W., Arvidson, R.E. 2004, Science 306, 1740.

Kral, T.A., Bekkum, C.R., McKay, C.P. 2004, Origins Life Evol. Biosphere. 34, 615.

Krasnopolsky, V.A., Maillard, J.P., Owen, T.C. 2004, Icarus. 172, 537.

Krasnopolsky, V.A. 2005, Icarus. 180, 359.

Leet, L.D. 1982, Physical Geology, 6th Ed. Prentice Hall, Englewood Cliffs, N.J.

Magingo, F.S.S., Stumm, C.K. 1991, FEMS Microbiol. Lett. 81, 273.

Malik, T. 2004, CNN.com.
http://www.cnn.com/2004/TECH/space/03/30/mars.methane/index.html
http://www.cnn.com/2004/TECH/space/03/30/mars.methane/index.html

McSween H. Y., Jr. 1994, Meteoritics 29 , 757.

Mitrofanov, I., Anfimov, D., Kozyrev, A., Litvak, M., Sanin, A., Tret'yakov, V., Krylov, A., Shvetsov, V., Boynton, W., Shinohara, C., Hamara, D., Saunders, R.S. 2002, Mars Odyssey. Science Online,
http://www.sciencemag.org/cgi/content/abstract/107361v1.

Murray, P.A., Zinder, S.H. 1984, Nature 312, 284.

Rieder, R., Geller, R., Anderson, R.C., Bruckner, J., Clark, B.C., Dreibus, G., Economou, T., Klingelhofer, G., Lugmair, G.W., Ming, D.W., Squyres, S.W., d'Uston, C., Wanke, H., Yen, A., Zipfel, J. 2004, Science 306, 1746.

Scherer, P., Lippert, H., Wolff, G. 1983, Biol. Trace Elem. Res. 5, 149.

Sowers, K.R. 1995, Methanogenic Archaea: An Overview. in Archaea: A Laboratory Manual, Methanogens, ed. K.R. Sowers and H.J. Schreier, Cold Spring Harbor Laboratory Press, Plainview, NY, 3, 93.

Sprott, G.D., Beveridge, T.J. 1993, in Methanogens, ed J.G. Ferry, Chapman and Hall, NY. 81.

Schonheit, P., Moll, J., Thauer, R.K. 1979, Arch. Microbiol. 123, 105.

Squyres, S.W., Arvidson, R.E., Bell, J.F., Bruckner, J., Cabrol, N.A., Calvin, W., Carr, M.H., Christensen, P.R., Clark, B.C., Crumpler, L., Des Marais, D.J., d'Uston, C., Economou, T., Farmer, J., Farrand, W., Folkner, W., Golombek, M., Corevan, S., Grant, J.A., Greeley, R., Grotzinger, J., Haskin, L., Herkenhoff, K.E., Hviid, S., Johnson, J., Klingelhofer, G., Knoll, A.H., Landis, G., Lemmon, M., Li, R., Madsen, M.B., Malin, M.B., McLennan, S.M., McSween, H.Y., Ming, D.W., Moersch, J., Morris, R.V., Parker, T., Rice, J.W., Richter, L., Rieder, R., Sims. M., Smith, M., Smith, P., Soderblom, L.A., Sullivan, R., Wanke, H., Wdowiak, T., Wolff, M., Yen, A. 2004a, Science 306, 1698.

Squyres, S.W., Grotzinger, J.P., Arvidson, R.E., Bell, J.F., Calvin, W., Christensen, P.R., Clark, B.C., Crisp, J.A., Farrand, W.H., Herkenhoff, K.E., Johnson, J.R., Klingelhofer, G., Knoll, A.H., McLennan, S.M., McSween, H.Y., Morris, R.V., Rice, J.W., Rieder, R., Soderblom, L.A. 2004b, Science 306, 1709.

Van Bruggen, J.J.A., Zwart, K.B., Hermans, J.G.F., van Hove, E.M., Stumm, C.K., Vogels, G.D. 1986, Arch. Microbiol. 144, 367.

Widdel, F. 1986, Appl. Environ. Microbiol. 51, 1056.

Winter, J., Lerp, C., Zabel, H-P., Wildenauer, F.X., Konig, H., Schindler, F. 1984, System. Appl. Microbiol. 5, 457.

Xun, L., Boone, D.R., Mah, R.A. 1988, Appl. Environ. Microbio. 54, 2064.

Zinder, S.H. 1993, in Methanogens, ed J.G. Ferry, Chapman and Hall, NY. 128.

Mike Mumma

Geronimo Villanueva

Bioastronomy 2007: Molecules, Microbes, and Extraterrestrial Life
ASP Conference Series, Vol. 420, 2009
K. J. Meech, J. V. Keane, M. J. Mumma, J. L. Siefert, and D. J. Werthimer, eds.

Radiation Environments on Mars and Their Implications for Terrestrial Planetary Habitability

S. I. Schneider and J. F. Kasting

The Pennsylvania State University, Department of Geosciences, University Park. 16802 PA. Email: sschneid@geosc.psu.edu

Abstract. The understanding of the surface and subsurface radiation environments of a terrestrial planet such as Mars is crucial to its potential past and/or present habitability. Despite this, the subject of high energy radiation is rarely contemplated within the field of Astrobiology as an essential factor determining the realistic parameter space for the development and preservation of life.

Furthermore, not much is known of the radiation environment on the surface of Mars due to the fact that no real data exist on this contribution. There are no direct measurements available as no surface landers/probes have ever carried nuclear radiation detection equipment to characterize the interactions arising from cosmic ray bombardment, solar particle events and the atmosphere striking the planetary surface. The first mission set to accomplish this task, the Mars Science Laboratory, is not scheduled to launch until 2011.

Presented here are some of such simulations performed with the HZETRN NASA code offering radiation depth profiles as well as a characterization of the diverse radiation environments. A discussion of the implications that these projected doses would have on terrestrial planetary habitability on Mars is presented as well as its implications for the habitability of terrestrial planets elsewhere. This work does not provide an estimate of the UV radiation fields on the Martian surface instead it focuses on the high energy radiation fields as composed by galactic cosmic rays (GCRs)

1. Introduction: Planetary Radiation Environments

Outer space is a high energy radiation environment where a constant background composed of GCRs pervades all space in an isotropic fashion. In addition to this, and depending on the proximity of a planet to its parent star, interplanetary space becomes sporadically filled by Solar Particle Events (SPE's) and Coronal Mass Ejections (CME's). The Earth is a particular case in that it is endowed with a powerful global magnetic field which deflects most of the incoming high energy radiation.

On the Earth, an average radioactive dose received by the public in the US, is approximately of ~0.36 rem/yr (including all man made radioactive sources such as radon concentrations in construction, atomic bomb fall-out, medical x-rays or nuclear waste from power reactors medical; Wilson et al. 1995) . Only 8% of this dose is coming from cosmic rays or approximately of 28.8mrem/yr[1].

[1]NRC Title 10 of the Code of Federal Regulations Part 2

Both Mars or Earth's Moon lack a global magnetic field and a well developed atmosphere. The moon has currently no magnetic field and no atmosphere, thus it is subjected to the full impact of free interplanetary space weather makes for an even worse case than Mars from the radiation standpoint.

2. The Martian Surface and Subsurface Radiation Environments

Mars present atmosphere has a column density at present of only 15-20 or an average surface pressure of 6 mbars ranging to 10mbrs. This thin atmosphere attenuates approximately half of the incoming high energy cosmic rays according to our simulations (Schneider 2007). The Martian surface is then subject to intensive charged particle bombardment which make of the Martian surface a potentially highly hazardous environment for earth-type biota, as well as for future human visitors. The evolution of the Martian surface radiation environment is directly coupled to the evolution of its radiation environments. This work simulates past radiation environments by transporting the incident radiation fields through the different stages of atmospheric development back into the Noachian, where we assume a thicker atmosphere capable of sustaining surface liquid water. We achieve this by increasing the atmospheric column density or, equivalently, by increasing surface pressure and transporting the radiation fields through it.

3. Results

Here we present some simulations performed with the NASA NZETRN code which include present surface and subsurface high energy radiation environments for Mars (Fig.1), the evolution of the Martian surface high energy radi-

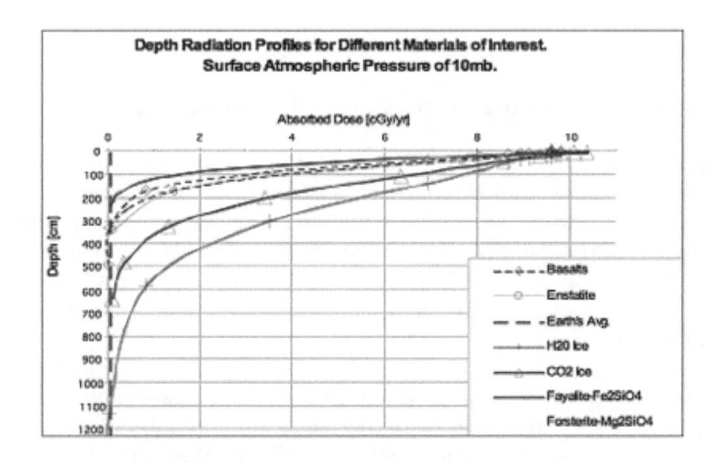

Figure 1. Simulation of the present Martian surface and subsurface radiation environments for a surface atmospheric pressure of 10mbars (Schneider 2007).

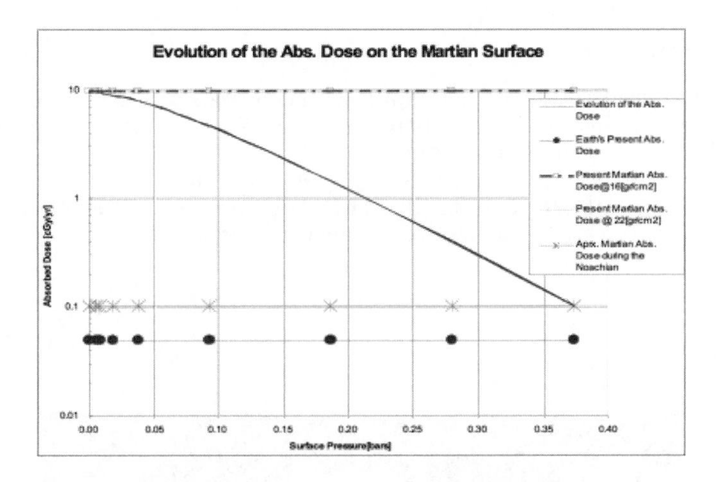

Figure 2. Evolution of the Surface Radiation Environment on Mars as a function of surface pressure (as to simulate the Geological timescales) in units of absorbed dose.

ation environments going back to the Noachian époque (Fig.2) as a function of atmospheric surface pressure (Fig.2). More detailed simulations can be found in (Schneider 2007).

4. Discussion and Conclusions

According to our simulations, present Mars surface radiation levels oscillate between 8 and 10 cGy/yr. This is an order of magnitude higher than the Earth's average surface absorption doses. However, communities of microorganisms are known to thrive at much higher doses on Earth. Regarding the subsurface radiation our simulations show that to match comparable subsurface Earth levels one must be below a depth of only 2 meters of basalt, under which water ice reservoirs may be found on Mars. When considering water ice as a transport media, the radiation fields reach comparable Earth dose levels at a depth of only 12 meters. This is highly relevant for potential life within the permafrost at locations above 60 deg latitude and at the poles on Mars, as well as other terrestrial planetary environments. When simulating an ancient Mars, we find that Mars should have had similar surface radiation environments as Earth even during the Hesperian epoch. The Noachian would have offered even better habitability prospects, in terms of radiation and especially when assuming the existence of surface liquid water. Life as we know it not only could have originated on Mars, but it is possible that it had sufficient time to evolve adapting to the changing environment perhaps by establishing shallow (~2m) subsurface habitats (Schneider 2007).

References

Schneider, S.I. 2007. 'An Astrobiological Evolution of Mars with a Focus on its Radiation Environments and the Implications for Planetary Habitability'. (Master Thesis). The Pennsylvania State University.

Wilson, J.W.; Cucinotta,F.A.; Badhwar,G.D.; Silberberg, R.; Tzao, C. H.; Townsend, L. W.; Tripathi, R. K. 1995. 'HZETRN: Description of a FREE-Space Ion and Nucleon Transport and Shielding Computer Program'. NASA TP-3495.

David DesMarais

Session VI

Prebiotic Chemistry

Antonio Lazcano

Reggie Hudson

Bioastronomy 2007: Molecules, Microbes, and Extraterrestrial Life
ASP Conference Series, Vol. 420, 2009
K. J. Meech, J. V. Keane, M. J. Mumma, J. L. Siefert, and D. J. Werthimer, eds.

Enigmatic Isovaline: Investigating the Stability, Racemization, and Formation of a Non-Biological Meteoritic Amino Acid

R. L. Hudson,[1,2] A. S. Lewis,[1] M. H. Moore,[2] J. P. Dworkin,[2] and M. P. Martin[2]

[1]*Department of Chemistry, Eckerd College, St. Petersburg, FL 33711 USA*
[2]*NASA Goddard Space Flight Center, Greenbelt, MD 20771 USA*

Abstract. Among the Murchison meteoritic amino acids, isovaline stands out as being both non-biological (non-protein) and having a relatively high abundance. While approximately equal amounts of D- and L-isovaline have been reported in Murchison and other CM meteorites, the molecule's structure appears to prohibit its racemization in aqueous solutions. We recently have investigated the low-temperature solid-phase chemistry of both isovaline and valine with an eye toward each molecule's formation, stability, and possible interconversions of D and L enantiomers. Ion-irradiated isovaline- and valine-containing ices were examined by IR spectroscopy and highly-sensitive liquid chromatography/time-of-flight mass spectral methods to assess both amino acid destruction and racemization. Samples were studied in the presence and in the absence of water-ice, and the destruction of both isovaline and valine was measured as a function of radiation dose. In addition, we have undertaken experiments to synthesize isovaline, valine, and their amino acid isomers by solid-phase radiation-chemical pathways other than the oft-invoked Strecker process.

1. Introduction

Meteorites provide the only direct evidence for extraterrestrial amino acids. To date, about 70 amino acids have been reported in Murchison and other objects, with the majority of the acids being non-biological (Pizzarello et al. 2008). Although many of these amino acids are racemic (i.e., equal abundance of D and L structures), reports of enantiomeric excesses are in the literature, with some values being as large as 10% (Cronin & Pizzarello 1997).

Carbonaceous chondrites, such as Murchison, are particularly attractive to astrobiologists as these meteorites harbor a rich variety of organic molecules and are among the most primitive material in the solar system. The amino-acid inventories of CM and CI carbonaceous chondrites have been examined, and distinct differences have been found between them (Ehrenfreund et al. 2001). CM types Murchison and Murray are enriched in, among other amino acids, isovaline and α-aminoisobutyric acid compared to CI types, such as Orgeuil and Ivuna.

Our own studies of Murchison samples have concentrated on isovaline and its isomers, such as valine, all having the formula $C_5H_{11}NO_2$. Figure 1 shows the result of a liquid chromatography/time-of-flight mass spectral measurement

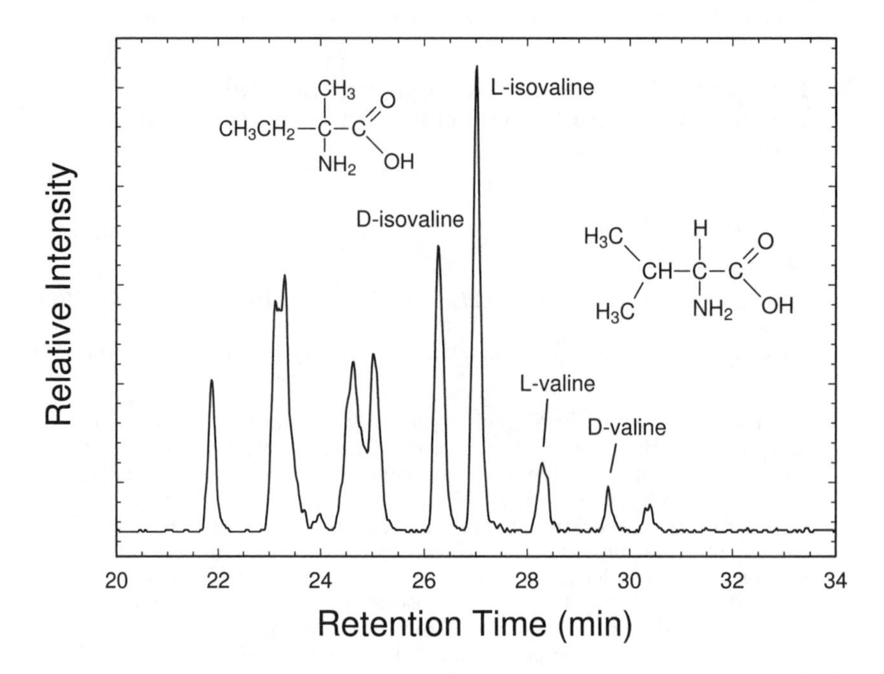

Figure 1. A liquid chromatography/time-of-flight mass spectral measurement of C_5 amino acids from a Murchison sample. Structures of isovaline and valine are shown in the left and right parts of the figure, respectively. About half of the unlabeled peaks also have been identified, but are not covered in this paper.

of a Murchison extract, looking only at isovaline, valine, and their isomers. The enhancement of isovaline over valine is obvious as are substantial apparent excesses of L-isovaline and L-valine over their D counterparts.

We recently have performed experiments on isovaline and valine to investigate their relative thermal and radiolytic stabilities in order to possibly explain their different CM and CI meteoritic abundances. We also have examined possible interconversions between these molecules, as well as racemization of each. Finally, we have undertaken experiments designed to synthesize isovaline and valine from known and suspected meteoritic compounds.

2. Experimental

Our experiments involved mid-infrared spectroscopy coupled with liquid chromatography and time-of-flight mass spectral (LC/ToF-MS) analyses. Relevant details for both techniques are in earlier papers from our laboratories (Hudson et al. 2005; Glavin et al. 2006). In brief, experiments began with the condensation of an ice sample onto a pre-cooled aluminum mirror in a vacuum chamber interfaced to the beamline of a van de Graaff accelerator. Mid-IR spectra of the resulting ices were recorded in situ in the 10 - 300 K range. Samples were irradiated to doses of 1 - 10 eV molecule^{-1} and then brought slowly up to room temperature overnight. Residual materials were removed by washing the metal substrate with water, and then analyzed by LC/ToF-MS without acid hydrolysis. Analytical and procedural blanks were performed at each step of our work. Reagents used were of the highest purity available from Sigma-Aldrich and Cambridge Isotopes.

3. Results

3.1. Thermal Studies

Our initial experiments concerned the thermal stability of isovaline and valine. In separate experiments, each amino acid (AA) was sublimed under vacuum from a small pyrex tube onto the 10 K aluminum mirror. Reflecting a mid-IR beam off the AA-coated metal surface produced an IR spectrum that could be followed as the sample was warmed. Variations in IR bands were seen near 100 K as the amino acid converted from a non-ionic to a zwitterionic form, the main change being loss of a C=O stretching feature near 1717 cm^{-1}. Further warming gave no additional changes and each amino acid was stable under vacuum up to the highest temperature examined. In short, there was no difference in thermal behavior between valine and isovaline at 10 - 300 K. Thus, temperature effects alone cannot explain the enhancement of one of these molecules over the other in carbonaceous chondrites.

3.2. Radiation Stabilities

Our second set of experiments involved the same preparation just described, but each amino acid sample was irradiated at 10 K under vacuum with a 0.8 MeV proton beam, to simulate exposure to cosmic radiation. For both valine and isovaline, a dose of about 1.4 eV $molecule^{-1}$ sufficed to destroy half the sample molecules; separate experiments with AA + H_2O-ice mixtures gave essentially the same result. When the temperature was raised to 80 K, the dose to destroy half the AA molecules dropped slightly, to about 1.0 eV $molecule^{-1}$. Our conclusion is that there is insufficient difference in the radiation stabilities of valine and isovaline to explain how one of these molecules might become enhanced over the other in an extraterrestrial radiation environment.

3.3. Amino Acid Reactions

Warming irradiated samples to room temperature, while under vacuum, left a residual material that could be washed from the aluminum substrate with water. These residues were analyzed with LC/ToF-MS for both isomerization and racemization of starting materials. We found no interconversion of isovaline and valine, excluding yet another possible means for an enhancement of one isomer over the other. However, we did find that in all cases racemization took place. A 10 K irradiation to a dose of 10 eV $molecule^{-1}$ converted pure L-valine into a 1:2 D:L valine mixture (i.e., D-to-L ratio of 0.5). A D-isovaline sample behaved almost exactly the same in an identical experiment. The radiation dose for Murchison fragments has been estimated as only about 530 Mrad, or about 1 eV $molecule^{-1}$ (Bonner et al. 1979). This means that the racemization we observed is probably much greater than background radiation will produce in the meteoritic case.

3.4. Amino Acid Syntheses

The experiments just described show that meteoritic abundance differences between isovaline and valine cannot be easily explained by an instability of either molecule relative to the other, or by a radiation-induced equilibrium that favors either isomer. Since an explanation is not possible on the basis of molecular destruction, we have begun to study formation paths for these two amino acids. We also are interested in simply being able to make isovaline. Although the oft-invoked Strecker reaction might lead to this compound in mixed molecular ices, a close inspection of the literature shows that laboratory detections of isovaline are based on the acid hydrolyses of reaction products (Nuevo et al. 2007). In other words, what has been shown is that amino-acid precursors, sometime called bound amino acids, can be formed and then, after strong-acid hydrolysis, isovaline is detected. What we seek is something more like the meteoritic case, the formation and direct detection of isovaline itself.

We knew from experience that molecules will fragment on irradiation and that the individual pieces will add to unsaturated molecules. For example, irradiation of H_2O produces H and OH radicals which add to acetylene to make vinyl alcohol (Hudson & Moore 2003), and irradiation of CH_3OH forms H and CH_2OH radicals which add to CO to make glycolaldehyde (Hudson et al. 2005). We thus reasoned that C_4-saturated alkylamines ($C_4H_{11}N$) would under fragmentation to form H and $C_4H_{10}N$ radicals which would add to CO_2 to make C_5 amino acids. Figure 2 summarizes the expected reaction paths. Supporting this strategy was the known presence of primary amines in both Murchison (Jungclaus et al. 1976) and Stardust (Sandford et al. 2006) samples.

On irradiation at 10 K of mixtures of $^{13}CO_2$ and either isobutylamine or *sec*-butylamine, an IR band immediately formed at 1717 cm^{-1}, the same feature mentioned earlier for valine and isovaline. This band persisted up to room

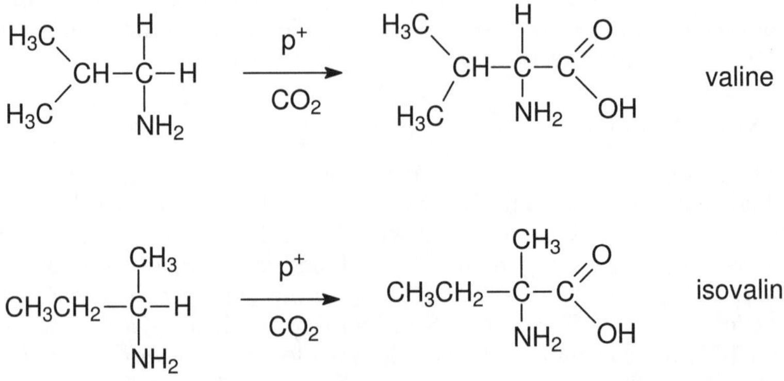

Figure 2. Reactions showing the formation of C_5 amino acids from proton-irradiated CO_2 and C_4 amines. The top reaction begins with isobutylamine, and the bottom reaction begins with *sec*-butylamine.

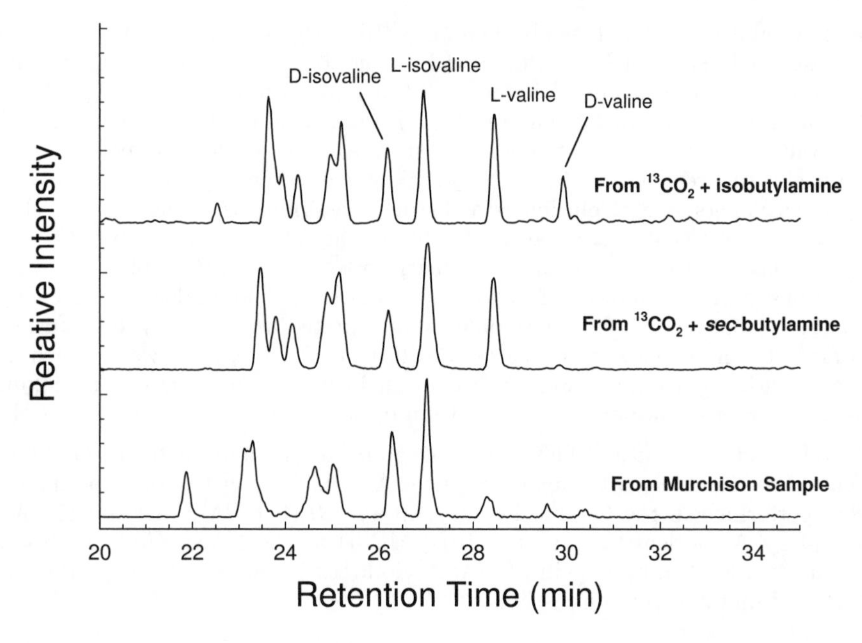

Figure 3. C_5 amino acids made in irradiated $^{13}CO_2$ + amine ices compared to C_5 amino acids in a Murchison sample (bottom trace).

temperature and could be seen regardless of the irradiation temperature. Our conclusion is that amino acids were formed in these low-temperature experiments, but that IR spectroscopy alone cannot identify the specific amino acids produced.

Irradiated $^{13}CO_2$ + amine ices were warmed under vacuum to room temperature, and the residues washed from the cold finger and analyzed by LC/ToF-MS. Figure 3 shows the results of two ice experiments and our Murchison analysis. Comparisons with blanks and standard amino acid samples left no doubt that valine and isovaline were formed in our experiments. The use of $^{13}CO_2$ also was important for confirming that the results were from the irradiation of the samples, and not atmospheric or other contamination.

4. Discussion

Our results show that isovaline's enhancement over valine in Murchison is not due to either a difference in thermal or radiation stabilities between the two molecules or a radiation-induced isomerization that favors one molecule over the other. As for these amino acids' formation, our work (Figure 3) shows unequivocal evidence, for the first time, of isovaline and valine syntheses by energetic processing of a cosmic-ice analog. Future work will examine the relative formation rates of these amino acids.

Our results also speak to the question of enantiomeric excesses of meteoritic amino acids. The amines we used were either achiral (isobutylamine) or a racemic mixture (1:1 D:L sec-butylamine). In both cases, Figure 3 shows that a substantial excess of L-isovaline and L-valine, over their D-counterparts, appears to have been produced by our reactions. An alternative and more-likely interpretation of Figure 3 is that another C_5 amino acid isomer is co-eluting with L-isovaline and L-valine, artificially boosting each molecule's LC peak. There are twelve different acyclic $C_5H_{11}NO_2$ amino acid isomers (twenty-three counting enantiomers), so co-elution is a distinct possibility that must be checked molecule by molecule. We plan to do this in the future, and to study the C_2, C_3, and C_4 radiation products for comparison to meteoritic results.

Finally, the bottom trace of Figure 3 shows a substantial L-isomer excess from a Murchison sample. This could be interpreted as a preference for nature to favor L amino acids, as is the case in terrestrial biology. However, such an interpretation must be viewed with caution in light of our success in making large apparent enantiomeric excesses from optically-inactive starting materials.

Acknowledgments. This work was supported by a grant to the Goddard Center for Astrobiology through the NASA Astrobiology Institute. Experiments were performed in the Cosmic Ice Laboratory (RLH, MHM, AL) and the Astrobiology Analytical Laboratory (JPD, MPM) at the NASA Goddard Space Flight Center. Danny Glavin of NASA Goddard is thanked for experimental work and many helpful discussions.

References

Bonner, W. A., Blair, N. E., & Lemmon, R. M. 1979, J. Amer. Chem. Soc., 101, 1049
Botta, O. & Bada, J. L. 2002, Surveys in Geophysics, 23, 411
Cronin, J. R. & Pizzarello, S. 1997, Science, 275, 951
Ehrenfreund, P., Glavin, D. P., Botta, O., Cooper, G., & Bada, J. L. 2001, Proc. Natl. Acad. Sci., 98, 2138
Glavin, D. P., Dworkin, J. P., Botta, O., Doty III, J. H., Martins, Z., & Bada, J. L. 2006, Meteoritics and Planetary Sci., 41, 889
Hudson, R. L. & Moore, M. H. 2003, Astrophys. J., 586, L107
Hudson, R. L., Moore, M. H., & Cook, A. M. 2005, Adv. in Space Res., 36, 184
Jungclaus, G., Cronin, J. R., Moore, C. B., Yuen, G. U. 1976, Nature, 261, 126
Nuevo, M., Meierhenrich, U. J., D'Hendecourt, L., Muñoz Caro, G. M., Dartois, E., Deboffle, D.,Thiemann, W. H.-P., Bredehöft, J.-H., & Nahon, L. 2007, Adv. Space Res., 397, 400
Pizzarello, S., Huang, Y. & Alexandre, M. R. 2008, PNAS, 105, 3700
Sandford, S. A., Aleon, J., Alexander, C. M. O'D., Araki, T., Bajt, S., Baratta, G. A., Borg, J., Bradley, J. P., et al. 2006, Science, 314, 1720

Bioastronomy 2007: Molecules, Microbes, and Extraterrestrial Life
ASP Conference Series, Vol. 420, 2009
K. J. Meech, J. V. Keane, M. J. Mumma, J. L. Siefert, and D. J. Werthimer, eds.

Oscillating Type of Prebiotic Models as an Intermediate Step to Life

V. N. Kompanichenko

Institute for Complex Analysis, 4 Sholom-Aleyhem Street, Birobidzhan 679016, Russia, email:kompanv@yandex.ru

Dept. of Chemistry and Biochemistry, University of California, 1156 High Street, Santa Cruz CA 95064, USA

Abstract. The principal transition of a prebiotic microsystem into a primary living unit proceeds through three stages. succession of transformations is theoretically substantiated. For 1st stage significant change of the external conditions turns a prebiotic microsystem at the point of bifurcation (the unstable intermediate position between the current and advanced stable states). During 2nd stage its unstable state relatively stabilizes by means of the balanced oscillations around the bifurcation point (the microsystem acquires the paradoxical bistate status). 3rd stage is characterized with the inversion of the ratio free energy/entropy contributions – it changes from the negative to positive, with appearance of the key properties of the biological organization (the ability to concentrate free energy and information, ability for intensified counteraction to external influences, expedient behaviour, continuous self-renovation). On the early Earth such successive transformation of organic matter occurred in changeable conditions of hydrothermal medium.

1. Introduction

The initial prebiotic model - a coacervate drop - possesses some signs of internal and external activity that are peculiar to a living organism as well (for instance, ability to selectively extract substance from the environment). Nevertheless, plenty of experimental attempts to transform coacervates into really living units failed. Similar obstacle appears in the experimental efforts to get a really living unit of much more complex prebiotic models - RNA-World macromolecules, liposomes, proteinoide microspheres, etc. It seems that a cause of the obstacle consists in the fundamental difference between non-living prebiotic microsystems and living microorganisms. Better understanding of this difference could facilitate future experimental research in the origin of life field. The attempt to systematize unique properties of biological systems and emphasize their distinction from any non-living systems (including prebiotic models) was a starting point of the systemic approach to the origin of life (Kompanichenko, 2002, 2004).

2. Properties of Biological Systems

About 230 biological properties distinguished by 73 competent world scientists in. the book Fundamentals of Life (Palyi et al eds., 2002) that is the latest summarizing in this field of fundamental biology. The properties very often reword each other. Their juxtaposition and integration allowed the author to formulate 31 fundamental biological properties. They are subdivided into two groups. According to the elaborated generalization, 19 properties can be considered as **unique** fundamental properties of biological systems, which are not peculiar to any other natural system. The remaining 12 out of 31 are attributed to the **non-unique** fundamental biological properties. The same or similar features characterize some non-biological systems as well, although they are devoid of any biological specificity (Kompanichenko, 2004). During the origin of life process on the ancient Earth the unique properties for the first time appeared, unlike the non-unique ones that transited from the maternal geological medium and acquired the biological specificity.

Nineteen unique properties were used as a basis for the further analysis and integration aimed to better comprehension of the essence of a living organism. According to the made generalization, the unique fundamental biological properties can be narrowed down to the following four integrated properties: a) Ability to concentrate free energy and information (by means of their extraction from the environment); b) Ability for the intensified counteraction to external influences; c) Expedient behavior (or expedient character of interaction with the environment); d) Regular self-renovation at different hierarchical levels. The most important of twelve non-unique fundamental biological properties are the following: thermodynamic and chemical nonequilibrium; integrity through cooperative events; capability for self-organization. These attributes characterize the phenomena that are investigated in the framework of nonequilibrium thermodynamics. The necessity of nonequilibrium conditions is remarkable. This requirement relates the origin of life process with the wide range of nonequilibrium events related with the bifurcate transitions and formation of dissipative structures.

3. 1st Stage of the Origin of Life: A Prebiotic Organic Microsystem at the State of Bifurcate Transition Under Nonequilibrium Conditions

Behaviour of a chemical system under conditions far from equilibrium radically differs from its behaviour under conditions near equilibrium. Nonequilibrium conditions are responsible for transformations that radically change of the structure of a system. The universal scheme of such transition is the following: stable existence of a system - rise of instability through the powerful fluctuations - the highest point of instability (bifurcation, or critical point), radical change of the structure of a system - choice of new way of development - the next period of its stable existence. At the bifurcation point a system undergoes a lot of accidental changes that may influence upon the choice of its further way of development. Finally the system has chosen one of plenty permissible ways, which can be united into two principal trends: a) to complication through self-organization, b) to simplification and degradation, up to full destruction. Back change of the external conditions initiates reverse transition of the system in a direction to the initial state. Investigation of diverse bifurcate transitions is a major aspect of the theory of dissipative structures (the founder I. Prigogine) and synergetics (the founder H. Haken). The transition was described in detail in their basic works with the co-authors (Nicolis, Prigogine, 1977; Prigogine, Stengers, 1984; Haken, 1978).

There exist some principal characteristics of a system being at the state of bifurcate transition. The most essential of these bifurcate, or critical properties are the following: 1) sharp heterogeneity with the intensive counter processes along and against the gradients; 2) continuous fluctuations (waves) and re-arrangement of molecules; 3) integrity through cooperative processes; 4) incessant exchange by matter and energy with the surroundings (open system). Each of four critical properties is distinctly reflected in the vital processes. Although the corresponding characteristics of living cells do not express unique essence of life, they are in its background. Meanwhile, the critical properties are temporal characteristic of a chemical system: they appear since the beginning of the bifurcate transformation and disappear with the completed transition to the advanced stable state. On the basis of these facts the following thesis is formulated: **Life on Earth arose of the bifurcate state** that inhered to still indeterminate kind of organic microsystems. This threshold of the origin of life process can be designated as its 1st stage.

The 1st stage ends with the paradoxical situation: the bifurcate state should be considered as a starting-point of living systems, but the bifurcation point is in principle unstable. A system cannot be at the bifurcation point for a long time, because a lot of happening accidental changes that inevitably turn its development whether into the initial state or the advanced one. But nature on Earth apparently found the way to prolong the bifurcate state of initial prebiotic microsystems about four billion years ago. The permissible opportunities for this way are investigated in the next chapter.

4. 2nd Stage of the Origin of Life: Oscillating Prebiotic Microsystem at the Balanced Bifurcate State

The period of a bifurcate transition is characterized by appearance of two opposite forces in a system. They strive to turn development of the system into the opposed states (initial and advanced). Action of these forces dualizes of the structure of a system, maintains the internal tension, and keeps the considered bifurcate properties. The bifurcation point designates peak of the bifurcate period, and it is unstable in principle. It seems, only permissible way for a system to be inside the bifurcate area for a ling time consists in balanced oscillations around the bifurcation point. The regular advanced and reverse transitions over the critical point really might occur under oscillating conditions in the surroundings, because a system at the critical state is open and very sensitive to the external changes. On this basis the following thesis can be formulated as a hypothesis: the critical properties in a prebiotic microsystem can be kept for a long time **through regular oscillations around the bifurcation point**.

The object for further theoretical investigation is a prebiotic microsystem that oscillates around the bifurcation point and, therefore, keeps the critical properties. Because behavior of various organic compounds and prebiotic models under oscillating conditions is still not explored, it is accepted conditional chemical composition of the microsystem: mixture of precursors of lipids, nucleic and amino acids, and other compounds with indefinite proportions. Relative stability of the prebiotic microsystem is supported by the balanced oscillations between two polar states. The system simultaneously strives to get both stable states. The opposite forces are approximately equal; as a result, the advanced force blocks its return into the initial state and the reverse force opposes its transition into one of the advanced states. This rare and unusual type of natural systems was called a **bistate system**, or shortly **bisystem** (Kompanichenko, 2004). It may appear in oscillating conditions in the maternal medium, when the regular inversions of the responsible parameters maintain the balance between the competing nonequilibrium states. Once forming, a bistate system has a chance to resist the destruction through the adequate respond to external actions. This opportunity follows of the Le Chatelier principle: any action to a chemical system due to change of conditions in the outside world initiates its counteraction.

Basing on the above, the following set of most essential properties inherent to a bistate system is theoretically argued. Acquisition of these properties achieves by a bistate prebiotic microsystems for 2nd stage of the origin of life process. *1. Latent biforked structure* consisting of two autonomous co-structures

divided by the plane of smooth sub-symmetry. The plane is maintained by relative equilibrium between the opposite forces.

2. Dichotomy at the end of normal cycle of the existence. This property follows of the paradoxical way of organization of a bisystem, i.e. simultaneous repulsion the nonequilibrium co-structures from the central point of instability and integration through maintenance of relative equilibrium between them.

3. *Oscillating character of existence* related with periodic change of the dominant states. Oscillations of all inner processes support stability of a bisystem. Ceasing of the oscillations constrains a bisystem to leave the bifurcate area with the following irreversible transition into one of the stable equilibrium states. Oscillations between the polar forces sustain the internal tension that is a source of incessant dynamic processes and intensive rearrangement of molecules in a bistate system.

4. Display of mutability and heredity maintaining the ability to self-evolve According to the theory of dissipative structures, each forward transition over the bifurcation point (from the initial into advanced state) brings some accidental changes in the chemical system. During the back transition the system may return only into the altered initial state due to the happened changes. In this context, regular oscillations around the bifurcation point occurring in a bistate chemical system step by step work towards accumulation of the accidental changes that provides its ability of to evolve. The forward transitions initiate new transformations in a bisystem (mutability), while the reverse ones retain its state (heredity).

5. 3rd Stage of the Origin of Life: Inversion of the Balance Free Energy/Entropy Contribution and Arising of the Key Biological Properties

As it was based, a prebiotic bistate microsystem possesses a lot of properties that are essential for life. However, it is still not characterized with four key unique biological properties (distinguished in chapter 2), although some of their dawns are present - for instance, the ability to dichotomy. The essence of these key properties could be boiled down to the following thesis: a viable organism concentrates free energy (F) and information (I) through the ability for intensified and expedient reaction to external changes. Rewording, it is able to keep the positive F and I gradients in respect to the environment. By means of the positive free energy gradient a viable living system is able to transform the environment more efficiently than the environment transforms it (i.e. the ability for intensified counteraction). Due to the positive information gradient, a viable biological system knows about the environment more than the environment about it (this is a basis for its expedient behaviour). To keep this way of existence, the inflow of free energy F (providing the ability to do work) into a biological system must be higher than inflow of entropy S (devaluing the ability to carry out work). Entropy contribution is connected with the total input of the spontaneous (basic) processes in a natural system, while free energy contribution is related with the total input of non-spontaneous (coupled) processes. That means the ratio free energy contribution/entropy contribution can be used to fix the approximate balance between these counter universal processes. The prevalence of F contribution over S contribution means the following: a) non-spontaneous processes proceed in a viable biological system faster than spontaneous ones; b) constructive transformations in a living organism are more efficient than destructive ones.

As it was based, the complex circulative physical and chemical processes proceed in a bistate system. Results of the local processes contribute into the common F/S ratio. Being a non-living system, a prebiotic bistate microsystem is characterized with the negative total F/S ratio. However, because of the oscillating character of existence the F/S ratio in a bistate prebiotic system all the time changes. Besides, sometimes amplitude of swinging of the ratio may extraordinary rise due to radical constructive or destructive transformation initiated by the accumulation of accidental changes. In case the high amplitude, from time to time free energy contribution exceeds entropy contribution, although the average long-term F/S balance is still negative. During these short periods, the physic-chemical processes in the bistate prebiotic microsystem proceed in inverse: internal energetic gradients multiply, free energy and information accumulate through the turned interaction with the surroundings. This turn can be considered as a step to life through appearance of the initial unique biological properties.

It was considered that oscillations of parameters in the host medium are necessary condition to sustain bistate status of prebiotic microsystems. There are main thermodynamic and physic-chemical parameters in a liquid geological medium: pressure, temperature, pH, Eh, concentrations of diverse compounds, including organic molecules. In the changeable conditions prebiotic bistate microsystems undergo periodic stress. The stress affecting to a bistate prebiotic microsystem for the period of negative F/S balance leads to its degradation because of the entropy contribution prevailing. However, in case the period of positive F/S ratio a stress stimulates intensive reorganization of the microsystem possessing excess of free energy. The surplus of free energy can be used for constructive transformations in the bistate microsystem as a respond to stress, with the following rise of free energy producing due to more efficient organization of the network of chemical reactions. The enhanced inflow of free energy in the microsystem increases the positive F/S balance further and facilitates new constructive transformations. In this way several successive actions from the outside world influenced for the favorable period may result in increscent avalanche-like constructive conversion of the prebiotic bisystem and turn the **average F/S balance** on the positive branch. This extraordinary radical constructive transformation creates new highly effective energy-giving ways of the chemical reactions. H. Baltscheffsky (1997) investigated such drastic constructive transformations in the process of biological evolution and called them anastrophes (antipode to catastrophes). As soon as this greatest anastrophe has happened, the microsystem is organized already in the way which provides faster free energy producing than entropy. The excessive free energy in probionts binds in energy rich compounds. There appears the ability to concentrate free energy and information. Just at this moment a prebiotic bistate microsystem is transforming into the simplest living unit – a probiont. This radical transformation gives rise to metabolism of the initial biological organism and transition of chemical reactions from *in Vitro* into *in Vivo*. Once formed, a probiont has opportunities to support existence because the Le Chatelier principle acts: under regular external actions a probiont persistently develops counteraction directed to maintenance of own structural and functional stability, i.e. the biological way of organization.

6. Brief Scenario of the Origin of Life on the Early Earth

Most of the scientific knowledge that was put in background of the elaborated systemic conception of the origin of life covers not only to Earth, but to processes in the vast explored Universe too. The used key notions free energy and entropy are universal and characterize ability or disability of a natural system to carry out work elsewhere. For instance, outbreak of ash by active volcanoes on Earth and Io is the same kind of work that can be approximately evaluated in terms of the spent free energy. The opposite spontaneous and non-spontaneous processes are universal as well. They are described the transition of a natural system to more or less probable state correspondingly. All stars, planets and other space bodies in the Universe are composed of about one hundred elements systematized in the Periodic Table. Taking these and some other reasons into account, the distinguished three stages in principle can be considered in the context of a general scenario of the origin of life in the Universe.

The following scenario of the origin of life on the early Earth can be outlined on the basis of the elaborated systemic conception. About four billion years ago intensive tectonic-volcanic processes maintained nonequilibrium conditions and heterogeneous distribution of substance in hydrothermal systems on the early Earth. In this way some local parts in hydrothermal media were enriched with organic matter. Organic compounds could as penetrate into hydrothermal systems from the surface (exogenous or cosmic source) as to be synthesized in hot solutions, for instance in the course of Strecker and Fischer-Tropsch synthesis reactions (Holm, Andersson, 2005). On the background of the gradual temperature decrease organic molecules segregated into numerous of microsystems disseminated in the local high temperature aquatic zones. Significant change of conditions in the hydrothermal medium (macrofluctuation) turned the microsystems into the bifurcate state, i.e. at the intermediate unstable state between stability and destroying. The following smaller microoscillations of thermodynamic and physic-chemical parameters in the optimum regime stabilized bifurcate state some of these microsystems. They were transformed into prebiotic bistate microsystems, while the rest part of this huge number of microsystems left the bifurcate area and whether transited to passive existence in the medium or destroyed. Under continuous optimum oscillations in the medium some of the prebiotic bistate microsystems transited into living probionts, others lost own bistate status and were eliminated of further evolution to life with the natural selection. This process resulted in the formation of a community of probionts interacting with the environment. Then habitats of the communities expanded from the submarine hydrothermal media or/and (hot) terrestrial groundwater systems into the hot ocean.

References

Baltscheffsky H.: 1997, Major anastrophes in the origin and early evolution of biological energy conversion, J. Theor. Biol., 187, 495-501.

Fundamentals of Life (Palyi G., Zucci C., and Caglioti L. eds.), 2002, Elsevier SAS, Paris.

Haken H.: 1978, Synergetics, Springer-Verlag. Berlin, New York:

Holm N., Andersson E., 2005. Hydrothermal simulation experiments as a tool for studies for the origin of life on Earth and other terrestrial planets: a review. *Astrobiology*, **5**, 4, 444-460.

Kompanichenko V.N.: 2002, Life as high-organized form of the intensified resistance to destructive processes, *Fundamentals of Life*, Elsevier SAS, Paris, 111-124.

Kompanichenko V.N.: (2004) Systemic approach to the origin of life. *Frontier Perspectives*, **13** (1), 22-40.

Nicolis G. and Prigogine.I.: 1977, Self-organization in nonequilibrium systems, Wiley, New York.

Prigogine I. and Stengers I.: 1984, Order out of chaos, Bantam, New York.

Bishun Khare and Kevin Hand

Bioastronomy 2007: Molecules, Microbes, and Extraterrestrial Life
ASP Conference Series, Vol. 420, 2009
K. J. Meech, J. V. Keane, M. J. Mumma, J. L. Siefert, and D. J. Werthimer, eds.

Prussian Blue as a Prebiotic Reagent

M. Ruiz-Bermejo, C. Menor-Salván, S. Osuna-Esteban, and
S. Veintemillas-Verdaguer

Centro de Astrobiología [Consejo Superior de Investigaciones Científicas-Instituto Nacional de Técnica Aeroespacial (CSIC-INTA)]. Carretera Torrejón-Ajalvir, Km. 4,2. E-28850 Torrejón de Ardoz, Madrid, Spain.

Abstract. Ferrocyanide has been proposed as a potential prebiotic reagent and the complex salt Prussian Blue, $Fe_4[Fe(CN)_6]_3$, might be an important reservoir of HCN, in the early Earth. HCN is considered the main precursor of amino acids and purine and pyrimidine bases under prebiotic conditions. Recently, we observed the formation of Prussian Blue in spark discharge experiments using saline solutions of ferrous chloride, $FeCl_2$. Using Prussian Blue as starting material in ammonium suspensions, we obtained organic compounds containing nitrogen. These results seem to indicate that Prussian Blue could have been first, a sink of HCN, and then in subsequent reactions, triggered by pH fluctuations, it might have lead to organic life precursors.

1. Introduction

Hydrogen cyanide (HCN) is ubiquitous in the Universe. HCN has been identified in the gas phase of the interstellar medium as well as in comets, in the atmosphere of planets and on the surface of moons. HCN is one of the main products obtained in prebiotic synthesis experiments with spark discharge, UV irradiation and laser shock on various gas mixtures (Chen and Chen 2005; Saladino et al. 2004). Several experiments suggest that HCN could have served as a prebiotic precursor of purines, pyrimidines, and amino acids, as well as of others compounds such as oxalic acid and guanine. Concentrated solutions of HCN produce, by polymerization, nucleic acid bases and amino acids, whereas in dilute solutions (< 0.01 M) hydrolysis becomes dominant. At lower concentration in the primitive ocean, hydrolysis of HCN should be predominant to form ammonia and formic acid. Therefore, if HCN polymerization was actually important for the production of the first and essential biomolecules, there must have been routes by which dilute HCN solutions were concentrated. Some iron derivatives, such as ferrocyanide $[Fe(CN)_6]^{-4}$ and the insoluble complex salt Prussian Blue, $Fe_4[Fe(CN)_6]_3$ [ferric hexacyaneferrate (II)], have been proposed as intermediates to store HCN in the primitive Earth (Arrhenius et al. 1994). In a recent work, we have demonstrated that Prussian Blue is easily formed under plausible prebiotic conditions using spark discharges, an $CH_4/N_2/H_2$ atmosphere and saline solutions of ferrous salts (Ruiz-Bermejo et al. 2007). At $4 < pH < 10$, Prussian Blue is a stable and insoluble compound but outside that interval it can undergo solubilization, releasing cyanide. Oxidizing or highly reductive mediums as well as ultraviolet light could also lead to unstabilization of Prussian Blue. Here, we report the first results on the synthesis of organic compounds from Prussian Blue in ammonium solutions in conditions-like Oró for the synthesis of adenine (Oró and Kimball 1961).

2. Experimental Process

Suspensions of $Fe_4[Fe(CN)_6]_3$ in NH_4OH were heated at 70 ^oC, 100 ^oC, 130 ^oC and 150 ^oC during 24 h, 48 h and 1 week. All samples were centrifuged after the time of reaction, and the supernatant separated from the insoluble solid (mixture of iron hydroxides plus unreacted Prussian Blue). The supernatants were chromatographied using flash grade silica gel. In all cases, two fractions were colleted. Both fractions were dried by reduced pressure, yielding a white solid for the first fraction and a yellow solid for the second one. The white solids were analyzed for nucleic acid bases and carboxylic acids by GC-MS of their trimethylsilyl derivatives (Ruiz-Bermejo et al. 2007). For each reaction, the IR spectra of both solids were recovered in CsI pellet on the reflectance mode of operation.

3. Results

In all experiments, the IR spectra of the reaction raws are similar and show strong differences with respect to the Prussian Blue IR spectrum (Figure 1).

The IR spectra of the fraction 1 and fraction 2 present bands at 2044-2032 cm^{-1} that can be related to -CN groups and R-C=C=NH. The signal at 2119 cm^{-1} can be assigned to N=C=O or N=C=N or -N≡C groups. The features at 1760-1660 cm^{-1} may be associated with C=O and C=N bonds, and the bands at 3128 and 1409 cm^{-1} with the ammonium cation ($NH_4{}^+$).

In addition, the fraction 1 and fraction 2 were analized by ICP-Mass spectrometry. The fraction 1 is an organic mixture free on iron but the fraction 2 contains a 31% of iron. Thus, regarding these results with the IR spectra, it seems that fraction 2 is formed, among others, by iron complexes containing -CN groups (for example for the soluble salt $K_4[Fe(CN)]_6$ the band related to nitrile groups appears at 2042 cm^{-1}) and that fraction 1 is an organic complex solid formed by compounds rich in nitrogen and unsaturations but not nitriles.

Under all conditions assayed, organic compounds containing nitrogen were detected by GC-MS (Table 1), in good agreement with the spectroscopy data. The production of urea, dimethylhydantoin, and some carboxylic acid seems to improve at 100 ^{o} C with reaction time between 24 and 48h.

It is interesting to point out the formation of hydantoin derivatives since these compounds are precursors of α-amino acids and potentially intermediates in the prebiotic synthesis of peptides (Taillades et al. 1998) and the detection of sugar related compounds such as glycerol and lactic acid.

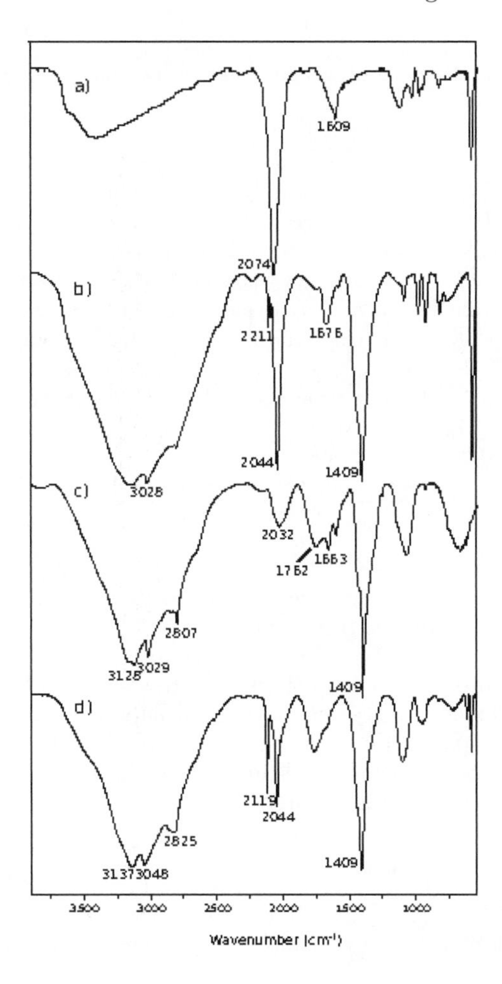

Figure 1. Figure 1. a) IR spectrum of Prussian Blue. IR spectra of the reaction heating at 100 C during 168 h. b) Reaction raw. c) Fraction 1. d) Fraction 2.

4. Conclusions

- Iron could act as a scavenger and concentrator of HCN in the prebiotic hydrophere.
- The pH-modulated release of cyanide from Prussian Blue in the presence of ammonia leads to the synthesis of organic compounds rich in nitrogen.
- The formation of Prussian Blue lowers the fixation of HCN like insoluble HCN polymers and favour the synthesis of single nitrogen organic molecules, such as hydantoin.

Table 1. Reaction conditions and organic compounds collected from Prussian Blue.

T (ºC)	t (h)	MSTFA (derivatives)
70	24	Urea, lactic acid
70	48	Urea, lactic acid, glycerol
70	168	Urea, lactic acid
100	24	Urea, lactic acid, glycerol, dimethylhydantoin, malic acid
100	48	Urea, dimethylhydantoin, 2-hydroxy-2-methyl-propionic acid
100	168	Urea, lactic acid, unknown amino acid, glycerol, glycine
130	24	Urea, dimethylhydantoin, 2-aminoisobutyric acid
130	48	Urea
130	168	Urea, lactic acid, glycerol, unknown amino acid
150	24	Urea, glycine
150	48	Urea, glycerol
150	168	Lactic acid, 2-hydroxybutiric acid, glycerol

- Sugar related compounds were obtained under relatively mild conditions using Prussian blue as precursor.

Acknowledgments. The authors have used the research facilities of Centro de Astrobiología (CAB) and have been supported by Instituto Nacional de Técnica Aeroespacial Esteban Terradas (INTA) and the project AYA2006-15648-C02-02 of the Ministerio of Educación y Ciencia (Spain). We also thank M.-P. Zorzano for her valuable comments.

References

Arrhenius T., Arrhenius G., Paplawsky W. 1994, Archean geochemistry of formaldehyde and cyanide and the oligomerization of cyanohydrin. Orig. Life Evol. Biosph. 24, 1-17.

Chen Q. W. and Chen. C. L. 2005, The role of inorganic compounds in the prebiotic synthesis of organic molecules. Current. Org. Chem. 9, 989-998.

Oró J. and Kimball P. 1961, Synthesis of purines under possible primitive Earth conditions. I. Adenine from hydrogen cyanide. Archives of Biochemistry and Biophysics, 94, 217-227.

Ruiz-Bermejo M., Menor-Salvn, C., Osuna-Esteban, S., Veintemillas-Verdaguer S. 2007, The effects of ferrous and other ions on the abiotic formation of biomolecules using aqueous aerosols and spark discharges. Orig. Life Evol. Biosph. 37, 507-521.

Saladino R., Crestini C., Costanzo G., DiMauro E. 2004, Advances in the prebiotic synthesis of nucleic acids bases: Implications for the origin of life. Current Org. Chem. 8, 1425-1443.

Taillades J., Beuzelin I., Garrel L., Tabacik V., Bied C., Commeyras A. 1998, N-Carbamoyl-amino acids rather than free-amino acids formation in the primitive hydrosphere: A novel proposal for the emergence of prebiotic peptides. Orig. Life Evol. Biosph. 28, 61-77.

Bioastronomy 2007: Molecules, Microbes, and Extraterrestrial Life
ASP Conference Series, Vol. 420, 2009
K. J. Meech, J. V. Keane, M. J. Mumma, J. L. Siefert, and D. J. Werthimer, eds.

Organic Material by Irradiation of HCN Solutions

M. Colin-García,[1,2] A. Negrón-Mendoza,[1] and S. Ramos-Bernal[1]

[1]*Instituto de Ciencias Nucleares, Universidad Nacional Autonoma de Mexico*

[2]*Current address:Instituto de Geología, Universidad Nacional Autónoma de México*

Abstract. Chemical evolution is the chemistry of formation and organization of bio-organic compounds under astronomical and geological primitive condition. In this context, comets have been considered a source of important molecules to chemical evolution: they could have brought molecules that contributed to the formation of the primitive atmosphere, and also organic material, necessary to the emergence of life on Earth. Ionizing radiation (as an inner and outer supply) could have been an important source of energy in the synthesis of species detected on comets. In this work we attend to understand the effect of ionizing radiation on a key molecule, hydrogen cyanide, which has been widely detected in comets, the ISM and other environments. The role of ionizing radiation in the decomposition of frozen and liquid solutions of HCN has been investigated, and the formation of bio-molecules has been monitored. For the analysis, gas chromatography-mass spectrometry and spectrophotometric techniques were used to detect the formed compounds. These experiments show that there are many organic compounds formed in the irradiated samples, such as carboxylic acids, urea, bases (adenine and cytosine), as well as amino acids. These compounds are formed even at low radiation doses in both systems.

1. Introduction

Comets are minor bodies; however, they give us unique information about chemical evolution processes that occurred at early stages of the Universe. Within the scenery of origin of life, Chamberlin and Chaberlin, in 1908, were the first to suggest that extraterrestrial material could have pla yed an important role into this process. Later, J. Oró reworked this hypothesis implying a specific role of comets, as raw material sources for chemical evolution on Earth. This idea has been widely investigated by others, and it has been reinforced (Hartman et al. 1985; Negrón-Mendoza et al. 1994; Irvine 1998; Stephton 2001). Now, it is widely recognized that not only local processes such as UV irradiation and electrical discharges on the primitive atmosphere were important in triggering necessary reactions on the early Earth, but extraterrestrial contribution was also important. Comets were formed in cold environments and they have an important amount of volatiles. It was thought that their formation conditions could have protected their components from chemical evolution (Cottin, Szopa, & Moore 2001). However, when comets were formed in the interstellar medium (ISM), UV photons and cosmic ions had already interacted with grains in (Strazzulla, Castorina, & Palumbo 1995). In fact, Hudson & Moore

(2001) state that all the icy Solar System objects are at some degree modified by ionizing radiation.

1.1. Comets, What Are They Made Of ?

Currently, the most accepted model suggests that comets are made of icy, rocky and organic components. Greenberg (1985) proposed that a cometary nucleus could contain up to 30% of its mass as water ice, 26% of silicates, 23% refractory material, and 9% in the form of carbonaceous molecules. Cometary ices are predominantly made of water; in fact, water-rich ices are important components of dense molecular clouds and are common throughout the Solar System. In fact, water is a common molecule in the interstellar medium and in astrophysical environments (Ehrenfreund & Charnley 2000) and this is corroborated in comets, where water ice constitutes the major component in which the others are diluted (Gerakines et al. 2001). Having in mind this hypothesis we tested the capability of formation of new and more complex products, by the over irradiation of a solution of hydrogen cyanide in different temperature regimen. Hydrogen cyanide was chosen because it is a molecule broadly detected on different environments such as cometary tails (Ip et al. 1999; Magee-Sauer et al. 1999) and the ISM (Irvine 1998; Boonman et al. 2001) and has a great capability to transform into new compounds. In a previous research we studied the behaviour of this solution in a range from 0 to 419 kGy, in this study we expand the observations, mainly in the case of liquid samples, up to 844 kGy.

2. Method

2.1. Reagents, HCN Production, and Sample Preparation

All chemicals used were of the highest purity commercially available. Aqueous solutions of HCN were prepared by adding diluted (25%) sulphuric acid (Aldrich reagent 95-98%) to potassium cyanide salt (Aldrich reagent 97%), and by bubbling and dissolving the released HCN (gas) into triple distilled water. The concentration of the HCN solution employed was 0.15 M. Aliquots of this solution (150 ml) were placed in special glass ampoules for irradiation. After that, the samples were saturated burbling argon to displace the oxygen dissolved. Two sets of samples were used: one irradiated at room temperature, the other one was placed in a Dewar and frozen with liquid nitrogen. These last samples were maintained at 77 K by adding liquid nitrogen.

2.2. Irradiation

Frozen and liquid solutions were irradiated at different doses. Irradiations were carried o ut in a high intensity gamma source of Co-60 (Gamma-beam 651-PT) at ICN, UNAM. The irradiation doses were from 0 to 844 kGy for room temperature irradiation, and in the case of frozen samples they were exposed to doses up until 419 kGy.

2.3. Analysis

After irradiation, frozen solutions were thawed. The pH was measured for all the samples, and they were analyzed by several techniques.

Urea production: Liquid aliquots were used to quantify urea. The production was followed by UV spectroscopy (Negrón-Mendoza et al. 1986), through the reaction of this molecule with diacetyl monoxime (DAM).

Biuret test: The chemical bond that links two amino acids is called amide bond; in the case of peptides and proteins this bond is called peptide bond. Amide unions react specifically with the copper ion in a strong alkaline medium, giving a blue color for positive test, which could be quantified spectroscopically at 263 nm.

Amino acids: An aliquot of each sample was used to detect free amino acids. The procedure, based in the reaction with ninhydrin, suggested by (Murariu et al. 2003), for complex biological mixtures, was followed.

Dry residue: The aqueous component was evaporated until dryness under controlled temperature conditions (40°C) in a rotary evaporator. The dry residue was weighted and an aliquot was derived in order to detect the formation of carboxylic acids.

Bases: A small fraction (50 mg) of the sample irradiated at 310 K and a dose of 672.2 kGy was disolved in 0.5 ml of acetic acid (0.05M). This solution was fractionated in an automatic equipment (FractoScan) through a SP-Sephadex C-25 column. Differential elution was carried out, first with an acid phase (acetic acid 0.05 M), then with water and finally with a base (amm onium hydroxide 0.02). Fractions were then analyzed in a UV-vis spectrophotometer and the fractions with high absorption were selected to analyze and identify the components by HPLC .

3. Results and Discussion

3.1. Dry Residue

The results show that the amount and aspect of the dry residue (organic matter) depends on the dose and the temperature of irradiation. In the case of liquid samples (irradiated at 310 K), the amount of dry residue is bigger than the one produced in samples irradiated at low temperature. It is important to notice that the irradiation of solutions, liquid or solid, produces the formation of free radicals. Reactions that take place, involving those radicals, are controlled by diffusion. In the case of the experiments performed at 310 K, formed radicals can combine freely at high rates; however, for the experiments at low temperatures (77 K) radicals are constrained in their movement. As a consequence the yield of dry residue changes dramatically as the dose increases for liquid solutions, whereas for frozen solutions the increase is smaller (Fig.1).

3.2. pH Change and Urea Production

pH is so important because it is an indirec measure of chemical changes. The measurements reveal that pH change as a function of dose, in both studied systems it becomes alkaline. This could be explained because of the accumulation of nitrogen containing compounds (amides, amines), such as urea. We confirmed

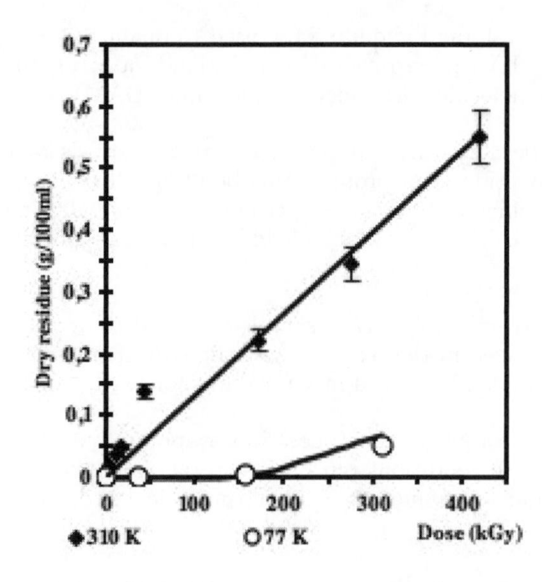

Figure 1. Dry residue production as a function of dose for the studied systems.

that the production of urea increases in both systems (Fig. 2). The production was higher in liquid solutions, than in frozen ones.

Production of amide bonds, detected by Biuret test, was higher for liquid experiments than in solid phase in both cases it was dose dependent (Fig. 3).

3.3. Production of Biomolecules

Amino acids: The production of amino acids was followed in free state. This production is a lso dose reliant and again the production was higher in the liquid system than in the solid one. Despite the fact that small amounts of these products are formed, the behaviour is close to be linear in both cases.

Nucleobases: A purine (adenine) and a pyrimidine (cytosine) bases were detected in the analyzed sample. The evidence of formation of those two molecules is very important, even if the detection was made in a single sample, due to their role as building blocks of nucleic acids and energy transfer molecules (Fig 4).

Carboxylic acids: This kind of molecules were detected after acid hydrolisis of dry residue. Several carboxylic acids were identified in both set of samples, such as: oxalic malonic, succinic, glutaric, carboxysuccinic, tricarballylic, and citric acids.

4. Conclusions

Ionizing radiation promotes chemical changes in the systems under study. Those changes are induced despite the temperature of irradiation and the low radiation doses. Irradiation of HCN solutions generates a dry complex residue (organic

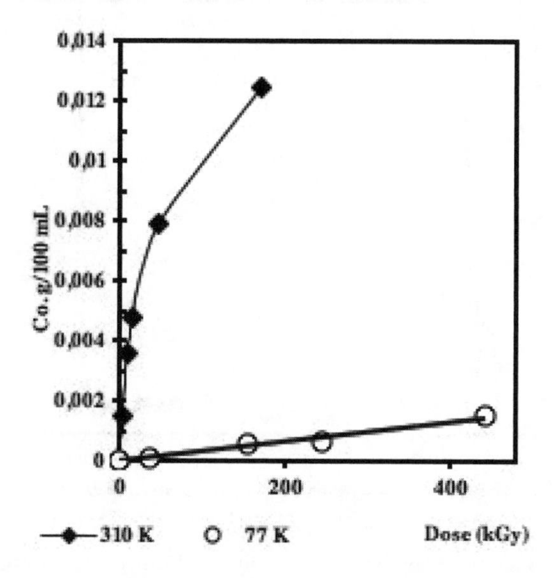

Figure 2. Urea production as a function of dose for both studied systems (77 and 310 K)

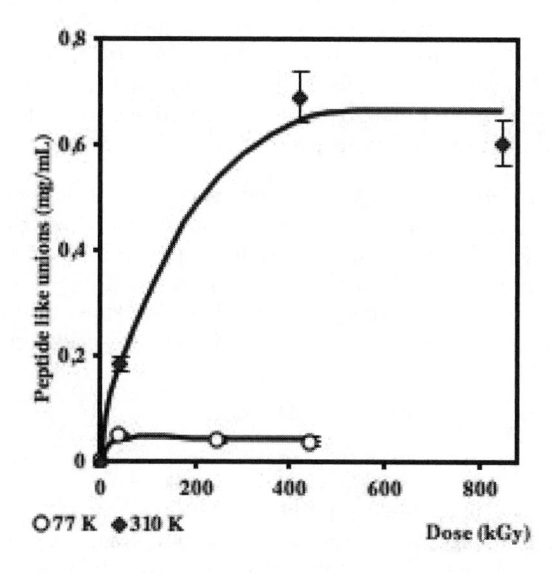

Figure 3. Detection of amide bonds in irradiated solutions of HCN at 310 and 77 K

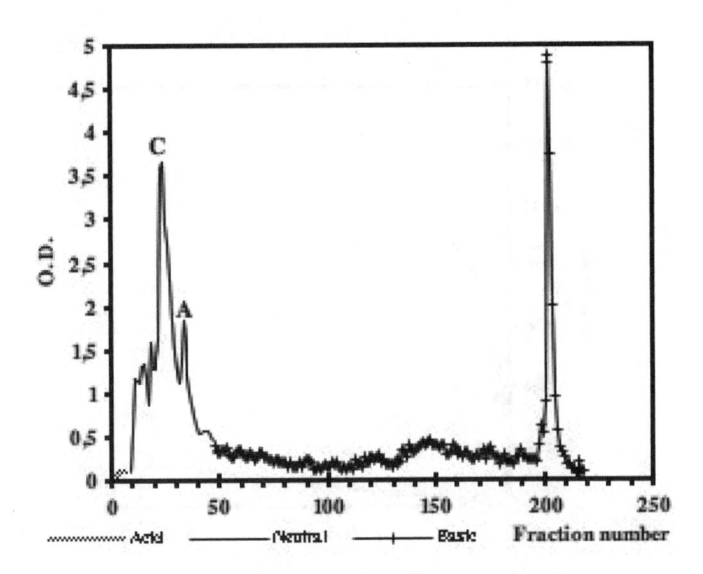

Figure 4. Analytic fractions of the polymer formed by a irradiation dose of 672 kGy of solutions of HCN at 310 K. Optical density was measured at 260 nm. The label peaks with C and A show the fractions were adenine and cytosine were detected.

in nature) and also water soluble products. The dry residue production is dose dependent and it is higher in liquid solutions. A great number of products such as amino acids, amides (urea), carboxylic acids (i.e. citric, succinic, etc.), bases (adenine and cytosine) are produced. This study confirms the role of HCN as a parental molecule and emphasizes the role of ionizing radiation as an effective energy source to promote reactions even in adverse conditions.

Acknowledgments. Travel Support was received from the NASA Astrobiology Institute and the National Science F oundation through the conference organizers (MCG). This work was partially supported by a grant (DGAPA-IN223406). Authors thank Physicist F. Garcia and M. in Sc. B. Leal for their technical assistance in irradiating samples. The experimental design and elaboration of special glassware was carried out by the expert manufacture of Mr. Ham.

References

Boonman, A.M.S., Stark, R., van der Tak, F.F.S., van Dishoek, E.F., van der Wal, P.B., Shăfer, F., de Lange, G.& Laauwen, W.M. 2001, Highly abundant HCN in the inner hot envelope of GL 2591: probing the birth of a hot core? Astrophys. J., 553, L63-L67

Cottin, H., Szopa, C. and Moore, M.H. 2001, Production of Hexamethylentetramine in photolyzed and irradiated interstellar cometary ice analogs, ApJ 561, L139-L142.

Goossens, M., Mittelbach, F., & Samarin, A. 1994,

Ehrenfreund, P.& Charnley S.B. 2000. Organic molecules in the interstellar medium, comets, and meteorites: A voyage from dark clouds to the early Earth. Annu. Rev. Astron. and Astrophys., 38, 427-483

Gerakines, P.A. Moore,M. H. & Hudson R.L. 2001, Energetic processing of laboratory ice analogs: UV photolysis versus ion bombardment. J. Geophys. Res., 106, 33381-33385

Greenberg, J.M. 1998, Making a comet nucleus. A & A 330, 375-380

Hartman, H.J., Lawless G., & Morrison P. 1985, Search for the universal ancestors, (National Aeronautic and Space Administration, Scientific and Technical Information Branch, Washington D.C.), 43-73

Hudson, R.L. & M.H. Moore 2001. Radiation chemical alterations in Solar System ices: an overview. Journal of Geophysical Research. 106: 33275- 33284.

Ip, W.H., Balsiger, H., Geiss, J., Goldstein, B.E., Kettmann, G., Lazarus, A.J., Mei er, A., Rosenbauer, H.,& Schelley, E. 1999, Giotto ISM measurements of the production rate of hydrogen cyanide in the coma of Comet Halley, Ann. Geophys., 8, 319-325

Irvine, W. 1998, Extraterrestrial organic matter: a review. Origins of Life and Evolution of the Biosphere., 28, 365-383

Magee-Sauer K., Mumma M.J., DiSanti M.A., Russo N.D. & Rettig T.W. 1999, Infrared Spectroscopy of the 1/3 Band of Hydrogen Cyanide in Comet C/1995 O1 Hale-Bopp, Icarus 142, 498-598

Murariu, M. Irim ia, M., Aelenei, N. & Drochiou, G. 2003, Spectrophotometric assay of amino acids in biological materials. Buletinul Institutului Politehnic din Iai XLIX (LIII), 219-225

Negrón-Mendoza, A. Chacón, E., Perezgasga, L.& Torres, J.L. 1986, Determinación de urea por el método de DAM . Informe Técnico Q-02-86, CEN-UNAM, México., 7 (Spanish)

Negrón-Mendoza, A. Albarrán G., Ramos S. & Chacón E. 1994, Some aspects of laboratory cometary models, J. Biol. Phys., 20, 71-76

Stephton, M.A. 2001, Life's sweet beginnings, Nature, 414, 857-858

Strazzulla, G., Castorina A.C. & M.E. Palumbo. 1995. Ion irradiation of astrophysical ices. Planetary and Space Science. 43: 1247-1251.

Nancy Kiang

Bioastronomy 2007: Molecules, Microbes, and Extraterrestrial Life
ASP Conference Series, Vol. 420, 2009
K. J. Meech, J. V. Keane, M. J. Mumma, J. L. Siefert, and D. J. Werthimer, eds.

Plausible Organic Synthesis on Titan's Surface

S. H. Abbas[1] and D. Schulze-Makuch[2]

[1]*Chemistry Department, Palomar Community College, San Marcos, CA 92069*

[2]*School of Earth & Environmental Sciences, Washington State University, Pullman, WA*

Abstract. Titan's nitrogen and hydrocarbon dominated atmosphere is the site of very complex organic chemistry leading to the production of numerous species including larger molecules from smaller molecular, ionic and radical precursors. Photochemical conversion of methane and other precursors produces smog in the middle to upper layers of the atmosphere, and an organic rain of methane, higher hydrocarbons, and nitrogen-containing aerosols falls onto Titan's surface, producing an Earth-like terrain of extended river channels that contain a "soup" of several important organic precursors. Here we describe plausible organic chemistry under Titan's surface and near surface conditions and present synthetic schemes based on acetylene, hydrocyanic acid, and other small organic precursors resulting in the production of larger molecules. Among the products of these reactions are important biological precursors such as nitrogenous bases and amino acids. We present details of several plausible prebiotic organic synthetic routes in Titan's surface environment and the implications of such chemistries to the overall process of prebiotic evolution on Titan.

1. The Environment of Titan

Titan's cold and reducing nitrogen and hydrocarbon-rich environment, which resembles early Earth's reducing atmosphere, qualifies Saturn's largest moon as a planet-size prebiotic laboratory. Complex hydrocarbon and nitrogen chemistry in Titan's atmospheric and surface environments may lead to the production of biologically-significant molecules such as amino acids. Data obtained from the Cassini-Huygens mission suggests a complex setting that contains hydrocarbon lakes, methane clouds, hydrocarbon rain, and very rich photochemistry. Thus, it is likely that Titan contains certain microenvironments of biological interest (Schulze-Makuch and Grinspoon, 2005).

2. Plausible Hydrocarbon Photochemistry

In the Titan environment, the presence of hydrocarbons, nitrogen containing precursors, hydrogen, electromagnetic radiations of various wavelengths, and cold temperatures, allows for addition and substitution type reactions to become dominant. Under the environmental conditions on Titan, atmospheric chemistry is dominated by radical reactions. One of the most significant roles that water plays is via the production of hydroxyl radicals, because OH radicals are very reactive precursors and water ice can be inferred to be a common compound on Titan's icy surface. The presence of methane in the atmosphere leads to photolysis during exposure to sunlight in the UV range. The photolysis of methane leads to the production of radical fragments including $.CH_3$, $:CH_2$, and others. Once produced, some of these radicals produce higher alkanes: For instance, the encounter between the CH_3 radicals produces ethane:

$$H_3C. + .CH_3 \rightarrow C_2H_6 \tag{1}$$

The encounter of two methylenes ($:CH_2$) may produce ethylene (C_2H_4):

$$H_2C: + :CH_2 \rightarrow C_2H_4 \tag{2}$$

Conditions prevailing in kinetically-slow environments such as Titan's atmospheric and near-surface environments (due to the low temperatures), result in the lowering of the enormous energy produced from the encounter of the methylenes and relieving the requirement for a third body absorber. The encounter of an ethyl radical with a methyl radical produces propane (C_3H_8):

$$H_3C. + .CH_2CH_3 \rightarrow C_3H_8 \tag{3}$$

3. Organic Synthesis of Amino Acids

Titan's rich organic inventory allows for various organic synthetic schemes to be envisioned (Abbas and Schulze-Makuch, 2002). The most promising of these schemes involve either acetylene or hydrocyanic acid. For temperature and pressure conditions prevalent on Titan's surface, the instability and available energy become an advantage because they allow for enhancement of reaction rates. Both, acetylene and hydrocyanic acid have the advantage of being common on Titan. Organic synthesis based on acetylene may be initiated via the catalytic self-assembly of three acetylenes (cyclotrimerization) to produce benzene:

$$3C_2H_2 + Catalyst \rightarrow C_6H_6 \tag{4}$$

The catalyst in reaction (4) is essential, because for the non-catalyzed reaction the energy barrier is too high. The catalyst allows for a reaction pathway that requires a lower energy of activation thus making it possible for reactants to overcome the energy barrier. A variety of substrates such as active metal surfaces (Pt, Pd, etc.), or certain silicates such as clay minerals or zeolites may act as surface catalysts that help facilitate reaction (4). High resolution spectra obtained from observations carried out by the Infrared Space Observatory confirm the presence of benzene in Titan's atmosphere with a column density of $\approx 2\times10^{15}$ molecules cm^{-2} . The detected atmospheric benzene may be produced by cyclotrimerization of acetylene on aerosol surfaces in the atmosphere. Some of the produced benzene may wash down and condense with hydrocarbon rain. If in reaction (4) one or more of the acetylenes is an alkyl-substituted acetylene, then the product will be a substituted benzene (C_6H_5R):

$$H - CaCR + 2C_2H_2 + Catalyst \rightarrow C_6H_5R \tag{5}$$

where R can be any alkyl or related group. Once the benzene or alkyl benzene is produced, the synthesis of more complex compounds can occur via simple substitution or addition reactions. For instance, alkyl-nitrobenzene ($RC_6H_4NO_2$) may be produced using NO+ , a highly reactive cation that is formed from ionizing radiation or the rearrangement of other precursors:

$$C_6H_5R + NO_2^+ \rightarrow RC_6H_4NO_2 \tag{6}$$

Alkyl nitrobenzene ($RC_6H_4NO_2$), the product of reaction (6) may in turn become the precursor for the synthesis of complex organic molecules such as amino acids. A simple illustration of this conversion is the catalytic hydrogenation of the nitro group in alkyl nitrobenzene (the metal surface catalyzed addition of H_2 to the nitro group). It converts the alkyl nitrobenzene to alkyl amino benzene ($RC_6H_4NH_2$)[†]:

$$RC_6H_4NO_2 + H_2 + Catalyst \rightarrow RC_6H_NH_2 \tag{7}$$

The oxidation of the R group in $RC_6H_4NH_2$ yields a carboxyl group (COOH), producing an amino acid product. If R in reaction (7) is a methyl group, the oxidation of $RC_6H_4NH_2$ using an appropriate oxidizing agent produces para-amino benzoic acid (HO-$COC_6H_4NH_2$), a promising chemical compound for a biological scheme as it has the backbone of an amino acid. The advantage of this pathway is that it should produce high yields of the products and that no liquid water is needed for these reactions to occur. The synthesis is limited by yields from several intermediary steps and the abundance of a suitable catalyst in Titan's surface environment. The HCN produced in the Titan environment may self-combine in aqueous alkaline solutions present on the surface to form amino acids with the assistance of UV photons:

$$3HCN + 2H_2O + h\nu \rightarrow CHCO_2HNH_2 + CN_2H_2 \tag{8}$$

In addition to glycine, reaction (8) yields cyanamide (CN_2H_2), which can link amino acids together as the first step in the formation of proteins. Reaction (8) is limited to areas on Titan's surface or interfaces that have liquid water. Liquid water may be temporarily available on Titan's surface due to cryovolcanism or meteorite impacts that lead to melting of large swaths of water ice. Liquid water might also have been more common earlier in Titan's history shortly after the formation of the outer Solar System. The HCN that forms due to atmospheric processes may be introduced into the surface environment via hydrocarbon precipitation. Keller et al. (1998) suggest that HCN can be produced in Titan's ionosphere via electron dissociative recombination reactions and conclude that ion neutral chemistry produces $HCNH^+$ as the major ion species at the ionospheric peak.

4. Discussion and Conclusions

A comparative evaluation of the plausible synthetic schemes can be made based on availability of reactants and reagents involved in each of the schemes; and the presence, or lack thereof, of the appropriate conditions that make the reaction scheme favorable. Among the several synthetic schemes leading to the production of amino acids, the first presented reaction sequence utilizing acetylene via benzene seems to be the most likely to occur due to the abundance of the required reactants and reagents, and because the Titan environment provides appropriate conditions for this reaction sequence. The recent detection of benzene, phenyls, and other complex organics on Titan supports the abovementioned discussion. As new chemical species are detected and identified in

---

[†]Amino acids from Hydrocyanic Acid

Figure 1. Artistic depiction of a possible scenario for a future mission to Titan, as conceived for the Titan Saturn System Mission (TSSM), a joint NASA/ESA proposal for the exploration of Saturn and its moons. While a Titan-dedicated orbiter provides global remote science, context information, and relaying communications to and from Earth, a comprehensive close and personal investigation is accomplished via a montgolfiere (hot-air balloon) circumnavigating Titan between 2km and 10 km in altitude and a probe with the capability to land on the surface of a northern lake to study the liquid composition. (Credit: Corby Waste, NASA JPL)

Titan's atmospheric or surface environments, more pieces are added to the puzzle of the complex web of chemical reactions occurring on Titan. Controlled laboratory experiments simulating microenvironments on Titan should prove invaluable, and theoretical predictions provided here and elsewhere should be tested. The Cassini-Huygens mission has enhanced our understanding of the chemistry occurring in both the atmospheric and surface environments and the forces that shape their nature. As more data from the Cassini-Huygens mission is analyzed, further insights will be gained on Titan's potential for supporting prebiotic evolution. Ultimately, however, a new mission is needed. The TANDEM (Titan and Enceladus Mission), also called Titan Saturn System Mission (TSSM) with its objective to understand the atmosphere, surface and interior, to determine the chemistry, and to derive constraints on the origin and evolution of Titan and the Saturnian System as a whole, would be ideal to gain further insights (Figure 1). The diversity and complexity of possible prebiotic compounds found will determine whether the course of the pre-biological process has progressed to the stage whereby chemical evolution is leading to the production of significant biomolecules or whether it was halted at a certain stage.

References

Abbas, S., & Schulze-Makuch, D. 2002, European Space Agency Special Publications 518, 345-348

Keller, C., Anicich, V., & Cravens T. 1998, Plan. Space Sci., 46, 1157-1174

Schulze-Makuch, D., & Grinspoon, D.H. 2005, Astrobiology, 5, 560-567

Bioastronomy 2007: Molecules, Microbes, and Extraterrestrial Life
ASP Conference Series, Vol. 420, 2009
K. J. Meech, J. V. Keane, M. J. Mumma, J. L. Siefert, and D. J. Werthimer, eds.

The Radiowave Synthesis in the Droplets of a Thunderstorm Cloud as a Precursor of Photosynthesis

V. A. Gusev

Sobolev Institute of Mathematics, Siberian Division, Russian Academy of Sciences, Koptyuga pr. 4, Novosibirsk, 630090 Russia, email: vgus@math.nsc.ru

Abstract. Oparins's coacervate droplets (Oparin 1924) have long served as a prototype of primary microbial cells that appeared in the Earth 4 to 4.5 billion years ago. At present, this concept is only of historical value because its constructive principles were formulated in the 1920s in terms of classical thermodynamics. According to the modern views, the transformation of chaos into order and the appearance of complex molecular structures are possible only under conditions far from thermodynamic equilibrium. However, no new paradigm commensurable with Oparin's idea of the natural origin of life has been formulated so far for these conditions, and the heuristic role of a coacervate droplet still remains attractive. This paper advances the hypothesis stating that low-molecular-weight organic compounds, precursors of living cell components, may be synthesized from inorganic oxides in the presence of alternating electromagnetic field as an energy source.

1. Organic Compund Synthesis in Water Droplets

The synthesis of organic compounds can be implemented in the water droplets hovering in a thunderstorm cloud of the Earth or another planet prebiotic atmosphere. A stroke of lightning is known to excite a broad spectrum of electromagnetic waves. These, in turn, can excite the Langmuir vibrations of protons in water droplets. The molecular mechanism of this process has been described in detail (Gusev 2001a,2001b).

For the convenience of simulation, we will consider the ideal case, namely, that the Langmuir proton vibrations possess, on average, a spherical symmetry. This idealization does not contradict the physics of the process: since the object is spherically symmetrical, the geometry of stationary vibration processes taking place in this object should also possess a spherical symmetry. We will assume that the form of the Langmuir vibrations is represented by periodic thickening and thinning of protons in the central area of the droplet. We will discuss processes in droplets whose radius $R_o=5\times10^{-5}$ cm, which corresponds to the average microbe size.

It can be easily shown that the average energy u_H acquired by a proton upon the formation of the spherical clot in the droplet central area having the radius R amounts to,

$$u_H = \frac{2\,\pi\, n_H\, R^2\, q^2}{\epsilon}\left(\frac{R_0^3}{R_3} - 1\right), \tag{1}$$

where q is the proton charge, ϵ is the dielectric constant, and n_H is the proton concentration. The numerical estimate of the proton energy at $R=R_o/2$ and for the proton concentration $n_H \geq 6\times10^{16}$ cm^{-3} (this corresponds to pH ≤ 4) gives $u_H \geq 3$ eV. The activation energies of most homogeneous chemical reactions fall in the 1-3 eV range; therefore, in the central area with the radius R=$R_o/2$, the energy of protons is sufficient both for activating the reactions and for the synthesis itself to proceed. The calculations carried out in Gusev (2001a,2001b) allow one to estimate the required amplitude E ($\approx$ 700 V/m) and frequency ($\nu \approx 6\times10^9$ Hz) for an electromagnetic wave able to excite the Langmuir vibrations of protons with an energy of about 3 eV.

Further we will consider the dynamics of the de novo synthesized molecules with allowance for their physicochemical properties, namely, the presence of a dipole moment and charged or aliphatic groups. The molecules possessing an intrinsic dipole moment will be drawn into the central area of the droplet due to the nonuniformity of the centrosymmetrical field of the Langmuir protons. The maximum size of molecules concentrated near the center can be estimated from their rotational relaxation time ? provided that the Langmuir frequency ω_L=q$(4\pi n_H/m\epsilon)^{1/2}$ and $T = \pi\, n\, a^3/6kT$ (where m is the mass of the proton, n is the water viscosity, kT is the thermal factor) are related as $\omega_L T < 1$. If this condition holds, the molecule dipoles have time to change orientation simultaneously with the change of the charge in the central area and, hence, they are all attracted to the droplet center.

$$a \leq \left(\frac{6\,k\,T}{\pi\,n\,q}\right)^{\frac{1}{3}} \left(\frac{\epsilon\,m}{4\,\pi\,n_H}\right)^{\frac{1}{6}} \tag{2}$$

The numerical estimate gives a < 6Å. The sizes of all most important low-molecular-mass polar molecules of microorganisms fit in this range. The steady-state concentration of these molecules in the central area of the droplet can be estimated from the conditions of equality of the osmotic and electrical pressures.

$$n = n_H \left(\frac{R_0}{R}\right)^3 \frac{4\,q\,d}{3\,\sqrt{2}\,\epsilon\,a^2\,k\,T}, \tag{3}$$

where d is dipole moment of the molecule. Thus for glycine, the equilibrium concentration at pH $= 4$ reaches the value n $\approx 10n_H \approx 1$ mM, which corresponds to the average concentration of biochemical substrates in a microbial cell. At the given pH, the frequency of proton vibrations is about 6×10^9 Hz. If we assume that at least one chemical synthesis event takes place during each vibration, about 10^{10} chemical steps occur throughout the whole bulk of the droplet per second. For comparison, the organic composition of the E coli bacterial cell contains, on the whole, NC $\approx 10^{10}$ carbon atoms. Assuming that CO_2 undergoes complete chemical transformation within the cell, we can estimate the time t of accumulation of organic carbon. If diffusion is considered as a steady-state process, then

$$t = \frac{N_c}{8\,\pi\,D_c\,n_c\,R_0}, \tag{4}$$

where n_C and D_c are the equilibrium concentration and the diffusion coefficient of CO_2 in water. The numerical estimate of time gives t $\approx$ 1sek (n_c 5×10^{17} cm^{-3}; here we take into account the fact that for pH $= 4$ in a water droplet, the

CO_2 concentration in the atmosphere should exceed the today's carbon dioxide content in the atmosphere by two orders of magnitude).

It follows from the given estimates that the rate of CO_2 ingress from the atmosphere is the rate-determining step in the synthesis of organic compounds in a droplet. In addition, it is obvious that even in the case where a proton interaction event with the dissolved oxide does not always result in a chemical reaction, the accumulation of organic matter in the drop reactor considered here takes place over periods of time much shorter than geological periods. The local increase in the electromagnetic-wave field strength at the resonance Langmuir frequency of protons (up to 4 kV/m) during a stroke of lightning, which lasts for up to several microseconds (Bazel et al. 1978), would force the whole pool of dipole molecules to the center of a sphere with a radius of 10^{-6} cm. The energy of interaction of these dipoles with the charged proton clot would exceed appreciably the thermal factor, $8\,\pi\, n_H\, qdR(R_o/R)^3/9\,\epsilon\, kT >> 1$, and all the dipoles will line up to form radially directed chains. The average distance between the dipoles in this state is close to their sizes. Thus, conditions favorable for the formation of covalent links between the monomers to give polymer chains, for example, polypeptides N^+H_3 - HCG_1 - COO- N^+H^3 - HCG^2 - COO^- (here G_i are the side groups of amino acids), are created for a period of several microseconds.

The peptide bonds can also arise using the energy of the proton Langmuir vibrations. Since prior to the covalent linking, the dipole monomers already form the "right" sequence of alternation of amino and carboxy groups, in the case of a favorable monomer arrangement, oligomers containing up to several dozens of amino acids can be formed in this way. The limiting scale factor is the active area size, $R \approx 10^{-6}$ cm. The rotational relaxation time for the dipole oligomers with a size of greater than $6\AA$ exceeds the critical value: hence, they no longer experience the action of the spherically symmetrical field of Langmuir protons, which have drawn them into the droplet center, and the droplet bulk is gradually filled with these oligomers. The secondary structure of peptide polymers consisting of homochiral monomers, no matter in the L- or D-form, is known to be thermodynamically more stable than a chain consisting of a random set of stereoisomers (Elliot & Elliot 1997; Avetisov & Gol'danski 1996). Thermodynamic stabilization is due to the formation of hydrogen bonds between the peptide groups of every fourth amino acid residue in the chain, which provide the formation of an α-helix. An oligopeptide consisting of, e.g., $N = 12$ amino acids can contain three bonds of this type. The energy of formation of one hydrogen bond is $\delta U \approx 30\,\text{kJ/mol}$. The entropy change can be estimated as $\delta S = k\ln 2^N$, where 2^N is the number of states of the amino acid residues in the racemic chain. Then the free energy of the formation of the α-helix can be written as $\delta G = N(\delta U/4 - kT\ln 2)$. For an oligopeptide consisting of 12 amino acids, the equilibrium constant at 300 K $\kappa_{eq} \approx 4\times10^{-10}$, i.e., the formation of a regular oligopeptide structure as an α-helix is thermodynamically justified. However, this transformation requires overcoming of an energy barrier comparable with the energy of a covalent bond, and this process does not take place in the temperature range in which this oligomer exists.

As shown previously (Gusev 2001a,2001b), a proton is able to gain energy under the action of an electromagnetic wave while moving along the integrated three-dimensional hydrogen bond network formed by water molecules. Upon collision with a break, the proton stops, and a local energy release takes place.

Thus, if a chain of hydrogen bonds in the oligopeptide structure passes continuously into the hydrogen bond network of the water environment, the proton would migrate along the chain without difficulty. Otherwise, an amount of energy sufficient to overcome the activation barrier to stereochemical inversion will be released at the break site. This process of oligomer local excitation will end after the formation of a perfect α-helical structure. It is evident that deracemization, i.e., the formation of a homochiral oligopeptide, can occur in the central area of the droplet. It should be noted that the overall oligomer would still be a racemate.

Hydrophobic (neutral or charged) molecules would be displaced to the droplet periphery. Indeed, the average electric field strength in a droplet is equal to zero; hence, the strength acting on charged (not dipole) aliphatic molecules is also zero. Due to the hydrophobic interactions with water molecules, aliphatic molecules are displaced to the surface of the droplet. Let us estimate the characteristic time of this process in terms of the diffusion mechanism of molecule movement using the formula R^2=6Dt, where D$\approx10^{-5}$cm^2/s; then t$\approx$40 mcs. The space between the central area and the droplet surface would be occupied by neutral, readily soluble nondipole molecules.

After the droplet surface has been entirely covered by a monolayer of hydrophobic hydrocarbon molecules, the rate of the ingress of water-soluble components of atmospheric gases would be significantly restricted. Most of the carbon dissolved in the droplet as CO_2 would be spent for the synthesis of organic matter. This would reduce the concentration of protons, i.e., increase the pH. The synthesis of organic matter and, correspondingly, the increase in the pH in the droplet would stop when the proton energy u_H is lower than the activation energy of CO_2 protonation. The numerical estimate shows that a 5-10-fold decrease in the proton concentration is sufficient for this. Thus, the limiting value is pH $\leq$5 in the droplet for the specified initial value pH $\leq$ 4. The energy of protons will also decrease if the droplet is carried away from the thunderstorm area. As the external field amplitude decreases from E$\approx$700 V/m to E$\approx$350 V/m (for n_H = const), the proton energy would also decrease below the threshold value, and the de novo synthesis of organic matter would stop. However, in both cases, the energy of protons is still sufficient to ensure the subsequent chemical transformations within the compartment.

Thus, after completion of the synthesis of primary organic matter, a water droplet would be a prototype of a microbe cell consisting of a hydrophobic shell enclosing a solution of organic molecules of diverse compositions in physiological concentrations. The presence of regular homochiral peptide oligomers in the compartment space creates conditions for the next stage of organic synthesis using their catalytic properties. The time spent for the whole process of synthesis of primary organic matter is much shorter than geological periods; under conditions formulated above, this time is only 1 s. An advantage of this model is the possibility of its real-time verification. As my mine over described processes can take place on the Earth at the nowadays.

Acknowledgments. The author is grateful to Prof. A. D. Gruzdev for fruitful discussion.

References

Avetisov, V.A. and Gol'danskii, V.I., Usp. Fiz. Nauk (Russia), 1996, vol. 166, no. 8, pp. 873–891.

Bazel', E.M., Gorin, B.N., and Levitov, V.I., Physical and Engineering Foundations of Lightning Protection, Leningrad: Gidrometeoizdat (Russia), 1978.

Elliot, W.H. and Elliot, D.C., Biochemistry and Molecular Biology, Oxford: Oxford Univ. Press, 1997.

Gusev, V.A., Biophysics (in Russia), 2001a, vol. 46, no. 5, pp. 862–878.

Gusev, V.A., Newslett. ISSOL, 2001b, vol. 28, no. 1/2, pp. 34–37.

Oparin, A.I., The Origin of Life, Moscow: Mosk. Rabochii (Russia), 1924.

Steven Rodgers

Maggie Turnbull

Session VII

Origins of Viruses and Cellular Function

Thomas Balonek

Andrew Pohorille

Bioastronomy 2007: Molecules, Microbes, and Extraterrestrial Life
ASP Conference Series, Vol. 420, 2009
K. J. Meech, J. V. Keane, M. J. Mumma, J. L. Siefert, and D. J. Werthimer, eds.

The Earliest Ion Channels

A. Pohorille,[1,2] M. A. Wilson,[1,2] and C. Wei[1,2]

[1]*Exobiology Branch, MS 239-4, NASA Ames Research Center, Moffett Field, CA 94035*

[2]*Department of Pharmaceutical Chemistry, University of California San Francisco, 600 16th St., San Francisco, CA 94143*

Abstract. Supplying protocells with ions required assistance from channels spanning their membrane walls. The earliest channels were most likely short proteins that formed transmembrane helical bundles surrounding a water-filled pore. These simple aggregates were capable of transporting ions with efficiencies comparable to those of complex, contemporary ion channels. Channels with wide pores exhibited little ion selectivity but also imposed only modest constraints on amino acid sequences of channel-forming proteins. Channels with small pores could have been selective but also might have required a more precisely defined sequence of amino acids. In contrast to modern channels, their protocellular ancestors had only limited capabilities to regulate ion flux. It is postulated that subsequent evolution of ion channels progressed primarily to acquire precise regulation, and not high efficiency or selectivity. It is further proposed that channels and the surrounding membranes co-evolved.

1. Introduction

It is commonly believed that the emergence of cells in the beginnings of life was preceded by protocells — evolving, membrane-bound structures that were endowed with some, but not all, cellular functions. The essential functionalities of protocells include transport of nutrients, waste products and simple ions across cell walls. Small neutral molecules readily permeate cell membranes. In contrast, nutrients with ionizable groups, such as amino acids or nucleotides permeate cell walls at much lower rates. Furthermore, membranes are essentially impermeable to small ions, such as Na^+, K^+, Cl^- or Ca^{2+}.

The small permeabilities of membranes to ions can be readily understood from electrostatic considerations. Ions interact favorably with water, which is a highly polar medium, but not with non-polar, oily interior of bilayer membranes. The free energy of transferring an ion from water to the center of the bilayer is therefore large and unfavorable, creating a barrier (often called the Born barrier) to ion transport. Even though membrane deformations in response to the presence of an ion reduce the Born barrier by allowing water to penetrate into the membrane interior (see Fig. 1), this barrier remains substantial, ~ 27 kcal/mol for Na^+ (Wilson & Pohorille 1996).

High barriers to unassisted permeation mean that this is not a viable mechanism for supplying protocells with small ions. This constitutes a serious, evolutionary problem because ion transport is needed to regulate osmotic equilibria in

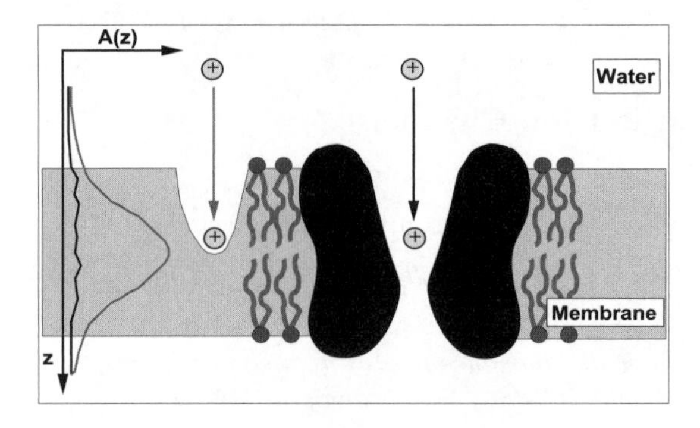

Figure 1. Schematic of unassisted (left) and assisted (right) ion transport through a membrane. The change in free energy associated with unassisted transport is shown in the curve to the left.

protocells, to provide charged species for the catalysis of chemical reactions, and to maintain macromolecular structures. Furthermore, ion transport is universally involved in bioenergetics and transmission of environmental signals. One way to reduce the Born barrier is to form water-filled pores (see Fig. 1), which act as conduits for ions by creating a locally polar environment across the membrane (Parsegian 1969). In nature, this is the function of protein ion channels.

In this paper, we review the basic properties of transmembrane ion channels, discuss a simple model of a protocellular channel, and draw some conclusions about the earliest evolution of assisted ion transport. Protobiologically relevant mechanisms for the synthesis of channel-forming proteins, their folding, insertion into membranes and aggregation into channels are not covered, but they were recently reviewed in the context of the origins of life (Pohorille, Schweighofer, & Wilson 2005) and membrane biophysics (White & von Heijne 2005).

2. Models of the Earliest Ion Channels

2.1. Structural and Functional Features of Contemporary Ion Channels

Ion channels are protein assemblies that span membranes to create ion-conducting pores. They are ubiquitous in all kingdoms of life. In higher organisms, they are most abundant in cells of the brain, the central nervous system and the heart. They are also quite common in microorganisms, although their functions are not always known. Ion channels are among the largest, functionally complex and most difficult to study proteins. In the last decade, however, we have learned a great deal about their structure from X-ray crystallography, which in turn shed light on their mechanisms of action. Despite the apparent complexity of ion channels, their transmembrane segments have remarkably simple architectures. Some have structures called β-barrels, but a majority are simply bundles of α-helices, which are the most common structural motifs in proteins. Helices are typically 16-25 amino acids in length, which is enough to span the membrane. The ion conducting pore is formed by 4 to 8 central helices, which often contain polar amino acids directed towards the pore in the channel.

Besides facilitating efficient ion transport, most ion channels exhibit two other, essential features. One is the ability to select ions that are being transported with respect to the charge (negative *vs.* positive ions) or valence (monovalent *vs.* divalent ions). Some channels show preferences for only one type of ions, e.g. Na^+ or K^+. For this purpose, they possess structural elements called selectivity filters, usually located near the mouth of the pore. The second feature is gating, which means that channels can exist in states that are either open or closed to ion flux. Channel opening is precisely regulated by environmental signals provided by small molecules, voltage, local pH or mechanical force exerted, for example, by the membrane. These types of channels are called, respectively, ligand-gated, voltage-gated, acid-sensing and mechanosensitive. To respond to signals, ion channels possess sensors or binding sites, which are frequently located in water-soluble domains, and fast, accurate and complex mechanisms of converting the signal into structural changes that lead to channel opening.

2.2. Antiamoebin as a Model of Protocellular Ion Channels

On the basis of phylogenetic analysis of amino acid sequences in different families of ion channels, considerable progress has been made towards understanding their evolutionary history. However, genetically coded channels from even the simplest microorganisms are still quite complex and shed relatively little light on their protocellular ancestors. Fortunately, there are naturally occurring or synthetic proteins that are good models of primitive ion channels. One group of such models are viral channel proteins, which are markedly smaller than their bacterial counterparts, and sometimes can be further truncated without appreciable loss of essential functions (Fischer & Sansom 2002). Some synthetic channels are even simpler, of which the best studied is built of just two types of amino acids - non-polar leucine and polar serine (Lear et al. 1997).

Especially interesting models are simple channels formed by natural, non-genomically coded proteins (peptides), which in nature exhibit antimicrobial activity. Here, we consider one member of this family, antiamoebin (AAM), a helical peptide 16 amino acids long. In response to transmembrane electric fields, AAM helices can insert into lipid bilayers and self-assemble into channels (Duclohier, Snook, & Wallace 1998). As shown in Figure 2, the AAM channel most likely consists of 8 monomers (O'Reilly & Wallace 2003), which surround a wide, water-filled pore. In the middle, the pore radius is 4.5 Å and widens at both ends to approximately 7-8 Å. Its size is sufficient for a Na^+ or K^+ ion to pass through most of the channel with its first hydration shell intact. In the narrowest region of the channel, a small fraction of the hydration shell is lost, but this is compensated by favorable interactions between the ion and carbonyl or hydroxyl groups of the protein. Thus, Na^+ and K^+ ions pass though the channel fully solvated by polar molecules or groups, encountering a free energy barrier of only $\sim$ 1 kcal/mol (Wilson et al. 2007). For Cl^-, a slight ion dehydration near the center of the channel is not compensated by favorable interactions with the peptide, which yields a somewhat larger barrier of $\sim$ 3.5 kcal/mol. In all cases, however, the barrier is greatly reduced, compared to the barrier to

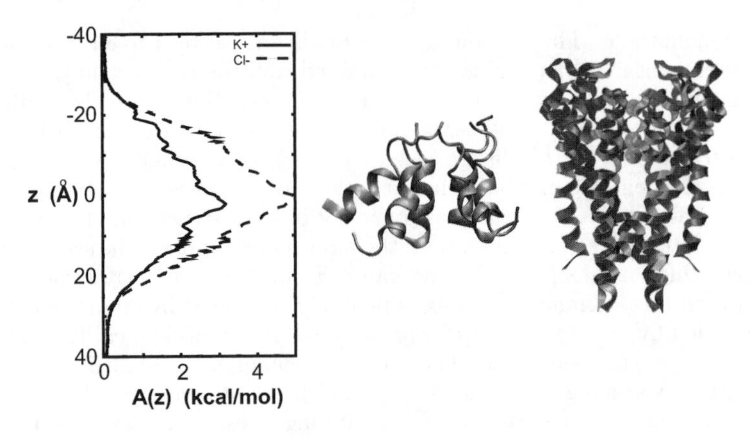

Figure 2. Free energy curves (left) for ion transport through the AAM channel (center). The pore-forming helices of KcsA are on the right side.

permeation through membranes. This indicates that the AAM channel, despite its simplicity, is very efficient at mediating ion transport.

An interesting feature of AAM is the presence of glutamine near the center of the helix. It was proposed that the highly polar side chain of this amino acid points towards the pore to assist directly ion transport (O'Reilly & Wallace 2003). Computer simulations, however, reveal that its role is different, but equally essential (Wilson et al. 2007). Glutamine side chains from each helix in the channel form hydrogen bonds with neighboring helices, which stabilize the channel structure by preventing individual helices from dissociating. This points to an evolutionarily convenient property of simple channels: their integrity can be controlled by local features of the amino acid sequence and does not require specific sequence of all residues in the helix.

2.3. From Simple to Complex Channels: Evolutionary Implications

In order to understand how simple ion channels might have worked in protocells it is helpful to compare basic properties of AAM channel with those of a voltage gated, bacterial channel, KcsA, which has been well studied both structurally and functionally (Doyle et al. 1998; Zhou & MacKinnon 2003). In both channels, the pore is surrounded by 8 helices and has a funnel shape (see Fig. 2). It has been suggested that these structural similarities may imply a similar mechanism of action (O'Reilly & Wallace 2003). This, however, does not appear to be the case. At the mouth of the pore KcsA has a narrow fragment built of five highly conserved amino acids, which acts as a selectivity filter. There are four binding sites for K^+ along the filter. At high ionic concentrations, two sites are occupied, which is the key to the high conductivity of this channel. Near the middle of the bilayer the pore widens to form a water-filled cavity. Interactions of ions with water in the cavity and polar groups from the surrounding helices markedly lower the Born barrier (Zhou & MacKinnon 2003). In contrast, the mechanism of ion transport through the AAM channel is much simpler. It does not involve multiple binding sites, concerted ion movement or direct participation of amino acid side chains. Yet, as shown from free energy calculations (Roux et al. 2000;

Wilson et al. 2007), ions pass through both channels without experiencing Born barrier. This indicates that the complex mechanism of ion conductance evolved in KcsA for reasons other than increasing efficiency of transport.

In contrast to AAM, KscA is ion selective. In a narrow selectivity filter, ions undergo substantial dehydration and favorable interactions with water molecules are substituted by equally favorable interactions with polar groups of the protein. Compatibility between pore and ion sizes is important to selectivity. Large ions are excluded from narrow pores or their segments whereas ions that are too small are selected against because they cannot interact optimally with all polar groups lining the pore (Gouaux & MacKinnon 2005). Flexibility of the pore and repulsive interactions between polar groups also influence selectivity (Noskov & Roux 2006). These considerations mean that sufficiently narrow simple channels can also be selective. One example is alamethicin, which is closely related to AAM. At high conductance levels it forms large pores made of 11 helices. Then, the channel exhibits no selectivity indicating that ions pass through in a fully hydrated state. In contrast, at low conductance levels the channel is markedly smaller and shows a selectivity sequence $K^+ > Na^+ > Li^+$ (Sansom 1993).

KcsA function requires it to be precisely gated. Only in the presence of an electric field a sequence of conformational transitions occurs that opens the channel to ions (Schmidt, Jiang, & MacKinnon 2006). The AAM and alamethicin channels, in contrast, exhibit a very primitive mechanism of regulation: the channel is open only so long as the helix bundle remains intact.

Membranes are not just passive barriers to ion permeation. Instead, channel gating might depend on membrane composition. For example, voltage-gated ion channels function properly in phospholipid bilayers, but not in membranes formed by other molecules. This is because of specific electrostatic interactions between the phospate groups and the protein. It has been suggested that this could be an evolutionary adaptation to the phospholipid composition of cell membranes (Schmidt et al. 2006). In another example, regulation of mechanosynthetic channels, which are activated by force exerted by membranes, requires the presence of anionic lipids that interact with positively charged residues in the protein (Powl, East, & Lee 2005). Membrane even plays a role in regulating a very simple AAM channel, which only forms in certain membranes (O'Reilly & Wallace 2003). This leads to a suggestion that ion channels and membranes, in which they were embedded, might have co-evolved.

3. Conclusions

The earliest ion channels might have been simply bundles of short, α-helical peptides spanning a membrane, as seen in AAM. If a pore was sufficiently wide for simple ions to move through the channel with their first hydration shell intact ions could pass the membrane at nearly diffusive rates. Since amino acid side chains played only a minor role in assisting transport, constraints on the amino acid sequence were rather modest. Perhaps the main requirement was the existence of sufficiently strong helix-helix interactions to ensure stability of the channel. Local interactions involving amino acids in the center of the helices might have been sufficient for this purpose.

Channels similar to AAM exhibit poor ion selectivity, mainly because of their large pore size. In this respect they are similar to the large mechanosynthetic channel, MscL, which protects cells from osmotic swelling. This similarity might not be accidental because osmotic regulation and supplying ions to protocells for biochemical processes were probably the earliest functions of ion channels, and neither required ion selectivity. Selective transport, needed for other functions, requires selectivity filters, as is the case for KcsA, but can also be exhibited by simple channels with narrow pores. This implies that ions undergo energetically unfavorable dehydration. To retain a low Born barrier ions must interact favorably with pore-lining groups of the protein. This almost always imposes constrains on the sequence of amino acids. Thus, we hypothesize that simple, ion selective channels evolutionarily followed AAM-like channels.

Although primitive channels could have been efficient and selective, they were not capable of precise regulation of ion transport. The main mechanism for deactivation ("closing") of these channels was dissociation of its helices. Thermodynamic equilibria between the "open" and "closed" states might have been influenced by environmental factors such as electric field due to charge imbalance between the interior and exterior of the protocell or mechanical forces due to cell swelling. These simple modes of regulation, however, were quite inadequate for more advanced functions (e.g. related to signaling or bioenergetics) acquired by protocells during their evolution. Thus, we propose that the impetus for evolution of ion channels to complex contemporary structures was provided not by the requirements for high efficiency or selectivity but for precise regulation.

Acknowledgments. This work was supported by grants from the NASA Exobiology Program and the NASA Astrobiology Institute.

References

Doyle, D. A., Cabral, J. M., Pfuetzner, R. A., Kuo, A., Gulbis, J. M.. Cohen, S. L., Chait, B. T., MacKinnon, R. 1998, Science, 280, 69
Duclohier, H., Snook, C. F., & Wallace, B. A. 1998, Biochim. Biophys. Acta, 1415, 255
Fischer, W. B., & Sansom, M. S. P. 2002, Biochim. Biophys. Acta, 1561, 27
Gouaux, E., & MacKinnon, R. 2005, Science, 310, 1461
Lear, J. D., Schneider, J. P., Kienker, P. K., & DeGrado, W. F. 1997, J. Am. Chem. Soc., 119, 3212
Noskov, S. Y., & Roux, B. 2006, Biophys. Chem., 124, 279
O'Reilly, A., & Wallace, B. A. 2003, J. Peptide Sci., 9, 769
Parsegian, V. A. 1969, Nat, 221, 844
Pohorille, A., Schweighofer, K., & Wilson, M. A. 2005, Astrobiology, 5, 1
Powl, A. M., East, J. M., & Lee, A. G. 2005, Biochemistry, 44, 5873
Roux, B., Berneche, S., & Im, W. 2000, Biochemistry, 44, 13295
Sansom, M. S. P. 1993, Q. Rev. Biophys., 26, 365
Schmidt, D., Jiang, Q. X., & MacKinnon, R. 2006, Nature, 444, 775
White, S. H., & von Heijne, G. 2005, Curr. Opin. Struct. Biol., 15, 378
Wilson, M. A., & Pohorille, A. 1996, J. Am. Chem. Soc., 118, 6580
Wilson, M. A., Wei, C., Bjelkmar, P., Wallace, B. A., & Pohorille, A. 2007, Biophys. J., in preparation
Zhou, Y., & MacKinnon, R. 2003, J. Mol. Biol., 333, 965

Bioastronomy 2007: Molecules, Microbes, and Extraterrestrial Life
ASP Conference Series, Vol. 420, 2009
K. J. Meech, J. V. Keane, M. J. Mumma, J. L. Siefert, and D. J. Werthimer, eds.

Theoretical Studies of Glutamate Enantiomers at Different pH Conditions

J. González, C. F. Corredor, and L. R. Lareo

Pontificia Universidad Javeriana, School of Sciences, Department of Nutrition and Biochemistry, Computational and Structural Biochemistry and Bioinformatics, Bogota, DC, Colombia

Abstract. Life, at least as we know it, is invariably linked to homochirality. Here, we present a new approach to understanding the homochirality using glutamic acid as a model. Glutamic acid is the key amino acid in living organisms. It was probably formed abiotically under conditions prevailing in the early Earth's atmosphere and its two enantiomers have been found in meteorites. The two enantiomers, however, do not exhibit biological activity to the same extent. The D-enantiomer usually has either very little or no activity as compared to the L enantiomer. In order to find an explanation for the differences, we have undertaken theoretical studies with classical and semi-empirical force field models on solvated L- and D-glutamate. Our results show strong interactions between charged atoms of the D-enantiomer that determine a very rigid spatial configuration resulting in a low capacity to adapt to different environmental conditions and to respond to energy perturbations. Our results might provide a possible explanation for the natural selection of the L-glutamate enantiomer as a basic constituent of living matter.

1. Introduction

One of the landmarks of terrestrial biochemistry is its homochirality, being based predominantly upon the L-amino acids and D-sugars to the virtual exclusion of their infrequent enantiomers D-amino acids and L-sugars. For the presence of racemic and scalemic mixtures of amino acids enantiomers in extraterrestrial materials, as meteorites, three main hypotheses were considered: a) they are products of extraterrestrial life, b) they are abiotic compounds from early stages of the solar nebula, and c) they are result of terrestrial contamination. The last hypothesis, the terrestrial contamination, will be discarded based on the data of concentrations identified in several of the meteorites and chondrites. McDonald & Bada (1995) based on the detection of amino acids, mainly of L-enantiomers in the Martian meteorite EETA 79001, considered these as terrestrial contaminants. The relative order of abundance found in this samples was *Glu* > *Ser* > *Gly* > *Ala* > *Asp*, considering only the amino acids commonly found in proteins. Brinton et al. (1998) studying Antarctic micrometeorites (AMMs) identifies an excess of L-enantiomers of amino acids and by this fact considered as terrestrial contamination. The abundance order found in these AMMs is: *Ala* > *Glu* > *Gly* > *Ser* > *Asp*, again presenting only the protein most common in amino acids. Similar results appears after the analysis of other samples, such as Allan Hill carbonaceous chondrite, meteorite 77306 (Kotra et al (1979)), where the abundance order found was: $Gly > Ala > Asp > Ser > Val > Thr > Leu > Ile$, in Martian meteorite Nakhla the abundance order is $Gly > Glu > Ser > Asp > Ala$. At last, as examples, in Murchinson carbonaceous meteorite the relative abundance order is: $Gly > Ala > Glu > Val > Ile > Leu > Ser$ (Glavin et al. (1999)).

From the previous examples, selected from all the reported data, emerge some clear facts: 1) the abundance of amino acids is not the same in all samples, 2) a significant number of amino acid not common on Earth are found in meteorites, and 3) the amounts of amino acids in extraterrestrial samples differ completely to the frequencies found in terrestrial samples for common amino acids in proteins. Based on the reports of SwissProt data bank these abundances are: $Leu > Ala > Gly > Ser > Val > Glu > Ile > Thr > Arg > Asp$. The only common fact is the presence of the same amino acids but the frequencies are different. A question that emerges is: does this fact depend on the modernity of the samples reported by the databank ? The data cited by Kvenvolden (1975) is useful for addressing this hypothesis. The order of amino acid abundances in Middle Middle Miocene (Mercenaria) and modern fossils are also different from the SwissProt and from the extraterrestrial samples, these are: for Miocene $Ala > Pro > Glu > Val > Leu > Lys$, and for modern fossils: $Asp > Gly > Ala > Ser > Glu > Leu > Pro$. With this data, it seems that the amino acids, in almost all extraterrestrial samples, belong to extraterrestrial sources and do not constitute terrestrial contamination. Following this, a new question arises:- How do these amino acids resist not only the different extreme conditions in space but also those inside Earth's atmosphere ? The studies of Abelson (1954, 1959) suggest a plausible answer. Abelson determined a stratification of stability to temperature by heating amino acid solutions at different times. It was found that the amino acids Ala, Gly, Glu, Leu, Ile, Pro and Val are relative stable molecules. Abelson found that Asp, Lys, and Phe were intermediary stable amino acids and the relatively unstable amino acids were Ser, Thr, Arg, Tyr. With this data Abelson concluded that these molecules are sufficiently stable to remain intact for more than a billion years. It is important to note the identified extraterrestrial amino acids consists mainly of the relative resistant group. This fact was confirmed in studies of the oxidative conditions (Khan & Souden (1971)), which showed the same stable amino acids. A further interesting fact in identifying meteoritic amino acids with having an extraterrestrial origin is the presence of amino acids that are uncommon on Earth's surface, such as norvaline, isovaline, norleucine, 2-methylnorvaline, pseudoleucine,amongst the amino acids identified in the Murchinson carbonaceous meteorite. It is still not easy to select the correct hypotheses presented above. However, the presence of these amino acids in the Murchinson meteorite, which is 3.4 billion years old, appears to suggest that abiotic extraterrestrial syntheses is the likeliest possibility but we can not exclude the possibility that extraterrestrial life is the origin (Cronin, Gandy & Pizzarello (1981)).

An important fact is that, in almost, all extraterrestrial samples there is a scalemic mixture between the two amino acids enantiomers. In general, the D/L enantiomers ratio in most samples range from 0.021 up to 0.670 (Botta & Bada (2002)). The origin of the enantiomers mixtures has several hypotheses. Bonner & Rubenstein (1987) suggest that ultraviolet circularly polarized light (CPL), of a particular handedness, emitted by neutron start could lead to either preferential syntheses or degradation of specific enantiomers of chiral compounds either in the interstellar molecular clouds or in planetary bodies. This is known as Rubenstein-Bonner hypotheses. Recently, it has been shown that amino are formed by irradiation of interstellar/circumstellar ice analogs with ultraviolet (UV) CPL, produced by a synchrotron radiation beamline, and thus allowed us to quantify the effect of such polarized light on the production of amino acids (Nuevo et al. (2007)). These results can be compared to the enantiomeric excesses measured in primitive meteorites such as Murchison. Another hypothesis of chirality is the parity violation energy differences (PVED's) that represents an essential property of particle and atomic handedness used to cope with the complex phenomenon of asymmetry in the universe. This proposal is based on the findings of Lee & Yang , in 1956, who proposed that parity is not always conserved. Their classical paper 'Question of parity conservation in weak interactions' analyzed numerous cases of strong and weak interactions and showed that the evidence for parity conservation applies only to the first interaction and not to the weak one. The absolute magnitudes of such PVED's are too small to be measured experimentally by our current available instruments, but they can be evaluated by theoretical calculations. This origin and the amplification of the effect by the living matter could explain the fact of the stereoselectivity in living matter on Earth (Avalos et al. (2000)).

Studies of D- and L- enantiomers of amino acids have added new dimensions to our understanding of the occurrence and fate of amino acids in terrestrial and extraterrestrial samples. The present study was developed with glutamic acid as example. Glutamic acid was chosen for a number of reasons. Practically all other amino acids can be metabolically derived from it (Morowitz (2002)), and following the results of Breslow & Levine (2006), it appears likely that there is partial transfer of enantioselective chiralities to all molecules derived from it. As mentioned previously, it is present in several meteorites and other extraterrestrial sources (Engel & Macko (2001)), and glutamate chronology studies show racemization in foraminifers fossils (Takano et al. (2006)). Moreover, it is present as scalemic mixtures in meteorites (Engel & Macko (2001)), which apparently is one of Earth's prebiotic amino acids (Miller (1974); Delaye & Lazcano (2005)), and was demonstrated recently to play an important role in the evolution of the hominoid brain (Burki & Kaesmann (2004)). It is also a key excitatory neurotransmitter in higher organisms (Dingledine et al. (1999)) and is utilized by the food industry as one of the major food additives and it is responsible for the umami flavor (McCabe & Rolls (2007)). In addition, just as with other amino acids, it occurs as two enantiomers: L and D-Glu; both enantiomers present different interaction capacities with their receptor and transporters (Moriyoshi et al. (1991)), and there are binding differences with calcium and other minerals by the two enantiomers (Hazen Filley & Goodfriend (2001)). Thus, using glutamic acid, and through theoretical and computational processes, we present a proposal that (in conjunction with other hypotheses) CPL and PDVE's explain the Earth living matter stereoselectivity for L- type enantiomers.

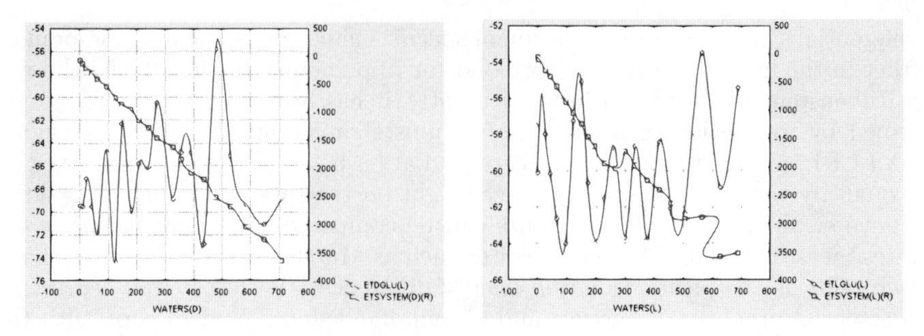

Figure 1. Energy landscapes, for the total system and for different D- and L-Glu in all established conformations.

2. Methods

Initially, L and D-Glu structures *in vacuo* was derived from Biopolymer program from InsightII, then the charges where modify, at different pH values, with charged hydrogen option. These molecules where solvated with layer from 0 to 21 Å that is from *in vacuo* up to about 500 water molecules by glutamate molecule, using Solvation program from InsightII. Using the program Discover 3 the total, potential and kinetic energies from the system, the glutamate and the solvent was calculated by, both, energy minimization and molecular dynamics following Charmm parameterizations. Once the structures where optimized, we calculated other molecular parameters as dipolar moments, using the program MM2, from classical molecular mechanics. For this optimized structures also all the internal variables where calculated with the program MOE. Using the same program partial atomic charges where calculated following the method of Gasteiger (PEOE). With these optimized structures was generated all the possible hydrogen bonds for each conformer and the same conformers where compared by the VMS programs using the RMS as spatial similarity indicator. All the statistical analysis where developed with the program Statistica 6.0 and graphs where developed with the programs Statistica 6.0 and Mathematica 3.0.

3. Results

Glutamic acid has partial charges, even if the total net charge is null, at different pH values. This fact has a strong effect on the results of conformers' calculation. Though the results shown here belong to the neutral pH value, the most abundant form in biological samples on earth, the trend is similar for all the pH range evaluated. The evaluation of energetic values, potential, kinetic and total energy, for all the conformations that both enantiomers can acquired reports the existence of different energy landscape for D- and L-Glu as appears represented in Figure 1. For L-Glu the minima has similar values, unlike the values found for D-Glu.

The potential energy values are -74.267 ± 3.061 and $-81.540 \pm 4.379 KJ/mol$ for L- and D-Glu respectively. The, roughly, 10% difference in mean values implies a possible major stability for D-enantiomer against L-Glu. This appears confirmed using a simple model as: $\Delta E(L-Glu, D-Glu) = E_t(L-Glu) -$

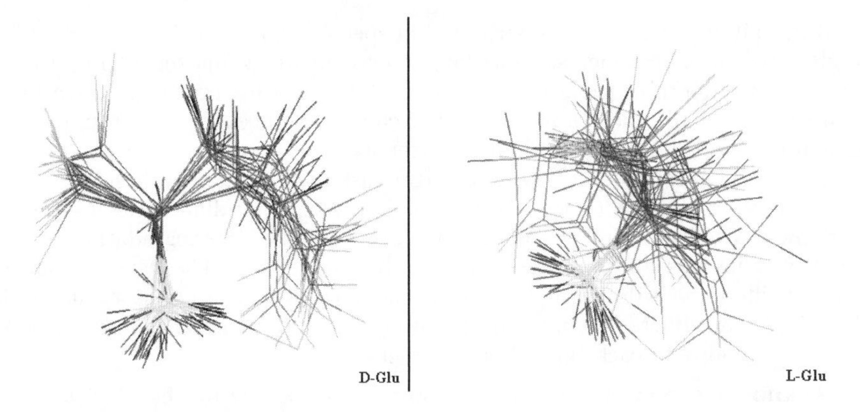

Figure 2. Representation of all found D- and L-Glu conformations with identical solvation levels for both enantiomers.

$E_t(D-Glu)$, where if $\Delta E(L-Glu, D-Glu) > 0$ then D-Glu is the more stable structure, and the average for all identical conditions is $\Delta E = 7.344 \pm 5.766$. The representation of all the conformations acquired by D- and L-Glu in identical environmental conditions are shown in Figure 2.

From this study, it is apparent that L-Glu is a more flexible molecule than D-Glu. After the calculation of several molecular features, with the goal understanding the factors that rule their behavior, we found that the partial charges of possible charged atoms at different environmental pH's is useful for generating a multiple correlation model that shows this as a major source of D-enantiomers low flexibility, implying low resilience and consequently it is significantly more fragile with respect to L-Glu. L-Glu is a more flexible and resilient molecule that can support strong environmental changes and energy stress, absorbing the energy differences thus changing the conformations by the flexible lateral chain. The equations describing these findings are: For L-Glu $Q_p(N) = -1.31 - 0.10Q_p(OXT) - 0.32Q_p(O) - \underline{1.30}Q_p(OE1) - \underline{1.60}Q_p(OE2)$, $R^2 = 0.7145$; $p = 0.0003$, and for D-Glu $Q_p(N) = -0.49 - \underline{0.33}Q_p(OXT) - 0.18Q_p(O) - \underline{0.70}Q_p(OE1) - \underline{0.73}Q_p(OE2)$, $R^2 = 0.5317$; $p = 0.0121$, where Q_i are the partial charges of the charged atoms in the molecules. Underlined values are statistically significant at $p = 0.05$ level.

4. Conclusions

From the present study, with glutamate as example, it is possible to present the following general conclusions: the D-Glu enantiomer is more stable, less flexible and it presents lower resilience than the L-Glu enantiomer. The lower flexibility of D-Glu limits its possibilities to interact with different ligands, while at the same time the higher stability of the D-Glu enantiomer makes it more susceptible to degradation by external energetic stress such as thermal perturbations and photolysis. If stability is related to energy values, the D-Glu enantiomer requires more energy for its synthesis, since it shows higher energy values. This trend is independent of pH due to the fact that in the entire pH range the glutamic acid exhibits partial charges distributed in specific regions of the molecule. The higher D-Glu enantiomer stability might explain living matter's "preference" for a more adaptable molecule such as L-Glu. This feature might have been transmitted to all other amino acids that could have been derived from it. The global proposal to explain the living matter stereoselectivity present two phases: enantiomeric excess (ee) following the Rubenstein-Bonner hypothesis for selective synthesis along with the parity violation of energy differences and latter the low flexibility of one of the enantiomers facilitates the degradation of this kind of molecules allowing the excess of the other type. The parity violation energy differences will be reflected, using the molecular orbitals theory, in subtle but significant differences in the overlapping matrix of bonding electrons among atoms belonging to each type of enantiomers.

Acknowledgments. The present work was possible by the financial support of Pontificia Universidad Javeriana.

References

Abelson, P. H. (1954) Carnegie Inst. Wash. Yearb. 53, 97-101.
Abelson, P. H. (1959) In: Reseraches in Geochemistry, 1, 79-103. New York, Wiley.
Avalos, M. R. Babiano, P. Antas, J. L. Jimenez & J. C. Palacios (2000) Tetrahedron: Assym. 11, 2845-2874.
Bonner, W. A. & E. Rubenstein (1987) Biosystems 20, 99-111.
Botta, D. & J. L. Bada (2002) Surv. Geophys. 23, 411-467.
Breslow, R. & M. S. Levine (2006) Tetrahedron Lett. 47, 1809-1812.
Brinton, K. L. F., C. Engrand, D. P. Glavin, J. L. Bada & M. Maurette 1998 Orig. Life Evol. Biosphere 28, 413-424.
Burki, F. & H. Kaesmann (2004) Nature Genetics 36, 1061-1063.
Cronin, J. R., W. E. Gandy & S. Pizzarello (1981) J. Mol. Evol. 17, 265-272.
Delaye, L. & A. Lazcano (2005) Phys. Life Rev. 2, 47-64.
Dingledine, R., B. Georges, K., Bowie, D. & Traynelis, S. (1999) *Pharm. Rev.* 51, 7-61.
Engel, M. H. & S. A. Macko (2001) *Precambrian Res.* 106, 35–45.
Glavin, D. P., J. L. Bada, K. F. L. Brinton & G. D. McDonald (1999) Proc. Nat. acad. Sci USA 96, 8835-8838.
Hazen, R. M. Filley, T. R. Goodfriend (2001) Proc. Natl. Acad. Sci. USA 98, 5487–5490.
Khan, S. U. & F. J. Souden (1971) Geochim. Cosmochim. Acta 35, 854-858.
Kotra, R. K., A. Shimogama, C. Ponnamperuma & P. E. Hare 1979 J. Mol. Evol. 13, 179-184.
Kvenvolden, K. A. (1975) Ann. Rev. Earth Planet. Sci. 3, 183-212.
Lee, T. D. & Yang, C. N. (1956) Phys. Rev. 104,254–258.
McCabe, C. & E. T. Rolls (2007) Eur. J. Neurosci. 25, 1855-1864.
McDonald, G. D. & J. L. Bada Geochim. Cosmochim Acta 1995 59, 1179-1184.
Miller, S. L. (1974) *Orig. Life* 5, 139-151.
Moriyoshi, K., M. Masu, T. Ishii, R. Shigemoto, N. Mizumo & S. Nakanishi (1991) Nature 354, 31-37.
Morowitz, H. J. (2002) *Manuscript.* Krasnow Institute for Advanced Studies. George Mason University, VA.
Nuevo, M., U. J. Meierhenrich, L. d'Hendecourt, G. M. Muñoz Caro, E. Dartois, D. Deboffle, W. H. P. Thiemann, J. H. Bredehöft & L. Nahon (2007) Adv. Space Res. 39, 400-404.
Takano, Y. Kobayashi, K. Ishikawa, Y. & Marumo, K. (2006) Organic Geochemistry 37, 334–341

Session VIII

Life in Extreme Environments

Susan Leschine

Jackie Denson

Bioastronomy 2007: Molecules, Microbes, and Extraterrestrial Life
ASP Conference Series, Vol. 420, 2009
K. J. Meech, J. V. Keane, M. J. Mumma, J. L. Siefert, and D. J. Werthimer, eds.

Sulfate Reduction: A Model for Subsurface Martian Life

J. Denson, V. Chevrier, D. Sears, and D. M. Ivey

Arkansas Center for Space and Planetary Sciences, University of Arkansas, Fayetteville, AR 72701

Abstract. ESA's Mars Express and NASA's Mars Exploration Rovers have discovered extensive deposits of sulfate minerals on the surface of Mars. These findings provide some of the most compelling evidence that there were once large deposits of liquid water on the martian surface. We have performed a kinetic study investigating the stability of $MgSO_4$ brine solutions under simulated martian conditions. This included precise control of temperature (O ° C), pressure (6-7 mbar) and atmospheric composition (100% CO_2). Our results indicate that highly concentrated brines (20-25 wt%) dramatically alter the evaporation/sublimation rates when compared to water/water ice, with a rate approximately 10 fold lower than expected. These concentrated brines also promoted the existence of metastable liquids, which suggests a unique niche for martian life. Sulfate reduction is thought to be one of the earliest forms of respiration to arise on the early Earth. This is based on a wide range of evidence, including phylogenetic studies of present day sulfate reducers and isotopic fractionation studies of sulfur bearing minerals. Therefore sulfate-reducing microorganisms are an intriguing group to serve as a model for potential martian life. The ability of sulfate-reducing microorganisms to survive and potentially grow under simulated martian surface and subsurface conditions is currently being investigated. This research will explore the utility of sulfate reduction as a form of respiration for martian microbes, give valuable insight into possible biosignatures produced by such organisms on Mars, and provide an initial glimpse into the energetics of such a system.

1. Introduction

1.1. Sulfates and Mars

Liquid water is known to be unstable on the surface of Mars, although gullies found on the valley and crater side suggest recent liquid water activity. The surface conditions on Mars are very close to the triple point of water, and under some conditions metastable liquid water could form (Heldmann *et al.* 2005). Moreover, solutes in the water may account for the stability of liquid water on Mars (Brass 1980), mainly because of diminution of the freezing point. The lower temperatures of brines enhance stability of the liquid phase by decreasing the evaporation rate (Sears & Chittenden 2005). Recent exploration of the martian surface has revealed abundant $MgSO_4$ salts (Gendrin *et al.* 2005; Squyres *et al.* 2004), suggesting that martian brines might contain high sulfate concentrations, which also could account for the presence of water in the regolith (Vaniman *et al.* 2004). Mg-sulfate brines present a eutectic temperature higher than halides, i.e.

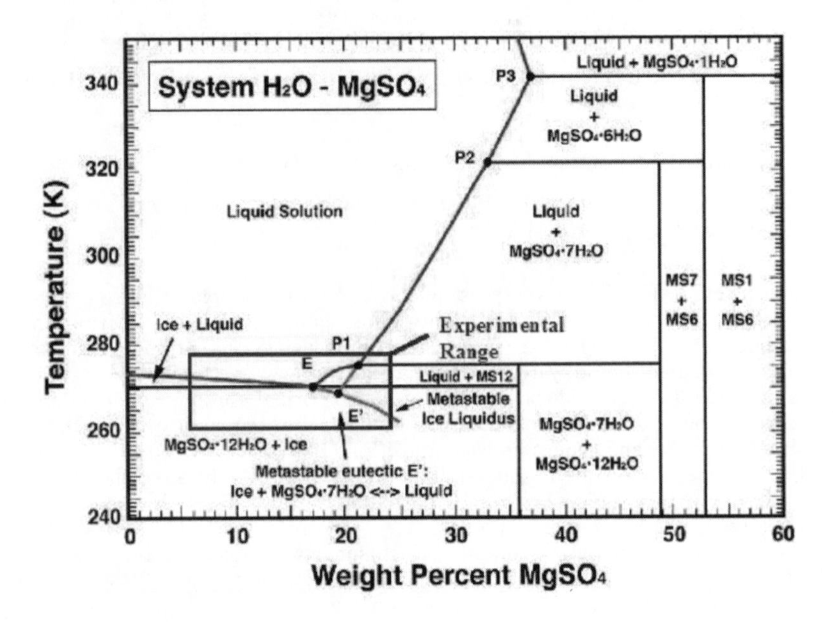

Figure 1. $MgSO_4$ phase diagram.

–3.6 ° C versus –21.2 ° C and –49.8 ° C for NaCl and $CaCl_2$ (Sears & Chittenden 2005), respectively, and can exist in a variety of hydrated states (Figure 1).

The specific goal of this series of experiments is to investigate the stability of $MgSO_4$ brines under simulated Mars surface conditions. This will contribute to the overall knowledge of discovering the potential of sulfates for serving as a possible source of subsurface water on Mars as well as their possible role in contributing to metastable water on the martian surface. A series of experiments were performed in a Mars simulation vacuum chamber, under a 100% CO_2 atmosphere with a constant pressure between 6 and 7 mbar, and a temperature of 0 ° C. Various brine concentrations ranging from 5 to 25 wt% were placed in petri dishes along with a thermal couple and placed onto an analytical balance located within the vacuum chamber. Mass loss was measured for a minimum of 2 h once the proper pressure was obtained. The first 20 min of data was excluded in order to allow the balance to adjust to the low-pressure environment. The observed evaporation rates for low brine concentrations (5-15 wt%) were in close agreement to data obtained for pure water under the same conditions. However the concentrated sulfate brines (20-25 wt%) evaporated more slowly than predicted even when the effects on water potential of the sulfates were considered (Figure 2).

Crystallization of the sulfates was routinely observed in the concentrated (20-25 %) brine solutions utilized in this study with the brine mixture consistently remaining as unfrozen slurry throughout our experiments. This study provides evidence supporting the hypothesis that sulfate minerals could conceivably

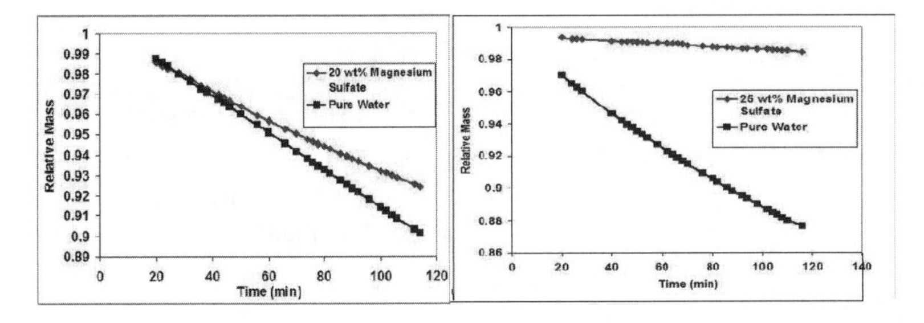

Figure 2. Mass loss of 20 wt% (*left*) and 25 wt% (*right*) $MgSO_4$ brines.

serve as a reservoir of subsurface water on the martian surface. This is intriguing from an astrobiology perspective in that two fundamental life-promoting components – a liquid medium (water or brine) and a metabolic energy sink – such as SO_4^{2-} as terminal electron acceptor - may be present, at least periodically, at or near the surface of Mars

1.2. Sulfate-Reducing Microorganisms

Several previous studies have investigated the ability of microorganisms to survive under martian surface conditions (Moll & Vestal 1992; Tauscher *et al.* 2006). Recently it was shown that under shallow subsurface conditions, given a minimal state of hydration, that even common bacteria can survive for limited times in a Mars-like environment (Diaz & Schulze-Makuch 2006). However, these studies have not taken into account the presence of sulfates at the surface, nor have they specifically investigated sulfate reduction as a model metabolic strategy for potential martian life.

It is thought that sulfate reduction was one of the earliest forms of respiration to arise on the early Earth. This is based on a wide range of evidence, including phylogenetic analysis of the enzymes utilized by present day sulfate reducers and evidence of biological (mass-dependent) isotopic fractionation of sulfur bearing mineral deposits dating to ~3.47 Gyr (Shen *et al.* 2001; Wagner *et al.* 1998). Therefore sulfate-reducing microorganisms are a valid system to consider as a model for potential martian life (Knoll *et al.* 2005). In this work, we propose to investigate the ability of selected representatives of sulfate-reducing microorganisms to survive under simulated martian conditions, identify the biosignatures indicative of sulfate reduction that these organisms would produce if they existed on Mars, and perform an analysis of how martian conditions - specifically low pressure - affect the biochemistry of these organisms. Psychrophilic sulfate reducing microorganisms, including *Desulfotalea psychcrophila*, *Desulfotalea arctica,* and the mesophilic *Desulfotomaculum halophilum* were initially chosen for experimentation. They represent chemolithoautotrophic, heterotrophic, and sporulating organisms. In addition, the availability of whole genome sequences for *D. psychrophila* make it an ideal candidate for elucidating the biochemical effects of survival and/or growth in a low-pressure environment such as the shallow subsurface of Mars.

Living organisms produce distinctive indicators of life. These include minerals of biological origin, gases, organics, and isotopic signatures in all of the

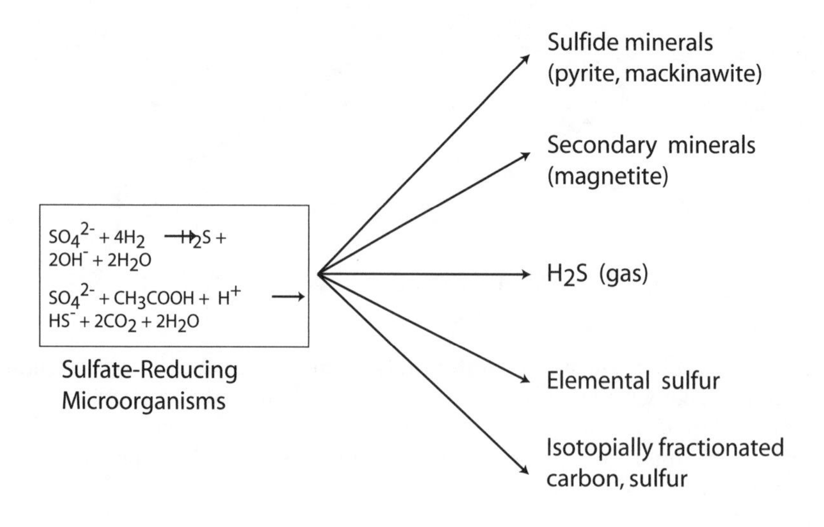

Figure 3. Potential biosignatures produced by sulfate-reducing microorganisms.

above. These signatures can be directly related to a biological process or be produced as a secondary effect of that process, such as mineral formation influenced by the geochemistry of the environment interacting with life's processes. The formation of biosignatures directly related to sulfate reduction can involve a wide variety of elements, including Fe, P, S, Ca, Mg and Mn, and compounds thereof, as depicted in Figure 3.

The analysis of mineralogical (elemental sulfur, sulfides such as pyrite, pyrrhotite and mackinawite) and gas (H_2S) signatures produced by biological sulfate reduction is a central focus of this study, this will include the identification of the biosignatures produced by living organisms under simulated martian conditions, as well as the assessment of the stability of these signatures under these conditions.

Organisms that survive or grow in the harshest environments on Earth tend to possess extraordinary, often unique physiologies, the study of which has broadened our understanding of life and provided valuable tools for biotechnology-based industry. Similarly, the physiologies that would be predicted of extraterrestrial organisms are likely to be extraordinary, particularly from an Earth-centered perspective, and have been the subject of considerable scientific curiosity. Future objectives of this research are to utilize a proteomics-based approach to determine novel biochemical changes associated with the growth or survival of these organisms under the low vacuum of the surface and shallow subsurface of Mars. In order to accomplish this task we will utilize two-dimensional gel electrophoresis, a biochemical technique that allows major changes in protein expression and/or modification to be detected. This will allow us to compare the proteins of pure cultures of the organisms grown under simulated martian conditions with those grown under defined terrestrial conditions, for example, where pressure is the only variable. Mass spectroscopy techniques will be utilized to precisely identify the proteins, thereby leading to a better understanding of how the physiologies of the microorganisms are being altered by the simulated Mars conditions. These experiments will provide initial insight into the biochemical pathways and energetics that might be associated with microbial survival and/or growth on Mars, or other low pressure systems.

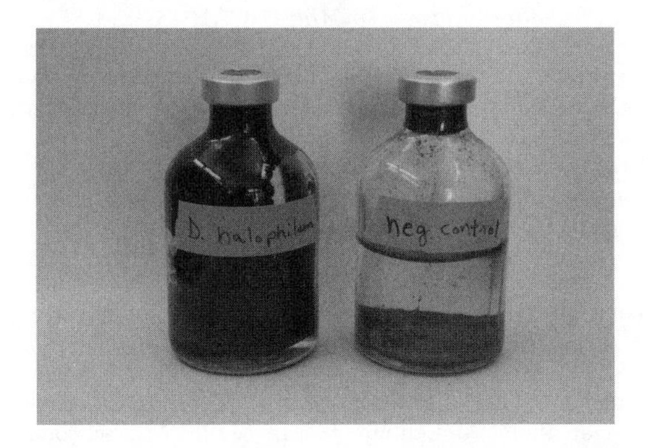

Figure 4. *Desulfotomaculum halophilum* inoculated culture (left) exhibiting growth, and uninoculated control vial after 7 d incubation at 50 torr/25 ° C. The black precipitate has been tentatively identified as an iron sulfide.

2. Discussion

Currently, experiments are underway utilizing the psychrophilic sulfate reducers *Desulfotalea psychcrophila*, *Desulfotalea arctica,* as well as the mesophilic spore forming *Desulfotomaculum halophilum.* A unique portable vacuum biochamber has been built that is entirely dedicated to this project. It will allow us to simulate martian surface and subsurface culture experiments under a wide variety of conditions, including psychrophilic conditions as well as subsurface conditions where positive pressures are encountered. Initial culture work utilizing the moderately halophilic, sporeforming bacteria, *Desulfotomaculum halophilium* has established that this organism can sustain growth under simulated shallow subsurface conditions (20-50torr) at 25 ° C. This was accomplished by placing serum vials inoculated with this organism beneath a layer (8 cm) of simulated martian regolith, JSC-1, while exposing the culture to the overlying 100% $C0_2$ atmosphere through dual syringe outlets. Interestingly this organism produces a mineral precipitate tentatively identified as an iron-sulfide under both terrestrial and simulated martian subsurface conditions (Figure 4).

The initial characterization of this mineralogical biosignature is currently being performed utilizing both X-ray diffraction and X-ray fluorescence. The future goals of this research are to further identify the biosignatures indicative of life that could exist on Mars, as well as to develop an understanding of the metabolic strategies and biology required to sustain such systems.

Acknowledgments. This research was funded by the W.M. Keck Foundation and the Department of Biological Sciences, University of Arkansas, Fayetteville.

References

Brass, G. W., 1980, *Stability of brines on Mars*, ICARUS 42, 1, 20–28

Diaz, B., & Schulze-Makuch, D., 2006, *Microbial survival rates of Escherichia coli and Deinococcus radiodurans under low temperature, low pressure, and UV-Irradiation conditions, and their relevance to possible martian life*, ASTROBIOLOGY 6, 2, 332–347

Gendrin, A., Mangold, N., Bibring, J. P., Langevin, Y., Gondet, B., Poulet, F., Bonello, G., Quantin, C., Mustard, J., Arvidson, R., LeMouelic, S., 2005 *Suffates in martian layered terrains: the OMEGA/Mars Express view*, SCIENCE 307, 5715, 1587–1591

Heldmann, J. L., Toon, O. B., Pollard, W. H., Mellon, M. T., Pitlick, J., McKay, C. P., Andersen, D. T., 2005, *Formation of Martian gullies by the action of liquid water flowing under current Martian environmental conditions*, JOURNAL OF GEOPHYSICAL RESEARCH-PLANETS 110, E5

Knoll, A. H., Carr. M., Clark, B., Marais, D. J., D., Farmer, J. D., Fischer, W. W., Grotzinger, J. P., McLennan, S. M., Malin, M., Schroder, C., Squyres, S., Tosca, N. J., and Wdowiak, T., 2005, *An astrobiological perspective on Meridiani Planum*, EARTH AND PLANETARY SCIENCE LETTERS 240, 1, 179–189

Mellon, M. T., and Phillips, R. J, 2001, *Recent gullies on Mars and the source of liquid water*, JOURNAL OF GEOPHYSICAL RESEARCH-PLANETS 106, E10, 23165–23179

Moll, D. M., and Vestal, J. R., 1992 *Survival of microorganisms in smectite clays-Implications for martian exobiology*, ICARUS 98, 2, 233–239

Sears, D. W. G., and Chittenden, J. D., 2005, *On laboratory simulation and the temperature dependence of the evaporation rate of brine on Mars*, GEOPHYSICAL RESEARCH LETTERS 32, 23

Shen, Y. A., Buick, R., and Canfield, D.E., 2001, *Isotopic evidence for microbial sulphate reduction in the early Archaean era*, NATURE 410, 6824, 77–81

Squyres, S.W., Grotzinger, J. P., Arvidson, R. E., *et al.*, 2004 *In situ evidence for an ancient aqueous environment at Meridiani Planum, Mars*, SCIENCE 306, 5702, 1709–1714

Tauscher, C., Schuerger, A. C., and Nicholson, W. L., 2006, *Survival and germinability of Bacillus subtilis spores exposed to simulated Mars solar radiation: Implications for life detection and planetary protection*, ASTROBIOLOGY 6, 4, 592–60

Vaniman, D. T., Bish, D. L., Chipera, S. J., Fialips, C. I., Carey, J. W., and Feldman, W. C, 2004 *Magnesium sulphate salts and the history of water on Mars*, NATURE 431, 7009, 663–665

Wagner, M., Roger, A. J., Flax, J. L., Brusseau, G. A., and Stahl, D. A., 1998 *Phylogeny of dissimilatory sulfite reductases supports an early origin of sulfate respiration*, JOURNAL OF BACTERIOLOGY 180, 11, 2975–2982

Jane Siefert

David Schwartzman

Bioastronomy 2007: Molecules, Microbes, and Extraterrestrial Life
ASP Conference Series, Vol. 420, 2009
K. J. Meech, J. V. Keane, M. J. Mumma, J. L. Siefert, and D. J. Werthimer, eds.

A Hot Climate on Early Earth: Implications to Biospheric Evolution

D. W. Schwartzman[1] and L. P. Knauth[2]

[1]*Department of Biology, Howard University, email:dschwartzman@gmail.com*

[2]*School of Earth and Space Exploration, Arizona State University*

Abstract. There is now robust evidence for a much warmer climate on the early Earth than now. Both oxygen and silicon isotopes in sedimentary chert and the compelling case for a near constant isotopic oxygen composition of seawater over geologic time support thermophilic surface temperatures until about 1.5-2 billion years ago, aside from a glacial episode in the early Proterozoic. This temperature scenario has important implications to biospheric evolution, including a temperature constraint that held back the emergence of major organismal groups, starting with phototrophs. A geophysiology of biospheric evolution raises the potential of similar coevolutionary relationships of life and its environment on Earth-like planets around Sun-like stars.

1. Introduction

A long standing debate in the Earth sciences has centered on the temperature history of the Earth's climate over geologic time, especially during the Precambrian. Was temperature then similar to the present or was it significantly higher, particularly for the Archean and early Proterozoic? Critical in this debate is the interpretation of the empirical record of oxygen isotopes in sedimentary chert, silica deposited on the seafloor. The fractionation of oxygen isotopes (i.e., O^{16} and O^{18}) between chert and seawater is a function of temperature at the time the isotopic record is established. If this fractionation is set during burial close to the temperature of the ocean near the seafloor then a climatic record can potentially be inferred if the seawater O^{16} / O^{18} ratio is known.

2. Discussion

The hot climate interpretation of the oxygen isotopic record of marine Archean cherts (Knauth, 2005; Knauth and Lowe, 2003) has been in the past commonly challenged by appeals to diagenetic resetting of the original isotopic ratios but this explanation is now considered implausible even by advocates of cooler climates (e.g., Kasting and Ono 2006) in view of the robust case made by Knauth and Lowe (2003). Silicon isotopes from the same cherts have been interpreted as supporting the hot climate scenario (Robert and Chaussidon, 2006). A commentary on the latter paper noted the correspondence of the upper temperature limit to photosynthesis to the estimated surface temperature derived from the chert oxygen and silicon isotopic record (De La Rocha, 2006).

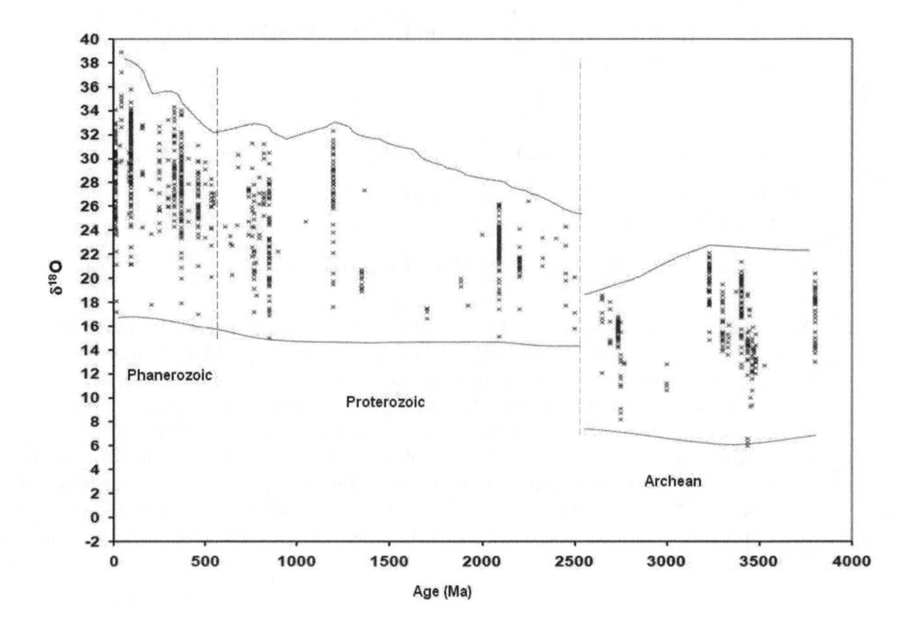

Figure 1. Oxygen iostope variations in cherts with time. Data from Knauth 2005

We argue that a compelling case for a very warm Archean/early Proterozoic (50-70 deg C) (aside from the Huronian) can now be made. The empirical foundation for this temperature history is derived from the sedimentary chert oxygen isotopic record, specifically from the highest O^{18}/O^{16} ratio at any age, since processes that would reset the climatic signature such as recrystallization with exchange with meteoric or higher temperature water during burial would decrease this ratio (Knauth, 2005). Alternatively, the shift in the range of O^{18}/O^{16} ratios of cherts downward from the late Phanerozoic to the Archean by 10 per mil (d O^{18} values; see Fig. 1) corresponds to a 45-50 deg C increase in climatic temperature or to the same shift in the seawater O^{18}/O^{16} ratio at a constant temperature or to a mixture of both effects.

Figure 1 shows this record with d O^{18} values being a laboratory standard-corrected measure of the O^{18}/O^{16} ratio. The upper bound corresponds to the ratio expressing the climatic signature. The silicon isotopic record from the same samples with highest O^{18}/O^{16} ratio at any age show the two isotopic systems are correlated back into the Archean (Robert and Chaussidon, 2006). This correlation rules out resetting of climatic record by reequilibration during burial, since the silicon concentrations are too low in any potential water that would reset the record. The inferred climatic temperature record from silicon isotopes agrees with the temperature history inferred from oxygen isotopes, assuming the seawater O^{18} /O^{16} ratios back into the Archean were close to the present.

Though the oxygen isotopic record of carbonates is more easily reset than cherts, the upper bound to the d O^{18} values of carbonates over geologic time, assuming fractionation from the same seawater as cherts, agrees with the inferred climatic record derived from cherts. Therefore it is misleading to argue that the average carbonate d O^{18} values express a progressive trend towards more negative seawater d O^{18} going back into the Archean (e.g., Jaffres et al.., 2007).

The high paleotemperatures derived from the oxygen and silicon isotopic record of Archean cherts are contingent on the assumption of near constant isotopic oxygen composition of seawater over geologic time. If Archean seawater were sufficiently isotopically lighter than modern values, the derived temperatures would be closer to a modern climate. A mechanism has been proposed for the necessary variation in the oxygen isotopic composition which requires shallower Archean ridgecrest depths than in the Phanerozoic, reducing high temperature seawater interaction with oceanic crust (Kasting and Ono, 2006; Kasting et al., 2006; Jaffres et al., 2007). However, the empirical record of ancient seawater-altered oceanic crust, including Archean eclogite proxies, appears to strongly support a near constant oxygen isotopic composition over geologic time.

In summary the evidence for near present seawater O^{18} /O^{16} ratio back into the Archean includes the following:

1. Paleozoic seawater is not significantly lighter than the present: direct measurement of fossil seawater from salt inclusions (Knauth and Roberts 1991). (Same conclusion from clumped isotope study on carbonate fossils, Came et al. 2007).

2. Inferences from the geologic record of ancient seawater-altered oceanic crust (ophiolites, their ecologite proxies, greenstones).

Buffering of oxygen isotopic ratio of seawater by reaction with oceanic crust over geologic time involves both low and high temperature reactions. At low temperatures the oceanic crust gains O^{18} relative to seawater, while at high temperatures the oceanic crust loses O^{18}, coming closer to seawater O^{18}/O^{16} ratio. This buffering has resulted in a near constant seawater O^{18}/O^{16} ratio going back to the Archean (Muehlenbachs et al., 2003).

Archean/Proterozoic ophiolites Joruna (2 Ga) (Muehlenbachs et al., 2004), Isua (3.8 Ga) (Furnes et al., 2007) and 3.3-3.5 Ga Onverwacht greenstone (Hoffman et al., 1986) have oxygen isotope patterns similar to Phanerozoic, with depleted oxygen isotopic sections observed in the Onverwacht and Joruma. Subduction loss of the depleted sections is a likely explanation for the lack of preservation, e.g., in the Isua ophiolites. This subducted crust is represented in Archean ecologite proxies for depleted/enriched seawater-altered oceanic crust, with abundant depleted values, clear evidence of high temperature interaction with seawater (Jacob, 2004; Jacob et al., 2005). Further, altered mineral assemblages imply a higher temperature in the Archean upper oceanic crust than the modern (Nakamura and Kato, 2004).

Similarly, high temperature alteration phases in the ophiolites show no evidence of equilibration with del O ^{18}-depleted seawater (Turner et al., 2007). Hence the buffer was similar to now, O^{18}/O ^{16} seawater ratio close to present value.

Other evidence for near present seawater O^{18} $/O^{16}$ ratio back into the Archean:

1. O^{18} is transferred to hydrosphere even by light meteoric water interaction in the case of Iceland, a subaerial, shallow ridgecrest (Hattori and Muehlenbachs, 1982; Gautason and Muehlenbachs, 1998), hence undermining the argument that postulated high ridgecrests would not permit loss of O^{18} by the Archean oceanic crust.

2. Archean and Proterozoic volcanogenic massive sulfide deposits with ore fluids on the order of 300 deg C and d O $^{18} = 0$ per mil, again no evidence of very light seawater (personal communication, Muehlenbachs).

3. As Robert and Chaussidon (2007) point out, the oxygen isotopic record from chert-phosphate pairs give seawater values close to present day (d O $^{18} = 0$ +/- 5 per mil) back into the Archean (Karhu and Epstein, 1986) , even if the temperatures of equilibration for some samples may be higher than the sediment/seafloor interface (note: the chert samples analyzed were not necessarily giving the heaviest d O ^{18} values for each age, hence diagenetic equilibration temperatures are probable for several samples).

4. A near present oxygen isotopic composition is inferred from hydrogen isotopic record measured from 2 Ga ophiolite-like complex (Lecuyer et al., 1996).

Other evidence for a very warm Archean/early Proterozoic (aside from the Huronian) includes:

1. Generally higher weathering intensities then, e.g., paucity of arkoses (Corcoran and Mueller, 2004). These higher intensities would not have occurred if seafloor weathering/carbon sink was the real player, pulling down the temperature, hence this empirical evidence supports a hot climate and high carbon dioxide/methane levels.

2. Highly stratified oceans apparently persisting until the mid Proterozoic (Lowe, 1994).

3. The sequence of inferred organismal emergence times have upper temperature limits for these metabolisms that are consistent with the chert paleotemperatures (see Figure 2). Oxygen levels alone cannot explain the big delay in the appearance of the Higher Kingdoms, since even atmospheric oxygen levels were likely high enough for their emergence long before they did, hence a temperature constraint is more plausible.

   If Archean temperatures were similar to the Phanerozoic, then some of the low- temperature prokaryotes should be grouped near the root on the rRNA phylogenetic tree with the hyperthermophiles/thermophiles. Rather, low temperature organisms are rooted further from the root than the hyperthermophiles/thermophiles (Schwartzman and Lineweaver, 2004-a,b). Further, melting temperatures of proteins resurrected from sequences inferred from robust molecular phylogenies give paleotemperatures at emergence consistent with the very warm early climate inferred from the oxygen isotopic record of ancient cherts (Gaucher et al., 2008).

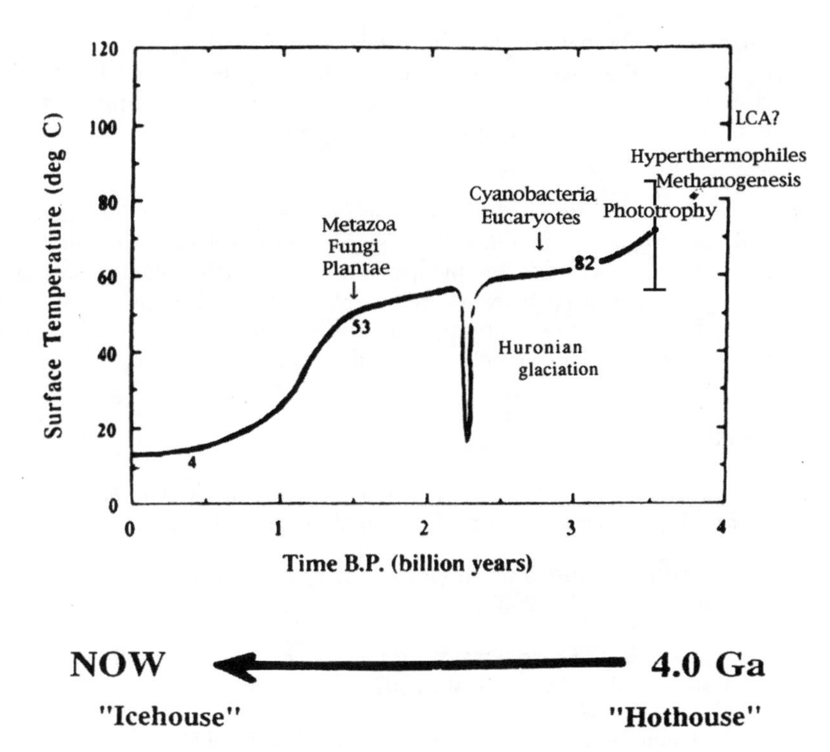

Figure 2. Temperature history of the Biosphere. A first order temperature curve is shown, with a likely uncertainty of about +/- 10 deg C. Emergence times of original groups are showm. The numbers on the curve are the model-derived ratios of the present biotic enhancement of weathering (BEW) to BEW values at indicated times (Schwartzman 1999, 2002). Trends are from 4.0 Ga to now

4. Evidence for high CO_2 levels prior to the first documented evidence of likely atmospheric methane at 2.8 Gya is consistent with the apparent requirements for the formation of Mn-bicarbonate clusters leading to oxygenic photosynthesis, i.e., high bicarbonate and therefore high CO_2 levels are necessary (on order of 1 bar, pH of about 6.5) (Schwartzman et al., 2008).

5. There are no apparent glaciations for over a billion years (2 Gya to Neo-Proterozoic snowball), with the Huronian plausibly explained by the rapid destruction of a methane-dominated greenhouse. (The apparent glaciation at 2.9 Gya may have a similar explanation, i.e., a transient rise of atmospheric oxygen).

6. The inferred biotic enhancement of weathering (BEW) for the present and its progressive increase to this value over geologic time emerges as a model result with reasonable assumptions regarding continental growth and outgassing rates (Schwartzman, 1999, 2002; see Figure 3); the Archean (and Hadean) was near abiotic with respect to BEW, so this factor must be subtracted out in modeling the long-term carbon cycle.

3. Conclusions

We conclude with a summary of implications of this climatic temperature history to biotic and biospheric evolution. These include the following:

1. A temperature constraint on emergence of major organismal groups (Schwartzman, 1999, 2002).

2. An atmospheric pO_2 constraint on macroeucaryotes, including metazoa in the Phanerozoic (Berner et al., 2007).

3. An atmospheric pCO_2 constraint on cyanobacterial emergence (Schwartzman et al., in press) and the emergence of leaves (megaphylls) in Devonian (Beerling et al., 2001).

4. A geophysiology of biospheric evolution, likewise on Earth-like planets around Sun-like stars (Schwartzman, 1999, 2002).

References

Beerling, D.J., Osborne, C.P., and Chaloner, W.G. 2001, Evolution of leaf-form in land plants linked to atmospheric CO2 decline in the Late Palaeozoic era. Nature, 410, 352-354.

Berner, R.A., VandenBrooks, J.M., and Ward, P.D., 2007, Evolution: Oxygen and evolution. Science 316: 557-558

Came, R.E., Eiler, J.M., Veizer, J., Azmy, K., Brand, U., and Weidman, C.R., 2007, Coupling of surface temperatures and atmospheric CO2 concentrations during the Palaeozoic era.. Nature 449, 198-201.

Corcoran, P.L. and Mueller, W. U. 2004, Aggressive Archaean weathering. In: The Precambrian Earth: Tempos and Events. Developments in Precambrian Geology 12, edited P.G. Eriksson, W. A. Altermann, D.R. Nelson, W.U. Mueller, O. Catuneaunu. Elsevier, pp. 494-504.

De La Rocha, C. L. 2006, In hot water. Nature, 443, 920-921.

Furnes, H., de Wit, M., Staudigel, H., Rosing, M. and Muehlenbachs, K., 2007, A Vestige of Earth's Oldest Ophiolite. Science 315: 1704-1707.

Gaucher, E. A., Govindaraja, S., and Ganesh, O.K. 2008, Palaeotemperature trend for Precambrian life inferred from resurrected proteins. Nature 451, 704-708.

Gautason, B. and Muehlenbachs, K., 1998, Oxygen isotopic fluxes associated with high-temperature processes in the rift zones of Iceland Chemical Geology 145: 275-286.

Hattori, K. and Muehlenbachs, K., 1982, Oxygen isotope ratios of the Icelandic crust. Journal Geophysical Research 87, No.B8: 6559-6565.

Hoffman, S.E., Wilson, M. and Stakes, D.S., 1986, Inferred oxygen isotope profile of Archaean oceanic crust, Onverwacht Group, South Africa. Nature 321: 55-58.

Jacob, D.E. 2004, Nature and origin of eclogite xenoliths from kimberlites. Lithos 77, 295– 316.

Jacob, D.E., Bizimis, M. and Salters, V. J. M. 2005, Lu–Hf and geochemical systematics of recycled ancient oceanic crust: evidence from Roberts Victor eclogites. Contrib. Mineral Petrol. 148, 707-720.

Jaffres, B.D., Shields, G.A. and Wallman, K., 2007, The oxygen isotope evolution of seawater: A critical review of a long-standing controversy and an improved geological water cycle model for the past 3.4 billion years. Earth-Science Reviews, 83: 83-122.

Karhu, J. and Epstein, S., 1986, The implication of the oxygen isotope records in coexisting cherts and phosphates. Geochimica et Cosmochimica Acta, 50: 1745-1756.

Kasting , J.F., Howard, M.T., Wallmann, K., Veizer, J., Shields, G. and Jaffrés, J. (2006) Paleoclimates, ocean depth, and the oxygen isotopic composition of seawater Earth and Planetary Science Letters 252, 82–93.

Kasting, J.F. and Ono, S. 2006, Palaeoclimates: the first two billion years. Phil. Trans. Royal Society B 361: 917-929.

Knauth, L.P. 2005, Temperature and salinity history of the Precambrian ocean: implications for the course of of microbial evolution. Palaeogeography Palaeoclimatology Palaeoecology, 219, 53-69.

Knauth L.P., and Lowe, D.R. 2003, High Archean climatic temperature inferred from oxygen isotope geochemistry of cherts in the 3.5 Ga Swaziland Supergroup, South Africa. Bulletin Geological Society of America,115, No.5, 566-580.

Knauth, L.P. and Roberts, S.K., 1991, The hydrogen and oxygen isotopic history of the Silurian-Permian hydrosphere as determined by measurement of fossil water. Geochemical Society Special Publication No. 3, 91-104.

Lecuyer, C., Gruau, G., Fruh-Green, G.L., and Picard, C., 1996, Hydrogen isotope composition of Early Proterozoic seawater. Geology 24: 291-294

Lowe, D.R., 1994. Early environments: Constraints and opportunities for early evolution. In: Early Life on Earth. Nobel Symposium No.84. (ed. S. Bengtson) pp. 24-35, Columbia University Press. N.Y.

Muehlenbachs, K., Banerjee, N.R., and Fumes, H. 2004, Oxygen isotopic composition of the Paleozoic and Precambrian ocean water. GSA Paper No. 205-3, 2004 Denver Annual Meeting (November 7–10, 2004).

Muehlenbachs, K., Fumes, H., Fonneland, H.C. and Hellevang, B. 2003, Ophiolites as faithful records of the oxygen isotope ratio of ancient seawater: The Solund-Stavfjord Ophiolite Complex as a Late Ordovician example. In: Ophiolites in Earth History. Edited by Y. Dilek and P.T. Robinson. Geological Society, London, Special Publication 218. pp. 401-414.

Nakamura, K. and Kato, Y., 2004, Carbonatization of oceanic crust by the seafloor hydrothermal activity and its significance as a CO_2 sink in the Early Archean. Geochimica et Cosmochimica Acta, 68: 4595-4618.

Robert, F. and Chaussidon, M. 2006, A palaeotemperature curve for the Precambrian oceans based on silicon isotopes in cherts. Nature 443, 969-972.

Robert, F. and Chaussidon, M., 2007, Reply. Nature 447, E1-2. doi:10.1038/nature05831.

Schwartzman, D., 1999, 2002, Life, Temperature and the Earth: The Self-Organizing Biosphere. Columbia Univ. Press.

Schwartzman, D.W., Caldeira, K., and A. Pavlov, 2008, Cyanobacterial emergence at 2.8 Gya and greenhouse feedbacks. Astrobiology 8, 187-203.

Schwartzman, D.W. and Lineweaver, C.H., 2004a, The hyperthermophilic origin of life revisited. Biochem. Soc. Transactions 32: 168-171.

Schwartzman, D. and C.H. Lineweaver, 2004b, Temperature, Biogenesis and Biospheric Self-Organization. In: Non-Equilibrium Thermodynamics and the Production of Entropy:Life, Earth, and Beyond (eds. A. Kleidon and R. Lorenz), chapter 16, Springer Verlag.

Turner, S., Tonarini, S., Bindeman, I., Leeman, W. P., and Schaefer, B.F., 2007, Boron and oxygen isotope evidence for recycling of subducted components over the past 2.5 Gyr. Nature 447: 702-705.

Bioastronomy 2007: Molecules, Microbes, and Extraterrestrial Life
ASP Conference Series, Vol. 420, 2009
K. J. Meech, J. V. Keane, M. J. Mumma, J. L. Siefert, and D. J. Werthimer, eds.

Survival of Purines and Pyrimidines Adsorbed on a Solid Surface in a High Radiation Field

A. Guzman-Marmolejo,[1] S. Ramos-Bernal,[1] A. Negrón-Mendoza,[1] and G. Mosqueira[2]

[1]*Instituto de Ciencias Nucleares, Universidad Nacional Autonoma de Mexico*

[2]*Direccion General de Divulgacion de la Ciencia, Universidad Nacional Autonoma de Mexico*

Abstract. According to astronomical data, organic molecules are abundant in interstellar space. These molecules have arisen from non-equilibrium processes driven by the energy of photons and cosmic rays. The presence of dirty ices show that a rich low temperature solid phase chemistry takes place in such environments. These chemical evolution reactions have been assumed to proceed mainly within solid surfaces of interstellar dust particles, as well as on macrobodies. Among solid surfaces for chemical processes, alumino-silicates are widely distributed in terrestrial and extraterrestrial bodies, such as meteorites, and the Martian soil, which showed the presence of carbonates and clays. Therefore, alumino-silicates are considered a likely inorganic material to promote organic reactions that might have played a role in the survival of organic molecules adsorbed on their surfaces. It is also known that they have a high surface area and a high affinity for organic compounds.Purines and pyrimidines are important organic compounds due to their role in biological processes. Their synthesis and stability are of paramount importance in chemical evolution. In this work we propose a mechanism to account for the survival of purines and pyrimidines adsorbed in a solid surface in a high radiation field.

1. Introduction

Organic molecules are a common element through the entire Universe. Up until now, more than 150 molecules have been detected in the ISM (Irvine 1998). Organic material is continuously exposed to high radiation fields and chemical reactions take place even in adverse temperature and concentration conditions, such as those of the ISM . In those places, the role of solid surfaces should be fundamental in the processes of complication of organic material. Among solid surfaces, clays are important mainly because they are widely distributed through the Universe, they can be a ubiquitous component of interstellar dust (Bowey & Adamson 2002), due to their strong affinity to organic molecules, and also because they could have played a possible role as protective agents against a source of energy (Ponnamperuma et al. 1982). On the other hand, purines and pyrimidines are fundamental in present living beings as part of genetic material, energetic molecules ATP, etc. So we decided to test the hypothesis of clays as protective agents, for biomolecules, against strong radiation fields.

2. Method

Triple distilled water and high purity reagents (Sigma) were employed in all the experiments. Water solutions (10^4 M) of different purines and pyrimidines were prepared: Xanthine (X), Guanine (G), Cytosine (C), Thymine (T) and Uracil (U). Na-Montmorillonite from the Geology Department (University of Missouri, Columbia) was employed as adsorbent surface.

2.1. Preparation of Samples

Samples were prepared by adding 0.1 grams of sodium montmorillonite to 3 mL of each base. Samples' pH was fixed in 2 with HCl. The samples were maintained in agitation for 30 minutes to allow adsorption. Solid was separated by centrifugation (10,000 rpm). Percentage of adsorption was determined by comparing the supernatant with a standard solution, both UV-vis and HPLC were employed. Desorption was carried out by changing pH (14), adding NaOH to the solid previously separated by centrifugation. The recovery of the bases was done after three cycles of treatment with NaOH. It was possible to desadsorb all the bases from the clays.

2.2. Radiolysis

Aliquots of 3 mL of free bases (in solution) were bubbled with argon to eliminate dissolved oxygen. They were irradiated in a high intensity gamma source (Gammabeam 651-PT). Irradiation doses were from 0.4 to 24.4 kGy. Samples of puric and pyrimidic bases were prepared with clay allowing adsorption. They were irradiated in suspension in the same conditions in the gamma source.

2.3. Analysis

To analyze samples, spectroscopic and chromatographic (High Performance Liquid Chromatography) techniques were employed. HPLC analysis was made in a special column (Restek Ultra IBD 150 x 4.6 mm) to separate the bases. UV-vis detector device (Varian, UV-50) was used to follow the decomposition.

3. Results and Discussion

There is a differential behavior in the adsorption percentage of the tested bases. Purines are completely adsorbed by the clay (adenine, guanine and xanthine). For the pyrimidines, only the cytosine is fully adsorbed; uracil and thymine are adsorbed 30 and 3.5% respectively. The order of adsorption on the clay is: adenine > xanthine > cytosine > guanine > uracil > thymine. The analyzes for irradiated samples in aqueous solutions show that the presence of clay reduces the decomposition of the bases. The decomposition behavior for all the tested compounds are shown in Figures 1 to 3. The most labile compound is thymine that in free state is decomposed at a dose of 3 kGk, and adenine is the one that can survive at higher doses (42 kGy).

The clay-bases systems show that the presence of Na-montmorillonite modifies the radiation decomposition of purines and pyrimidines. For those bases that are adsorbed more than 55% the decomposition deceases significantly, that is the case of adenine, guanine, cytosine and xanthine. However, for bases that

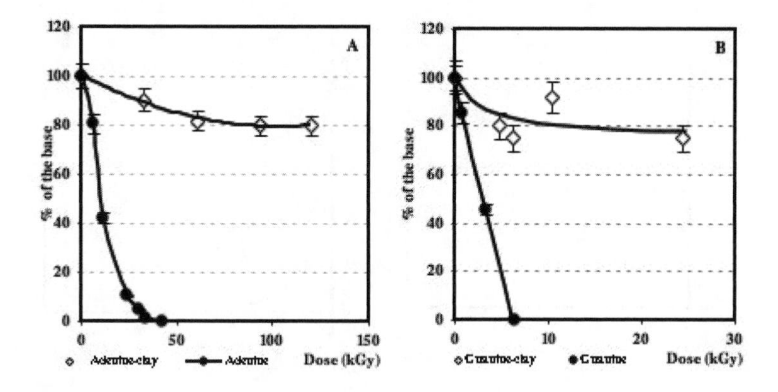

Figure 1. Decomposition of bases as a function of irradiation dose. Results for free and adsorbed bases are presented. A)Adenine. B)Guanine.

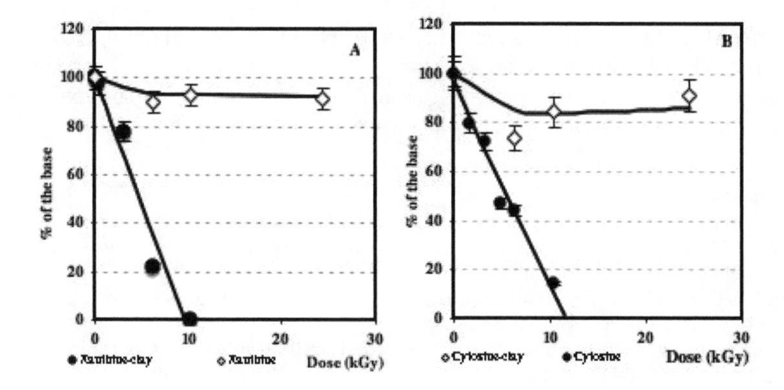

Figure 2. Decomposition of bases as a function of irradiation dose. Results for free and adsorbed bases are presented. A)Xanthine. B)Cytosine.

are poorly adsorbed, such as thymine and uracil, the decomposition is higher. In those cases just the small fraction that is adsorbed in the clay survives.

4. Conclusions

The tested clay exhibits a concentrator effect in the studied molecules. The presence of the montmorillonite modifies the decomposition of purines and pyrimidines. Decomposition diminishes in presence of clay. The bases are more resilient to the degradation induced by radiation when they are absorbed in montmorillonite. These results show that clays may have played an important role as protective agents toward degradation of molecules exposed to a high radiation field.

Acknowledgments. Travel Support was received from the NASA Astrobiology Institute and the National Science Foundation through the conference or-

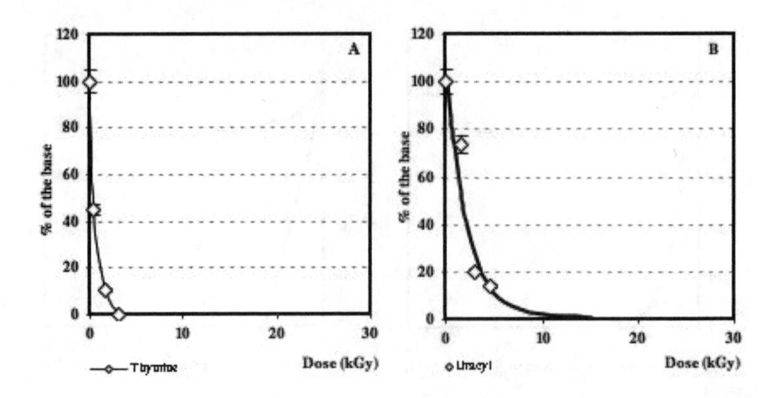

Figure 3. Decomposition of bases as a function of irradiation dose. Results for free and adsorbed bases are presented. A)Thymine. B)Uracil.

ganizers (AG). This work was partially supported by a grant (DGAPA-IN223406). We thank the help of Physicist F. Garcia and M. in Sc. B. Leal for their technical assistance in the irradiation of samples, and S. Ham for the experimental design and elaboration of special glassware.

References

Bowey, J. & Adamson A.J. 2002, *A mineralogy of extrasolar silicate dust from 10mm spectra.* Mon.Not. R. Astron. Soc. 334, 94-106

Irvine, W. 1998, *Extraterrestrial organic matter: a review.* Origins of Life and Evolution of the Biosphere. 28, 365-383

Ponnamperuma,C. Shimoyama A. & Friebele E. 1982, *Clay and the Origin of Life.* Origins of Life 12, 9-40

Bioastronomy 2007: Molecules, Microbes, and Extraterrestrial Life
ASP Conference Series, Vol. 420, 2009
K. J. Meech, J. V. Keane, M. J. Mumma, J. L. Siefert, and D. J. Werthimer, eds.

Patterned Growth in Extreme Environments

J. Curnutt, E. Gomez, and K. E. Schubert

California State University, San Bernardino
Department of Computer Science
5500 University Parkway
San Bernardino, CA 92407-2397

Abstract. In this paper, cellular automata are used to model patterned growth of organisms in extreme environments. A brief introduction to cellular automaton modeling is given to assist the reader. Patterned growth of soil surface cyanobacteria and biovermiculation microbial mats in sulfuric acid caves are modeled and simulations conducted. Simulations are compared with actual systems, and future directions are discussed.

1. Introduction

In resource limited environments, organisms grow in patterns that are self-enforcing and exhibit hysteresis, see Catena (1999); Meron et al. (2004); von Hardenberg et al. (2001). Among the techniques that have been used to model these patterns are evolutionarily stable strategies in game theory and differential equations, see Klausmeier (1999); HilleRisLambers et al. (2001); Meron et al. (2004); Thiéry et al. (1995); von Hardenberg et al. (2001). While good results have been generated using differential equations, they require tuning of the parameters and experience in mathematical and numerical techniques to obtain valid results.

In this work we developed cellular automata that produce similar predictions to the differential equation models, while preserving the rapid modeling and hypothesis testing of cellular automata. Similar models can also be applicable to group animal behavior, see Couzin et al. (2002); Hoare et al. (2004); Krause & Tegeder (1994); Pitcher et al. (1996). While cellular automata have been used to show patterning due to slope considerations, see Dunkerley (1997), our method for deriving rules for cellular automata from observed data in organism growth patterns accounts for soil nutrients, water, root growth patterns, and geology allowing scientists to easily examine the effects of modifying conditions without damaging the environment. We apply this model to identify factors affecting patterning with respect to growth, die-out, and stabilization in extreme environments. We compare the results of our model with biovermiculation microbial mats growth in acid caves, and cyanobacteria growth in Zzyzx, CA.

2. Cellular Automata

A cellular automaton (CA) is a computational model that is discrete in both space and time. Essentially we divide space into boxes called cells, and only calculate their values at discrete time (you could consider this as dividing space-time into boxes if you like). CA use a set of fixed rules to transition from one state to another.

A state could represent anything, and in our case it represents the amount of water, nutrients and the biomass. The rules describe how the organism grows or dies in the presence of water, nutrients and competition from other organisms. CA rules are not usually expressed as formulas, rather they are visual, such as drawing pictures of the neighboring cells and then labeling the next state. To avoid tedium, we group similar cell patterns using heuristics discerned from discussions with biologists and borne out in practice.

Typically CA are deterministic, do not use the current value of the state in the calculation of the next state, and use only one state variable. We consider a generalization of this in which these constraints are relaxed. For this paper, we only relax the determinism constraint, as this allows for more realistic looking results.

Consider the following simple transition rule.

> If 3 or 4 of the cells in a neighborhood (± 1 row and/or column) of a cell are 1 then the next state for the cell is 1.

This is illustrated by the transition of cell S:

Time t_0

0	1	1
1	S	0
0	0	0

$\Rightarrow$

Time t_1

0	1	1
1	1	0
0	0	0

Thus we only need to specify the necessary value for biomass to enable organism growth, which value of the sum allows organism survival, and which value will result in organism death.

Such rules are deceptively simple, but can lead to very complicated results. We allowed the neighborhood to vary to simulate plant root systems and other nutrient effects. We also allowed the value of the sum to be specified as a list of numbers or a single interval.

To show how this can relate to a case of crowding, consider the following simple rules. An organism is in a hostile environment. It can live if it has no more than 6 neighbors. Above that number, resource competition is too high. A new organism will grow if 2 or 3 of the neighboring squares have organisms (it takes at least 2 to generate a new organism, but the new one needs room- i.e. few neighbors). Start with a row of five organisms and an interesting thing happens, as you can see in Figure 1. The ten small figures show the first 10 time periods, the last one is the area after 40 time periods, when it is stable. If you allow life to be supported then the lines continue to fill the region. None can grow in the "dead zones" once they are established, and placing a living organism there will cause them to die (it will have too many neighbors). This self-enforcing nature of the pattern is the basis of the maze-like growth of plants in extreme environments. More complicated rules that fit more realistic situations generate patterns that imitate the striking structures found in nature. This paper presents first steps toward a general model.

3. Cyanobacteria Growth

Cyanobacteria are aquatic, photosynthetic bacteria and are notable for many reasons, including being the oldest fossils, the original producers of atmospheric oxygen, the source of much of our oil, and an ability to grow in extreme environments (including Antarctica). The cyanobacteria in Zzyzx are fossils, but preserve the structured growth we are considering.

Figure 2 shows a typical cyanobacteria fossil. These patterns were fairly dense, so we used a neighborhood with a range of 1 square (1 square in all directions for a total of 8 neighbors). Cyanobacteria often grow in blooms, so we modeled this using growth on having 3 neighbors only. We also allowed random death with probability of 10%. The resulting simulation is shown on the right in Figure 2, and shows much of the same structure as is seen in the actual fossil on left in Figure 2.

4. Biovermiculation Growth

Biovermiculations are microbial mats composed of bacteria, extracellular polysaccharide slime, embedded clay and other particles, in situ precipitated minerals (e.g. sulfur and gypsum), and even some small invertebrates like mites and nematodes.

Investigators have identified them from sulfuric acid caves, see Hose et al. (2000). More recently, observers have begun to see them in a wide variety of chemical and physical subsurface settings including mines, carbonate caves, lavatubes, and even Mayan ruins (P.J. Boston, personal communication(2007)).

These structures are interesting because of their intrinsically intriguing biology and geochemistry, the distinctive growth patterns that they exhibit, and

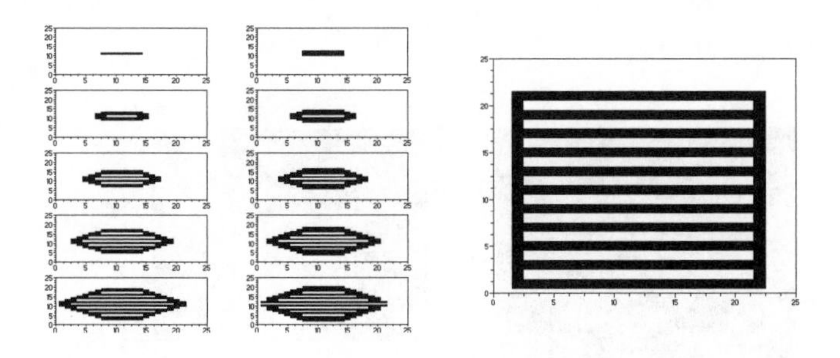

Figure 1. Left graphic shows the first ten time periods for the rules: if neighbors=2 or 3 then grow, if neighbors > 6 then die, else stay same. Right graphic shows the final pattern for these rules.

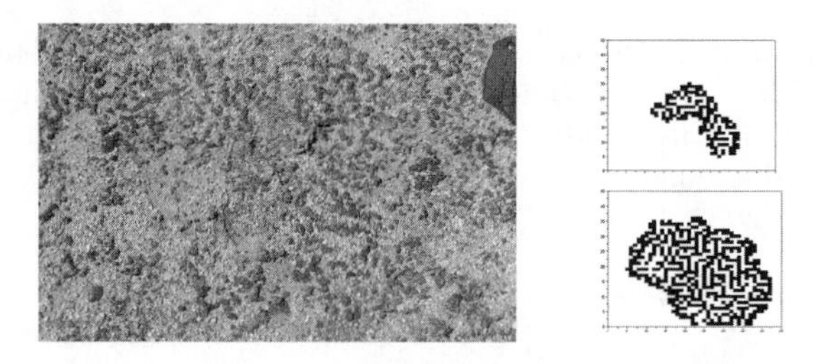

Figure 2. Left Picture is of cyanobacteria fossils from Zzyzx, CA. Right graphics are simulated cyanobacteria at $t = 5$ (top) and $t = 40$ (bottom).

also because they may be a highly distinctive biosignature that could be interpretable in extraterrestrial settings based on gross morphology, see Boston (2001). Interest in the biovermiculations has grown as better methods of studying such structures has become available. They also provide a model system of a biomat that might occur on the interiors of various cave types on Mars. Lavatubes have been identified on Mars, see Boston (2004), and more recently confirmed in a more elaborate study, see Cushing et al. (2007). Mechanisms to create solutional caves in evaporite mineral deposits on Mars have also been proposed, see Boston (2006). Such potential subsurface habitats could conceivably house or have housed microbial populations on Mars and left traces similar to those found in geomicrobiological communities in Earth's subsurface.

The patterning of biovermiculation growth is still a mystery. To understand it will require simulations that test different sets of rules enabling us to arrive at a good pattern match for the microbial mat growth that we are observing in nature. Such simulations will then be correlated with actual pattern examples from cave walls and other occurrences. Figure 3 shows an area that has biovermiculations so thick that they become the solid mat, but there are weird uninhabited areas that follow the rock curvatures. These are not simple water pathways that have prevented growth, so their origin is unknown. In figure 3, no nutrient or water differentiation was induced, only crowding rules, and an ability to pull nutrients from surrounding cells. The result was the depleted

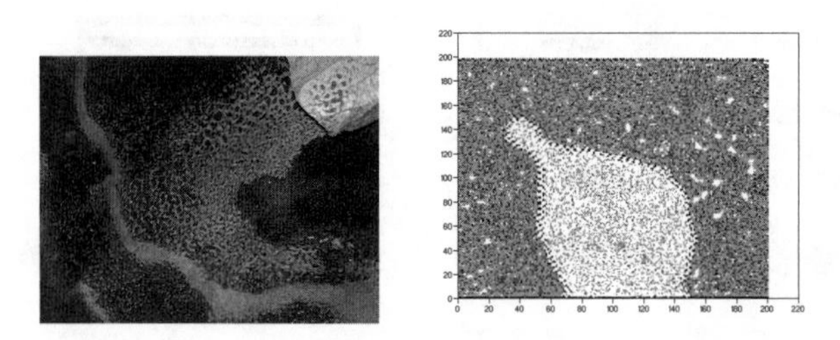

Figure 3. Left picture is biovermiculation with discontinuities. Right graphic is simulated biovermiculation growth.

region in the middle. Note the shape has many indents, like the actual system in Figure 3.

5. Conclusions

Cellular automata show great potential for simulation of patterned organism growth in extreme environments. In this paper an introduction to the method was provided. Further we examined the use of cellular automata to model patterned growth of cyanobacteria and biovermiculations. Early results are quite promising, and further work is going on to systematize the rule generation, and to connect differential models with cellular automaton models.

Acknowledgments. The authors gratefully acknowledge the assistance of Penny Boston, for her expertise with biovermiculation, numerous pictures of biovermiculation patterns, and for proofreading the manuscript. The authors also gratefully acknowledge Chris McKay and Jim Nienow for their assistance with the cyanobacteria portion. This paper is dedicated to Norval Curnutt, Jane's wonderful and supportive husband, who has finished his fight with cancer. Norval is the best of the best and will be sorely missed by all of us.

References

Boston, P. J. *Encyclopedia of Cave and Karst Science*, chapter Extraterrestrial Caves, pages 355–358. Fitzroy-Dearborn Publishers, Ltd., London, UK, 2004.

Boston, P. J., Hose, L. D., Northup, D. E., and Spilde, M. N. The microbial communities of sulfur caves: A newly appreciated geologically driven system on Earth and potential model for Mars. In R. Harmon, editor, *Karst Geomorphology, Hydrology, & Geochemistry*, pages 331–344. Geological Soc. Amer., 2006.

Boston, P. J., Spilde, M. N., Northup, D. E., Melim, L. A., Soroka, D. S., Kleina, L. G., Lavoie, K. H., Hose, L. D., Mallory, L. M., Dahm, C. N., Crossey, L. J., and Schelble, R. T. Cave biosignature suites: Microbes, minerals and mars. *Astrobiology Journal*, 1(1):25–55, 2001.

Catena vol. 37, 1999. Special issue on banded vegetation.

Couzin, I. D., Krause, J., James, R., Ruxton, G. D., and Franks, N. R. Collective memory and spatial sorting in animal groups. *J. Theor. Biol.*, (218):1–11, 2002.

Cushing, G. E., Titus, T. N., Wynne, J. J., and Christensen, P. R. Themis observes possible cave skylights on mars. *Geophysical Research Letters*, 34(L17201), 2007.

Dunkerley, D. L. Banded vegetation: development under uniform rainfall from a simple cellular automaton model. *Plant Ecol.*, (129):103–111, 1997.

HilleRisLambers, R., Rietkerk, M., van den Bosch, F., Prins, H. H. T., and de Kroon, H. Vegetation pattern formation in semi-arid grazing systems. *Ecology*, (82):50–62, 2001.

Hoare, D. J., Couzin, I. D., Godin, J. G. J., and Krause, J. Context-dependent group size choice in fish. *Animal Behavior*, 67:155–164, 2004.

Hose, L. D., Palmer, A. N., Palmer, M. V., Northup, D. E., Boston, P. J., and Duchene, H. R. Microbiology and geochemistry in a hydrogen sulphide-rich karst environment. *Chemical Geology*, 169:399–423, 2000.

Klausmeier C. A. Regular and irregular patterns in semiarid vegetation. *Science*, (284):1826–1828, 1999.

Krause, J. and Tegeder, R. W. The mechanism of aggregation behavior in fish shoals: individuals minimize approach time to neighbours. *Anim. Behav.*, (48):353–359, 1994.

Meron, E., Gilad, E., von Hardenberg, J., Shachak, M., and Zarmi, Y. Vegetation patterns along a rainfall gradient. *Chaos, Solutions and Fractals*, (19):367–376, 2004.

Pitcher, T. J., Misund, O. A., Fernö, A., Totland, B., and Melle, V. Adaptive behaviour of herring schools in the norwegian sea as revealed by high-resolution sonar. *ICES Journal of Marine Science*, (53):449–452, 1996.

Thiéry, J., d'Herbès, J. M., and Valentin, C. A model simulating the genesis of banded vegetation patterns in niger. *J. Ecol.*, (83):497–507, 1995.

von Hardenberg, J., Meron, E., Shachak, M., and Zarmi, Y. Diversity of vegetation patterns and desertifcation. *Phys Rev Lett*, (87:198101-14), 2001.

Abel Méndez

Bioastronomy 2007: Molecules, Microbes, and Extraterrestrial Life
ASP Conference Series, Vol. 420, 2009
K. J. Meech, J. V. Keane, M. J. Mumma, J. L. Siefert, and D. J. Werthimer, eds.

Planetary Habitability Under Dynamic Temperatures

A. Méndez

Department of Physcis and Chemistry, University of Puerto Rico at Arecibo, PR 00614 (amendez@upra.edu)

Abstract. One of the main environmental factors affecting the viability, growth and distribution of microbial life is temperature. Microorganisms are generally exposed to extreme diurnal temperature variations in the surface layer of planetary bodies where temperature is mostly in phase with solar radiation. This is epecially the case of bare soils in terrestrial deserts and the surface of Mars where potential microbial life might not only have to survive low nutrients and water levels but also to dynamic temperatures. Although there is extensive research about microbial growth kinetics at constant temperatures, there is little knowledge about microbial growth kinetics at dynamic temperatures, especially those associated with natural environments. Therefore, this work presents a new dynamic growth kinetics model based on a thermal habitability function. The model is being use in understanding the growth, distribution and potential habitats for microbial life in planetary bodies with extreme surface temperature fluctuations.

1. Introduction

Planetary surface environments such as Earth and Mars can experience extreme daily temperature variations. Even on Earth soils temperatures can change over 30° C between day and night times (McKay et al. 2003). Any potential microbial life in natural environments is therefore subjected to wide dynamic temperatures. The field of predictive microbiology deals with the construction of mathematical models to predict inactivation or growth of microorganisms as a function of some environmental variables (i.e. temperature, water activity and pH). The models of Van Impe et al. (1992), Fujikawa et al. (2004) and Baranyi and Roberts (1994) are currently used to predict microbial growth under dynamic temperatures. All three models have been used by the food industry to predict growth of bacteria at different variations of temperatures (usually step changes) (Baranyi and Roberts 1995). However, no model is being used to predict growth under the dynamical temperatures of natural environments. Our understanding of microbial growth in natural environment temperatures comes usually from the extrapolation of their behavior at constant temperatures. One of the main reasons for the limited research in this area is the experimental difficulty in measuring microbial growth under dynamical temperatures.

This work proposes a mathematical model to predict microbial growth under dynamic temperatures, especially those associated with natural environments. Such models are important to understand the growth, distribution and interactions of microbial life not only on Earth but also in planetary environments such as Mars where terrestrial contamination might be possible (National

Academy of Sciences 2007). It is the final goal of this work to apply the models obtained in this work to the Martian surface environment to quantify the potential risk of forward contamination of Mars by space probes.

2. Thermal Habitability Function

The suitability for life of any environment depends on complex environmental chemical and physical interactions. The qualitative concept of the habitability of an environment can be measured indirectly in many ways but there is no standard definition to measure it. This work proposes to use the specific growth rate as a function of the environment as a practical way to measure and quantify the habitability of an environment. Under this definition, environments with higher growth rates for a particular species or community are more habitable than others. The habitability of an environment for a particular individual or community is defined here as the ratio between its specific growth rate in the environment μ and its potential maximum growth rate μ_m (usually measured in vitro under steady conditions). The general form of the proposed habitability equation is

$$H(p_1, p_2, ...) = \frac{\mu(p_1, p_2, ...)}{\mu_m(p_1, p_2, ...)} \tag{1}$$

where $H(p_1, p_2, ...)$ is the habitability as a function of some physical or chemical parameters p_i of the environment. The habitability is a unitless scale that extends from zero to one, where zero ($H = 0$) means no growth and one ($H = 1$) means maximum relative growth rate. The habitability can be computed from local to global scales for individuals or communities.

It has been shown that the relation between growth rate as a function of environmental parameters can be conveniently expressed by a cardinal model (Rosso et al. 1993). In the cardinal model of Rosso et al. (1995) the gamma function gives the relative growth rate as function of growth temperatures. The gamma function is what is redefined in this work as the thermal habitability, a normalized relative growth function. Although, the model of Rosso et al. has been tested and used extensively to describe growth as function of temperature, it fails to explain some experimental data (under preparation). Therefore, this work uses a new expression to define growth rate as function of temperature given by

$$\mu = \begin{cases} 0 & \text{if } T \leq T_{min} \\ \mu_m H(T) & \text{if } T_{min} < T < T_{max} \\ 0 & \text{if } T \geq T_{max} \end{cases}$$
$$H(T) = \left(\frac{T - T_{min}}{T_{opt} - T_{min}}\right)^s \left(\frac{T_{max} - T}{T_{max} - T_{opt}}\right)^{sp} \quad p = \frac{T_{max} - T_{opt}}{T_{opt} - T_{min}} \tag{2}$$

where μ is the specific growth rate, μ_m is the maximum growth rate, s is a shape parameter (usually close to 2), T is temperature and T_{min}, T_{opt} and T_{max} are the minimum, optimum and maximum growth temperatures, respectively. This empirical expression is similar to the one proposed by Rosso et al. but with the advantage of being simpler and able to explain more microbial growth data than previous models (under preparation). In fact, it was noted later during this work that the same expression was already proposed in the field of crop science to explain development rate of crops as function of temperature (Xin et al. 1995). In equation 2 the shape parameter s for plants is closer to one (Yan and Hunt 1999) while in bacteria is closer to two. This means that the thermal habitability function $H(T)$ can be used to predict relative growth rate as function of temperature not only for microbial life but also for more complex life like plants (all primary producers).

3. Conclusion

A dynamic microbial growth model was constructed by combining the diferential equation for logistic growth with the thermal habitability function of equation 2. The biomass is given by the differential equation

$$\frac{dy}{dt} = \mu_m H(T)\left(1 - \frac{y}{A}\right) \tag{3}$$

where y is a direct or indirect measure of microbial mass (i.e. optical density) as function of both time t and temperature T, and A is the maximum biomass. Equation 3 is solved numerically by the Runge-Kutta method using a program created in the programming language IDL (Interactive Data Language).

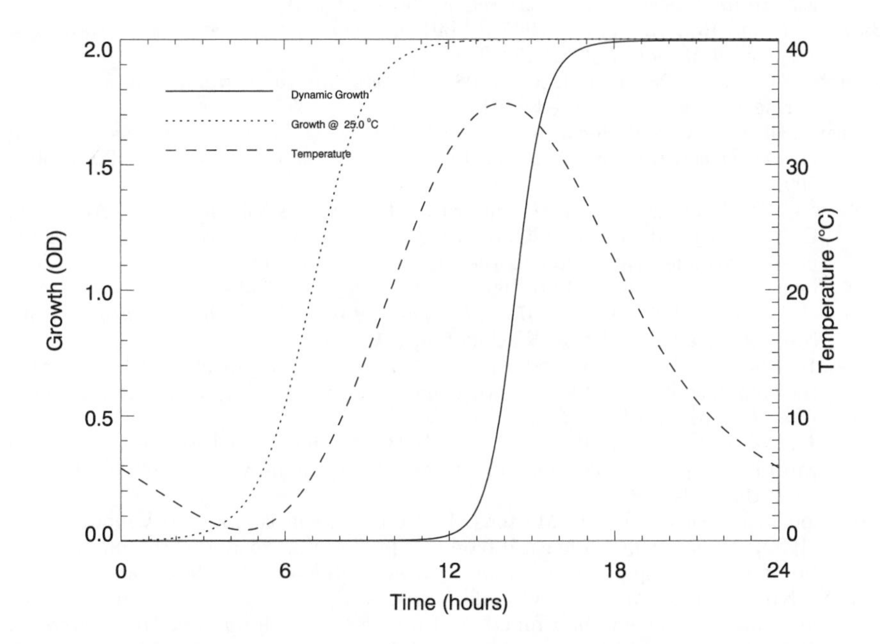

Figure 1. Comparison of the modeled growth of *B. subtilis* at constant temperature (25° C) with growth in a desert environment with daily temperature oscillations between 0° C and 35° C. The effect of the dynamic temperatures environment affected both the lag time and growth rate of the bacteria. Growth starts after noon with higher growth rates.

The common terrestrial bacterium *Bacillus subtilis* in a desert environment (McKay et al. 2003; Campbell and Norman 1998) was used as a sample case to solve the equations. The thermal habitability for *B. subtilis* is defined by its minimum, optimum and maximum growth temperatures: 11° C, 39° C and 53° C, respectively (Zwietering et al. 1990). The solution of the model predicts the potential growth as function of time of day of *B. subtilis* in a desert environment as compared to growth at a fixed temperature (Figure 1).

The presented model can be used to predict microbial biomass as a function of time of day in environments with large temperatures oscillations. Various experiments are in progress to test and validate the model but initial results with *E. coli*, *B. subtilis* and soil communities agree with the model. The model will be used to understand microbial growth in extreme thermal environments and to evaluate the potential thermal contamination risk of martian surface environments by common terrestrial bacteria.

Acknowledgments. Support for this work was received by the NASA Astrobiology Institute MIRS Program, the University of Puerto Rico at Arecibo and by Puerto Rico Space Grant Consortium.

References

Baranyi, J. and Roberts, T. A. (1994). A dynamic approach to predicting bacterial growth in food. *Int J Food Microbiol*, 23(3-4):277–294.

Baranyi, J. and Roberts, T. A. (1995). Mathematics of predictive food microbiology. *Int J Food Microbiol*, 26(2):199–218.

Campbell, G. S. and Norman, J. M. (1998). *Introduction to environmental biophysics.* Springer, New York, 2nd ed edition.

Fujikawa, H., Kai, A., and Morozumi, S. (2004). A new logistic model for escherichia coli growth at constant and dynamic temperatures. *Food Microbiol*, 21(2004):501–509.

McKay, C. P., Friedmann, E. I., Gomez-Silva, B., Caceres-Villanueva, L., Andersen, D. T., and Landheim, R. (2003). Temperature and moisture conditions for life in the extreme arid region of the atacama desert: four years of observations including the el nino of 1997-1998. *Astrobiology*, 3(2):393–406.

National Academy of Sciences (2007). *The limits of organic life in planetary systems.* National Academies Press, Washington, D.C.

Rosso, L., Lobry, J., Bajard, S., and Flandrois, J. (1995). Convenient model to describe the combined effects of temperature and ph on microbial growth. *Appl Environ Microbiol*, 61(2):610–616.

Rosso, L., Lobry, J. R., and Flandrois, J. P. (1993). An unexpected correlation between cardinal temperatures of microbial growth highlighted by a new model. *J Theor Biol*, 162(4):447–463.

Van Impe, J. F., Nicolai, B. M., Martens, T., De Baerdemaeker, J., and Vandewalle, J. (1992). Dynamic mathematical model to predict microbial growth and inactivation during food processing. *Appl Environ Microbiol*, 58(9):2901–2909.

Xin, X., Kropff, M. J., McLaren, G., and Visperas, R. M. (1995). A nonlinear model for crop development as a function of temperature. *Agrig. and Forest Meteor.*, 77:1–16.

Yan, W. and Hunt, L. A. (1999). An equation for modeling the temperature response of plants using only the cardinal temperatures. *Annals of Botany*, 84:607–614.

Zwietering, M., Jongenburger, I., Rombouts, F., and van 't Riet, K. (1990). Modeling of the bacterial growth curve. *Appl Environ Microbiol*, 56(6):1875–1881.

Bioastronomy 2007: Molecules, Microbes, and Extraterrestrial Life
ASP Conference Series, Vol. 420, 2009
K. J. Meech, J. V. Keane, M. J. Mumma, J. L. Siefert, and D. J. Werthimer, eds.

Chromatic Complementary Adaptation (CCA) for the Exploration of Exoplanetary Life

M. V. Tarasashvili and N. G. Alexidze

Dept. of Biochemistry and Biotechnology, I. Javakhishvili Tbilisi State University, I Chavchavadze ave. 1. 0128. Tbilisi, Georgia

Abstract. The purpose of our research is develop an astrobiological model of possible living processes on exoplanets. Imitations of exoplanetary systems have significant theoretical and practical value. Modeling in astrobiology involves: Selection of exoplanets suitable for the origin of life; Theoretical modeling of exoplanetary environment with polarization-holography methods; Imitation of stellar spectra and experiments with Chromatic Complementary Adaptation; Laboratory modeling of the ecosystem and alien life. The main problem is the integration of investigations both in astrophysics and biotechnology. We have showed for the first time the possibility and expediency of such approach, for the solution of astrobiological problems.

1. Introduction

Research is based on the modeling of possible biochemical processes on exoplanets. Imitations of exoplanetary systems have significant theoretical and practical values. Modeling in astrobiology involves: 1) Selection of exoplanets suitable for the origin of life; 2) Theoretical modeling of exoplanetary environment with polarization-holography methods; 3) Imitation of stellar spectra; 4) Experiments with Chromatic Complementary Adaptation; 5) laboratory modeling of alien ecosystems and life. Theoretical discussions and results obtained from our experiments demonstrate that potentially inhabitant planetary bodies reveal unusual environmental conditions and provide unique evolutionary scenario and biodiversities. Modeling of star spectra, selection of adapter pigments and other lightharvesting complexes and synthesis of the artificial photosynthetic film, in future could provide oxygen supply for space stations, space ships, Lunar and Martian colonies.

2. Strategy and Methods

First task was to identify exoplanets or their hypothetic satellites suitable for life. After analyzing astronomical parameters for 200 exoplanets and their parent stars, we have found at least of them suitable for life and to be possibly suitable. During these calculations we have determined limits of the ecospheres around each star processing data for L-luminosity, S-star classes and R-radiation parameters. Calculations were followed by the experiments. We have constructed an experimental chamber that allows us to simulate actual ecological and illumination characteristics of the selected exoplanetary body and conduct environmental modeling. For the illumination purposes, actual stellar spectra were analyzed and broken down into Gaussian components. Results were used for the selection of appropriate optical filters. In the experimental chamber, combinations of optical glass are located under the primary light-source that summarizes discrete spectra and give back actual emission spectra of the stars of selected classes, illuminating exoplanetary bodies. At the same time, the chamber is equipped with lx-meter for the measurement of the radiation intensity and thermometers for the measurement of temperature.The experimental Chamber has enabled us to create modeled environment that might present of selected exoplanets. Following parameters have been imitated, measured and investigated: 1) Star spectra and surface illumination conditions; 2) Mean daily temperature; 3) Rate and preferable type of photosynthesis; 4) Expected prevalent coloring of photosynthetic organisms; 5) Suitable combinations of pigments and light-harvesting systems; 6) Selection and adjustment mechanisms of various photosynthetic organisms to preferred spectral shifts within integral imitated spectra. Some of our important results are presented below. Star classes are sequenced in accordance to the order of our investigations:

2.1. Life at A Class Stars

If radiation waves shorter than 400nm enter the planetary atmosphere, then organisms should develop increased numbers of carotene-like pigments. Artificial coating of leaves with carotene envelope or extracts, and solutions added to protozoa colonies, increase their resistance to UV-radiation from 280 to 400nm. Phodoxantin gave best results so far.

2.2. Life at B Class Stars

Our experiments have revealed plants: Rosmarinus officinalis, Hyssopus officinalis, Tarragon, Foeniculum Vulgare and others containing aromatic hydrocarbons (benzo[a]anthracene, chrysene, fluoranthene, b-caryophyllene, citronellol and geraniol), bicyclic monoterpene derivatives (L-pinocamphene, isopinocamphone) in main bodies - leaves and stem - to be unusually highresistant against incoming radiation waves shorter then 280nm, including gamma and x-ray (short exposure). Thus, protection from UV and even gamma- and X-rays can be maintained with the increase vaporization of aromatic hydrocarbons (cystamine, benzol-derrivatives, ets). During our investigations, experimental plants were excreting clouds of various aromatic hydro-carbonaceous substances. Similar clouds and viscose colloidal yellowish polymer cover on plants are expected on planets at highradioactive stars.

2.3. M Class Stars and Brown Dwarfs: Beware of IR Life(?)

Ontogenesis of living creatures is the resemblance of their evolutionary development. In this case, insect-eater carnivorous plants are of special interest. Life at warm and low-radioactive red and Infra-Red stars does not necessarily require development of the variety of photosynthetic pigments. Instead, heterotrophic competition takes place. Members of such ecosystems most likely tend to develop as consumers predator and parasites, grow larger and evolve faster. For example, carnivorous plants have evolved in warm, wet climate under the thick plant cover, where preferably Red and IR waves (and maybe UV waves, if they are not cut by the planetary atmosphere) reach the surface. Although these plants are still partially photosynthetic, heterotrophic nutrition and digestion is still preferred. These results are based on investigations of metabolic rates, occurring under the artificial illumination of M-class and Brown Dwarf stars.

2.4. Life at K and G Class Stars

These stars are subjects of our future investigations - dependence of Chromatic Complementary Adaptation on Ecosystem parameters (temperature, radiation intensity, illumination, humidity, etc.) and its photochemical foundation is of interest. Initial results suggests that if an atmosphere (or other living matrix aerial or liquid) does not prevent the entrance of incoming radiation waves, then co-pigmentation and other biochemical adaptations take place in order to defend living organism for damaging extra-intensities. This is especially important for the inhabitants of such planets that the cross inner border of the ecosphere at periapsis.

3. Conclusions

In order to keep focusing on CCA, this research highlights evolutionary aspects of optical adaptation in living systems. Chromatic Complementary Adaptation is the consequence of the adjustment of chromo-protein composition of photosynthetic light-harvesting antennae system. Changes in light properties and light-harvesting structures help the living organisms to maximize or minimize the absorption of prevalent wavelength of light in the environment and maintain sufficient rates of natural artificial photosynthesis as well as various ways of Carbon conversion.

It is obvious that the variety of forms and coloring is expected at any inhabited body; however, the prevalence of green-colored plants on Earth was taken into consideration, accepted as objective fact and used as initial research point. Thus, we have concluded the following:

1. Life can occur at different stages of stellar and planetary evolution. Stars of O, B, A classes can have planets with life (consider evolution of stars in multiple systems).

2. Size and luminosity of stars varies, but if there are planets or their satellites within ecosphere limits, life can evolve.

3. Biodiversity is the result of combinations between the adaptations to ecosystems, circadian events and CCA phenomena. Still, variety (biodiversity) is dependant on high-quality life, entering planetary atmosphere.

4. Genetic regulation of CCA makes possible the creation of transgenic organisms adapted to various illumination conditions that can be used for the implementation of Terraformation (of the Moon and/or Mars).

5. CCA is an excellent tool for the prediction of some physical and biochemical properties of extraterrestrial life, but terrible and almost useless for the visual detection of life through the optical instruments, especially on those planetary bodies covered with atmosphere.

It now remains to investigate the precise metabolic mechanisms involved in CCA and photosynthesis under the illumination of different stars. Using the phenomena of CCA, photochemical properties of pigments and research in photosynthesis, I. Javakhishvili Tbilisi State University, Biochemical Association of Georgia, together with the Institute of Cybernetics, are making efforts to synthesize bionic film-covers for alternative oxygen-supply of space stations. This method involves artificial initialization of the processes similar to bacterial adhesion in bio-films and synthesis of crystal-enriched glass for the energy transformations (UV and short-waved light are transformed into extra amount of Red and IR waves) and used in photosynthesis.

Acknowledgments. We thank Abastumani Astrophysical Observatory staff for the stellar spectra and expert advice; Georgia Institute of Cybernetics for the supplement of Optical Devices; Prof Shalva Sabashvili and Prof. Elene Davitashvili for assistance.

References

Calvin, M. & Gazenko, O. 1975. Foundations in Space Biology and Medicine. Vol. 1 (pp. 217-316), Vol. 2 (pp 78-87); Vol. 3 (pp 318-324) M. Nauka & NASA STIO WD.C.

Goodwin, T. W. 1965. Chemistry and Biochemistry of Plant pigments. Aberrwyth, Wales. Academic Press. NY. London.

Pickelner S.B. 1976. Physics in Space (pp 66-67, 538-539, 559-564) SE. Moscow.

White, D. 2000. The physiology and Biochemistry of prokaryotes. Indiana University, Oxford University Press, Second ed.

Bioastronomy 2007: Molecules, Microbes, and Extraterrestrial Life
ASP Conference Series, Vol. 420, 2009
K. J. Meech, J. V. Keane, M. J. Mumma, J. L. Siefert, and D. J. Werthimer, eds.

Exobiology at Southern Brazil: Spore Dosimetry and the UV Solar Radiation

P. H. Rampelotto,[1] M. B. Rosa,[1] N. J. Schuch,[1] D. K. Pinheiro,[2] A. P. Schuch,[3] and N. Munakata[4]

[1]*Exobiology and Biosphere Laboratory, Southern Regional Space Research Center, National Institute for Space Research, Santa Maria, RS, Brazil*

[2]*Space Science Laboratory of Santa Maria, Federal University of Santa Maria, Santa Maria, RS, Brazil*

[3]*Department of Microbiology, Institute of Biomedical Sciences, University of São Paulo, SP, Brazil*

[4]*Faculty of Science, Rikkyo University, Tokyo, Japan*

Abstract. The ultraviolet - UV is considered the range of solar radiation most immediately lethal to the life organisms on the Earth's surface. In this context, since 2000, the monitoring of the biologically-effective solar radiation using spore dosimeter at the Southern Space Observatory (29.4°S, 53.8°W), South of Brazil, has been performed. The biological dosimeter is based in the spore inactivation doses of *Bacillus subtilis* strain TKJ6312, who is sensitive to the UV solar radiation. Monthly expositions of biological dosimeter have been compared with solar irradiance obtained by Brewer spectrophotometer. Correlations indices about $r > 0.86$ shows the potential applicability of the biosensor in the monitoring of biologically-effective solar radiation. Since spores are stabile microorganisms, considering extreme environment variations, the biosensor may be used for studies of the effects of the solar radiation in others planetary environments for future work.

1. Introduction

The study of solar ultraviolet radiation (UVR) at different latitudes of the planetary surfaces has been discussed in the last decades, because the UV radiation plays a decisive role on the course of the biological evolution. The understanding of photobiological processes under different planetary climates like the early Earth, Mars and the comprehension of the role of the ozone leer protecting the biosphere for the UVB radiation is a very important aspects for better understanding the life evolution (Rontó et al. 2004).

Considering the exobiology context, a special attention has been given to Mars, due mainly the similarities between the present Mars and the early Earth, when considering the life appearance on the Earth (Jakosky 1998). Although the climatology effects of the UV on the early Earth is still relatively speculative, the study of the biological effects of UVR on Mars can help us to understand a possible biological evolution on Mars, as well as the transference mechanism of organisms from Mars to spacecraft and still this reasoning could be applied to understand the potential UV stress generated on the molecules of essential microorganisms presents in the early Earth's surface (Rontó et al. 2003). Therefore, studies have been focused in understanding of how it is possible to adapt the biological UV dosimetry's concepts and technologies developed under terrestrial conditions for the Mars environment; temperatures from -80°C to +20°C, atmospheric pressure about 7 mbar and a mixture atmospheric gases about 95.3 % CO_2, 2.7% N_2, 1.6% Ar, 0.2% O_2, 0.03% H_2O (Fajardo-Cavazos et al. 2007). In this work, the potential applicability of spores as biosensors for biologically-effective solar radiation on Earth is presented, aiming to open the discussion in terms the potential applicability of the spore dosimetry for and in Mars environmental conditions.

2. Methodology

Bacillus subtilis TKJ 6312. Due the deficiency of both DNA repair mechanisms, Nucleotide Excision Repair (NER) and Spore Photoproduct Lyase (SP lyase), the spores were genetically modified to be sensible to UVR, but maintaining the resistance of extreme environment conditions (Munakata et al. 2000). In this method, four spots of about 10^6 spores of *B. subtilis* strain TKJ 6312 are spotted on a membrane filter, which two spots were covered with cardboard to serve as unexposed controls. After, the samples are covered and dually wrapped with blue polyethylene sheets (Umeya Sangyo, Tokyo) and placed in a plastic slide holder (Munakata et. al. 2006). The samples are characteristically stable, when stored in dark either before or after exposure. After monthly exposures, the SID (Spore Inactivation Dose) is calculated from the natural logarithm of surviving fraction; SID = -ln (Ne/Nc), where Ne and Nc represent the average number of colony-formers recovered from exposed and control spots, respectively. The spore's action spectrum is obtained by "*Okazaki Large Spectrograph*" comprising from 254 nm to 400 nm (Munakata et. al. 1996). This is used in the calculation of SID by the Brewer spectrophotometer. The SID's values were compared with the integrals of monthly UVB as well the SID calculated by the Brewer Spectrophotometer (Schuch et. al. 2006). All exposures were performed at the National Institute For Space Research – INPE's, Southern Space Observatory (SSO) (29.4°S, 53.8°W), in Southern Brazil.

3. Results

A seasonal variation of Biologically-Effective Solar Radiation (BESR), in terms of SID and UV-B is presented in **Figure 1**. As expected, the temporal variation of BESR and UV-B presents a seasonal profile associated to the annual cycle of the zenithal solar angle. The data profiles shows higher correlation indices, 0.86 (2000), 0.86 (2001), 0.96 (2002), 0.88 (2003), 0.98 (2004) and 0.90 (2005). Considering the SID calculated, pondered by the Brewer and the SID observed, obtained by the spores (**Figure 2**), an elevated correlation index about $r \sim 0.91$ ($p < 0.0001$) is also observed.

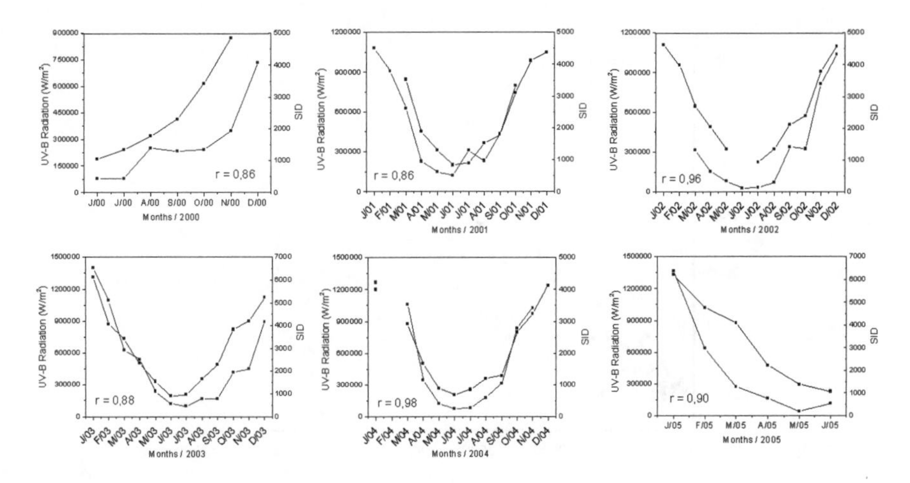

Figure 1. Seasonal profile of BESR obtained by Spores (blue line) and UV-B radiation, by Brewer (black line) at the INPE's Southern Space Observatory (SSO) for: 2000 - 2005

4. Discussion

Since the 70's the biological dosimetry has been developed, which studies have tried to apply it irrefutably as UV-biosensor and obeying the criterions established by BIODOS project from the European Commission (Horneck, 2000). The simplicity, facility of use and transport, long term storage, and well defined action spectrum is the principal requisites when field experiments are wished. The exposures have been made under natural sunlight by several minutes, days or weeks. However, for long term application, some problems need be circumvented including the reduction of the dose and the protection of the samples from adverses climatic conditions. This new stage, long term exposures, began in 2000 in our station at the SSO. The results observed in Figure 1 and 2 indicate the potential applicability of spores as biosensors of biologically-effective solar radiation on the Earth surfaces. To adapt this biosensor technology for Mars UV environments some considerations need to be considered. Mars receives characteristically 44% less of the total solar radiation compared with the Earth. However, the UVC and UVB presents in comparison higher levels, due the similarity lack between the Earth's and the Mars atmosphere (Mancinelli & Klovstad 2000). In the spores, NER and SP lyase appear sufficiently to explain the great majority of repair of damage by UVC and UVB (Nicholson et. al. 2005). As they are deficient of both repair mechanisms mentioned and considering a well defined action spectrum, they could be an alternative to study the biological implication of UVR on Mars surface. Although the UVR on Mars surface is higher in comparison with the observed values on the Earth, the dusts, rocks and ice shows that Mars surface presents physical substrates adequate to protect biomolecules against the harsh radiation regimen and allow the long-term survival of microbes under these substrates (Rontó et al. 2003). To ratify

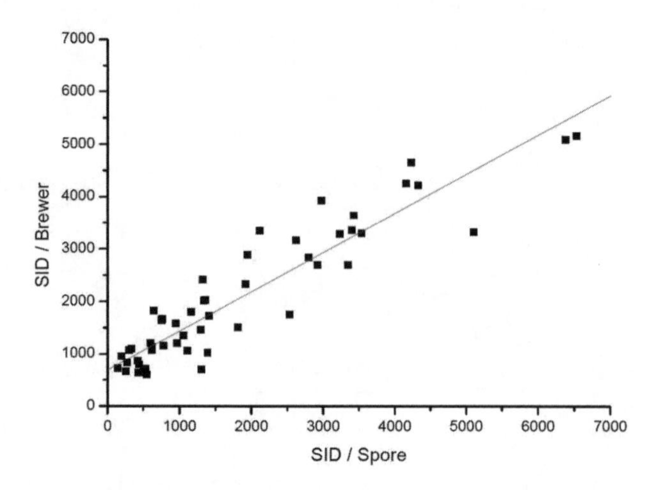

Figure 2. SID observed (by the Spores) versus SID calculated (by Brewer) for 2000-2005

the possible application of biosensor under others planetary environmental, simulations of Mars radiation conditions have been performed, where a layer of dust of 1 mm thickness shows to be completely effective in the prevention of spore inactivation, when submitted a exposure about 12 kJ (Mancinelli & Klovstad 2000).

5. Conclusions

The results presented in this paper indicate the potential applicability of spores as biosensors of biologically-effective solar radiation on the Earth surfaces. This biosensor technology may be adapted for application on the Mars surface but with some considerations for the difference of the UV environments.

Considering the interests of the international exobiology groups to study the spatial solar radiation under planetary environments, the application of *Bacillus subtilis* TKJ 6312 has been also considered. However, some problems are recognized to applications under extreme environments and further studies to elucidate these questions are still necessary. The current interests have been to find appropriate filter materials with good optical properties, reasonable and uniform transmittance for the UVC-UVB bands (Mars condition) with a same comparatively performance with the used under Earth's scenarios.

References

Fajardo-Cavazos, P., et al. 2007, Astrobiology, 7, 512
Horneck, G. 2000, Adv. Space Res., 12, 1983
Jakosky, B., 1998, Cambridge University Press, Cambridge, UK
Mancinelli, R.L. & Klovstad, M. 2000, Planet. Space Sci., 48, 1093
Munakata, N., et al. 1996, Photochem. Photobiol., 63, 74
Munakata, N., et al. 2000, Photochem. Photobiol., 72, 739
Munakata, N., et al. 2006, Photochem. Photobiol., 82, 689
Nicholson, W.L., Schuerger, A.C. & Setlow, P. 2005, Mutat. Res., 571, 249.
Rontó, G., et al. 2003, Photochem. Photobiol., 77, 34
Rontó, G., et al. 2004, Adv. Space Res., 33, 1302.
Schuch, A.P., et al. 2006, Adv. Space Res., 37, 1784

David Harrington

John Rummel and Karen Meech

Bioastronomy 2007: Molecules, Microbes, and Extraterrestrial Life
ASP Conference Series, Vol. 420, 2009
K. J. Meech, J. V. Keane, M. J. Mumma, J. L. Siefert, and D. J. Werthimer, eds.

Astrobiology in the Environments of Main-Sequence Stars: Effects of Photospheric Radiation

M. Cuntz,[1] L. Gurdemir,[1] E. F. Guinan,[2] R. L. Kurucz[3]

[1]*Department of Physics, University of Texas at Arlington, Arlington, TX 76019-0059, USA*

[2]*Department of Astronomy and Astrophysics, Villanova University, Villanova, PA 19085, USA*

[3]*Harvard-Smithsonian Center for Astrophysics, Cambridge, MA 02138, USA*

Abstract. We explore if carbon-based macromolecules (such as DNA) in the environments of stars other than the Sun are able to survive the effects of photospheric stellar radiation, such as UV-C. Therefore, we focus on main-sequence stars of spectral types F, G, K, and M. Emphasis is placed on investigating the radiative environment in the stellar habitable zones. Stellar habitable zones are relevant to astrobiology because they constitute circumstellar regions in which a planet of suitable size can maintain surface temperatures for water to exist in fluid form, thus increasing the likelihood of Earth-type life.

1. Theoretical Approach

The centerpiece of all life on Earth is carbon-based biochemistry. It has repeatedly been surmised that biochemistry based on carbon may also play a pivotal role in extraterrestrial life forms, if existent (e.g., Bennett & Shostak 2007). This is due to the pronounced advantages of carbon, especially compared to its closest competitor (i.e., silicon), which include: its relatively high abundance, its bonding properties, and its ability to form very large molecules as it can combine with hydrogen and other molecules as, e.g., nitrogen and oxygen in a very large number of ways (Goldsmith & Owen 2002).

In the following, we explore the relative damage to carbon-based macromolecules in the environments of stars other than the Sun using DNA as a proxy. We focus on the effects of photospheric radiation from main-sequence stars, encompassing the range between F0 and M0. Our models consist of the following components:

1. The radiative effects on DNA are considered by applying a DNA action spectrum (Horneck 1995). It shows that the damage is strongly wavelength-dependent, increasing by more than seven orders of magnitude between 400 and 200 nm. The different regimes are commonly referred to as UV-A, UV-B, and UV-C.

2. The planets are assumed to be located in the stellar habitable zone (HZ). Following the concepts by Kasting et al. (1993) and Underwood et al.

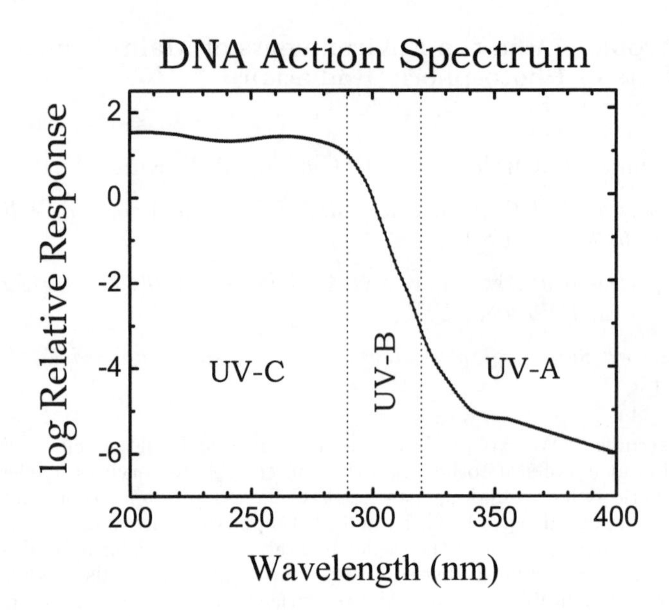

Figure 1. DNA action spectrum following Horneck (1995).

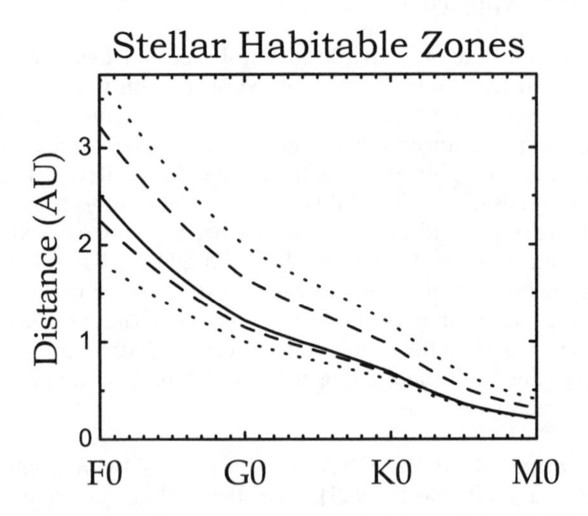

Figure 2. Stellar habitable zones for main-sequence stars of different spectral types. The dashed lines indicate the conservative HZ, whereas the dotted lines indicate the generalized HZ. The solid line denotes the Earth-equivalent position.

(2003), we distinguish between the conservative and generalized HZ (see Fig. 2). The inner and outer edge of the conservative HZ are given by the onset of water loss and CO_2 condensation, respectively, whereas the inner and outer edge of the generalized HZ are given by the runaway greenhouse effect and the breakdown of greenhouse heating, respectively, needed to permit the existence of fluid water on the planetary surface.

3. Stellar photospheric radiation is represented by using realistic spectra, which take into account millions or hundred of millions of lines for atoms and molecules (Castelli & Kurucz 2004, and related publications). Clearly, significant differences emerge between the different spectral types, both concerning the total amount of radiation and their spectral distribution.

4. We also consider the effects of attenuation by a planetary atmosphere. The following cases are considered: Earth as today, Earth 3.5 Gyr ago, and no atmosphere at all (Cockell 2002). For general discussions see, e.g., Guinan & Ribas (2002) and Guinan et al. (2003).

Our results are presented in Figs. 3 and 4. They show the relative damage to DNA due to photospheric radiation from stars between spectral type F0 and M0. The results are normalized to today's Earth, placed at 1 AU from a star of spectral-type G2V. We also considered planets at the inner and outer edge of either the conservative or generalized HZ as well as planets of different atmospheric attentuation.

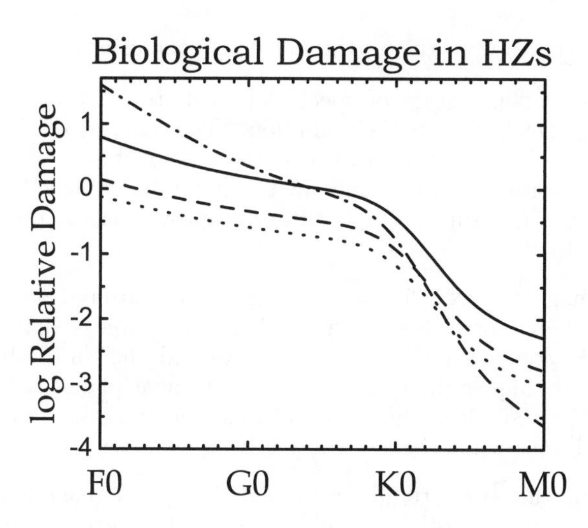

Figure 3. Biological damage to DNA for a planet at an Earth-equivalent position without an atmosphere (solid line), an atmosphere akin to Earth 3.5 Gyr ago (dashed line) and an atmosphere akin to Earth today (dotted line). The dash-dotted line refers to a planet without an atmosphere at 1 AU.

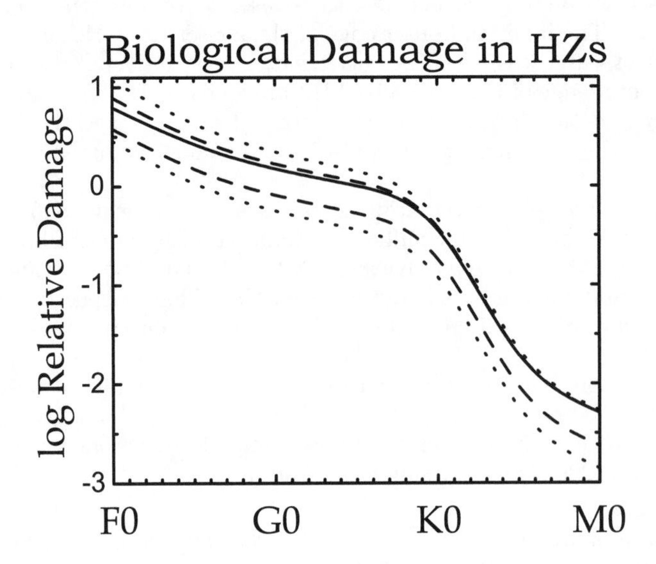

Figure 4. Biological damage to DNA for a planet (no atmosphere) at an Earth-equivalent position (solid line), at the limits of the conservative HZ (dashed lines) and at the limits of the generalized HZ (dotted lines).

2. Conclusions

Based on our studies we arrive at the following conclusions:

1. All main-sequence stars of spectral type F to M have the potential of damaging DNA due to UV radiation. The amount of damage strongly depends on the stellar spectral type, the type of the planetary atmosphere and the position of the planet in the habitable zone (HZ). Our results constitute a quantitative update and improvement of previous work by Cockell (1999).

2. The damage to DNA for a planet in the HZ around an F-star (Earth-equivalent distance) due to photospheric radiation is significantly higher (factor 5) compared to planet Earth around the Sun, which in turn is significantly higher than for an Earth-equivalent planet around an M-star (factor 180). Small modifications of this picture occur for different planetary positions inside their respective HZs.

3. Regarding the cases studied, we found that the damage is most severe in the case of no atmosphere at all, somewhat less severe for an atmosphere corresponding to Earth 3.5 Gyr ago, and least severe for an atmosphere like Earth today.

Our results are of general interest for the future search for planets in stellar HZs, the chief goal of future NASA search missions (e.g., Turnbull & Tarter

2003). Our results also reinforce the notion that habitability may at least in principle be possible around M-type stars, as previously discussed by Tarter et al. (2007). Note, however, that a more detailed analysis also requires the consideration of chromospheric UV radiation, especially stellar flares (e.g., Robinson et al. 2005).

References

Bennett, J. & Shostak, S. 2007, Life in the Universe, 2nd Ed. (San Francisco: Addison Wesley)
Castelli, F. & Kurucz, R. L. 2004, in Modelling of Stellar Atmospheres, IAU Symp. 210, ed. N. E. Piskunov, et al. (San Francisco: ASP), CD-ROM, Poster 20
Cockell, C. S. 1999, Icarus, 141, 399
Cockell, C. S. 2002, in Astrobiology: The Quest for the Conditions of Life, ed. G. Horneck & C. Baumstark-Khan (Berlin: Springer), 219
Goldsmith, D. & Owen, T. 2002, The Search for Life in the Universe, 3rd Ed. (Sausalito: University Science Books)
Guinan, E. F. & Ribas, I. 2002, in The Evolving Sun and Its Influence on Planetary Environments, ed. B. Montesinos, et al. (San Francisco: ASP), 85
Guinan, E. F., Ribas, I., & Harper, G. M. 2003, ApJ, 594, 561
Horneck, G. 1995, J. Photochem. Photobiol. B: Biology, 31, 43
Kasting, J. F., Whitmire, D. P., & Reynolds, R. T. 1993, Icarus, 101, 108
Robinson, R. D., et al. 2005, ApJ, 633, 447
Tarter, J. C., et al. 2007, Astrobiology, 7, 30
Turnbull, M. C. & Tarter, J. C. 2003, ApJS, 145, 181
Underwood, D. R., Jones, B. W., & Sleep, P. N. 2003, Int. J. Astrobiology, 2, 289

Rebecca Turk

Bioastronomy 2007: Molecules, Microbes, and Extraterrestrial Life
ASP Conference Series, Vol. 420, 2009
K. J. Meech, J. V. Keane, M. J. Mumma, J. L. Siefert, and D. J. Werthimer, eds.

Biological Evolution on the Earth Influenced by Astronomical Objects: Especially Gamma-ray Sources

J. Ponert[1] and P. Příhoda[2]

[1]*Czech Botanic Society, Department of Science, Wintrova 6 160 00, Praha, Czech Republic*

[2]*Czech Planetarium Prague, Czech Republic*

Abstract. Taking in to account 20,000 explosions of intragalactic supernovae per million years, the sources estimated at 10^{56} - 10^{57} MeV producing the high intensity of gamma- and xray-radiation even after its reduction through the Earth atmosphere, may have a significant mutagenic action. During the time period of the last 4 billion years not less than one hundred explosions up to the mean distance 126 pc from the Earth. All such explosions were able to evoke a genetic revolution among most taxonomic groups of terrestrial organisms. For mountain organisms, the more frequent supernova explosions in distance up to 400-900 pc are of importance, maritime organisms could be influenced mainly by secondary radiation products, rather than directly by the gamma and X-rays from the supernovae. The mechanisms of macroevolution depending on supernovae is elucidated. Smaller genetical revolutions in the macroevolutional process (formation of genera) took place on the average once every 10 millions or more years, fundamental genetic revolutions once in 100 millions or more years. Also other newly discovered astronomical gamma-ray sources have to be taken in account.

1. Organic Evolution on Earth and Astronomical Events

While the question of the origin of life seems to be the whole - space problem and the origin of life on the Earth only its subquestion, the evolution of organisms directly on the Earth is the fact proved by several scientific disciplines with paleontology as the central one. But we are interested in the factors and mechanisms of this evolutionary process. While the microevolutionary mechanisms can be studied even in the experimental way (extensive literature of case studies and even the monograph about the whole system of microevolutionary mechanisms effective in the frame of large taxonomical units (e.g. Ponert 1979) exists), the difference of such microevolutionary systems in comparison of different large taxonomical units are pointed out (Ponert 1980). The system of reticular evolution of bacteria leads some scientists even to unjustifiable rejection of the evolution in whole. But it is clear that the evolution as the process of changes is the fact, even if the concept of evolution as the process of improvement can be dependent on the scientists point - of - view and the criteria used.

By looking for broader effective factors on evolutionary processes we find out the geological processes as origin of continents from Pangea and their drift, the geophysical processes as the change of position of the earth axis effective on the way and place of fall of electric active elementar particles of non - earth origin and the changes of earth magnetism, the position of the Earth in the Sun system, the changes of local climates and distribution of climatic zones. Climatic cycles differently affect the populations of one - years and long living plents on the other hand. The changes of atmosphere composition originating from organisms activity, lastly - though in lower extend - the industrial activity of man are also of importance.

In the evolution of organisms effects also one stabile factor - the earth gravity. The gravity constant different from the earth one can influence in the case of biological evolution on the other astronomical objects of the Earth - type even to different models of successful biological species.

Astronomical energetic sources of gamma - radiation, X - radiation and / or neutrons can influence the organic evolution on the Earth in the case when the intensity of doses coming to the Earth are so high to excite some mutations in the organisms.

Other astronomical events as meteorits falls or colisions of the Earth with small planets are more of destructive consequences though in the case of lager extend the catastrophs are able to change the spectrum of surviving species in favour of the successful colonizers.

As shown by the author (Ponert & Příhoda 1979) even the deep hit to the biological evolutionary process is possible in the case of special situation in which not only the origin of individual mutations, but transformation of the whole biological system on the Earth takes place.

At expermental mutagenesis by gamma - rays, X - rays, fast or slow neutrons we are able to increase the frequency of mutations compered with the values of spontaneous situation. But in natural populations even the reparation of primary mutations and further elimination of some persisting ones by original genotypes takes place.

The previously mentioned special evolutionary situation on the Earth developed after some explosions of near intergalactic Supernovas is characteristic with total transformation of genotypes to multiple mutants so that the next competition, hybridization, genetic drift and selection can be realized without the previous original genotypes, so exclusively between the new mutants in the case of restoration of their reproductive ability. In such a way the astronomical event starts to be not only a microevolutionary factor. In a similar way, of course, we can imagine further situations, when the ratio of original genotypes is substantially reduced but not fully eliminated.

For the origin of 1 mutation probably 15 - 20 ionizations are necessary. The different sorts of radiation have different ionization abilities. The number of fragmentations of chromosomes and point - mutations is in linear dependence on the dosage, while number of chromatide and chromosomal translocations and inversion is of quadratic dependence. Higher intensity of irradiation even at the same radiation dosage does not change the number of point- mutations obtained, but through the larger disturbation of reparative mechanisms increases the ratio of finally obtained chromosomal mutations.

E. g., the dosage of 2 krad of gamma - radiation evokes chromosomal aberations at 50% of metaphases of plants of the genus Rumex. The question is if such values can be available as the result of some explosions of intragalactic supernovae or some other sources of gamma - radiation?

If we concentrate our attention to the intragalactic supernovae of 2nd type (Ponert &Příhoda 1979) and take in account the diameter of the disc of the subsystem of supernovae of the 2nd type as 27.6 kpc (Poveda & Woltjer 1968) and its surface $600(kpc)^2$, then the explosion on 1 $(kpc)^2$ surface on the average is realized once a 3 x 10^4 years. The mean distance from the Earth of 10 nearest explosions of supernovae of 2nd type is 93 pc during 10^9years period, for 100 of such explosions 200 pc, respectively. Energy of explosion is estimated (Tucker 1970) as 10^{44} J. At the moment of the maximum of explosion gamma - rays prevail. The tables of obtained dosage of organisms in rentgens R in the dependance on the distance of Supernova from the Earth and the size of organism are given by Ponert & Příhoda 1979. Possible changes of chemical composition of Earth atmosphere are taken in account.

Based on the recent knowledge of SN2001g and close double - stars the supernovae of the 1st type seem to be of importance, too. Mutagenic active gamma - rays, X - rays and neutrons have to be obtainable on the Earth surface after penetration through the atmosphere, eventually also through some layer of soil or water. For this reason only the penetrating gamma - rays are taken in account. The velocity of gamma - rays from the source comparable with light - velocity is of similar importance. Biological importance of supernovae is as follows: 1) in chemical - element evolution as source of elements heavier than Fe - prebiotic material formation - cosmic importance 2) as gamma-sources for biological mutagenesis - Earth type importance

References

Dubinn N.P. (1976): Obšaja genetika, ed. 2. - "Nauka" (Moskva), 590 pp.

Ponert J. (1979): Microevolutionary mechanisms in Magnoliophyta (Angiosperms). - Publishing House of Georgian Academy of Science "Mecniereba" (Tbilisi), 338 pp.

Ponert J. (1980): Taxonomic peculiarities of microevolutionary mechanisms and problems of special macroevolutionary mechanisms. - Proceedings of Symposium "Micro - and macro - evolution," Academy of Science of Estonia (Tartu), 106 - 109

Ponert J., Příhoda P. (1979): Explosions of intragalactic supernovae - apossible external factor for macroevolution of organisms on the Earth. - Biologické listy (Praha),44 (1) : 27-44.

Poveda A., Woltjer L. (1968), Astron. J. 73 : 65

Ting - Ting Gao, Yong - Feng Huang (2006): On the evolution of the apparent size of gamma - Ray burst remnants. - Chin. J. Astron. Astrophys. 6(3): 305 - 311.

Tucker W. H. (1970), Astroph. and Space Sci, 9 : 315 a. Astrophys. J. 161:1161

Kensei Kobayashi

Bioastronomy 2007: Molecules, Microbes, and Extraterrestrial Life
ASP Conference Series, Vol. 420, 2009
K. J. Meech, J. V. Keane, M. J. Mumma, J. L. Siefert, and D. J. Werthimer, eds.

Determination of Oxygen Production by Cyanobacteria in Desert Environment Soil

J. E. Bueno Prieto
Department of Biology, Universidad Nacional de Colombia,
email:jebuenop@unal.edu.co

Abstract. The cyanobacteria have been characterized for being precursor in the production of oxygen. By means of photosynthetic reactions, they provide oxygen to the environment that surrounds them and they capture part of surrounding dioxide of carbon. This way it happened since the primitive Earth until today. Besides, these microorganisms can support the harmful effects of ultraviolet radiation. The presence of cyanobacterias in an environment like a dry tropical bioma, such as the geographical location called Desert of The Tatacoa (Huila - Colombia), is determinant to establish parameters in the search of biological origin of atmospheric oxygen detected in Mars. In that case, I work with a random sample of not rhizospheric soil, taken to 15 cm of depth. After determining the presence of cyanobacterias in the sample, this one was in laboratory to stimulate the oxygen production. The presence of oxygen in Mars is very interesting. Since oxygen gas is very reactive, it disappear if it is not renewed; the possibility that this renovation of oxygen has a biological origin is encouraging, bearing in mind that in a dry environment and high radiation such as the studied one, the production of oxygen by cyanobacterias is notable. Also it is necessary to keep in mind that the existence of cyanobacterias would determine water presence in Mars subsoil and the nutrients cycles renovation. An interesting exploration possibility for some future space probe to Mars might be the study of worldwide distribution of oxygen concentration in this planet and this way, indentify zones suitable for microbian life.

1. Introduction

Life developed on early Earth in extreme conditions similar to Mars. Recent evidence shows that there used to be water on Mars. Some barren landscapes on Earth that are similar to Mars harbor life today. A thin atmosphere and harsh conditions are not a deterrent to the survival of some bacteria. Their activities made it possible for the more complex forms of life we see today to evolve. Microbes also do most of the recycling of dead things, so that life can continue. A type of photosynthetic bacteria called cyanobacteria (originally misnamed blue-green algae) put the first molecular oxygen in the atmosphere about 2 billion years ago, raising the oxygen level to about 10% of what it is today, allowing the ozone layer to form and oxygen-using creatures to evolve.

Much later, plants took over the job of producing oxygen, but they can do this only because of organelles called chloroplasts, which evolved from cyanobacteria.

The Martian atmosphere is much thinner than that of Earth. It consists mostly of carbon dioxide, or CO_2 (about 95%), and contains virtually no oxygen

(O_2). Because many bacteria, archaea, and algae can use inorganic carbon dioxide as their source of carbon (used to build proteins and other cell components), the predominance of carbon dioxide would be a plus. Also, as noted earlier, many of Earth's microbes do not require O_2, so the lack of O_2 does not preclude life.

A more troubling feature of the Martian atmosphere is the very low level of nitrogen (N_2). On Earth, N_2 makes up 78% of atmospheric gases. On Mars it only composes 3% and oxygen 0.5-1%. Many bacteria can use N_2 as a sole source of the nitrogen they need for proteins, nucleic acids, and other cell components, but the low level of N_2 would certainly limit the amount of microbial growth. The presence of cyanobacteria in an environment like a dry tropical bioma, is critical to establish parameters in the search of biological origins of atmospheric oxygen detected on Mars.

2. Methodology

Work with a random sample of non-rhizospheric soil in the desert of the Tatacoa (Huila-Colombia), taken from 15cm of depth the samples were conserved in a hermetic container and to temperature of 20 degrees the ground. After determining the presence of cyanobacteria in the sample with a culture in nutritive agar, this was paced in a laboratory to stimulate the oxygen production in a container that allowed receiving constant ultraviolet light by 12 hours continuous.

3. Results

A rank of 15-20 percent in production of oxygen in conditions of desert dry atmosphere in a subsoil layer with depth of 15 cm, in a non rhizospheric soil (Soil without presence of roots), the presence of this type of microorganisms in a soil with very poor water presence is a sample of adaptation to extreme conditions, in addition to important a physiological and metabolic development.

4. Discusion

Life depends on chemical reactions that cannot occur spontaneously under the conditions where they are found to occur. On Earth, many microbes make a living by catalyzing reactions, such as the oxidation of minerals or the production of methane, that either do not occur abiotically or occur at a much lower rate.

Different compositions and concentrations of gases may exist in some areas under the Martian surface. Such a possibility would be difficult to prove—unless it is proved indirectly the presence of life in the subsurface regions and in greater abundance than expected.

Also it is necessary to keep in mind that the existence of cyanobacteria would indicate the presence of water in the Mars subsoil an the nutrient cycle's renovation.

5. Conclusion

An interesting exploration possibility for some future space probe to Mars might be the study of the world wide distribution of oxygen concentration on this planet and this way, identify zones suitable for microbial life.

Acknowledgments. My mother, my University, teachers and Avianca airlines in Colombia.

References

Amann, R.I., Ludwig, W. & Schleifer, K.H. (1995). Phylogenetic identication and in situ detection of individual microbial cells without cultivation. Microbiol. Rev. 59, 143–169.

Conrad, P.G., and K. H. Nealson. 2001. A non-Earthcentric approach to life detection. Astrobiology1(1):15–24.

Friedman, E. L. 1986. The Antarctic cold desert and the search for traces of life on Mars. AdvancesinSpaceResearch6(12):265–268.

Jakosky, B. M., K. H. Nealson, C. Bakermans, R. E. Ley, and M.T. Mellon. 2003. Subfreezing activity of microorganisms and the potential habitability of Mars' polar regions. Astrobiology 3(2): 343–350.

Nealson, K. H. 2001. Searching for life in the universe. Annals of the New York Academy of Sciences 950: 241–258.

Salyers, A. A., and D. W. Whitt. 2000. History of life on Earth. Chapter 2 in Microbiology: Disease, Diversity and the Environment. Washington (DC): Fitzgerald Science Press (original publisher); New York: John Wiley & Sons.

Wharton, R. A., C. P. McKay, R. L. Mancinelli, and G. M. Simmons. 1989. Early martian environments: The Antarctic and other terrestrial analogs. Advances in Space Research 9(6): 147–153.

Daniel Altschuler

Bioastronomy 2007: Molecules, Microbes, and Extraterrestrial Life
ASP Conference Series, Vol. 420, 2009
K. J. Meech, J. V. Keane, M. J. Mumma, J. L. Siefert, and D. J. Werthimer, eds.

Evaluation of Biological and Enzymatic Activity of Soil in a Tropical Dry Forest: Desierto de la Tatacoa (Colombia) with Potential in Mars Terraforming and Other Similar Planets

A. N. Moreno

Departament of Biology, Universidad Nacional de Colombia, Columbia, email: andmorenom@unal.edu.co

Abstract. Desierto de la Tatacoa has been determined to be a tropical dry forest bioma, which is located at 3° 13" N 75° 13" W. It has a hot thermal floor with 440 msnm of altitude; it has a daily average of 28°C, and a maximum of 40°C, Its annual rainfall total can be upwards of 1250mm. Its solar sheen has a daily average of 5.8 hours and its relative humidity is between 60% and 65%. Therefore, the life forms presents are very scant, and in certain places, almost void. It was realized a completely ramdom sampling of soil from its surface down to 6 inches deep, of zones without vegetation and with soils highly loaded by oxides of iron in order to determine the number of microorganisms per gram and its subsequent identification. It was measured the soil basal respiration. Besides, it was determined enzymatic activity (catalase, dehydrogenase, phosphatase and urease). Starting with the obtained results, it is developes an alternative towards the study of soil genesis in Mars in particular, and recommendations for same process in other planets. Although the information found in the experiments already realized in Martian soil they demonstrate that doesnt exist any enzymatic activity, the knowledge of the same topic in the soil is proposed as an alternative to problems like carbonic fixing of the dense Martian atmosphere of CO_2, the degradation of inorganic compounds amongst other in order to prepare the substratum for later colonization by some life form.

1. Introduction

Well it is known that all the biochemical reactions are catalyzed by enzymes, that are proteins that have catalytic properties due to their power of specific activation. The enzymes are specific for the type of reaction in which they participate, that is to say, are specific for a certain substrate. Within the Mars experiments search, they have considered technical that can not be accurate at the time of the detection of life, as it were it the case of the gas-chromatographic spectrometer of mass (GCMS)

The determination of the enzymatic activity has like advantage the high sensitivity, in addition it is like a useful tool in the knowledge of cyclization of the different important elements like nitrogen, phosphorus. If the colonization of another planet thinks about future, is necessary to know and to understand the processes that occur in the formation and ground maintenance, that will be the substrate that has supported the life like since it does in the Earth.

Total the biochemical activity of the ground is constituted by a series of reactions catalyzed by enzymes (Skujins, 1967). The enzymes are soluble proteins, of organic nature and colloidal state, elaborated by the alive cells, that act independently of these, must be able catalytic specific and they are destroyed by 100 the humid heat to ^oC. It is considered that the enzymatic activity is one of the numerous forms to measure the quality of the ground (Kennedy, et al., 1995). The cycles of nutrients in the ground imply a series of biochemical, chemical reactions and physical-chemistries, being the biochemical processes mediated by the microorganisms, roots and animals of the soil.

The enzymes of the ground can come as much from microorganisms as of plants and animals, although it is considered that the enzyme main source is the microorganisms. Some enzymes, for example deshydrogenase, only is in viable cells, but most of enzymes they can exist secretadas by microorganisms, or originated by microbial remainders or remainders of plants. Its more useful physical form in the ground is when they are immobilized in mineral colloids (clays) either organic (humic molecules) since then they are the more resistant to the degradation and remain active in the ground by the more time. We have to indicate that, due to the difficulty in its location, we must have present which they are difficult to extract of the ground, reason why indirectly studied enzymes by means of its activity.

The enzymes play a fundamental role in the cycles of as important elements as nitrogen (urease and protease), phosphorus (phosphatase) or carbon (-glucosidasas), of there, the importance of the study of this type of enzymes. It agrees to indicate that the activities enzymatic that we can measure, must of being considered like potential measures, since they are moderate "in vitro" under optimal conditions of temperature, pH, etc. Nevertheless, the information that offers these measures allows to contribute knowledge on the biochemical processes that occur in the same one, and therefore they behave like biological indices (Frakenbeyer and Dick, 1983). Phosphatase constitutes an enzyme group that catalyzes anhídridos ester hydrolysis and of phosphoric acid (Schmidt and Laskowski, 1961).

The total activity of dehydrogenase in the ground reflects the global oxidativas reactions of the edaphic microflora and, consequently, a good indicator of the microbiological activity is considered him (Skujins, 1973). Urease is an enzyme of the cycle of nitrogen; it catalyzes the hydrolysis reaction of the urea in ammonium and carbon dioxide, and one is present in superior plants and microorganisms (particularly bacteria). Basal breathing of the ground: Anderson (1982), defined several terms related to the breathing: Breathing: yield of the power processes in which reduced the organic and inorganic products can act like electron givers. Breathing of the ground: Consumption of O2 or CO_2 loosening by alive organisms.

In this study, I determine the enzymatic activity of the ground in a barren zone with the purpose of obtaining data on the global metabolic activity of the same one. Also enzymatic activities of the group of hydrolasae have been determined, to be able to have a better knowledge of the influence of the treatment made on the processes that affect the cycle of nitrogen (urease), phosphorus (phosphatase). It is tried to contribute a general concept of the important metabolic processes that they occur in the ground, being the this substrate that supports the life in this or another planet. To know the formation the ground in the planet earth, is a tool that surely will have much validity in the planning of terraformation of another planet.

2. Methodology

I try myself to obtain the samples in totally lacking places of vegetation, thus to guarantee that the enzymatic activity outside total bacterial origin. Several samples were taken from different zones and they met in a total sample to guarantee the homogeneity of the sample. Work to a depth of 7,5 cm El ground had been incubated aerobic, in conditions of laboratory. For the different treatments I sift the ground in a mesh of 1 mm.

For the determination of the breathing of the ground use a camera of static breathing. This technique consists of catching CO_2 involved during the breathing of a ground sample, in a solution of NaOH located within the air chamber. A cubical bottle of wide mouth with 16 cm of side, was used like incubation camera, while a plate of plastic Petri was used like container of the NaOH. Once the samples were deposited in a layer of 5 cm in the incubation camera, a tube of PVC with 5 cm of diameter was installed vertically within the ground to support the plate of Petri. Hollows through the tube were made to allow the free circulation of CO_2 within the camera. The cover was covered with paraffin immediately after 10 mililiter of NaOH 1N were spilled within the plate of Petri.

For the quantification of the activity of the enzyme dehydrogenase use the method raised by (Casida ET al.1964), which is based on the colorimétrica determination of the released product 2,3,5-triphenilformazan (TFF) that is originated after incubating the sample with chloride of 2,3,5-tripheniltetrazolio (37 CTT) to ^oC by 24 hours. The activity of phosphotase was determined acid (phosphomonoesterasae), according to is described in (Eivazi and Tabatabai 1977), by means of later colorimetric determination with p-nitrophenol released to the incubation of the ground with a solution buffer of (P-nitrophenil Sodium phosphate and toluene). The enzymatic activity of ureasa was determined by means of the method of (Tabatabai and to bremner, 1972), which consists of the determination of ammonium produced by urease after cover the ground in a solution buffer and toluene to 370 centigrade ones during two hours. The activity catalase I determine myself after the incubation of the ground with peroxide of I hydrogenate (HÒ2), volume the measurement of the amount of peroxide of I hydrogenate myself disintegrated in the time unit.

3. Results and Discussion

The values of the found enzymatic activities in this zone are relatively low with respect to the values found in similar works with other types of ground, for example the average of the CO_2 production is minor who half of the breathing of a ground in the high valley of rio black and Neuquén in Argentina in where a basal breathing of 6 is divided equally ($mgCO_2/g.hr$), according to (P, Gili ET al.2004). This reflected one which the environmental conditions are unfavorable for the biological activity in the ground of the desert of the Tatacoa, but that somehow is managed to catch the little activity of the present microorganisms. The result is encouraging since in spite of the almost null presence of life in the sampled zones, a result in the enzymatic activity was obtained that gives account of the life existence. The results do not have differences between the different samples, which guarantee a good uniformity of the total sample. The

Table 1. Enzymatic activities in sampled zones.

CO_2 mgCO_2/g.hr	Urease μ mol of NH_3/g	Catalase μmol of H_2O_2 g.min^{-1}	Dehydrogenase μg of TPF/g.hr	Phosphase mg p-pitrofenol mg.hr
0.4	1.4	0.05	0.03	1.5
0.2	2.2	0.03	0.02	1
0.1	1.1	0.02	0.01	0
0.1	2	0.01	0.03	1.6
0.2	0.9	0	0.01	2.3

found total breathing is without single place to doubt the emanated one by the microorganisms of the ground, since as previously it were described, the samples were taken from zones totally deprived of vegetation.

The activity ureasa of the ground of the desert is again relatively low with respect to works made in other sites, for example in (Gil-Sotres, 2002).El average of the activity urease is from 3 to 5mol of NH_3/g, this is almost 5 times greater than in this work, this difference can be cradle in the difference of density of microorganisms between the two zones. In addition the availability to Nitrogen is much more low in the desert that in another type of ecosystem. Catalase like dehydrogenase is enzymes that show the metabolic reactions of viable cells, in the desert the conditions are not propitious for many forms of life even for many sorts of microorganisms, reason why it was to be expected a very low activity of these enzymes, in this work were almost null values, in the case of catalase thinks that the results are rather product of the chemical decomposition of peroxide of I hydrogenate, as it suggests it (Tabatabai and Bremner.1972), reason why the found data are not really reliable in this aspect. In the case of the activity dehydrogenase, does not exist a chemical reaction in the ground that manages to replace it, reason why it becomes only and infallible in the detection of life, by the found results is possible to be obtained like conclusion that if a microbiological activity in the this places.

4. Conclusions

The enzymes are potentially useful in the determination of the existence of life in other places, having the advantage to detect processes that are characteristic of the life like the metabolism. It is important to know the processes biochemical that they happen in the ground, since to these reactions they are those that determine their origin and transformation. For the case of the colonization of Mars or another planet, of recognizing the substrate ground it is of fundamental character, because it is this that represents the support for all the human activities. It is needed to extend our knowledge of the enzymatic activity of the ground and to directing this same one towards the application to the astrobiology. The enzymes are constituent main of the Earth life and its origin I determine the change and the adaptation of the life to different conditions, in addition they are the more complex organic molecules and with greater antiquity in the planet.

References

Casida, l.E. 1977. Microbial metabolic activity in soil as measured by dehidrogenasa determinations.Appl. Environ. Microbiol. 34: 630-636.

Eivazi, F. and M.A. Tabatabai. 1977. Phosphatases in soils. Soil Biol.. Biochem. 9:167-172.

Frankenberger, W.T. and W.A. Dick. 1983. Relationships between enzyme activities and microbial Growth and activity indices in soils: Method of assay. Soil Sci. Soc. Am. J. 44:282-287.

GILI, P., MARANDO, G., IRISARRI, J. *et al.* Actividad biológica y enzimática en suelos afectados por sales del Alto Valle de Río Negro y Neuquén. *Rev. Argent. Microbiol.*, oct./dic. 2004, vol.36, no.4, p.187-192. ISSN 0325-7541.

Kennedy, A.C., and K.L Smith. 1995. Soil microbial diversity and the sustainability of agricultural soils. Plant Soil 170:75-86.

Schmidt, G. and M. Laskowski, 1961. Phosphate ester cleavage (survey). p. 3-35. In P.D. Boyer et al (ed.) The enzymes. 2nd ed. Academic Press, Inc., New York.

Skujins, J. 1973. Dehidrogenase: An indicator of biological activities in soils. Bull. Ecol. Res. Commun. 17: 233-241.

Andrew Gould

Session IX

Co-Evolution of Life and the Environment on Early Earth

John Baross

Colin Goldblatt

Bioastronomy 2007: Molecules, Microbes, and Extraterrestrial Life
ASP Conference Series, Vol. 420, 2009
K. J. Meech, J. V. Keane, M. J. Mumma, J. L. Siefert, and D. J. Werthimer, eds.

Bistability of Atmospheric Oxygen and the Great Oxidation: Implications for Life Detection

C. Goldblatt,[1,2] A. J. Watson,[1] and T. M. Lenton[1]

[1]*School of Environmental Sciences, University of East Anglia, Norwich, NR4 7TJ, U.K.*

[2]*Now at: Space Science and Astrobiology Division, NASA Ames Research Center, MS 245-3, Moffett Field, CA 94035, U.S.A. Email: colin.goldblatt@nasa.gov*

Abstract. Earth's atmospheric evolution was punctuated by a rapid and nonlinear transition in oxygen inventory 2.4 – 2.3 billion years ago, from $< 2 \times 10^{-6}$ atm to $> 10^{-3}$ atm, known as the Great Oxidation. The cause of the Great Oxidation has been a major problem in understanding the evolution of the Earth system. In particular, oxygenic photosynthesis is thought to have evolved by 2.7 billion years ago, at least 300 million years before the Great Oxidation. We have shown that the origin of oxygenic photosynthesis gave rise to two simultaneously stable steady states for atmospheric oxygen. The existence of a low oxygen steady state explains how a reducing atmosphere persisted long after the onset of oxygenic photosynthesis. The Great Oxidation can be understood as a switch to the high oxygen steady state. The bistability arises because ultraviolet shielding of the troposphere by ozone becomes effective once oxygen exceeds 2×10^{-6} atm, causing a nonlinear increase in the lifetime of atmospheric oxygen. Identification of oxygen or ozone in an extrasolar planetary atmosphere has been proposed as an indicator of life, but the existence of a low oxygen stable steady state introduces an important false negative case of planets with stable low (undetectable) oxygen but thriving oxygenic photosynthesis. An oxygen rich atmosphere, detectable remotely and capable of supporting complex life, is likely to occur only on planets of similar size to Earth; too small and bulk atmospheric loss occurs, too large and energetic limitation of hydrogen escape prevents atmospheric oxidation.

1. Introduction

The evolution of the redox state of the atmosphere during the Precambrian (< 0.542 billion years ago (Ga)) can be characterised as reducing conditions until 2.4 – 2.3 Ga, then a rapid transition to oxidising conditions (Fig. 1). The strongest indicator of anoxia is evidence of mass independent fractionation (MIF) of sulphur isotopes in rocks 2.4 Ga and older (Farquhar *et al.* 2000, 2007; Papineau *et al.* 2007), but absent since 2.32 Ga (Bekker *et al.* 2004). Evidence of MIF has generally been taken to indicate $fO_2 < 2 \times 10^{-6}$ atm, but Zahnle *et al.* (2006) show that sufficient methane levels and volcanic sulphur input are equally important in causing a MIF signal. The transition to an oxic atmosphere is termed the "Great Oxidation". This was probably the largest chemical transition in Earth history, causing reorganisations of the major geochemical cycles,

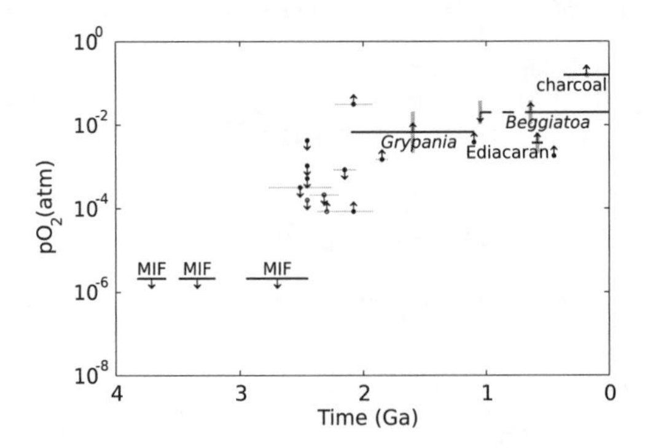

Figure 1. Compilation of oxygen constraints (Goldblatt *et al.* 2006), plus additional MIF data (Farquhar *et al.* 2007; Papineau *et al.* 2007). Upward arrow indicates lower bounds and downward arrows indicate upper bounds. Solid black lines are duration of the constraint. Circles represent palaeosols. Ediacaran, Grypania and Beggiatoa are biological constraints. Charcoal requires sufficient oxygen for combustion.

changing the dominant modes of microbial respiration and revolutionising atmospheric chemistry. The Great Oxidation would be most easily understood as the immediate consequence of the origin of oxygenic photosynthesis (Kopp *et al.* 2005), but evidence from biomarkers (Brocks *et al.* 1999, 2003; Summons *et al.* 1999), suggests this innovation had occurred at least 300 million years (My) earlier, by 2.7 Ga. This time lag has been a major puzzle in understanding the evolution of the Earth system. In section 2 we develop the solution offered by Goldblatt *et al.* (2006) to this problem. Whilst an alternate view suggesting high oxygen throughout Earth history exists (e.g. Ohmoto 1997; Ohmoto *et al.* 2006) this is a minority view (e.g. Holland 1999; Farquhar *et al.* 2007).

The early Earth is used as an analogue for possible inhabited extrasolar planets. The only practical method for assessing whether an extrasolar planet is inhabited is by remote analysis of its atmospheric content for signs of life (Lovelock 1965; Hitchcock and Lovelock 1967; Des Marais *et al.* 2002); understanding the evolution of Earth's atmosphere is thus a prerequisite for interpreting data from extrasolar planets. In section 3 of this paper, we discuss the implications of our new understanding of the dynamics of oxygen evolution for the search for extrasolar life.

2. Biogeochemical Modelling

2.1. Model Description

A three box model of the global redox system is briefly described here, following Goldblatt *et al.* (2006), to which the reader should refer for a full description. The model comprises a box for the atmosphere and surface ocean containing O_2 ($\mathbb{O}$) and CH_4 ($\mathbb{M}$) and a box of organic carbon in the crust ($\mathbb{C}$). Reservoir sizes are in moles.

A key driver of the model is the marine biosphere; this is driven mainly by a net primary productivity from oxygenic photosynthesis ($CO_2 + H_2O + hv \longrightarrow CH_2O + O_2$), N, which is set as a boundary condition. Decomposition of organic matter is by aerobic respiration (reverse of oxygenic photosynthesis) if there is sufficient ambient oxygen and by fermentation and acetotrophic methanogenisis otherwise. If there is sufficient oxygen, methane oxidising bacteria (methanotrophs) will utilise the methane produced ($CH_4 + 2O_2 \longrightarrow CO_2 + 2H_2O$). When oxygen availability is limited, the net effect of the biosphere is to deliver oxygen and methane to the atmosphere in a 2:1 stochiometric ratio ($CO_2 + H_2O + hv \longrightarrow \frac{1}{2}CH_4 + \frac{1}{2}CO_2 + O_2$), thus promoting the strong chemical disequilibrium in the atmosphere which is a characteristic of life (Lovelock 1965; Hitchcock and Lovelock 1967). Ω is the fraction of N which has this fate.

The major process in restoring equilibrium is atmospheric methane oxidation. This takes place as a series of reactions, some of which are photochemically mediated, but can be represented as an effective bimolecular reaction ($CH_4 + 2O_2 \longrightarrow CO_2 + 2H_2O$). We empirically parameterise this flux as a function of methane and oxygen abundances, fitting it to the results of detailed photochemical models (Pavlov *et al.* 2001; Pavlov and Kasting 2002; Pavlov *et al.* 2003), obtaining oxidation rate $\Psi\mathbb{M}^{0.7}$, where Ψ is a polynomial function of oxygen concentration. An important property of the atmospheric chemistry embodied in Ψ is that, once oxygen has increased to a certain level, an ozone layer forms shielding the troposphere from UV radiation. This dramatically decreases the rate of photolysis of water vapour in the troposphere, reducing hydroxyl radical availability, thus suppressing the rate of methane oxidation (Kasting and Donahue 1980; Pavlov and Kasting 2002).

Hydrogen loss to space is the source of long-term global oxidation (Hunten and Donahue 1976). For the Earth, H loss is diffusion limited, so depends on total H mixing ratio at the homopause (Hunten 1973). With methane as the only H bearing species in our model atmosphere, the loss rate can be represented $s\mathbb{M}$, where s is a constant. The stochiometry of methane derived H loss can be represented as $CH_4 + h\nu \longrightarrow 4H + C$ followed by $C + O_2 \longrightarrow CO_2$, net $CH_4 + O_2 + h\nu \longrightarrow 4H + CO_2$ (Catling *et al.* 2001).

Geologic forcing is set as a boundary condition with net input of reduced material, r, to the atmosphere–ocean system, from the solid Earth. The crustal organic carbon cycle is represented by burying a fraction, β, of the net primary productivity, N. It is assumed that the weathering rate of organic carbon is controlled by uplift and exposure, so is proportional to $\mathbb{C}$ with constant w.

The model is written as a set of ordinary differential equations:

$$\frac{d\mathbb{M}}{dt} = \frac{1}{2}\Omega(N+r) - s\mathbb{M} - \frac{1}{2}\Psi\mathbb{M}^{0.7} - \frac{1}{2}\Omega(\beta(N+r) - w\mathbb{C}) \quad (1)$$

$$\frac{d\mathbb{O}}{dt} = \Omega N - (1-\Omega)r - s\mathbb{M} - \Psi\mathbb{M}^{0.7} + (1-\Omega)(\beta(N+r) - w\mathbb{C}) \quad (2)$$

$$\frac{d\mathbb{C}}{dt} = \beta(N+r) - w\mathbb{C} \quad (3)$$

2.2. Steady State Solutions

Goldblatt *et al.* (2006) solved Eqs. 1 – 3 for steady state. However, the crustal organic carbon cycle has a timescale $\sim$ 500 Myr, so it is more appropriate to

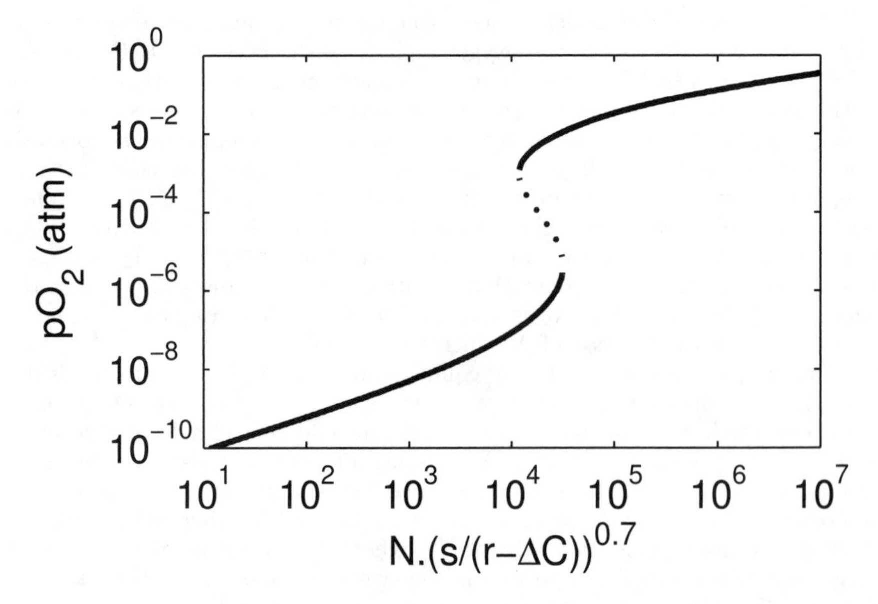

Figure 2. Steady state solutions for oxygen against composite forcing. Solid line indicates stable steady state and dotted line unstable steady state.

consider steady state for the atmosphere (Eqs. 1 and 2) with respect to some given $\frac{d\mathbb{C}}{dt} = \Delta\mathbb{C}$. Directly we find

$$\mathbb{M} = \frac{r - \Delta\mathbb{C}}{s} \tag{4}$$

and with the simplifying assumptions $N \gg r$ and $N \gg |\Delta\mathbb{C}|$ we find

$$\frac{\Psi}{\Omega} = N\left(\frac{s}{r - \Delta\mathbb{C}}\right)^{0.7} \tag{5}$$

where Ψ/Ω incorporates all oxygen dependency. Eq. 5 is solved numerically (Fig 2). There are two sets of stable steady state solutions for oxygen in the presence of oxygenic photosynthesis; a low oxygen solution, consistent with the MIF constraint, and a high oxygen, consistent with oxygen constraints after the Great Oxidation. The cause of the bistability is a strong nonlinear positive feedback on oxygen levels. Once oxygen exceeds 2×10^{-6} atm an ozone layer starts to form, decreasing the rate of methane oxidation. This reduced oxygen sink causes oxygen levels to increase, promoting the formation of the ozone layer. Oxygen continues to rise until the photochemical loss again balances the biospheric source. Inclusion of oxygen dependency in the biosphere (methanotrophy and aerobic respiration) is essential to obtain realistic oxygen levels on the high oxygen solution.

The discovery of this bistability led Goldblatt *et al.* (2006) to hypothesise that (1) the Great Oxidation can be understood as a switch between the two

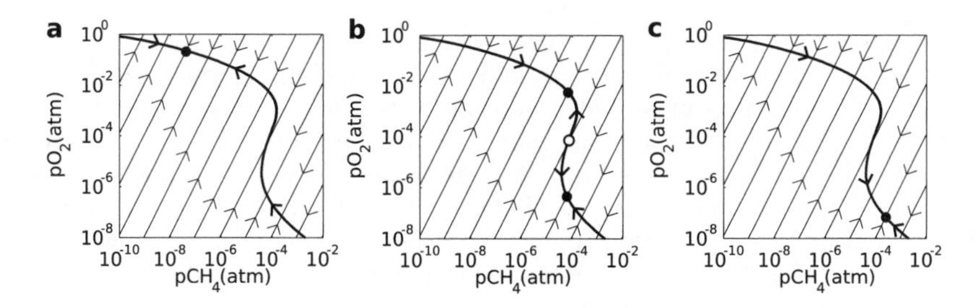

Figure 3. Methane–Oxygen phase space portraits. Filled circles indicate asymptotically stable improper nodes and the open circle an unstable saddle node. **a** Monostable high oxygen, $N(\frac{s}{r-\Delta\mathbb{C}})^{0.7} = 3.7 \times 10^6$. **b** Bistable region, $N(\frac{s}{r-\Delta\mathbb{C}})^{0.7} = 1.2 \times 10^4$. **c** Monostable low oxygen, $N(\frac{s}{r-\Delta\mathbb{C}})^{0.7} = 8.3 \times 10^3$.

stable states (2) the time lag between the origin of oxygenic photosynthesis and the Great Oxidation can be explained as the atmosphere residing in the low oxygen stable steady state (3) there was an earlier transition from prebiotic oxygen levels ($< 2 \times 10^{-13}$ atm, Kasting *et al.* 1979) to the low oxygen steady state when oxygenic photosynthesis evolved. Hypothesis (2) is supported by new geochemical data: trace metal (Anbar *et al.* 2007; Wille *et al.* 2007) and sulphur isotope (Kaufman *et al.* 2007) evidence indicates the existence of trace quantities of oxygen at 2.5 Ga, before the Great Oxidation, as predicted by our model.

We emphasise that the Great Oxidation was a rapid event defined by a strong nonlinearity in the atmospheric chemistry: formation of the ozone layer. It is not necessary to invoke major contemporaneous changes in boundary conditions to explain the transition.

2.3. Phase Portraits

The time dependent behaviour of the system can be illustrated with sample phase space portraits (Fig. 3). Note that, during the transition from the low to high oxygen stable steady states, the path through the phase space is sinuous. Whilst oxygen increases monotonically, methane will first decrease, then increase as UV shielding becomes effective and finally decrease once the high oxygen stable steady state has been reached. Explicit time dependent runs are described in Goldblatt *et al.* (2006) and Goldblatt (2008).

2.4. Climatic Implications

It has previously been suggested a crash in methane concentrations associated with the Great Oxidation could have caused drastic global cooling, triggering the low latitude (possibly "Snowball Earth") glaciations in the Palaeoproterozoic (Pavlov *et al.* 2000; Kasting 2005). This hypothesis was based on three premises which now appear incorrect. (1) Climate model results suggested that methane could contribute up to 30°C warming (Pavlov *et al.* 2000), but this is now known to have been a very large overestimate (Goldblatt 2008; Haqq-Misra *et al.* 2008). (2) The assumption that the Great Oxidation was accompanied by a methane decrease. Our results suggest otherwise. Ozone, which becomes abundant at the Great Oxidation is a greenhouse gas and the UV shielding it provides will have increased the photochemical lifetimes of methane and nitrous oxide. After the Great Oxidation, a sulphidic ocean may have provided a large biogenic source of nitrous oxide (Buick 2007). Thus, we suggest that the Great Oxidation caused significant warming due to increased ozone, methane and nitrous oxide levels. (3) The notion that glaciation followed the Great Oxidation. New MIF data (Papineau *et al.* 2007) show that the atmosphere remained reducing *after* the first Palaeoproterozoic glaciation.

We suggest two alternate hypotheses for the glaciations: (1) The Archean greenhouse may have been from a carbon dioxide, methane, ethane mixture (Haqq-Misra *et al.* 2008). A decrease in r earlier than the Great Oxidation, caused by changing patterns of volcanism (Kump and Barley 2007) or crustal oxidation (Claire *et al.* 2006), may have caused a step decrease in methane and ethane concentrations, triggering glaciation whilst oxygen remained low. Continued decrease in r may then have caused the switch to high oxygen at around 2.3 Ga, after the final glaciation, when the first geologic evidence of high oxygen is found. (2) Changes in the carbonate–silicate cycle might have caused glaciation: Rino *et al.* (2004) find a gap in the age distribution of detrital zircons between 2.5 and 2.3 Ga, implying low rates of crustal formation so low volcanic carbon dioxide emissions. In either case, the Great Oxidation probably assisted recovery from glaciation rather than causing glaciation.

3. Implications for Life Detection

Life detection by atmospheric analysis was initially proposed by Lovelock (1965) and Hitchcock and Lovelock (1967) as a search for disequilibrium in atmospheric chemistry generated by life. This was successfully demonstrated for the Earth with Galileo data by Sagan *et al.* (1993), but the necessary resolution would make this rigorous approach difficult for extrasolar planets (Kasting 1996). Hence recent discussion has focused on "biosignature" gases (e.g. Des Marais *et al.* 2002). Oxygen or ozone are seen as the most important as they are taken to indicate extant oxygenic photosynthesis. Also, Catling *et al.* (2005) argue that abundant oxygen is necessarily required by complex life (metazoa). Several possible false positives have been identified; abiotic production of O_2 in high H_2O or CO_2 atmospheres, or a frozen surface preventing oxidation of reduced rocks, but these can generally be excluded by other planetary properties (Des Marais *et al.* 2002) or by taking co-identification of O_3, CO_2 and H_2O as the indicator of life (Selsis *et al.* 2002). Also, O_3 may not be detectable in some O_2 rich atmospheres (Selsis *et al.* 2002).

Concentrations of oxygen or ozone detectable by proposed Terrestrial Planet Finder and Darwin missions correspond to those found after the Great Oxidation (Kaltenegger *et al.* 2007). Thus our identification of a low oxygen stable steady state introduces an important potential false negative. Undetectable levels of O_2 and O_3 are an *expected* property of some planets with extant oxygenic photosynthesis. Depending on planetary conditions, this could persist for any length of time, possibly longer than the > 300 My for Earth. Observable biogenic oxygen would require not only oxygenic photosynthesis, but the right combination of geochemical conditions to have caused a Great Oxidation on the planet in question.

The early Earth is a useful analogue for the properties of yet unexplored extrasolar worlds, but it is important to consider how different physical characteristics of such planets would affect both their habitability and the chances of detecting life. Consider a planet smaller than Earth, for example Mars. Most of Mars's atmosphere was lost early due to a combination of impact erosion, hydrodynamic loss and stripping by the solar wind (Catling 2009). These processes operated effectively because of Mars's small mass and corresponding low escape velocity, so bulk atmospheric loss should be expected for a small planet. A thin atmosphere would be unlikely to support abundant life (or contain sufficient atmospheric oxygen for aerobic metazoa) and would be a poor candidate for a positive detection of life by atmospheric analysis.

Conversely, the problem for a larger planet would relate to weaker loss to space. Hydrogen escape, the source of long term oxidation, is diffusion limited on Earth. This can be considered a buoyancy flux, so increases with planetary mass as g increases. However, at some mass greater than Earth's there will be a regime transition to energetic limitation of hydrogen escape (Watson *et al.* 1981). In this regime, the escape flux decreases with increasing mass as the gravitational binding energy which must be overcome is higher. Thus a "super-Earth" planet would be unlikely to oxidise sufficiently to form an oxygen rich atmosphere due to weak hydrogen escape, regardless of whether oxygenic photosynthesis exists.

In looking for an extrasolar planet with detectable oxygen (or, arguably, complex life), we must follow a 'Goldilocks' principle of neither too big nor too small. Conversely, a lack of observable oxygen or ozone does not indicate the absence of oxygenic photosynthesis.

Acknowledgments. C.G. received travel support from the NASA Astrobiology Institute through the conference organisers and from Roberts Funds from the Faculty of Science at the University of East Anglia.

References

Anbar, A. D. *et al.* 2007, A whiff of oxygen before the Great Oxidation Event? *Science*, 317:1903–1906

Bekker, A. *et al.* 2004, Dating the rise of atmospheric oxygen. *Nature*, 427:117–120

Brocks, J. J., Logan, G. A., Buick, R., and Summons, R. E. 1999, Archean molecular fossils and the early rise of eukaryotes. *Science*, 285(5430):1033–1036

Brocks, J. J., Buick, R., Summons, R. E., and Logan, G. A. 2003, A reconstruction of Archean biological diversity based on molecular fossils from the 2.78 to 2.45 billion-year-old Mount Bruce Supergroup, Hamersley Basin, Western Australia. *Geochim. Cosmochim. Acta*, 67(22):4321–4335

Buick, R. 2007, Did the Proterozoic 'Canfield ocean' cause a laughing gas greenhouse? *Geobiology*, 5:97–100

Catling, D. C. 2009 Atmospheric evolution of Mars. In V. Gornitz, editor, *Encyclopedia of Paleoclimatology and Ancient Environments.* Springer, pp. 66-75

Catling, D. C., Zahnle, K. J., and McKay, C. P. 2001, Biogenic methane, hydrogen escape, and the irreversible oxidation of early Earth. *Science*, 293:839–843

Catling, D. C., Glein, C. R., Zahnle, K. J, and Mckay, C. P., 2005, Why O_2 is required by complex life on habitable planets and the concept of planetary "oxygenation time". *Astrobiology*, 5(3):415–438

Claire, DM.W., Catling, D. C., and Zahnle, K.J. 2006, Biogeochemical modelling of the rise in atmospheric oxygen. *Geobiology*, 4(4):239–270

Des Marais, D. J. *et al.* 2002, Remote sensing of planetary properties and biosignatures on extrasolar terrestrial planets. *Astrobiology*, 2:153–181

Farquhar, J., Bao, H., and Thiemens, M. 2000, Atmospheric influence of Earth's earliest sulfur cycle. *Science*, 289:757–758

Farquhar, J., *et al.* 2007, Isotopic evidence for mesoarchaean anoxia and changing atmospheric sulphur chemistry. *Nature*, 449:706–710

Goldblatt, C. 2008, *Bistability of atmospheric oxygen, the Great Oxidation and climate.* PhD thesis, University of East Anglia

Goldblatt, C., Lenton, T. M., and Watson, A. J. 2006, Bistability of atmospheric oxygen and the Great Oxidation. *Nature*, 443(7112):683–686

Haqq-Misra, J. D., Domagal-Goldman, S. D., Kasting, P.J., and Kasting, J. F. 2008, A revised, hazy methane greenhouse for the Archean earth. *Astrobiology* , 8 (6): 1127-1137

Hitchcock, D. R. and Lovelock, J. E. 1967, Life detection by atmospheric analysis. *Icarus*, 7(2):149–159

Holland, H. D. 1999, When did the Earths atmosphere become oxic? A reply. *The Geochemical News*, 100:20–23

Hunten, D. M. 1973 The escape of light gasses from planetary atmospheres. *J. Atmos. Sci.*, 30:1481–1494

Hunten, D. M. and Donahue, T. M. 1976, Hydrogen loss from the terrestrial planets. *Annu. Rev. Earth Pl. Sc.*, 4:265–292

Kaltenegger, L., Traub,W. A., and Jucks, K. W. 2007, Spectral evolution of an Earth-like planet. *Astrophys. J.*, 658(1 I):598–616

Kasting, J. F. 2005, Methane and climate during the Precambrian era. *Precambrian Res.*, 137:119–129

Kasting, J. F. 1996, Planetary atmosphere evolution: do other habitable planets exist and can we detect them? *Astrophys. Space Sci.*, 241:3–24

Kasting, J. F. and Donahue, T. M. 1980, The evolution of the atmospheric ozone. *J. Geophys. Res.*, 85:3255–3263

Kasting, J. F., Liu, S. C., and Donahue, T. M. 1979, Oxygen levels in the prebiological atmosphere. *J. Geophys. Res.*, 84(C6):3097–3107

Kaufman, A. J., *et al.* 2007, Late Archean biospheric oxygenation and atmospheric evolution. *Science*, 317:1900–1903

Kopp, R. E., Kirschvink, J. L., Hilburn, I. A., and Nash, C. Z. 2005, The Palaeoproterozoic snowball Earth: A climate disaster triggered by the evolution of oxygenic photosynthesis. *Proc. Natl. Acad. Sci. U. S. A.*, 32:11,131–11,136

Kump, L. R. and Barley, M. E. 2007, Increased subaerial volcanism and the rise of atmospheric oxygen 2.5 billion years ago. *Nature*, 448(7157):1033–1036

Lovelock, J. E. 1965, A physical basis for life detection experiments. *Nature*, 207(4997): 568–570

Ohmoto, H. 1997, When did the Earth's atmosphere become oxic? *The Geochemical News*, 93:12–13 & 26–27

Ohmoto, H., Watanabe, Y., Ikemi, H., Poulson, S. R., and Taylor, B. E. 2006 Sulphur isotope evidence for an oxic Archaean atmosphere. *Nature*, 442:908–911

Papineau, S., Mojzsis, S. J., and Schmitt, A. K. 2007, Multiple sulfur isotopes from Palaeoproterozoic Huronian interglacial sediments and the rise of atmospheric oxygen. *Earth Planet. Sci. Lett.*, 255:188–212

Pavlov, A. A and Kasting, J. F. 2002, Mass-independent fractionation of sulfur isotopes in Archean sediments: strong evidence for and anoxic Archean atmosphere. *Astrobiology*, 2(1):27–41

Pavlov, A. A., Kasting, J. F., Brown, L. L., Rages, K. A., and Freedman, R. 2000, Greenhouse warming by CH_4 in the atmosphere of early Earth. *J. Geophys. Res.*, 105(E5):11,981–11,990

Pavlov, A. A., Brown, L. L., and Kasting, J. F. 2001, UV shielding of NH_3 and O_2 by organic hazes in the Archean atmosphere. *J. Geophys. Res.*, 106(E10):23267–23287

Pavlov, A. A., Hurtgen, M. T., Kasting, J. F., and Arthur, M. A. 2003, Methane-rich Proterozoic atmosphere? *Geology*, 31(1):87–90

Rino, S., *et al.* 2004, Major episodic increases of continental crustal growth determined from zircon ages of river sands; implications for mantle overturns in the early Precambrian. *Phys. Earth Planet. In.*, 146(1-2):369–394

Sagan, C., Thompson, W. R.., Carlson, R., Gurnett, D., and Hord, C. 1993, A search for life on Earth from the Galileo spacecraft. *Nature*, 365:715 – 721

F. Selsis, D. Despois, and J. Parisot. Signature of life on exoplanets: Can Darwin produce false positive detections? *Astron. Astrophys.*, 388(3):985–1003, 2002.

Summons, R. E., Jahnke, L. L., Hope, J. M., and Logan, G. A. 1999, 2-Methylhopanoids as biomarkers for cyanobacterial oxygenic photosynthesis. *Nature*, 400:554–557

Watson, A. J., Donahue, T. M., and Walker, J. C. G. 1981, The dynamics of a rapidly escaping atmosphere: applications to the evolution of Earth and Venus. *Icarus*, 48:150–166

Wille, M., et al. 2007, Evidence for a gradual rise of oxygen between 2.6 and 2.5 Ga from Mo isotopes and Re-PGE signatures in shales. *Geochim. Cosmochim. Acta*, 71:2417–2435

Zahnle, K., Claire, M., and Catling, D. 2006, The loss of mass-independent fractionation in sulfur due to a Palaeoproterozoic collapse of atmospheric methane. *Geobiology*, 4(4):271–283

David Morrison

Bioastronomy 2007: Molecules, Microbes, and Extraterrestrial Life
ASP Conference Series, Vol. 420, 2009
K. J. Meech, J. V. Keane, M. J. Mumma, J. L. Siefert, and D. J. Werthimer, eds.

Non-sparse Sampling of the Genomic Phase-Space of a Proto-cell in the Young Earth

D. J. Mullan

Dept. of Physics and Astronomy, Univ. of Delaware, Newark, DE 19716

Abstract. We estimate the numbers of reactions which occurred between certain molecules in liquid water during an interval of 200 My on the young Earth. We compare this with the size of genomic phase-space for a primitive cell whose genetic material consists of single-strand RNA. We find that the reactions which occurred in 200 My are sufficiently numerous to allow non-sparse sampling of a genome containing $p \approx 15$ genes, provided the genetic code used doublet codons. We argue on general grounds that $p \geq 10$ *could* give rise to a viable cell.

1. Introduction

Could the genome of a proto-cell have been assembled by chemical reactions in the young Earth within the time interval $t_{life} \approx 0.2$ Gy which elapsed between the end of heavy bombardment and the emergence of the earliest known cell? The answer depends on how densely the genomic phase-space was sampled by the number of reactions that actually occurred during the interval t_{life}.

In this work, we do not attempt the (essentially impossible) task of assembling even the smallest modern cell, with its 150 genes (Forster & Church 2006). Instead, we invert the task and ask: given the number N_r of reactions that *did* occur during t_{life}, what is the largest genome that *could have been* sampled during t_{life}? We estimate N_r and compare it with the size of phase space Φ_g for the genome of a cell containing p genes. We evaluate the maximum p_{max} for which N_r exceeds Φ_g, and ask: is a cell viable if it contains only p_{max} genes?

2. Size of Phase-Space

First, we consider protein phase-space Φ_p. For a protein containing a chain of n amino acids (AA's), Φ_p contains N_a^n elements, where N_a is the number of distinct AA's used by proteins. It is essential to note that many of the AA's in a protein can be replaced with other AA's without impairing protein function. That is, the protein is "neutral" to substituting many of its AA's. Let Q ($\equiv 10^q$) be a "neutrality" (or "non-specificity") factor, i.e. the number of distinct polypeptides which can perform the function of any given protein. Then that protein's phase-space actually contains only $\Phi'_p = N_a^n/Q$ distinct elements.

How large can Q be? For iso-1-cyto-c, with $n = 113$, $Q = 10^{35}$ (Yockey 2005). Thus, to assemble a given protein by random choices from an AA reservoir, there is no need to sample the vast phase-space Φ_p: instead, sampling need extend only to the much reduced Φ'_p.

Turning to genomic phase-space, we assume that proto-cells relied on the simplest genetic material: (+)single-strand RNA. In the Baltimore virus classification, (+) denotes RNA which can function as mRNA for protein synthesis. Some modern viruses rely on (+)ssRNA: these are not self-contained replicators in today's world, but in the pre-biotic Earth, (+)ssRNA could have given a "proto-cell" a survival advantage. RNA can also serve for catalysis.

The size of genomic phase-space can be estimated by noting that, in order to encode for a protein with n residues, the number of NT's in one gene is $(n+2)M_c$. Here, the codon multiplicity M_c=3 (or 2) for triplet (or doublet) codons, and +2 refers to start-stop codons. The genetic code relies on 4 bases, a number that is required for stable information carrier systems (Ganti 1979). For a single gene which encodes for one protein with neutrality factor Q, the gene phase-space contains $\Phi_{g1} = 4^{(n+2)M_c}/Q$ elements. For a proto-cell consisting of p genes, genomic phase-space contains $\Phi_g = \Phi_{g1}^p$ elements, i.e.

$$\log(\Phi_g) = p(n+2)M_c \log(4) - pq \quad (1)$$

3. Sampling of Genomic Phase-Space in the Young Earth

Our view is that, in the young Earth, RNA was assembled by reactions among nucleotides (NT's) which were present with a finite density in liquid water. Let us estimate the number of reactions N_r between NT's during t_{life}. At first, we estimate the number of collisions between NT's. The collision frequency between any molecule and NT's is $f_c = v_{th}\,\sigma\,d_n(NT)$ per sec, where v_{th} is speed, $d_n(NT)$ is the NT number density, and σ is cross-section. In liquid water, NT's (with molecular weights of 300-400) have $v_{th} \approx 10^4$ cm/sec. With linear scales of 0.34 nm (Wang et al. 1998), NT's have $\sigma \approx 10^{-15}$ cm^2. Thus, during $t_{life} = 0.2$ Gy, any molecule undergoes some $C_1 = 10^5 d_n(NT)$ collisions with NT's.

What $d_n(NT)$ values are appropriate in the oceans on the young Earth? Using a photochemical reaction network, Dose (1975) derived AA concentrations of 10^{-7} M, corresponding to $d_n(AA) \approx 10^{14}$/cc. In Miller-Urey experiments, the DNA/RNA bases are formed in conjunction with AA's. Moreover, the Murchison meteorite is observed to contain both AA's and bases, and with relative abundances which do not differ by many orders of magnitude. Thus, in an environment where AA's are generated, the RNA bases may be present in comparable quantities. Formation of NT's from the bases requires reactions with ribose and phosphate: this will certainly diminish the number densities of NT compared to AA's. Nevertheless, for present purposes, we assume that $d_n(NT)$ was not greatly different from $d_n(AA)$. We shall show that our conclusions are not affected significantly even if this assumption is in error by up to 20 orders of magnitude.

Setting $d_n(NT) = 10^{14}$/cc, we find that during 0.2 Gy, the number of collisions experienced by any particle with NT's was $C_1 \approx 10^{19}$. In particular, each NT undergoes C_1 collisions with other NT's in 0.2 Gy. Estimates of the total number $N_t(AA)$ of AA's in the young Earth are 10^{42-44} (Shklovskii & Sagan 1966; Bar-Nun & Shaviv 1975). These values are not implausible since the Earth's current biomass ($\approx 10^{19}$ gm) would contain some 10^{41} molecules if the mass were entirely in the form of AA's. Assuming that $N_t(NT)$ is not much less than $N_t(AA)$ in the young Earth, we find that the total number of collisions that NT's undergo with one another in 0.2 Gy is $C_{tot} = 10^{61-63}$.

In NT collisions, only a fraction (f_a) leads to a reaction where the NT's combine as the first step in forming RNA. As certain motifs emerge in RNA fragments, some reactions may be catalyzed to proceed faster. But to be conservative, we estimate N_r by relating f_a to the activation energy ΔE_a: $f_a \approx \exp(-\Delta E_a/kT)$. What value should be used for ΔE_a? In "typical" chemical reactions where the rate doubles for every increase in T by 10 K (at $T{\approx}300$ K), $\Delta E_a \approx 53$ kJ/mol, i.e. 0.56 eV per molecule. At T=273-373 K, this leads to $f_a \approx 10^{-(8-10)}$. Such values for f_a in biological contexts are not inconsistent with empirical results that enzymes (which effectively reduce ΔE_a) can enhance reaction rates by $\geq 10^5$. Combining f_a=10^{-9} with C_{tot}, we estimate that the number of reactions between NT's during 0.2 Gy is

$$N_r = C_{tot} f_a \approx 10^{52-54} \tag{2}$$

If our estimates of $d_n(NT)f_a$ are too large by 10^{20}, replace 52-54 in (2) by 32-34.

4. Sampling of Genomic Phase-Space: Was it Sparse or Non-Sparse?

If $N_r{<}\Phi_g$, sampling of genomic phase-space was sparse, and the probability of genome assembly $p_g{\approx}N_r/\Phi_g$ was small. But if $N_r{\geq}\Phi_g$, the sampling was non-sparse, and genome assembly was highly probable ($p_g{\to}1$). Comparing eqs. (1) and (2), we find that p_g=1 requires $q{\geq}q_c$, where $q_c = (n+2)M_c\, log(4) - (52-54)/p$. Noting that the number of distinct codons N_c= 4^{M_c}, we can write:

$$q_c = n\log(N_c) + 1.2M_c - (52-54)/p \tag{3}$$

Can the q of a "real protein" ever be as large as q_c? Q has an absolute upper limit of N_a^n. However, in order to ensure viability of a cell containing p genes, a minimum amount of specificity is required: there should be at least p distinct regions of phase-space. Otherwise, genes might be expressed in inappropriate locations in the cell. This suggests $Q_{max} = N_a^n/p$, i.e. $q_{max} = n\log(N_a) - \log(p)$.

If $q_c > q_{max}$, the Q required for genome assembly is so large as to be mathematically forbidden. The key quantity in determining whether the probability of genome assembly p_g is large (during an interval of 0.2 Gy) is:

$$K(M_c) = q_c - q_{max} = n\log(N_c/N_a) + \log(p) + 1.2M_c - (52-54)/p \tag{4}$$

For a cell with p genes, p_g can be no larger than $10^{-pK(M_c)}$. If $K(M_c)$ is positive, p_g for a viable cell (with $p{\geq}10$) approaches zero. But if $K(M_c)$ is negative, p_g *can* approach unity: the genome *could have been* assembled with high probability.

5. RNA Assembly: Triplet Codons *Versus* Doublets

For triplets, M_c=3, $N_c = 64$, and $N_a = 20$. This leads to $K(3) = 0.5n + 3.6 + \log(p) - (52-54)/p$. Viable proteins may require $n \geq 25$ (Dahiyat & Mayo 1997). This leads to $K(3){<}0$ only if $p \leq 3$. Such a small genome could hardly give rise to a viable cell. If triplet codons were operative in the young Earth, p_g for a viable cell would have been vanishingly small.

The current genetic code (based on triplet codons) may have evolved from an earlier code based on doublet codons (Yockey, 2005). Various mechanisms have been proposed as to how this code evolution might have occurred (Wu et al. 2005; Copley et al. 2005). A dynamical theory for the evolution of an optimal genetic code has been reported based on collective mechanisms which likely occurred in early communal life (Vetsigian et al. 2006): a finite time interval is required for the optimal triplet code to emerge. Inspection of the analysis suggests that for a code based on doublets, the reading-probability parameter $T_{cc'}$ would have the value $\nu/6$ (where ν is a mistranslation rate), i.e. 1.5 times larger than for triplet codons. This suggests that an optimal *doublet* code could have emerged in a shorter time interval than a *triplet* code.

With doublets, M_c=2, $N_c = 16$, and the maximum $N_a \leq 16$. In the limit $N_a \rightarrow 16$, the term $\log(N_c/N_a)$ in eq. (4) vanishes, and $K(2) \rightarrow 2.4+\log(p)$-(52-54)$/p$. Remarkably, $K(2)$ is negative, *independent of protein length n*, provided $p \leq p_{max}$=15. Thus, in 0.2 Gy, the genome for a proto-cell with up to 15 genes, *each of arbitrary length, could* have been assembled if doublet codons were at work. Even if our reaction rates are overestimated by 10^{20}(!), i.e. if in eq. (4) 52-54 $\rightarrow$ 32-34, $K(2)$ remains negative, i.e. p_g remains large, provided $p \leq 10$.

6. Conclusion

Doublet codons permit assembly of the genome for a proto-cell with as many as 15 genes in the young Earth in 0.2 Gy. In principle, a minimally viable cell requires, in addition to information storage, 5 necessary "house-keeping" systems: (i) energy generation, (ii) enclosure, (iii) digestion, (iv) waste ejection; (v) ability to replicate information in the genome (Simpson 1964). With a regulatory gene for each, the absolute minimum appears to be $p \approx 10$ genes. The proto-cell we consider here satisfies this absolute minimum. Our conclusion remains valid even if we have overestimated reaction rates by 10^{20}.

Acknowledgments. Thanks to Delaware Space Grant for partial support of this work.

References

Bar-Nun, A. & Shaviv, A., Icarus (1975), 24:197

Copley, S.D., Smith, E., & Morowitz, H. J., PNAS (2005), 192:4442

Dahiyat, B. & Mayo, S., Science (1997), 278:82

Dose, K. (1975), Biosystems, 6:224

Forster, A. C., & Church, G. M., Mol. Syst. Biol. (2006) doi:10.1038/msb4100090

Ganti, T. (1979), Theory of Biomechanical Supersystems, (Univ Park Pr., Balt.), p. 47

Shkolvskii, I. S. & Sagan, C. (1966), Intelligent Life in the Universe (Dell Publ. NY)

Simpson, G. G., Science (1964), 143:769

Vetsigian, K., Woese, C., & Goldenfeld, N., PNAS (2006), 103:10696

Wang, H.-Y., Elston, T., Mogilner, A., & Oster, G. (1998), Biophys. J., 74:1186

Wu, H.L., Bagby, S., & Van den Elsen, J. M., Mol. Evol. (2005), 61:54

Yockey, H. P. (2005), Information Theory, Evolution, and the Origin of Life, (Cambridge University Press), pp. 68-84

Lori Walton

Hiroshi Ohmoto

Session X

Extrasolar Planets / Planet Formation

Alan Boss

David Latham

Bioastronomy 2007: Molecules, Microbes, and Extraterrestrial Life
ASP Conference Series, Vol. 420, 2009
K. J. Meech, J. V. Keane, M. J. Mumma, J. L. Siefert, and D. J. Werthimer, eds.

Toward Earth-Like Planets

D. W. Latham

Harvard-Smithsonian Center for Astrophysics, 60 Garden Street, Cambridge, MA 02138, USA

Abstract. NASA's Kepler mission has the goal of detecting transiting planets as small as the Earth, even some that might be habitable. Kepler was launched 2009. After transiting planet candidates are detected, the challenge will be to sort out the real planets from the stellar imposters masquerading as planets. The ultimate test is a spectroscopic orbit that confirms the planet's mass. A key facility for this follow-up work will be HARPS-NEF, a collaboration between the Geneva Observatory and Harvard University to build a version of HARPS for a telescope in the North. Kepler stares at a single field of view in Cygnus and Lyra for the entire four-year mission, to allow the detection of planets with orbital periods as long as a year. As a result the systems discovered by Kepler will be relatively far away, faint, and difficult for follow-up observations. To complement Kepler we need an all-sky survey that can identify the nearest and brightest transiting systems. For the photometry to be good enough to push to smaller planets we need to get above the Earth's atmosphere. TESS is a proposal for such a mission.

1. Introduction

The design of NASA's Kepler mission[1] has been optimized for the detection of Earth-sized planets that transit solar-type stars in the zone where water could be liquid. The satellite will monitor more than 100,000 stars for at least 3.5 years with a limiting photometric precision of 20 ppm, which is sufficient to detect a transit by a planet like the Earth in a one-year orbit around a $V = 12$ magnitude star like the Sun. The mission is now scheduled for launch in February 2009.

The Kepler photometer is a 1-meter class Schmidt telescope feeding a focal plane with a mosaic of 42 2K×2K CCDs. More than 100 square degrees of sky in the constellations Cygnus and Lyra will fall onto the detectors. An Earth-trailing orbit has been chosen for Kepler, to allow continuous viewing of the target field, and to avoid the fluctuating environment of a low-Earth orbit. Thus the transmission of full images back to Earth is limited, and in normal operation only postage stamps centered on selected targets will be returned, for subsequent analysis on the ground.

[1] http://kepler.nasa.gov/

2. The Kepler Input Catalog

To allow the selection of optimum targets, the Kepler Input Catalog (KIC) is being prepared by a team led by the Smithsonian Astrophysical Observatory. All previously known objects in the Kepler target field will be included in the KIC, for example all the stars from the USNO-B Catalog (Monet et al. 2003). This will help with the identification of targets where contamination of the large Kepler images by nearby stars could be a problem. Magnitude-limited samples are dominated by luminous stars, such as distant giants, so considerable effort is being invested in the preparation of the KIC so that solar-type dwarfs can be selected with a minimum of contamination by stars that are too large to show transits by Earth-sized planets. The goal is to maximize the number of small stars from spectral type F down into the M dwarfs. The primary resource for the stellar classification reported in the KIC is multi-band ground-based photometry. Near-infrared JHK_s photometry from the 2MASS Catalog (Skrutskie et al. 2006) is being supplemented by new photometry in the Sloan Digital Sky Survey $griz$ bands plus a custom intermediate band centered on the Mg b features (designed to provide surface gravity information). The 2MASS photometry is critical for our photometric classification of stars. Fortunately the 2MASS Catalog goes deep enough to include all the likely targets for Kepler – more than a million 2MASS stars that fall on the Kepler CCDs are brighter than $K_s = 14.5$ magnitude. The new photometry is being obtained with the 1.2-m telescope at the Smithsonian's Fred Lawrence Whipple Observatory. KeplerCam, a new camera based on the Fairchild 486 4K $\times$ 4K CCD, was built for this purpose.

To refine the photometric stellar classifications, we are using the KIC to select promising target candidates for spectroscopic follow-up with Hectochelle on the MMT. This multi-fiber echelle spectrograph allows us to obtain single-order spectra of more than 200 stars in a 1-degree field in a single fiber configuration. Two high-speed robots can reconfigure the fibers in a matter of minutes. This allows us to obtain spectra of more than a thousand stars per night. We then use a new library of synthetic spectra to derive surface gravities, metallicities, and radial and rotational velocities, adopting the photometric temperatures from the KIC. To calibrate our photometric and spectroscopic classifications we are using observations of well-studied clusters, especially M67. More than 8000 stars in the Kepler field of view have been observed with Hectochelle. The astrophysical quantities derived from these spectra have precisions of better than 0.2 in both $\log g$ and [Fe/H].

3. Spectroscopic Follow-Up Observations

Since 1999 we have been using the CfA Digital Speedometers (Latham 1992) for the initial spectroscopic reconnaissance of transiting-planet candidates identified by wide-angle ground-based photometric surveys, initially for the Vulcan project, and more recently for TrES, HAT, and KELT (Latham 2003, 2007). We soon learned that the vast majority of candidates actually involved eclipsing stellar systems masquerading as planets. A common impostor is an F-star primary eclipsed by an M dwarf companion. Stars near the bottom of the main sequence have radii similar to Jupiter, so their light curves are indistinguishable from transiting planets. However, M dwarfs induce a reflex motion of many $\mathrm{km\,s^{-1}}$ on the primary star, which is easy to detect with spectroscopy on a modest telescope. For example, the velocity precision of nominally 0.5 $\mathrm{km\,s^{-1}}$ achieved by the CfA Digital Speedometers for slowly-rotating solar-type stars is sufficient to reveal companions down to a few Jupiter masses for periods less than a week.

Another configuration that can mimic a transiting Hot Jupiter is a grazing eclipsing binary. These are often revealed by two sets of lines in the very first spectroscopic observation. Although grazing eclipsing binaries are relatively common impostors among the transiting giant-planet candidates identified by wide-angle ground-based photometric surveys, we expect them to be much less common among the shallower light curves detected by Kepler. The most difficult stellar systems to distinguish from planets are triples (and even quadruples), where the light of an eclipsing binary is diluted by a third star (Mandushev et al. 2005), which itself can be a binary. The spectral lines of the eclipsing binary can be difficult to detect, not only because the third star often dominates the total light, but also because the lines of the eclipsing binary are usually very broad and shallow due to synchronization of the stellar rotation with the short orbital period.

For the initial spectroscopic reconnaissance of the transiting-planet candidates identified by Kepler we have recently put into operation the new Tillinghast Reflector Echelle Spectrograph (TRES) on the 1.5-m Tillinghast Reflector at the Whipple Observatory. Not only is TRES 2.5 magnitudes more sensitive than the CfA Digital Speedometer, but it also has the stability to deliver velocities down to a precision of perhaps 10 $\mathrm{m\,s^{-1}}$ if enough photons are available. A major advantage is that TRES gives full spectral coverage compared to the 45 Å window of the CfA Digital Speedometers.

Kepler will be able to detect planets as small as the Earth in orbits as large as 1 AU, where the reflex motion of a Sun-like star will be on the order of 0.1 $\mathrm{m\,s^{-1}}$. This is well beyond the capability of HIRES on Keck and HRS on the Hobby-Eberly Telescope, the two existing facilities that will contribute very precise radial-velocity observations for the most interesting planets found by Kepler. Furthermore, many of the Kepler planets will be fainter than $V =$ 12 magnitude, and even 1 $\mathrm{m\,s^{-1}}$ velocity precision will require exposure times measured in hours. Thus the ability to characterize the smallest and most interesting planets found by Kepler will be limited by the existing capabilities in the North to obtain very precise velocities. To address this serious limitation, Harvard University and the Geneva Observatory have joined in a collaboration to build a copy of the HARPS spectrograph [2] now in operation on the 3.6-m telescope at the European Southern Observatory on La Silla. HARPS has already demonstrated the ability to measure velocities with a precision of 1 $\mathrm{m\,s^{-1}}$ on well-behaved solar-type stars, and recent work shows some promise that the limit with HARPS can be extended to the level of 0.2 $\mathrm{m\,s^{-1}}$. However, this level of precision requires bright stars where lots of photons are available, and large amounts of telescope time are needed to average out and/or correct for intrinsic velocity variations due to astrophysical phenomena on the star, such as stellar oscillations.

[2]http://www.eso.org/sci/facilities/lasilla/instruments/harps/index.html

To allow follow up of Kepler targets, the copy of HARPS needs to be located on a telescope in the Northern Hemisphere, preferably one with good weather and excellent seeing during the summer months when Cygnus and Lyra are overhead. Site visits to potential telescope hosts in Arizona, California, and La Palma led to the choice of the William Herschel 4.2-m telescope, where the HARPS-North project has been received with enthusiasm. The spectrograph itself will be a close copy of the original HARPS, which was optimized for a 4-m class telescope. The interface to the telescope will be permanently mounted at a broken-Cassegrain focus, to which the beam can be redirected in a matter of just a few minutes. Thus HARPS-North will be well suited for queue observing and rapid response to changing observing conditions. Experience with HARPS has shown that very precise guiding of the stellar image on the center of the science fiber is important for the ultimate velocity precision, so careful attention is being paid to this issue in the design of the adaptor to the William Herschel Telescope. Another limit to the performance of HARPS has been the stability and lifetime of the Thorium-Argon comparison lamps. Therefore we are developing a comb laser for possible use in the wavelength calibration on HARPS-North.

HARPS-North is one of three new research facilities supported by the Harvard University Origins of Life Initiative, an interdisciplinary collaboration between five departments and centers, including the Harvard-Smithsonian Center for Astrophysics. The Geneva Observatory is an equal partner with Harvard on HARPS-North. The primary motivation for this new facility has been the characterization of the transiting planets discovered by the Kepler mission, but it will also be made available to the user community of the Isaac Newton Group of Telescopes.

4. Future Initiatives

Exciting advances in our understanding of extrasolar planets have come from follow-up observations with space missions, such as HST and Spitzer. Transmission spectroscopy with HST has detected atmospheric constituents during transit (Charbonneau et al. 2002), and photometry and spectroscopy in the infrared with Spitzer have detected secondary eclipses, thus measuring the thermal emission from planets (Charbonneau et al 2005; Deming et al. 2005). The thermal emission has even been mapped as a function of orbital phase (Knutson et al. 2007), thus revealing the way that energy is transported from the planet's day side to the night side. This line of research holds great promise for the future, especially after the launch of the James Webb Space Telescope [3]. However, the observations need vast numbers of detected photons, and the best progress will come with observations of the nearest and brightest transiting planets. What we need now is an all-sky survey to identify all the nearest and brightest transiting planets, down to planets the size of the Earth and out to periods of at least a few weeks. The Transiting Exoplanet Survey Satellite (TESS) is a project to carry out just such a survey.

---

[3]http://www.jwst.nasa.gov/

TESS was conceived in 2006 as a collaboration between MIT and the Harvard-Smithsonian Center for Astrophysics. The design concept is an array of CCD cameras covering 2000 square degrees on the sky in a single pointing. The optics are custom lenses with 125 mm clear aperture and 190 mm focal length. Each lens feeds a quad array of MIT/Lincoln Lab CCID-20 frame-transfer CCDs, thus providing a 4K$\times$4K array of 15μ pixels. The frame transfer is accomplished in 5 ms, so the duty cycle is very efficient and no shutter is needed. Because we want to emphasize our sensitivity to cool dwarfs, we have selected 600 to 1000 nm as the passband. Elimination of the blue also allows us to reduce the number of optical surfaces in the custom lenses compared to typical photographic lenses, and to reduce the vignetting. The sensitivity is estimated to be a signal-to-noise ratio of 100 in an exposure of 10 seconds at $I = 14.2$ magnitude.

The target list for TESS will mainly consist of solar-type dwarfs (mid F through K spectral types), amounting to about two million stars down to $I = 12$ magnitude, but we will also make a special effort to include all the accessible M dwarfs within 30 pc, amounting to about ten thousand stars. It is around M dwarfs where we expect to detect transiting planets as small as the Earth in orbits where water could be liquid.

The intention is to build and operate TESS in much the same style that was used for HETE-2 (Ricker et al. 2003). A low Earth equatorial orbit will be used so that the existing HETE-2 ground stations can be utilized for operations and data return. A major advantage of this orbit is that it avoids the Earth's radiation belts, and radiation damage is not an issue. This allows the use of many modern components without the need for traditional space qualification. NASA's Ames Research Center has joined the collaboration and will provide the spacecraft and launch.

Acknowledgments. Many people have contributed to the efforts outlined in this paper, and I thank them all. In particular I wish to acknowledge the critical contributions of Tim Brown, David Monet, Mark Everett, Gil Esquerdo. and Soeren Meibom to the preparation of the Kepler Input Catalog; of John Geary, Andy Szentgyorgyi, Steve Amato, Ted Groner, Wayne Peters, and Emilio Falco for their contributions to the success of KeplerCam; and of Andy Szentgyorgyi, Gabor Furesz, and Mark Ordway for their work on the Tillinghast Reflector Echelle Spectrograph. The Harvard University Origins of Life Initiative is ably led by Dimitar Sasselov, with active participation by Dave Charbonneau, Matt Holman, Charles Alcock, and many others at CfA and Harvard. The HARPS-North initiative would not be possible without the enthusiastic collaboration with Michel Mayor, Stephane Udry, Francesco Pepe, Didier Queloz, and Michel Fleury at the Geneva observatory, and of Rene Rutten and his staff at the Isaac Newton Group of Telescopes. I especially thank George Ricker for his leading role in the TESS project, and also Sara Seager, Josh Winn, Adam Burgasser, Jim Elliott, and Jackie Hewett at MIT, Pete Worden, Pete Klupar, and Robbie Schengler at NASA Ames, and all the usual suspects involved in transiting-planet research at CfA. This work received partial support from the Kepler mission under Cooperative Agreement NCC2-1390.

References

Charbonneau, D., Allen, L. E., Megeath, S. T., Torres, G., Alonso, R., Brown, T. M., Gilliland, R. L., Latham, D. W., Mandushev, G., O'Donovan, F. T., & Sozzetti, A. 2005, ApJ, 626, 523

Charbonneau, D., Brown, T. M., Noyes, R. W., & Gilliland, R. L. 2002, ApJ, 568, 377

Deming, D., Seager, S., Richardson, L. J., & Harrington, J. 2005, Nat, 434, 740

Knutson, H. A., Charbonneau, D., Allen, L. E., Fortney, J. J., Agol, E., Cowan, N. B., Showman, A. P., Cooper, C. S., & Megeath, S. T. 2007, Nat, 447, 183

Latham, D. W. 1992, in ASP. Conf. Ser. Vol. 32, IAU Colloq. 135: Complementary Approaches to Double and Multiple Star Research, ed. H. A. McAlister & W. I. Hartkopf, (San Francisco, ASP), 110

Latham, D. W, 2003, in ASP. Conf. Ser. Vol. 294, Scientific Frontiers in Research on Extrasolar Planets, ed. D. Deming & S. Seager (San Francisco: ASP), 409

Latham, D. W. 2007, in ASP. Conf. Ser. Vol. 366, Transiting Extrapolar Planets Workshop, ed. C. Afonso, D. Weldrake, & T. Henning, (San Francisco: ASP), 203

Mandushev, G., Torres, G., Latham, D. W., Charbonneau, D., Alonso, R., White, R. J., Stefanik, R. P., Dunham, E. W., Brown, T. M., & O'Donovan, F. T, 2005, ApJ, 621, 1061

Monet, D. G., Levine, S. E., Canzian, B., Ables, H. D., Bird, A. R., Dahn, C. C., Guetter, H. H., Harris, H. C., Henden, A. A., Leggett, S. K., Levison, H. F., Luginbuhl, C. B., Martini, J., Monet, A. K. B., Munn, J. A., Pier, J. R., Rhodes, A. R., Riepe, B., Sell, S., Stone, R. C., Vrba, F. J., Walker, R. L., Westerhout, G., Brucato, R. J., Reid, I. N., Schoening, W., Hartley, M., Read, M. A., & Tritton, S. B. 2003, AJ, 125, 984

Ricker, G. R., Atteia, J. L., Crew, G. B., Doty, J. P., Fenimore, E. E., Galassi, M., Graziani, C., Hurley, K., Jernigan, J. G., Kawai, N., Lamb, D. Q., Matsuoka, M., Pizzichini, G., Shirasaki, Y., Tamagawa, T., Vanderspek, R., Vedrenne, G., Villasenor, J., Woosley, S. E., & Yoshida, A. 2003, in AIP Conf. Ser. Vol. 662, Gamma-Ray Burst and Afterglow Astronomy 2001: A Workshop Celebrating the First Year of the HETE Mission, ed. G. R. Ricker & R. K. Vanderspek, (College Park: AIP), 3

Skrutskie, M. F., Cutri, R. M., Stiening, R., Weinberg, M. D., Schneider, S., Carpenter, J. M., Beichman, C., Capps, R., Chester, T., Elias, J., Huchra, J., Liebert, J., Lonsdale, C., Monet, D. G., Price, S., Seitzer, P., Jarrett, T., Kirkpatrick, J. D., Gizis, J. E., Howard, E., Evans, T., Fowler, J., Fullmer, L., Hurt, R., Light, R., Kopan, E. L., Marsh, K. A., McCallon, H. L., Tam, R., Van Dyk, S., & Wheelock, S. 2006, AJ, 131, 1163

Avi Mandell

Hannah Jang-Condell

Bioastronomy 2007: Molecules, Microbes, and Extraterrestrial Life
ASP Conference Series, Vol. 420, 2009
K. J. Meech, J. V. Keane, M. J. Mumma, J. L. Siefert, and D. J. Werthimer, eds.

Observable Differences Between Core Accretion and Disk Instability

H. Jang-Condell

Carnegie Insitution of Washington, Dept. of Terrestrial Magnetism

Abstract. There are two competing paradigms for giant planet formation: core accretion and disk instability. In core accretion, solid particles in a protoplanetary disk coagulate into larger and larger bodies until a core massive enough to accrete a gaseous envelope forms, on time scales of millions of years. In disk instability, the disk fragments into a self-gravitating giant planet-sized clump, within tens of thousands of years. How and when giant planets form affect whether or not other planets in that system are habitable. To address this problem, I present predictions of what planet-forming disks might look like by modeling patterns of shadowing and illumination from the central star on the disks surfaces. I model the signature of core accretion in an analytic disk model of a quiescent disk that is perturbed by a 10-20 Earth mass body. The perturbation in the disk image follows the position of the planet. I model the signature of disk instability in a 3D hydrodynamic simulation of a disk in which a planet-sized clump is forming (Boss 2001). In this case, the disk is optically thick and stirred up by the forming planet, so the exact location of the protoplanet is indeterminate. However, the high variability and corrugated structure in the disk images are in and of themselves indicators that gravitational instability is occuring. In either case, milliarcsecond resolution is necessary to image the perturbations. This might be achievable by the GMT or TMT in optical/near-IR, or by ALMA in radio.

1. Introduction

Giant planet formation is important in the context of astrobiology because giant planets dominate the dynamics of the planetary systems that they inhabit. How, when, and where giant planets form affect whether or not a system can support a habitable planet. It is therefore important to understand the formation processes of giant planets to predict the frequency of habitable systems.

The two main paradigms for giant planet formation are core accretion and disk instability. In either case, planet formation takes place in the circumstellar disk of gas and dust that remain from the planet formation process. Giant planet formation must be complete before this disk dissipates in $10^6 - 10^7$ years, otherwise there would be insufficient gas to accrete onto the planet.

In the core accretion scenario, solid particles coagulate into larger and larger bodies until a body that is massive enough to capture and accrete a gaseous envelope forms, typically around 10 or more $M_\oplus$ (Earth masses) (Hubickyj et al. 2005). The advantage of this paradigm is that it readily explains the assembly of the rocky inner planets of our Solar System, which formed in the same way that a giant planet core would. However, the predicted timescales are the same or longer than disk lifetimes, so it is questionable as to whether or not core accretion can act fast enough to create giant planets.

In contrast, disk instability acts very quickly, within thousands of years (Boss 2001). In this scenario, the circumstellar disk is massive and cold enough to be gravitationally unstable and rapidly fragments to form a self-gravitating body of roughly a Jupiter mass. However, simulations are unclear about the eventual fate of these clumps, whether they eventually collapse into planets or get disrupted by tidal forces in the disk. Also, disk instability predicts giant planets with metallicity the same as that of the star, but Jupiter has a much higher metallicity than the sun.

In this paper, we explore disk models representative of both core accretion and disk instability. We model radiative transfer in these disks and determine their observable properties over a range of wavelengths. By understanding the different observable signatures for core accretion versus disk instability, we can guide and interpret observations of protoplanetary disks.

2. Radiative Transfer

The method used for modeling radiative transfer in the disk models is described in detail in Jang-Condell & Sasselov (2003, 2004) and Jang-Condell (2008). The surface is defined to be where the disk becomes optically thick to stellar irradiation, or where $\tau_s \equiv \int_\ell \chi_P^* \rho \, dl = 2/3$ where χ_P^* is the total extinction averaged over the stellar spectrum, ρ is the density of disk material, and the integral is calculated along the line from the star to the surface. A fraction of incident stellar irradiation is scattered off the surface and the remainder is absorbed by material in the disk, reprocessed, and re-radiated at thermal wavelengths. The amount of stellar flux that is absorbed is strongly dependent on the angle of incidence at the surface of the disk, which is why depressed regions are shadowed and cooled while uplifted regions get brightened and heated.

We calculate images of the disk assuming that it is at 0° inclination, or face-on. If the star is spherical blackbody of radius R_* and temperature T_*, then its surface brightness is $B_0 = \sigma_B T_*^4/\pi$ where σ_B is the Stefan-Boltzmann constant. The flux received at distance r from the star and with angle of incidence $\cos^{-1}\mu$ is $\mu\sigma_B T_*^4 R_*^2/r^2$. A fraction α of this light is absorbed and the remainder scattered isotropically. Hence, the intensity of scattered light seen by the observer is $F = (1-\alpha)\mu B_0 R_*^2/(4r^2)$.

We calculate the emitted thermal emission from the internal temperature structure of the disk. For each pixel or point in the image plane, we calculate radiative transfer along the line of sight to find the emitted surface brightness: $dL_\lambda/dA = \int_0^\infty \exp(-\tau_\lambda) B_\lambda(T) \, d\tau_\lambda$ where B_λ is the Planck function expressed as a function of T the temperature in the disk, and τ_λ is the wavelength-dependent optical depth expressed as $\tau_\lambda = \int_\infty^z \int \chi_\lambda \rho \, dz'$ where χ_λ is the extinction coefficient and ρ is the density of the disk. We adopt values for the extinction coefficient from those tabulated in D'Alessio et al. (2001) at $\lambda = 0.8$, 769, and 1300 μm, interpolating between the values as a broken power law. Generally speaking, opacities increase at shorter wavelengths. Thus, shorter wavelengths probe the surface layers of the disk while longer wavelengths probe deeper in the disk.

Figure 1. Protoplanetary disks with embedded planets in scattered light.

3. Simulated Images

To interpret the observability of the features we model, it is useful to keep in mind that a distance of 100 parsecs, a feature 1 AU across subtends an angle of 0.01" (arcseconds). The closest protoplanetary disks are in the Taurus star-forming region, which is 140 pc away.

3.1. Core accretion

The model we adopt for core accretion is described in detail in Jang-Condell (2008). We begin with the standard viscous accretion disk model with stellar heating at the surface (Chiang & Goldreich 1997; Calvet et al. 1991; D'Alessio et al. 1998). The disk has a flared geometry, so that the ratio of scale height to the radius, h/r, increases with r rather staying constant as it would in a flat disk. We treat a planet core embedded in the disk as a point mass at the midplane that gravitationally compresses the disk in the vertical direction, creating a dimple at the surface of the disk. This creates a paired shadowing and brightening near the position of the planet as the side of the dimple closer to the star is shadowed and cools while the far side of the dimple is exposed to starlight and heats up.

In Figs. 1, 2 and 3 we display simulated images of a disk perturbed by a planet located at 4 AU (top row) and 8 AU (bottom row) with masses of 10, 20 and 50 $M_\oplus$ (left, middle, and right, respectively). In each image, the star is at the center, the planet location is marked by a cross (+), and the tick marks are spaced by 1 AU. Fig. 1 shows images in scattered light, in units of 10^{-7} of the surface brightness of the star. Fig. 2 shows images in Janskys per square arcsecond at infrared wavelengths, 40 and 70 μm for planets at 4 and 8 AU, respectively. These wavelengths are chosen to be close to the peak in thermal

Figure 2. Infrared images of protoplanetary disks with embedded planets at 40 μm at 4 AU and 70 μm at 8 AU.

Figure 3. Radio images of protoplanetary disks with embedded planets.

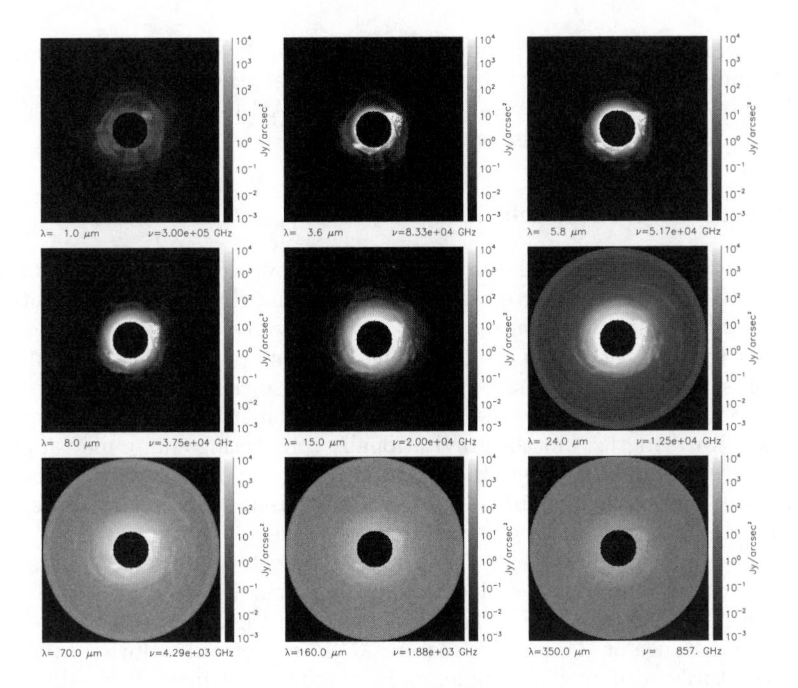

Figure 4. Simulated images of a disk undergoing gravitational instability

emission of the unperturbed disk at those distances. Fig. 3 shows images at 0.35 mm, radio frequency. In scattered light and infrared, the paired shadowing and brightening are clearly seen, particularly for larger planet masses. The feature is much less prominent in the radio. This is because cooling and heating due to the planet perturbation is confined mostly to the surface layers. Longer wavelengths probe deeper into the disk, where the effect is less pronounced.

3.2. Disk instability

To model disk instability, we apply our radiative transfer methods to a 3D hydrodynamic simulation of a gravitationally instable disk by Boss (2001), at a time step at which a self-gravitating clump has formed. The calculation of the scattered light image was carried out in Jang-Condell & Boss (2007). Here, we incorporate heating to the disk from stellar irradiation at the surface as well as shock heating from the hydrodynamic simulation to calculate the total disk temperature structure.

In Fig. 4 we show the synthesized images at various wavelengths, incorporating both scattered light and thermal emission. Wavelength increases from left to right, and top to bottom. At 1 μm, the shortest wavelength, scattered light dominates, so the image is very similar that shown in Jang-Condell & Boss (2007). As wavelength increases, we probe deeper into the disk, and thermal emission increases. At 24 μm, we begin to see thermal emission from the entire disk. The change in the morphology of the images from 1 to 350 μm reflects the change in the density structure of the disk from the surface to the midplane. The forming planet is not visible at short wavelengths, but can be seen as a bright spot at the 4 o'clock position at about 10 AU from the central star. For reference, the radius of the inner hole is 4 AU.

4. Discussion and Conclusion

Core accretion in a protoplanetary disk should show a relatively quiescent overall disk structure with a paired shadow/brightening that tracks the position of the embedded planet. In contrast, disk instability gives rise to large scale corrugations in the structure of the disk. The signature of a planet core embedded in a disk should fade at longer wavelengths because the shadowing and cooling effects. With disk instability, the position of the planet will not be determinable at optical wavelengths, but the gravitationally contracting clump may appear at longer wavelenghts. The age of the disk may also be an indicator: core accretion should still be occuring in $10^6 - 10^7$ year old, passively accreting disks, while disk instability occurs in very young, massive disks.

The keys to finding these signatures of planet formation in disks are a multi-wavelength approach and extremely high resolution. The scale of the perturbations produced by core accretion shown in Figs. 1–3 is on the order of 1 AU. At a distance of 100 pc, this requires 0.01" or better resolution to observe. The structures seen in disks undergoing gravitational instability will require similar resolution (Jang-Condell & Boss 2007). This will require either large-diameter telescopes or interferometry with long baselines. ALMA (Atacama Large Millimeter Array) or VLTI (Very Large Telescope Interferometer) should achieve this resolution in radio and visible interferometry, respectively. Large optical telescopes such as the GMT (Giant Magellan Telescope) or TMT (Thirty Meter Telescope) should also achieve this resolution.

Acknowledgments. This work was supported by the NASA Astrobiology Institute under Cooperative Agreement NNA04CC09A.

References

Boss, A. P. 2001, ApJ, 563, 367
Calvet, N., Patino, A., Magris, G. C., & D'Alessio, P. 1991, ApJ, 380, 617
Chiang, E. I. & Goldreich, P. 1997, ApJ, 490, 368
D'Alessio, P., Calvet, N., & Hartmann, L. 2001, ApJ, 553, 321
D'Alessio, P., Canto, J., Calvet, N., & Lizano, S. 1998, ApJ, 500, 411
Hubickyj, O., Bodenheimer, P., & Lissauer, J. J. 2005, Icarus, 179, 415
Jang-Condell, H. 2008, ApJ, 679, 797
Jang-Condell, H. & Boss, A. P. 2007, ApJ, 659, L169
Jang-Condell, H. & Sasselov, D. D. 2003, ApJ, 593, 1116
Jang-Condell, H. & Sasselov, D. D. 2004, ApJ, 608, 497

Bioastronomy 2007: Molecules, Microbes, and Extraterrestrial Life
ASP Conference Series, Vol. 420, 2009
K. J. Meech, J. V. Keane, M. J. Mumma, J. L. Siefert, and D. J. Werthimer, eds.

Preliminary Results in Dynamical Simulations of Water Origins for Earth-Like Planets

K. de Souza Torres[1] and O. C. Winter[2]

[1]*Instituto Nacional de Pesquisas Espaciais - INPE, São José dos Campos, Brazil*

[2]*Unesp - São Paulo State University, Faculdade de Engenharia de Guaratinguetá, Brazil*

Abstract. In its gaseous and solid forms, water is present in the most distant galaxies, among the stars, in the Sun, in its planets and their satellites and ring systems, and in comets. In its liquid form, it has played an essential part in the appearance, development and maintenance of terrestrial life. The origin of water on Earth remains one of the most important subjects of debate and controversy in solar system formation science. Possible sources of water can be divided into endogenous and exogenous. The most accepted endogenous source is the direct absorption of water from gas onto grains in the accretion disk, and the exogenous one is that the bulk of the Earth's water may have come from the asteroid belt in the form of planetary embryos with up to 10% of water. However, none of them alone is enough to explain Earth's water as a whole. In the present work, we use dynamical simulations of planetary formation and water delivery to investigate the implications of both of the main composition theories using chemical constraints like D/H ratio as discriminator. The goal is to understand how terrestrial planets got their water in the solar system and expand it to extrasolar systems. From the results we can conclude that the composition model with both main theories better explains the D/H ratio of Earth's water. Future work will add a cometary component in the water quantity and D/H ratio of terrestrial planets.

1. Introduction

Water is present in our planet in three different states: vapour, liquid and solid. In its liquid form, it has played an essential part in the appearance, development and maintenance of terrestrial life. In its gaseous and solid forms, water is omnipresent in the Universe: in the most distant galaxies, among the stars, in the Sun, in its planets and their satellites and ring systems, and in comets. Is there extraterrestrial life? We still await the answer, and the search for liquid water is an indispensable aspect of that answer (Encrenaz 2006). We are interested in finding how terrestrial planets in any planetary system get their water. In this research we are beginning to tackle this question by using numerical simulations to try to reproduce Earth with its oceans. To understand how water arrived on Earth we must examine the early stages of its formation.

2. Sources of Water

According with the most accepted theory of planetary formation (Encrenaz 2006), the protosolar nebula was hotter and denser toward its center and cooler and less dense farther out. These gradients influenced the chemical composition of different regions of the early solar system, including the distribution of water. Close to the nebula's center, high temperatures and pressures vaporized ice crystals and the light elements. The solar wind blew these materials toward the outskirts of the nebula, leaving mainly grains of rock behind to form the inner planets. Farther out, debris formed the carbonaceous chondrites which carry up to 10 percent of their mass in water ice (Morbidelli et al., 2000). Beyond the giant planets, water condensed in large quantities and formed comets, which are about half ice. Compared with these icy objects, Earth contains little water: only about 0.02 percent of its mass in its oceans, and somewhat more water sits beneath the surface. Nevertheless, Earth has substantially more water than could be expected at 1 AU from the Sun.

Over the years, several possible answers have been proposed for that question but, until recently, we have had little data for testing the hypothesis. In principle, it should be possible to determine the main sources of and relative contributions to Earth's water if they have distinct chemical and isotopic signatures. Signatures that are used as discriminators include the D/H ratio, the ratio of noble gases and the isotopic composition of the highly siderophile element Os. Among them, the D/H ratio discriminator is the one for which most data are available and so has been used most. Possible sources of water can be divided into endogenous and exogenous (Drake et al., 2005).

- Endogenous Theories: Scientists that support the endogenous sources argue there was enough water to be absorbed by the terrestrial planets locally. They created a simple model based on the temperature gradient that suggests that grains accreted to Earth could have absorbed 1-3 Terrestrial Oceans (1 $O_T = 1.4 \times 10^{24}$g) of water. However, the absorbed water has a protosolar nebula D/H ratio value (2.1×10^{-5}), while the D/H ratio in Earth's oceans is six times the protosolar value.

- Exogenous Theories: On the other hand, the exogenous theorists argue that, at 1AU from the Sun, the temperature is too high for enough water to exist to form all of Earth's oceans and an external source was therefore necessary. As comets contain a greater proportion of water than other known celestial objects do, they make natural candidates as a source of Earth's water. But spectral analyses of the chemical compositions of three comets - Halley, Hyakutake, and Hale-Bopp - revealed the D/H value in comets is about 2 times higher than the terrestrial value.

3. Numerical Simulations

So, it seems likely that the water of Earth's oceans cannot have come entirely from comets, nor could it have been entirely absorbed. The D/H value in Earth's water can be explained via a combination of an absorbed component (of value

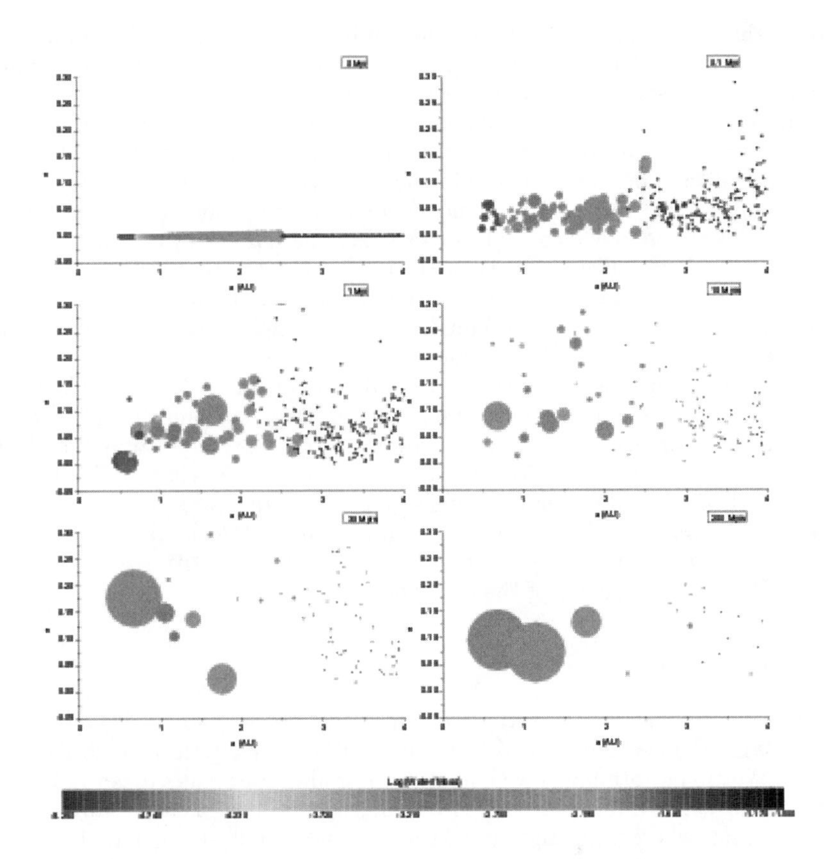

Figure 1. Snapshots for one of the simulations; surface density of 8 g/cm^2 at 1 UA.

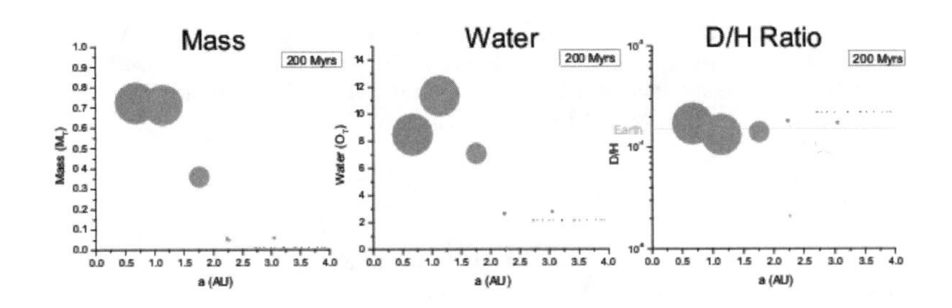

Figure 2. (a) Final Mass, (b) final water quantity and (c) final D/H ratio of the survivors (Simulation 2, $\Sigma_1 = 8$ g/cm^2).

close to the protosolar), a (certainly predominant) component of carbonaceous chondrites type and a cometary component. So, in this research we simulate the late stage planetary accretion, analysing the delivery of water, using the D/H ratio as discriminator. The simulations include both planetary embryos and planetesimals simulating the asteroids like in (Raymond et al., 2004). Also, one Jupiter-like planet is used to influence the accretion dynamics. We created a compound model that uses the endogenous and exogenous theories to define how much water and which D/H ratio value the embryos and planetesimals have initially. For embryos, we use a D/H value like the protosolar nebula (2.1 x 10^{-5}) and initial water defined by the endogenous theory model. For the planetesimals we give a water quantity (5%) and D/H value (a median value, 2.2 x 10^{-4}) representative of carbonaceous chondrites.

Simulations are performed using a modified version of Mercury(Chambers 1999). Mercury is a general-purpose software package for doing N-body integrations and we reprogrammed it to include water content. The total time in all the simulations is 200Myrs. We use 3 values of surface density at 1 AU (Σ_1), 6, 8 and 10 g/cm^2. Each surface density is simulated three times. Figure 1 shows some snapshots of one of the simulations. In all the simulations the same analyses are made: final mass (Figure 2a), final water quantity (Figure 2b) and final D/H ratio (Figure 2c) of the survivors.

4. Conclusions

In the above simulation, the water content of the largest embryos is comparable to that of Earth in both mass and D/H ratio. Other simulations produced similar results. We are performing now an analysis over all these results to get how much water could come from asteroids, according this model. But we already can conclude that, whereas any individual source of water fails, the compound model is better able to explain the mass of the Earth, the mass of Earth's water and the water's D/H ratio. Future work will include other chemical discriminators, simulations of other initial conditions (for other planetary systems) and add a cometary component in the mass, water's quantity and D/H ratio of terrestrial planets.

Acknowledgments. command. The authors are grateful to CAPES (Coordenação de Aperfeiçoamento de Pessoal de Nível Superior), CNPq (Conselho Nacional de Desenvolvimento Científico e Tecnológico) and FAPESP (Fundação de Amparo a Pesquisa do Estado de São Paulo) for supporting this research.

References

Encrenaz, Thérèse 2006. A la recherche de l'eau dans l'universe; Praxis. Publishing Ltd, Paris Observatory, Meudon, France.

Drake, Michael J., Campins, Humberto 2005. Origin of water on the terrestrial planets. Proceedings IAU Symposium No. 229.

Chambers, J.E 1999. A hybrid sympletic integrator that permits close encounters between massive bodies. Mon. Not. R. Astron. Soc. 304, 793-799.

Morbidelli, A., Chambers, J., Lunine, J. L., Petit, J. M., Robert, F., Valsecchi, G. B., &, Cyr, K. E. 2000. Source regions and timescales for the delivery of water on Earth. Meteorit. Planet. Sci. 35, 1309-1320

Raymond et al. 2004. Making other earths: dynamical simulations of terrestrial planet formation and water delivery, Icarus, 168, 1-17

Karla de Souza Torres

Leonardo Lareo

Bioastronomy 2007: Molecules, Microbes, and Extraterrestrial Life
ASP Conference Series, Vol. 420, 2009
K. J. Meech, J. V. Keane, M. J. Mumma, J. L. Siefert, and D. J. Werthimer, eds.

Transiting Extrasolar Planets and SuperWASP

D. R. Anderson[1] and The SuperWASP Consortium[2]

[1]*Astrophysics Group, University of Keele, Staffordshire, ST5 5BG, UK*

[2]*Comprises of University of Cambridge, Keele University, University of Leicester, The Open University, Queen's University Belfast, St Andrews University, the Isaac Newton Group, the Instituto de Astrofisica de Canarias and the South African Astronomical Observatory.*

Abstract. Since 1995 hundreds of planets have been discovered around main sequence stars, some in multiple planet systems. A range of search methods are currently being used or developed. The most successful of these so far has been radial velocity but transit, microlensing, and direct-imaging have also produced results. The transit method is powerful in that many stars can be surveyed at once, and much information is available for transiting planets. This method is now being used extensively on the ground and in space.

SuperWASP (Wide Angle search for Planets) is a ground-based transit survey made up of UK universities. There are two instruments, one on La Palma, Canary Isles, and the other in Sutherland, South Africa, giving near-all-sky coverage. The first two SuperWASP planets were announced at in September 2006. Candidates are currently being identified and followed up using the radial velocity method. It is hoped that there will be more planets to announce soon. COROT began the space-based transit survey era in December 2006 and is already producing results. Over the next few years more satellites will join the effort and with that it is likely Earth-mass planets will be found in the habitable zones of Sun-like stars. The search for life as we know it, and as we don't, will soon gather pace.

1. Extrasolar Planets

Since 1995 hundreds of planets have been discovered around main sequence stars, some in multiple planet systems[1]. Many known planets are ≈Jupiter-mass and in close orbits. This is likely to be a selection effect which will reduce as sensitivity increases and a common goal is achieved: to find less massive planets at greater orbital distances. Ultimately, for many the aim is to find Earth-like planets in the habitable zone (liquid water) of Sun-like stars. So far the most successful detection methods have been radial velocity and transit. Microlensing and direct imaging have also produced results. Astrometry will begin to compete in the near future. The radial velocity method has so far proved the most successful, but most instruments can focus on only one star at a time. So stars thought likely to harbor planets are chosen and then monitored for a characteristic variation in the radial velocity that could indicate a planetary companion. Recently this

[1]http://exoplanet.eu

method produced a 5 Earth-mass planet, Gliese 581b(Udry et al. 2007), that is thought to be in the habitable zone of an M-Dwarf; this is the most promising candidate to date for life. When a planet is detected a shortfall of the method is revealed as the inclination angle of the planets orbit is not known then only the minimum mass can be estimated. Microlensing has drawbacks too: observations are not repeatable, there have been few events so far, and follow-up observations are very difficult as the sources are so faint.

2. Transiting Extrasolar Planets

The transit method has some clear advantages(Charbonneau et al. 2007). Many stars can be monitored at the same time. If bright enough, a candidate planet can be confirmed using the radial velocity method. The planet radius and mass are then known, so, therefore, is the average density and hence something can be said about the composition of the planet. A transit occurs when an extrasolar planet passes in front of its parent star along our line of sight. For a hot-Jupiter orbiting a solar type star the effect of this is a $\sim$1% dip in the light received from the star. For a star with a hot-Jupiter, the probability of a transiting orientation is $\sim$10%. The possibilities of following up transiting planets are rich. By observing the secondary eclipse (as the planet passes behind the star) the planet-to-star brightness ratio can be determined(Deming et al. 2005). Knutson et al. (2007) used the Spitzer Space Telescope to produce a temperature distribution map of transiting extrasolar planet HD 189733; the increase in brightness as the dayside of the planet rotated into view was detected. Transmission spectroscopy can reveal atmospheric composition signatures. Sodium was the first constituent to be detected(Charbonneau et al. 2002) and more recently water vapor has been detected(Tinetti et al. 2007).

3. SuperWASP - A Transit Survey

3.1. Background

The Super Wide Angle Search for Planets survey is a ground-based, wide-angle search for transiting extrasolar planets(Pollacco et al. 2006). The consortium comprises a number of UK universities and three astronomical institutions.
The consortium built, maintains, and operates two robotic observatories: a northern observatory on La Palma, Canary Isles, and a southern observatory near Sutherland, South Africa.

3.2. Operations

Each clear night the roof of each observatory opens and observations begin. A small number of fields are observed frequently so that any transit signal is recorded with reasonable time resolution. The instrument is extremely wide-field: a single camera observes a portion of sky 7.8°x7.8°, whereas the Moon is only 0.5° in diameter. In a single frame there can easily be a quarter of a million stars. It is a numbers game: a fraction of stars will have planets, a fraction

Table 1. Parameter table for SuperWASP's first two planets

Planet	$M_{\rm pl}$ $[M_{\rm J}]$	$R_{\rm pl}$ $[R_{\rm J}]$	Period [days]	M_* $[M_{\rm Sun}]$	Sp. Type
WASP-1b	0.89 ± 0.2	1.358 ± 0.1	$2.51997 \pm 1.6{\times}10^{-4}$	1.24 ± 0.17	F7V
WASP-2b	0.88 ± 0.11	1.017 ± 0.15	$2.152226 \pm 4{\times}10^{-5}$	0.84 ± 0.11	K1V

of those will be hot-Jupiters, a fraction of those will be in a favorable orientation.

3.3. Data Reduction and Exploitation

A large amount of data (up to 80GB) are produced each night. These data are reduced, de-trended and quality-checked. An algorithm (Collier Cameron et al. (2006)) then searches the millions of lightcurves for the characteristic signal of a small dip with a regular frequency. Candidates are prioritized and filtered(Collier Cameron et al. 2007b) before being examined by eye. For promising candidates radial velocity measurements of the parent star are performed to determine if the companions are indeed of planetary mass.

3.4. Results

From a short, initial season running only the northern observatory, and with only five of its current complement of eight cameras, we have so far detected two confirmed planets: WASP-1b and WASP-2b (Collier Cameron et al. 2007a). Table 1 is a table of parameters of the two planets. As we continue to identify and follow up more candidates from the initial season and the expanded subsequent seasons (two observatories with eight cameras each) we expect to find many more planets.

4. Future Prospects

Most planets known to date are massive and in close orbit around their parent stars. The aim is to find lower mass planets ($\sim$ Earth-mass) at greater orbital separation (within the habitable zone).

4.1. Transit Timing Variations

By studying transiting systems other planets could be discovered by using the method of transit timing variations(Agol et al. 2005; Holman and Murray 2005). The effect of another body in a transiting system will be to cause small variations in the period of the transiting planet; the other body need not be in a transiting configuration. If the system is in a certain (resonant) configuration this effect will be large enough to detect with current technology, even for an Earth-mass companion.

4.2. Space Missions

COROT[2] is a European transit and astroseismology satellite launched at the very end of 2006. The first announcement of a detected transiting extrasolar planet (a hot-Jupiter) came in May 2007. COROT is sensitive to planets around twice the mass of Earth. Kepler[3] is a US transit survey satellite scheduled for launch in November 2008, and will be sensitive to Earth-mass planets in orbit at habitable distances around Sun-like stars.

There are two further planned space missions that do not rely on the transit method, but are worth mentioning anyway. Darwin[4] is a European space mission scheduled for launch in 2015. It aims to directly image Earth-like planets orbiting nearby stars. This will be achieved using nulling interferometery which will cancel out the starlight allowing the much fainter emission from planets to be detected. Spectra will be taken of detected planets and analyzed for possible signatures of life (e.g. oxygen, water, ozone). Terrestrial Planet Finder[5] is a similar US mission, but with budget problems. An additional planned feature is a coronagraph that would attempt to detect the reflected light from planets.

Acknowledgments. The author expresses gratitude to the organizers of Bioastronomy 2007 and AbGradCon 2007 for a wonderful conference, and acknowledges travel support that enabled attendance.

References

Agol, E., et al., 2005, MNRAS, 359, 567
Collier Cameron, A., et al., 2006, MNRAS, 373, 799
Collier Cameron, A., et al., 2007a, MNRAS, 375, 951
Collier Cameron, A., et al., 2007b, MNRAS, 380, 1230
Charbonneau, D., et al., 2002, ApJ, 568, 377
Charbonneau D., et al., 2007, Protostars and Planets V, 701
Deming, D., et al., 2005, Nature, 434, 740
Holman, M.J., Murray, M. W., 2005, Science, 307, 1288
Knutson, H., et al., 2007, Nature, 447, 183
Pollacco, D., et al., 2006 PASP, 118, 1407
Tinetti, G., et al., 2007, Nature, 448, 169
Udry, S., et al., 2007, A. & A. , 469 , 43

---

[2]http://smsc.cnes.fr/COROT

[3]http://kepler.nasa.gov

[4]http://sci.esa.int/home/darwin

[5]http://tpf.jpl.nasa.gov

Bioastronomy 2007: Molecules, Microbes, and Extraterrestrial Life
ASP Conference Series, Vol. 420, 2009
K. J. Meech, J. V. Keane, M. J. Mumma, J. L. Siefert, and D. J. Werthimer, eds.

Transit Spectroscopy of the Extrasolar Planet HD 209458b

P. Rojo,[1] J. Harrington,[2] D. Deming,[3] and J. Fortney[4]

[1]*Departamento de Astronomía, Universidad de Chile*

[2]*Department of Physics, University of Central Florida*

[3]*Planetary System Laboratory, NASA Goddard Space Flight Center*

[4]*Space Sciences and Astrobiology Division, NASA Ames Research Center*

Abstract. An attempt to detect water in the atmosphere of the extrasolar planet HD 209458b using transit spectroscopy is presented. A radiative transfer model, designed and built specifically for this project, predicts the dependence in wavelength of the stellar spectrum modulation due to a transiting planet, given a planetary temperature/pressure/composition profile. A total of 352 spectra around 1.8 microns were obtained during three transit events using ISAAC at the Very Large Telescope.

Correlating the modeled modulation with the infrared spectra yields a non-detection of water in the atmosphere of HD 209458b due to overwhelming contamination from telluric water lines.

Original data-reduction techniques that were developed during this work are also presented.

At the time of writing this article, around 30 of the almost 270 known extrasolar planets transit in front of their host stars. Despite the small number, those transiting planets have proved invaluable by providing an excellent opportunity to accurately measure planetary properties. The unexpected variety found even in this small sample makes it clear that planets are not just small stars, since not even their bulk properties can be solely determined by their masses.

Strong limits on molecular abundances in exoplanetary atmospheres are eagerly awaited by theorists to validate their models. In addition, conclusions of atmospheric abundances over a statistically-significant sample will undoubtedly help selecting appropriate targets to maximize the science return from ambitious space-based missions in the coming decade. Missions whose ultimate goal will be to detect and characterize Earth-like planets.

Here we present an attempt to use the transit spectroscopy technique on the system HD 209458. This technique uses the fact that the area a transiting planet blocks from its host star varies with wavelength according to the unique spectral features of molecules in the exoatmosphere, allowing identification and abundance measurements from appropriately-timed observations.

1. The Model

We developed an efficient radiative transfer code that predicts the modulation (which is defined as the in-transit spectra divided by the stellar, or out-of-transit,

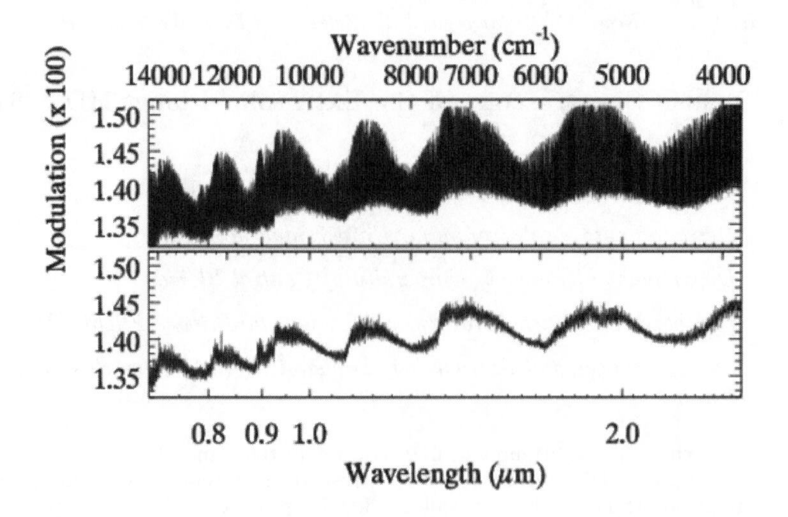

Figure 1. Infrared modulation of HD 209458 due to water, convolved to wavelength resolutions of $R = \lambda/\Delta\lambda = 30000$ and $R = 5000$, for the top and bottom frame, respectively. This modeled modulation includes close to 22 million lines from Partridge & Schwenke (1997) water line database, more than half a million wavenumber points, and over 50 planetary layers. The result was obtained in just a couple of hours with a desktop computer. Planetary profile from Fortney et al. (2005)

spectra) given molecular line data, system geometry, and thermal and abundance profiles for any transiting planet. For each wavelength and projected planetary radius, the code follows a ray through its trajectory into the atmosphere. Besides calculating molecular line opacities from up-to-date molecular line databases, it also calculates continuum opacity for clouds and collision-induced absorption from H2-He and H2-H2 pairs (Borysow et al. 2001; Jørgensen et al. 2000). Fig. 1 shows sample modulation spectra.

2. The Data

Three transit events were observed with ISAAC at the VLT (Moorwood et al. 1998). Chosen wavelength range was believed to be at the right distance from an absorption band in order to obtain an exoplanetary spectrum that is both strong enough for detection and is not overwhelmed by the telluric absorption. The integration time of each spectral frame was limited to 60 seconds and we use the standard ABBA nodding sequence.

Due to the small modulation amplitude (Fig. 1), it was necessary to correct for previously ignored systematic errors. For instance, we developed a new method to remove the appearance of fringes on flat fields that did not appear on data frames (Rojo & Harrington 2006).

We use an optimal extraction algorithm (Dermody and Harrington, 2006, private communication) to obtain the spectra. Then, we tried 3 different methods to remove the telluric water. This is the most critical step and it is only

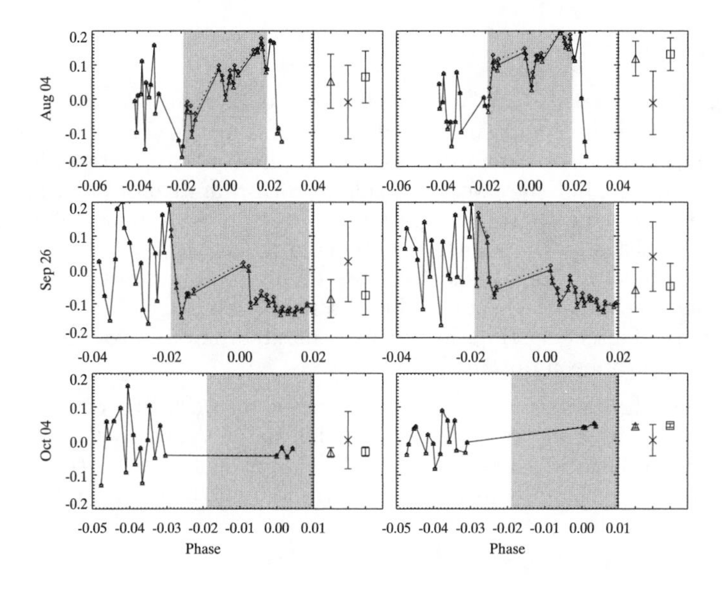

Figure 2. Correlation result for each night and beam (ABBA beams A and B, on the left- and right-side frames, respectively). The left side of each frame shows the correlation value for each spectra. The solid line indicates the results from the real data. The dotted line indicates the result for the synthetic data. The shaded areas indicate the in-transit phases. On the right side of the frame, the symbols triangle, diamond, and square are the average correlation of the out-of-transit, in-transit-real, and in-transit-synthetic spectra, respectively.

possible due to the temperature difference between the telluric and the exoplanetary atmosphere. Wavelength channels that still remained variable were masked out from further analysis. The corrected spectra was divided by an out-of-transit average to obtain the modulation.

3. The Result

The modeled modulation was correlated with each observed modulation. A successful detection would translate in a correlation value of the out-of-transit spectra that is distinctly closer to zero than the correlation values of the in-transit spectra. This is not the case in any of the observing nights. Furthermore, creating a synthetic dataset (by adding the modeled modulation to each in-transit spectra immediately after extraction) does not change the correlation result by a significant amount (Fig. 2) indicating that, even if the signal were there, our analysis would have not detected it.

After performing several tests to check the pipeline (including scaled planetary atmospheres, where detection was indeed attained), the only explanation left is that in the chosen wavelength range the telluric water absorption was too strong and variable, overwhelming any signal there may be from the extrasolar planet.

4. Future Work

Recently, Tinetti et al. (2007) and Barman (2007) identified water in the atmosphere of an exoplanet using broad-band photometry. Their results seem to confirm the existence of water on an extrasolar planet atmosphere with a high degree of confidence, but they were not able to constrain its molecular abundance. A strong result using IR transit spectroscopy will provide more information as it is more sensitive to abundance variations in models.

To overcome the biggest shortcomings on new observations, we developed a new method to choose the best possible observing wavelength from ground-based observations. Observing time has been granted, where improved observations have just recently been obtained.

Acknowledgments. This material is based on work supported by the National Aeronautics and Space Administration under grant NAG5-13154 issued through the Science Mission Directorate. The presentation on this conference was possible by the support from a BioAstronomy 2007 grant, and the Fondap Astrophysics, Chile

References

Barman, T. 2007, ApJ, 661, L191

Borysow, A., Jørgensen, U. G., & Fu, Y. 2001, Journal of Quantitative Spectroscopy and Radiative Transfer, 68, 235

Fortney, J. J., Marley, M. S., Lodders, K., Saumon, D., & Freedman, R. 2005, ApJ, 627, L69

Jørgensen, U. G., Hammer, D., Borysow, A., & Falkesgaard, J. 2000, A&A, 361, 283

Moorwood, A., Cuby, J.-G., Biereichel, P., Brynnel, J., Delabre, B., Devillard, N., van Dijsseldonk, A., Finger, G., Gemperlein, H., Gilmozzi, R., Herlin, T., Huster, G., Knudstrup, J., Lidman, C., Lizon, J.-L., Mehrgan, H., Meyer, M., Nicolini, G., Petr, M., Spyromilio, J., & Stegmeier, J. 1998, The Messenger, 94, 7

Partridge, H. & Schwenke, D. W. 1997, J. Chem. Phys. , 106, 4618, [P&S]

Rojo, P. M. & Harrington, J. 2006, ApJ, 649, 553

Tinetti, G., Vidal-Madjar, A., Liang, M.-C., Beaulieu, J.-P., Yung, Y., Carey, S., Barber, R. J., Tennyson, J., Ribas, I., Allard, N., Ballester, G. E., Sing, D. K., & Selsis, F. 2007, Nat, 448, 169

Bioastronomy 2007: Molecules, Microbes, and Extraterrestrial Life
ASP Conference Series, Vol. 420, 2009
K. J. Meech, J. V. Keane, M. J. Mumma, J. L. Siefert, and D. J. Werthimer, eds.

Stellar Properties and Rocky Planet Habitability: Is There a Correlation?

J. A. Robles and C. H. Lineweaver

Planetary Science Institute, Research School of Astronomy & Astrophysics and Research School of Earth Sciences, The Australian National University, Canberra Australia

Abstract. The properties of a rocky planet's host star may be associated with the origin of life and the evolution of observers on that planet. By comparing the Sun to other stars in our Galaxy, we may be able to identify such properties. If our Sun is a typical star, stellar conditions appropriate for life, and life itself, may be common in the Universe. On the other hand, if the Sun is atypical or even unique, then the life orbiting the Sun may also be atypical or unique. Here we present the comparison between the Sun and stars for three important parameters that may be associated with the origin and evolution of life. We find that:

i) The Sun is more massive than $95 \pm 2\%$ of nearby stars.
ii) The Sun is younger than $53 \pm 2\%$ of stars in the Galactic disk.
iii) The Sun is more metal rich than $65 \pm 2\%$ of nearby stars.

1. Introduction

Quantifiable relationships should exist between some properties of a host star and the properties of any rocky planets in orbit around it. For example, the ages of a host star and its planets will be nearly identical. The ratios of refractory elements, e.g. Mg/Fe, in a planet will mirror the observable ratio in its host star (Lineweaver and Robles. 2009). It is also plausible that more massive host stars have more massive terrestrial planets.

The relationships between host star properties and the habitability of its orbiting planet are more subtle and difficult to quantify. We only know of one star that harbours a life-bearing planet. However, the hypothesis that there are a number of stellar properties (or combination of properties) that are optimal for the formation of habitable planets, is testable. As an extreme example, if the Sun were anomalously old (let us say the oldest star in the Galaxy) this would suggest that the evolution of observers takes an extremely long time and this would be evidence that observers like us are rare in the Universe (Carter 1983). Nothing as obvious as this example is showing up in our analysis. However, the basic concept is the same.

2. Methodology

By comparing the Solar properties to those of a stellar sample representative of the local stellar population, we hope to identify one or more habitability-related properties. Solar properties showing up as an anomaly are probably correlated with the habitability of the Earth. One must be extremely careful when dealing with statistics of one, for example, if we choose a large number of solar properties, then, any possible anomalous property would fade away when weighted against the number of properties analysed (Robles et al. 2009). If on the other hand we choose only a few parameters in which we know *a priori* the Sun is somewhat of an outlier, the outcome of the analysis would be a foregone conclusion.

Is our current knowledge of the Sun and stars consistent with the idea that the Sun is a typical star? This question has been previously addressed with apparently conflicting results — while Gustafsson (1998) and Allende Prieto (2006) have suggested that the Sun is a typical star, other studies (Gonzalez 1999*a*,*b*; Gonzalez et al. 2001) have suggested the opposite.

The most delicate part of such an analysis is the selection of an adequate stellar sample for each property. Ideally, the Sun needs to be compared to a large stellar sample free of selection effects — therefore, representative of the 'whole' stellar population for every property. The assembly of such a stellar sample is close to impossible, so great effort should be spent assembling a minimally biased stellar sample for each parameter and quantifying its biases.

3. Results

3.1. How Massive is the Sun?

The mass of a star largely determines its luminosity, temperature, main sequence life-time and circumstellar habitable zone dimensions. Mass is probably the most important property of a star. In Figure 1 we compare the Sun (denoted by "$\odot$") to the 125 nearest stars within 7 parsecs $\sim$ 23 lightyears. The distance limit of the selected stellar sample (RECONS (Research Consortium on Nearby Stars), Henry 2006), permits the observation of the faint end of the stellar population (M-dwarfs) is still observable. The Sun is more massive than $95\pm2\%$ of the stars. The distribution of stars as a function of stellar mass (Initial Mass Function, IMF), is represented by the thick grey line and its associated uncertainty by the hashed-shade (Kroupa 2002). For the IMF, the Sun is more massive than $94 \pm 2\%$ of the stars. There is good agreement between the histogram and the IMF model.

3.2. How Old is the Sun?

The left-hand panel in Figure 2 compares the age of the Sun to two different stellar age distributions: the Milky Way Galaxy and the cosmic stellar ages. The histogram represents the Milky Way Star Formation History (SFH) derived by Rocha-Pinto et al. (2000). The Sun is younger than $53 \pm 2\%$ of the stars in the Galaxy. The grey represents the cosmic SFH with its associated uncertainty (Hopkins and Beacom 2006) — the Sun is younger than $86 \pm 5\%$ of the stars in the Universe. Why are these two distributions so different? Most stars in the Universe reside in elliptical galaxies. These galaxies had initial bursts of star formation, ran out of gas and can no longer form stars. Our Galaxy's SFH, on the other hand, features bursts of star formation as satellite galaxies interacted or fell into the Milky Way.

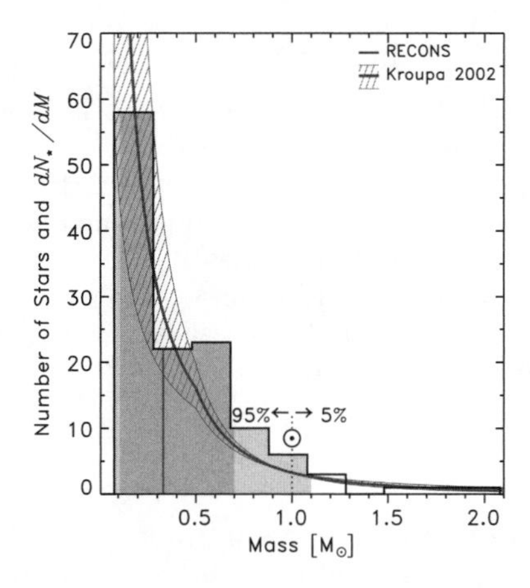

Figure 1. Mass histogram of the 125 nearest stars (RECONS Henry 2006). The median mass ($\mu_{1/2} = 0.33\ M_\odot$) of the distribution is indicated by the vertical grey line. The 68% and 95% confidence intervals around the median are indicated by the vertical dark grey and light grey bands respectively. We also use these conventions in Fig. 2. The solid curve and hashed area around it represents the IMF and its associated uncertainty (Kroupa 2002). The Sun, indicated by "$\odot$", is more massive than $95 \pm 2\%$ of these stars.

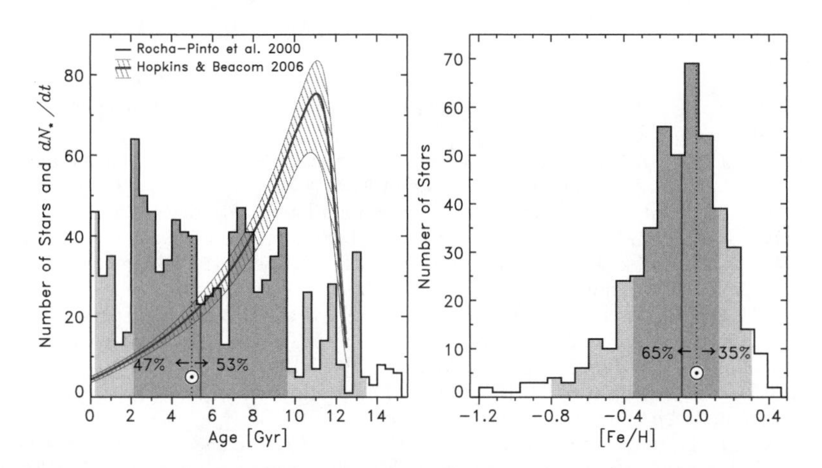

Figure 2. Left-hand panel: The galactic stellar age distribution from Rocha-Pinto et al. (2000). The Sun is younger than $53 \pm 2\%$ of the stars in the Galaxy. The median age is $\mu_{1/2} = 5.4$ Gyr. In contrast to the Galaxy's stellar ages, the cosmic SFH (Hopkins and Beacom 2006), indicates that the Sun is younger than $86 \pm 5\%$ of the stars in the Universe. Right-hand panel: Stellar metallicity histogram of the 453 FGK *Hipparcos* stars within 25 pc (Grether and Lineweaver 2007). The median metallicity $\mu_{1/2} = -0.08$. The Sun is more metal-rich than $65 \pm 2\%$ of the stars.

3.3. How Metal Rich is the Sun?

The heavy element content of a star is related to its ability to form planets. The abundances of oxygen, carbon and refractory elements scale approximately with iron abundance ([Fe/H]). Therefore, a star's [Fe/H] abundance has a direct impact on the abundances of elements that determine planet characteristics, e.g. heat budget provided by radioactive elements. The right-hand panel in Figure 2 shows the stellar metallicity distribution of 453 F,G and K stars within 25 parsecs $\sim$ 80 lightyears (Grether and Lineweaver 2006). The Sun is more metal rich than $65 \pm 2\%$ of these nearby stars.

4. Discussion

In our search for extrasolar worlds, we want to know how special are stars with habitable planets. Comparing the Sun to other stars can be used as a fishing expedition for any property that may be associated with the prerequisites for life. Astrobiological constraints on life requirements and their relation with stellar parameters will enable us to follow up any candidate parameter this study may uncover.

Acknowledgments. We would like to thank Mount Stromlo Observatory at The Australian National University, the Bioastronomy 2007 and ABGradCon local organising committee for travel support. Special thanks to Karen Meech.

References

Allende Prieto, C. (2006), 'Solar Chemical Peculiarities?', *ArXiv Astrophysics e-prints astro-ph/0612200* .

Carter, B. (1983), 'Anthropic Bias: Observation Selection Effects in Science and Philosophy', *Philos. Trans.R. Soc. London* A, 310–347.

Gonzalez, G. (1999*a*), 'Are stars with planets anomalous?', MNRAS 308, 447–458.

Gonzalez, G. (1999*b*), 'Is the Sun anomalous?', *Astronomy and Geophysics* 40, 25.

Gonzalez, G., Brownlee, D. and Ward, P. (2001), 'The Galactic Habitable Zone: Galactic Chemical Evolution', *Icarus* 152, 185–200.

Grether, D. and Lineweaver, C. H. (2006), 'How Dry is the Brown Dwarf Desert? Quantifying the Relative Number of Planets, Brown Dwarfs, and Stellar Companions around Nearby Sun-like Stars', ApJ 640, 1051–1062.

Grether, D. and Lineweaver, C. H. (2007), 'The Metallicity of Stars with Close Companions', ApJ 669, 1220–1234 .

Gustafsson, B. (1998), 'Is the Sun a Sun-Like Star?', *Space Science Reviews* 85, 419–428.

Henry, T. J. (2006), 'RECONS catalog of the 100 nearest stellar systems', *RECONS database* .

Hopkins, A. M. and Beacom, J. F. (2006), 'On the Normalization of the Cosmic Star Formation History', ApJ 651, 142–154.

Kroupa, P. (2002), 'The Initial Mass Function of Stars: Evidence for Uniformity in Variable Systems', *Science* 295, 82–91.

Lineweaver, C. H. and Robles, J. A. (2008), 'The Chemical Compositions of Other Earths', *ASP* .

Robles, J. A., Lineweaver, C. H., Grether, D., Flynn, C., Holmberg, J., Pracy, M. and Gardner, E. (2009), 'How typical is the Sun?', *ApJ, 684, 691* .

Rocha-Pinto, H. J., Scalo, J., Maciel, W. J. and Flynn, C. (2000), 'An Intermittent Star Formation History in a "Normal" Disk Galaxy: The Milky Way', ApJ 531, L115–L118.

Session XI

Habitable Planets and Their Stars

Adrian Mellot

Nader Haghighipour

Bioastronomy 2007: Molecules, Microbes, and Extraterrestrial Life
ASP Conference Series, Vol. 420, 2009
K. J. Meech, J. V. Keane, M. J. Mumma, J. L. Siefert, and D. J. Werthimer, eds.

Extreme Habitability: Formation of Habitable Planets in Systems with Close-in Giant Planets and/or Stellar Companions

N. Haghighipour

Institute for Astronomy and NASA Astrobiology Institute, University of Hawaii-Manoa, Honolulu, Hawaii, USA

Abstract. With more than 260 extrasolar planetary systems discovered to-date, the search for habitable planets has found new grounds. Unlike our solar system, the stars of many of these planets are hosts to eccentric or close-in giant bodies. Several of these stars are also members of moderately close (<40 AU) binary or multi-star systems. The formation of terrestrial objects in these "extreme" environments is strongly affected by the dynamics of their giant planets and/or their stellar companions. These objects have profound effects on the chemical structure of the disk of planetesimals and the radial mixing of these bodies in the terrestrial regions of their host stars. For many years, it was believed that such effects would be so destructive that binary stars and also systems with close-in giant planets would not be able to form and harbor habitable bodies. Recent simulations have, however, proven otherwise. I will review the results of the simulations of the formation and long-term stability of Earth-like objects in the habitable zones of such "extreme" planetary systems, and discuss the possibility of the formation of terrestrial planets, with significant amounts of water, in systems with hot Jupiters, and also around the primaries of moderately eccentric close binary stars.

1. Introduction

An analysis of the orbital properties of the currently known extrasolar planets indicates that many of these objects have dynamical characteristics that are profoundly different from those of the planets in our solar system. While around the Sun, planets keep nearly circular orbits with smaller planets at closer distances and the larger ones farther way, from the 263 exoplanets that were discovered at the time of the writing of this article, 68 were at distances smaller than 0.1 AU from their host stars, and 126 revolve in orbits with eccentricities larger than 0.2 (figure 1). A deeper look at the parent stars of these planets indicates that the differences between their planetary systems and the Solar System are not limited to the semimajor axes and eccentricities of their planetary companions. Several of these stars (approximately 20% of currently known extrasolar planet-hosting stars) are members of binaries or multi-star systems.

Since the masses of the majority of extrasolar planets are within the range of a few Neptune- to several Jupiter-masses, the eccentric and close-in orbits of these bodies, and the fact that several of them exist around the components of binary stars, raise questions on the formation of such *extreme* planetary systems, and also on the possibility of the existence of smaller planets, such as terrestrial-

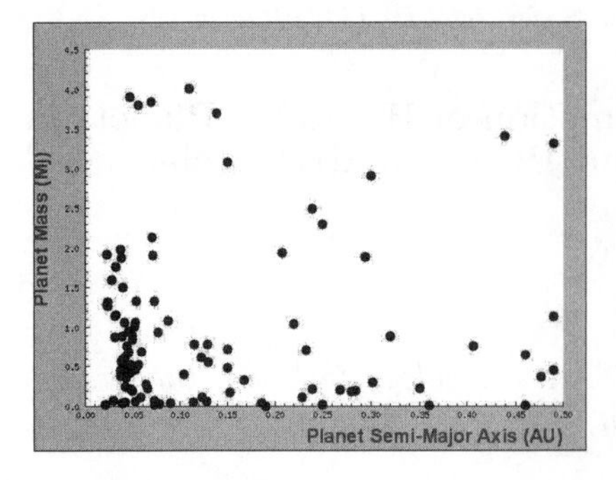

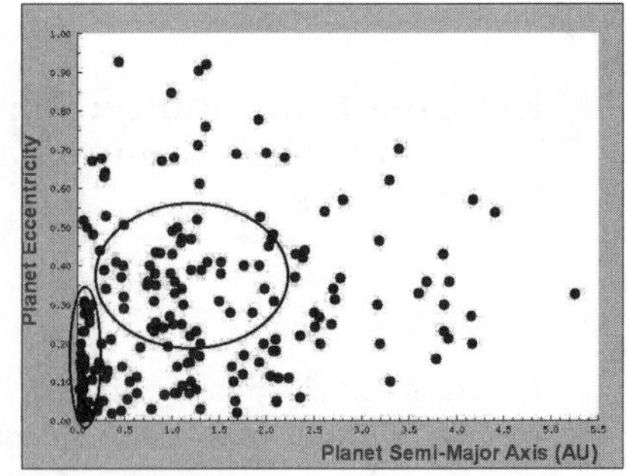

Figure 1. Graphs of the mass and eccentricity of some of the extrasolar planets. The graph on the left shows close-in giant planets. The graph on the right shows planets with large eccentricities. The two encircled areas depict extreme planetary systems. Graphs from exoplanet.eu.

class objects, around their host stars, and particularly in their systems' habitable zones. In this paper, I review the current status of research on the formation of habitable planets in these extreme planetary environments, and discuss the results of the simulations of Earth-like planet formation in systems with close-in giant planets and binary companions.

2. Systems With Close-in Giant Planets

It is widely accepted that the close-in giant planets have formed at larger distances and have migrated to their current positions. During their migrations, these objects perturb the orbits of smaller bodies on their paths and affect the interactions among these objects. In general, the perturbative effect of a migrating giant planet causes the orbital eccentricities and/or inclinations of planetesimals and protoplanets to rise to higher values. Whether these dynamically hot objects can cool down and re-interact to form terrestrial bodies depends on the mass of the migrating planet and its rate of migration. While, as shown in figure 2, simulations by Armitage (2003) indicate that a giant planet migration will disrupt a disk of planetesimals and reduces its surface density to very low values (unless the migration occurs in a very short time), a systematic search of the parameter-space by Lufkin et al (2006) have resulted in identifying the ranges of the mass and rate of the migration of a giant planet for which dynamically excited planetesimals can return to low inclination and low eccentricity orbits (through gas-drag and dynamical friction) and increase the possibility of terrestrial planet formation (figure 2). As shown by the four-panel graph in figure 2, for a low-mass (top left), or a fast migrating (bottom left) giant planet, the majority of planetesimals acquire eccentricities smaller than 0.4, implying that they may survive planetary migration and return to dynamically cool orbits in short times.

Recent numerical integrations by Fogg & Nelson (2005, 2007), Raymond et al. (2006), and Mandell et al. (2007) have shown that it is indeed possible for many planetesimals to survive giant planet migration and form terrestrial bodies (figure 3). As shown by Raymond et al. (2006) and Mandell et al. (2007), under

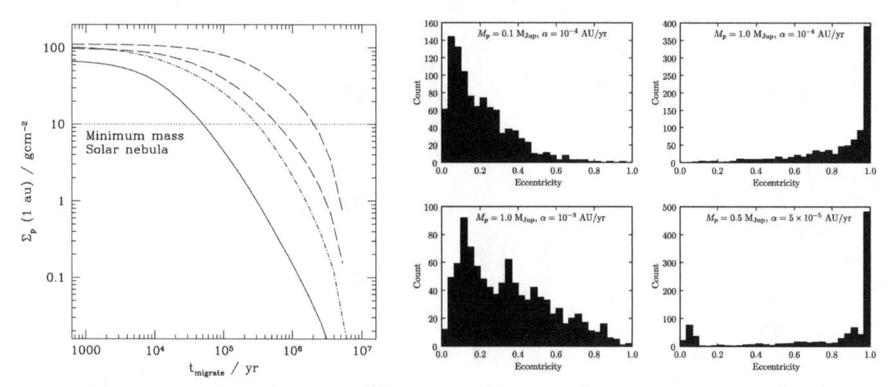

Figure 2. Left: the variation of surface density of a disk of planetesimals in term of the time of the migration of a giant planet (Armitage 2003). As shown here, unless the migration is fast, giant planet migration destabilizes planetesimals resulting in their removal from the disk. Right: eccentricity of planetesimals during giant planet migration. As shown in the two left graphs, for a low mass, or a fast migrating giant planet, the increase in planetesimals' eccentricity may not be too high implying that they may be able to return to low eccentricity orbits.

certain circumstances, not only do planetesimals survive giant planet migration, they can also form Earth-like objects, with substantial amounts of water, in the habitable zone of their central stars. By integrating the orbits of 1200 planetesimals and 80 Moon- to Mars-sized objects distributed in a protoplanetary disk with a mass of 17 Earth-masses, these authors have shown that an Earth-like planet can form in the habitable zone of a star after 200 Myr, when a Jupiter-size object migrates from 5 AU to 0.25 AU in 10^5 years (figure 3). The initial distribution of water in the disk of planetesimals is similar to those of our solar system's primitive asteroids, with a 0.5 water-to-mass ratio for protoplanetary objects beyond the location of the giant planet.

3. Habitable Planets in Binary Star Systems

Given that a large fraction of main and pre-main sequence stars are formed in binaries or clusters (Abt 1979; Duquennoy & Mayor 1991; Mathieu et al. 2000), it is not surprising that approximately 20% of the currently known extrasolar planet-hosting stars are members of binary systems. When the separation of these binaries are larger than 100 AU, the perturbative effect of the gravitational force of one stellar component on the formation and dynamics of planets around the other star is negligible. However, when the separation of a binary is smaller than 100 AU, the perturbative effect of the farther companion becomes more pronounced, and strongly affects the long-term stability and the possibility of the formation of planets in binary systems.

Planet formation in binaries has been the subject of debate for a long time. Artymowicz & Lubow (1994) have shown that a circumstellar disk around the

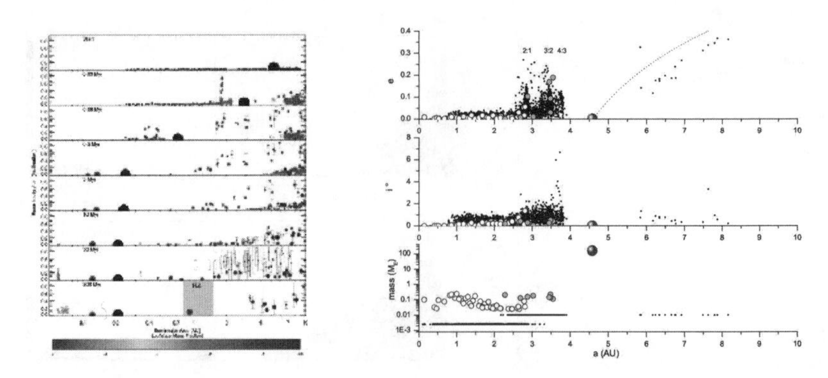

Figure 3. Habitable planet formation during giant planet migration. Graphs from Raymond et al. (2006) left, and Fogg & Nelson (2007) right.

primary of a binary system may lose a large portion of its planet-forming material due to the interaction with an eccentric secondary star. Thébault et al. (2004) have shown that such an interaction may also increase the orbital eccentricities of smaller objects inhibiting giant/terrestrial planet formation by destabilizing the orbits of their building blocks. However, the detection of circumstellar disks in systems such as L1551, in which the binary separation is ~45 AU, and the stars of the system maintain circumstellar materials with masses of approximately 0.05 solar-masses (comparable to the minimum mass of our solar system's nebula) in disks with radii of 10 AU (Rodriguez et al. 1998), and the recent detection of Jovian-type planets in three moderately close (<40AU) binaries of γ Cephei (Hatzes et al. 2003), GL 86 (Els et al. 2001), and HD 41004 (Zucker et al. 2004; Raghavan et al. 2006), with separation smaller than 20 AU, have shown that it is possible for binary systems to maintain enough material in their circumstellar disks to trigger planet formation in the same fashion as around single stars. These *binary-planetary* systems, with their small separations, and with hosting Jupiter-like planets, present another case of extreme planetary systems. Whether such systems can harbor habitable planets depends on the degree of the interactions between their secondary stars, their giant planets, and their disks of embryos. As shown by Haghighipour (2006), terrestrial-class objects in binary-planetary systems can maintain long-term stability in orbits close to their host stars and outside the influence zones of their systems' giant planets. That means, in order for habitable planets to have stable orbits in such systems, the habitable zones of their central stars have to be much closer to them than the orbits of their giant planets. In a recent article, we studied this topic by integrating the orbits of 120 Moon- to Mars-sized planetary embryos, with water contents similar to those of our solar system's primitive asteroids, in a system consisting of the Sun, Jupiter, and a farther stellar companion (Haghighipour & Raymond 2007). Figure 4 shows the results for different values of the mass of the secondary star and its orbital parameters. As shown on the right graph, it is possible to form terrestrial-class objects, with substantial amounts of water, in the habitable zone of the primary star. The left graph of figure 4 shows a case in which a 1.17 Earth-masses object is formed at 1.16 AU from the primary star, with an orbital eccentricity of 0.02, and water to mass ratio of 0.00164 (Earth's water to mass ratio is approximately 0.001).

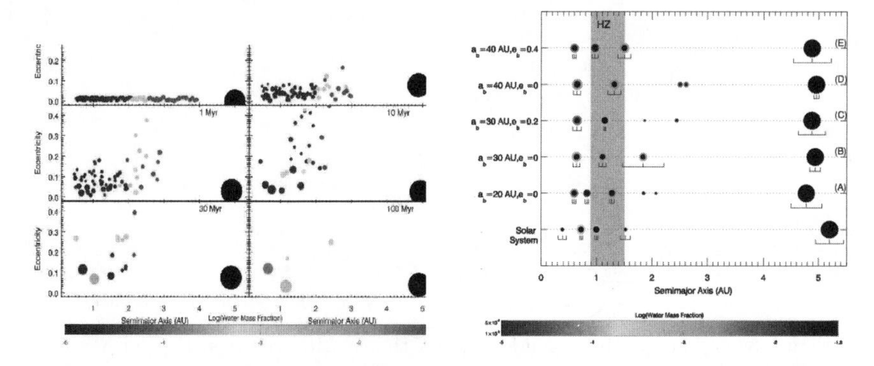

Figure 4. Habitable planet formation in a binary star system with Sun-like stars. Graph from Haghighipour & Raymond (2007).

The results of our simulations also indicate a relation between the perihelion of the binary (q_b) and the semimajor axis of the outermost terrestrial planet (a_{out}). As shown in the left graph of figure 5, similar to Quintana et al. (2007), simulations with no giant planets favor regions interior to $0.19q_b$ for the formation of terrestrial objects. That means, around a Sun-like star, where the inner edge of the habitable zone is at ~ 0.9 AU, a stellar companion with a perihelion distance smaller than $0.9/0.19 = 4.7$ AU would not allow habitable planet formation. In simulations with giant planets, on the other hand, figure 5 shows that terrestrial planets form closer-in. The ratio a_{out}/q_b in these systems is between 0.06 and 0.13. A detailed analysis of our simulations also indicate that the systems, in which habitable planets were formed, have large perihelia. The right graph of figure 5 shows this for simulations in a binary with equal-mass Sun-like stars. The circles in this figure represent systems with habitable planets.The numbers on the top of the circles show the mean eccentricity of the giant planet. For comparison, systems with unstable giant planets have also been marked. Since at the beginning of each simulation, the orbit of the giant planet was considered to be circular, a non-zero eccentricity is indicative of the interaction of this body with the secondary star. As shown here, Earth-like objects are formed in systems where the interaction between the giant planet and the secondary star is weak and the average eccentricity of the giant planet is small. That implies, habitable planet formation is more favorable in binaries with moderate to large perihelia, and with giant planets on low eccentricity orbits.

Acknowledgments. Support for N.H. through the NASA Astrobiology Institute under Cooperative Agreement NNA04CC08A with the Institute for Astronomy at the University of Hawaii-Manoa is acknowledged.

References

Abt, H. A., 1979, The Frequencies of Binaries on the Main Sequence. AJ, 84, 1591

Armitage, P. J., 2003, A Reduced Efficiency of Terrestrial Planet Formation Following Giant Planet Migration. ApJ, 582, L47

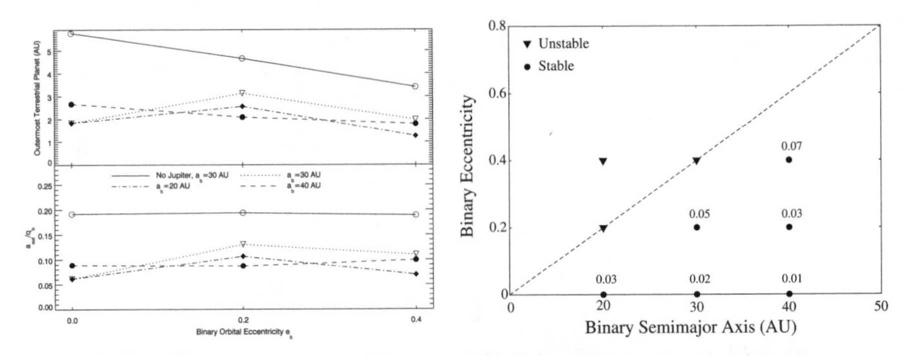

Figure 5. The graph on the left shows the relation between the perihelion of an equal-mass binary and the location of its outermost terrestrial planet. The graph on the right shows the region of the (e_b, a_b) space for a habitable binary-planetary system. Figures from Haghighipour & Raymond (2007).

Artymowicz, P., & Lubow, S. H., 1994, Dynamics of Binary-Disk Interaction. 1: Resonances and Disk Gap Sizes. ApJ, 421, 651

Duquennoy, A., & Mayor, M., 1991, Multiplicity Among Solar-Type Stars in the Solar Neighborhood. II - Distribution of the Orbital Elements in an Unbiased Sample. A&A, 248, 485

Els et al., 2001, A Second Substellar Companion to Gliese 86 System. A Brown Dwarf in an Extrasolar Planetary Systems, A&A, 370, L1

Fogg, M. J., & Nelson, R. P., 2005, Oligarchic and Giant Impact Growth of Terrestrial Planets in the Presence of Gas Giant Planet Migration. A&A, 441, 791

Fogg, M. J., & Nelson, R. P., 2007,The effect of Type I migration on the Formation of Terrestrial Planets in Hot-Jupiter Systems. A&A, 472, 1003

Haghighipour, N., 2006, Dynamical Stability and Habitability of the γ Cephei Binary-Planetary System. ApJ, 644, 543

Haghighipour, N., & Raymond, S. N., 2007, Habitable Planet Formation in Binary-Planetary Systems. ApJ, 666, 436

Hatzes et al., 2003, A Planetary Companion to γ Cephei A. ApJ, 599, 1383

Lufkin, G., Richardson, D. C., & Mundy, L. G., 2006, Planetesimals in the Presence of Giant Planet Migration. ApJ, 653, 1464

Mandell, A. M., Raymond, S. N., & Sigurdsson, S., 2007, Formation of Earth-like Planets During and After Giant Planet Migration. ApJ, 660, 823

Mathieu et al., 2000, Young Binary Stars and Associated Disks. In: Protostars and Planets IV, ed by V. Mannings, A. P. Boss & S. S. Russell, Univ. Arizona Press, Tucson, pp 703

Quintana et al., 2007, Terrestrial Planet Formation Around Individual Stars Within Binary Star Systems. ApJ, 660, 807

Raghavan et al., 2006, Two Suns in The Sky: Stellar Multiplicity in Exoplanet Systems. ApJ, 646, 523

Raymond, S. N., Mandell, A. M, & Sigurdsson, S., 2006, Exotic Earths: Forming Habitable Worlds with Giant Planet Migration, Science, 313, 1413

Rodriguez et al., 1998, Compact Protoplanetary Disks Around the Stars of a Young Binary System. Nature, 395, 355

Thébault et al., 2004. Planetary Formation in the γ Cephei System. A&A, 427, 1097

Zucker et al., 2004, Multi-order TODCOR: Application to observations taken with the CORALIE echelle spectrograph. II. A planet in the system HD 41004. A&A, 426, 695

Edwin Turner

Siegfried Franck

Bioastronomy 2007: Molecules, Microbes, and Extraterrestrial Life
ASP Conference Series, Vol. 420, 2009
K. J. Meech, J. V. Keane, M. J. Mumma, J. L. Siefert, and D. J. Werthimer, eds.

Dynamic Habitability of Extrasolar Planetary Systems

S. Franck, W. von Bloh, and C. Bounama

Potsdam Institute for Climate Impact Research (PIK), P.O. Box 601203, 14412 Potsdam, Germany

Abstract. We estimate the likelihood to find habitable Earth-like planets on stable orbits for 86 selected extrasolar planetary systems, where luminosity, effective temperature and stellar age are known. For determining the habitable zone (HZ) an integrated system approach is used taking into account a variety of climatological, biogeochemical, and geodynamical processes. Habitability is linked to the photosynthetic activity on the planetary surface. We find that habitability strongly depends on the age of the stellar system and the portion of land / ocean coverages. The conditions for orbital stability are estimated with the help of a method based on the Hill radius. Almost 60 % of the investigated systems could harbour habitable Earth-like planets on stable orbits. In 18 extrasolar systems we find even better prerequisites for dynamic habitability than in our own solar system.

1. Introduction

The existence of Earth-type planets around stars other than the Sun is strongly implied by various observational findings including (1) the steep rise of the mass distribution of planets with decreasing mass, which implies that more small planets form than giant ones; (2) the detection of protoplanetary disks (with masses between ten and one hundred times that of Jupiter) around many solar-type stars younger than ~3 Myr; and (3) the discovery of "debris disks" around middle-aged stars, the presumed analogs of the Kuiper Belt and zodiacal dust (Marcy *et al.* 2005; Santos *et al.* 2005, and reference therein). Lineweaver & Grethel (2003) conclude that 25 %–100 % Sun-like stars harbour planets.

Even if it seems today beyond the technical feasibility to detect Earth-mass planets we can apply computer models to investigate known exoplanetary systems to determine whether they could host Earth-like planets with surface conditions sufficient for the emergence and maintenance of life on a stable orbit. Such a configuration is described as dynamic habitable. Jones *et al.* (2001) have investigated the dynamic habitability of several exoplanetary systems. They used the boundaries of the habitable zone (HZ) originating from Kasting *et al.* (1993). To test the intersection of stable orbits and the HZ, putative Earth-mass planets were launched into various orbits in the HZ and a symplectic integrator was used to calculate the celestial evolution of the extrasolar planetary system.

In this paper, we adopt a somewhat different definition of the HZ already used by Franck *et al.* (1999, 2000a,b). Here habitability (i.e., presence of liquid water at all times) does not just depend on the parameters of the central star, but also on the properties of the planet itself. In particular, habitability is linked to the photosynthetic activity of the planet, which in turn depends on the planetary atmospheric CO_2 concentration, and is thus strongly influenced by the planetary geodynamics. This leads to additional spatial and temporal limitations of habitability, as the stellar HZ (defined for a specific type of planet) becomes narrower with time due to the persistent decrease of the planetary atmospheric CO_2 concentration.

The stability of orbits of hypothetical Earth-like planets is calculated by a method of Jones *et al.* (2005). They evaluated the dynamic habitability using nR_H derived from giant's orbital eccentricity without carrying out time-consuming orbital integrations, where R_H is the Hill radius of the giant planet and n is a multiplier that depends on the giant's orbital eccentricity. In the present paper we calculate the dynamic habitability of 86 extrasolar planetary systems in dependence of the relative continental area of a putative Earth-like planet.

2. Model Description

In our calculation of the HZ we are following an integrated system approach. On Earth, the carbonate-silicate cycle is the crucial element for a long-term homeostasis under increasing solar luminosity. In most studies (see, e.g., Caldeira & Kasting 1992), the cycling of carbon is related to the present tectonic activities and to the present continental area as a snapshot of the Earth's evolution. On the other hand, on geological time-scales the deeper parts of the Earth are considerable sinks and sources for carbon. In addition, the tectonic activity and the continental area change noticeably. Therefore, we favour the so-called geodynamical models that take into account both the growth of continental area and the decline in the spreading rate (Franck *et al.* 2000a). Our numerical model couples the stellar luminosity, the silicate-rock weathering rate, and the global energy balance to allow estimates of the partial pressure of atmospheric and soil carbon dioxide, the mean global surface temperature, and the biological productivity, as a function of time (Fig. 1). The main feedback loop stabilising the planetary climate is given by the silicate rock weathering: An increase of the luminosity leads to a higher mean global temperature causing an increase in weathering. Then more CO_2 is extracted from the atmosphere weakening the greenhouse effect. Overall the temperature is lowered and homeostasis is achieved.

The HZ around an extrasolar planetary system is defined as the spatial domain of distances R where the planetary surface temperature stays between 0°C and 100°C and where the atmospheric CO_2 partial pressure is higher than 10^{-5} bar to allow photosynthesis. This is equivalent to a non-vanishing biological productivity, $\Pi > 0$, i.e.,

$$HZ := \{R \mid \Pi(P_{\text{atm}}(R,t), T_{\text{surf}}(R,t)) > 0\}. \tag{1}$$

According to the definition in Eq. 1 the boundaries of the HZ are determined by the surface temperature extrema, $T_{\text{surf}} = 0°\text{C} \vee T_{\text{surf}} = 100°\text{C}$, or by the minimum CO_2 partial pressure, $P_{\text{atm}} = 10^{-5}$ bar. Therefore, the specific parameterisation of the biological productivity plays a minor role in the calculation of the HZ. In our approach habitability is linked to the photosynthetic activity of the planet. In particular photosynthesis is most relevant for the direct detection of life on extrasolar terrestrial planets. The TPF and Darwin space missions

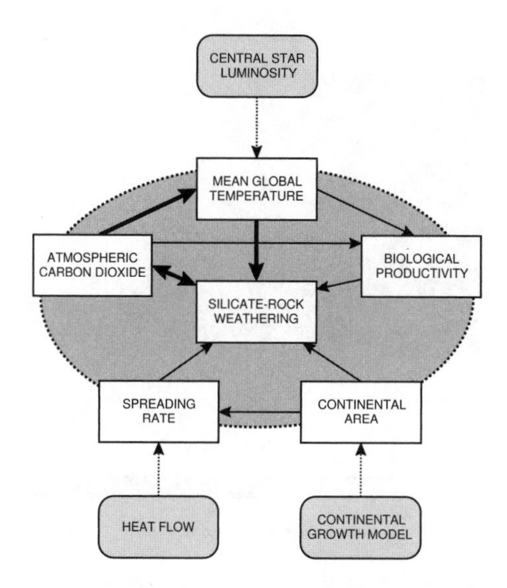

Figure 1. Box model of the integrated system approach (Franck *et al.* 2000a). The arrows indicate the different forcings (dotted lines) and feedback mechanisms (solid lines). The main feedback loop stabilising the climate is marked with bold arrows.

of NASA and ESA are planning to detect O_2 or its photolytic product O_3 as a biosignature of life (Des Marais *et al.* 2003) built up by oxygenic photosynthesis. Furthermore, habitability is strongly affected by the planetary geodynamics. In principle, this leads to additional spatial and temporal limitations of habitability.

In order to calculate the orbital stability we apply an approximation derived by Jones *et al.* (2005): if the Earth-like planet approaches three Hill radii of the giant planet severe orbital perturbations of the terrestrial planet occur. The Hill radius R_H is defined as

$$R_H = \left(\frac{m}{3M}\right)^{1/3} a, \tag{2}$$

where m is the mass of the giant planet, M the central star mass and a the semi-major axis. Then the inner and outer boundaries for unstable orbits around a giant planet are defined by

$$R_{\rm int} = a(1-e) - n_{\rm int}(e) R_H, \tag{3}$$
$$R_{\rm ext} = a(1+e) + n_{\rm ext}(e) R_H, \tag{4}$$

where e is the eccentricity of the giant planet. The values of the functions $n_{\rm int}(e)$ and $n_{\rm ext}(e)$ are taken from Table 4 in Jones *et al.* (2005). They are in the range of $[2\ldots3]$ for $n_{\rm int}(e)$ and $[3\ldots16]$ for $n_{\rm ext}(e)$. In our study we use results of Espresate (2005) for the stellar parameters of 133 extrasolar planetary systems. Missing values for stellar ages were taken from Jones *et al.* (2005). For 86 of these 133 systems the stellar luminosities, effective temperatures and ages are given. It must be pointed out that the determination of the age of low mass stars is error-prone and has to be taken with care. Together with the corresponding orbital

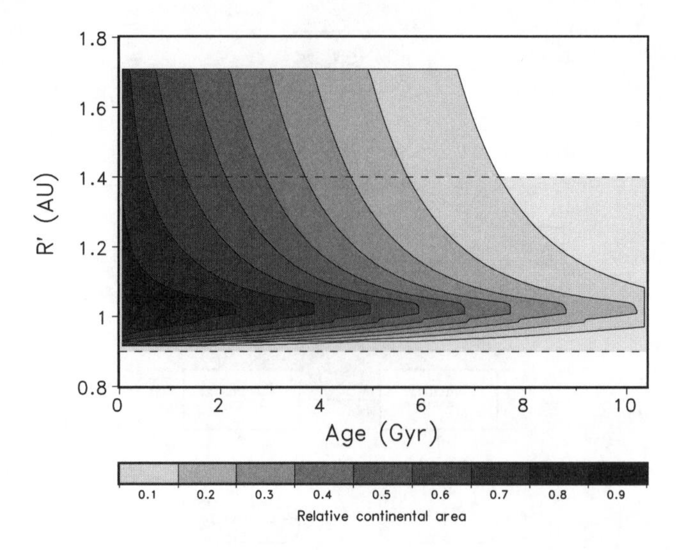

Figure 2. The habitable zone around a central star for a luminosity of $L = 1\mathrm{L}_{\odot}$ and fixed effective temperature of the central star ($T_{\mathrm{eff}} = 5700$ K) as a function of the age of the Earth-like planet. The colour shaded areas indicate the extent of the HZ for different relative continental areas. The horizontal dashed lines indicate the HZ defined by climatic constraints only (Kasting *et al.* 1993).

parameters and masses of the giant planets (data taken from Jean Schneider's extrasolar planets encyclopaedia http://www.obspm.fr/encycl/encycl.html) the habitable zones and the orbital stability domains can be calculated.

3. Results and Discussion

The HZ for a fixed central star luminosity of $L \equiv 1\,\mathrm{L}_{\odot}$ and fixed effective temperature of $T_{\mathrm{eff}} = 5700$ K is shown in Fig. 2 as a function of the age of the planetary system. For the investigation of an Earth-like planet under the external forcing we adopt a model planet with a prescribed continental area. The fraction of continental area to the total planetary surface is varied between 0.1 and 0.9. According to Eq. 1 the inner and outer limits of the HZ are given by vanishing biological productivity. For a central star with different luminosity and effective temperature the limits of the HZ have to be rescaled (von Bloh *et al.* 2007).

The likelihood that an Earth-like planet is both on a stable orbit and also within the HZ can be quantitatively estimated from the width of the HZ excluding the interval of orbital instability:

$$\Delta R = \max(HZ \setminus \{R_{\mathrm{int},i}, R_{\mathrm{ext},i}\}_{i=1,n_p}) - \min(HZ \setminus \{R_{\mathrm{int},i}, R_{\mathrm{ext},i}\}_{i=1,n_p}), \quad (5)$$

where n_p is the number of detected planets in the extrasolar planetary system and $R_{\mathrm{int},i}$ and $R_{\mathrm{ext},i}$ are the inner and outer limits for unstable orbits, respectively. The widths ΔR of the dynamic habitable zones are shown in Fig. 3

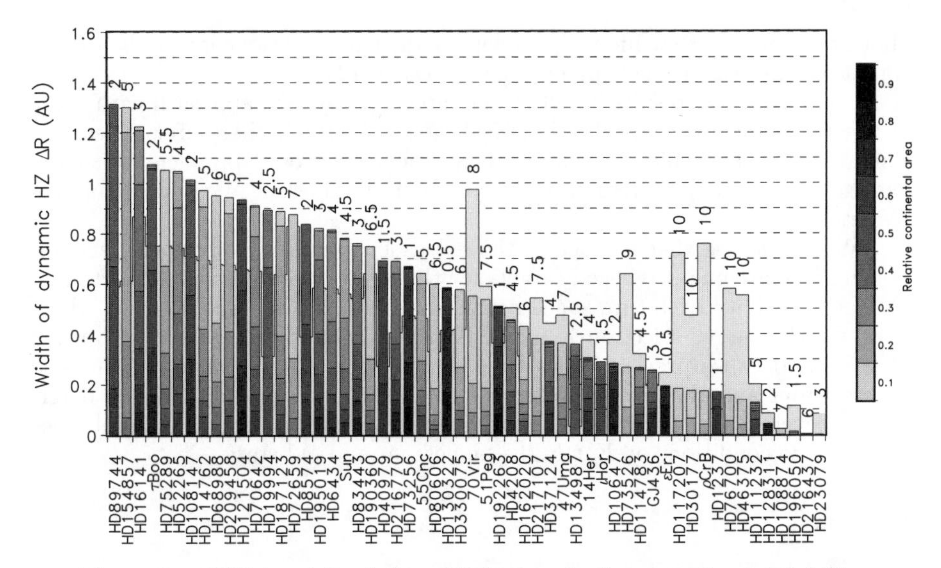

Figure 3. Widths, ΔR, of the orbital range both warranting habitability and orbital stability for the solar system and extrasolar planetary systems for different continental areas. The gray shaded area denotes the results for the HZ defined by climatic constraints only (Kasting *et al.* 1993). The numbers at the horizontal bars are the corresponding ages of the planetary systems.

ordered by the width of the HZ for the 55 extrasolar planetary systems with $\Delta R > 0$ for different relative continental areas. For comparison the width of the dynamic HZ for the solar system is additionally plotted. We find that 18 systems have a larger width of the HZ than the solar system. These systems are characterised by a more luminous star hosting a hot Jupiter not perturbing the orbit of an Earth-like planet. On the other hand rather old systems like, e.g., 70 Vir, HD117207 and ρ CrB have much smaller HZs than the solar system. The habitable zone differs significantly from the HZ defined by climatic constraints only (Kasting *et al.* 1993). In between are systems with a giant planet outside the HZ partially destroying stability inside the HZ. Two examples are the systems 47 UMa and ε Eri.

4. Conclusions

We studied the principle possibility of orbitally stable Earth-like habitable planets in 86 extrasolar systems. In particular, we considered Earth-like planets with different ratios of land / ocean coverages and applied an integrated system approach. We found that in 54 out of 86 systems (63 %) potential dynamic habitable Earth-like planets can exist. According to our results the solar system is a relatively ordinary planetary system as shown in Fig. 3. There are at least 18 extrasolar planetary systems with better prerequisites to harbour dynamic habitable Earth-like planets. This supports the so-called "Principle of Mediocrity." This principle proposes that our planetary system and life on Earth are about average and that life will develop by the same rules wherever the proper conditions and the needed time are given (Darling 2001).

Comparing our results based on the integrated system approach with those of the HZ determination following Kasting *et al.* (1993) there are, at least in some cases, remarkable differences. In particular, in our approach rather old extrasolar planetary systems are less probable for hosting dynamic habitable planets.

We emphasise that the age of the extrasolar planetary system is very important in searching privileged targets for the remote spectroscopic detection of biological activity by future space borne missions like TPF and Darwin. However, for the target selection it might be useful to apply a broader definition of the HZ because precise values for the inner and outer boundaries are still controversial and the correct values of the stellar ages are difficult to determine. Nevertheless, our results can give an explanation for a possible absence of habitable conditions on Earth-like planets around old stars.

References

Caldeira, K. & Kasting, J. F. 1992, Nature 360, 721

Darling, D. J. 2001, Life everywhere: the maverick science of astrobiology (New York, Basic Books)

Des Marais, D. J., Harwitt, M. O., Jucks, K. W., Kasting, J. F., Lin., D. N. C., Lunine, J. I., Schneider, J., Seager, S., Traub, W. A. & Woolf, N. J. 2003, Astrobiology 2, 153

Espresate, J. 2005, Catalog of 156 confirmed extrasolar planets and their 133 parent stars. Astro-ph/0508317 v1 15 Aug 2005

Franck, S., Kossacki, K. & Bounama, C. 1999, Chem. Geol. 159, 305

Franck, S., Block, A., von Bloh, W., Bounama, C., Schellnhuber, H.-J. & Svirezhev, Y. 2000a, Tellus 52B,94

Franck, S., von Bloh, W., Bounama, C., Steffen, M., Schönberner, D. & Schellnhuber, H.-J. 2000b, JGR 105 (E1), 1651

Jones, B. W., Sleep, P. N. & Chambers, J. E. 2001, A & A 366,254

Jones, B. W., Underwood, D. R. & Sleep, P. N. 2005, ApJ 622, 1091

Kasting, J. F., Whitmire, D. P. & Reynolds, R. T. 1993, Icarus 101, 108

Lineweaver, C. H. & Grether, D. 2003, ApJ 598, 1350

Marcy, G. W., Butler, R. P., Fischer, D., Vogt, S., Wright, J. T., Tinney, C. G. & Jones, H. R. A. 2005, Progress Theoret. Phys. Suppl. 158, 24

Santos, N. C., Benz, W. & Mayor, M. 2005, Science 310, 251

von Bloh, W., Bounama, C. & Franck, S. 2007, PSS 55, 651

CAMBRIDGE
JOURNALS
International Journal
of Astrobiology
For the study of the origin, evolution and
distribution of life in the universe

Andy Flower

Gustavo Porto de Mello

Bioastronomy 2007: Molecules, Microbes, and Extraterrestrial Life
ASP Conference Series, Vol. 420, 2009
K. J. Meech, J. V. Keane, M. J. Mumma, J. L. Siefert, and D. J. Werthimer, eds.

Astrobiologically Interesting Stars Near the Sun: Galactic Orbits and Mass Extinctions

G. F. Porto de Mello,[1] J. .R Lépine,[2] and W. da Silva Dias[3]

[1] *Universidade Federal do Rio de Janeiro/Obs. do Valongo, Ladeira do Pedro Antonio 43, 20080-090, Rio de Janeiro, Brazil, e-mail: gustavo@ov.ufrj.br*

[2] *Instituto Astronômico, Geofísico e de Ciências Atmosféricas, USP, Rua do Matão, 1226, Cidade Universitária, 05508-090, São Paulo, Brazil, e-mail: jacques@astro.iag.usp.br*

[3] *Instituto de Ciências Exatas, Departamento de Física/Química, Av. BPS 1303, 37500-903, UNIFEI, Itajubá, Brazil, e-mail: wilton@unifei.edu.br*

Abstract. Life based on carbon chemistry, water oceans and planetary surfaces relies upon stellar properties, such as mass, age and chemical composition, which can be well constrained for nearby stars. Our goal is to produce a catalogue of Sun-like stars, nearer than 15 parsecs, which may have harbored Earth-like planets within their continuously habitable zone (CHZ) for the last ∼3 Gyr (giga-years). We determined their masses, ages and galactic orbits. These objects should be prime targets for future space-based missions aimed at detecting, interferometrically, spectral infrared biomarkers and thereby deducing the presence of oxygenated atmospheres and photosynthesizing life in extra-solar Earth-like planets. We show that ∼7 % of the nearby stars have optimum astrobiological interest, and that only ∼2 % have characteristics truly approaching the Sun's. The solar galactic orbit is shown to submit Earth's biosphere to long passages through the galactic spiral arms. If these passages offer a biological hazard, most stars of the neighborhood offer more benign environments than the Sun.

1. Introduction

Here we extend the study of Porto de Mello et al. (2006), presenting preliminary results of a thorough compilation of the temperatures, luminosities, chemical composition, multiplicity and degree of chromospheric activity (an age indicator) for all Sun-like stars within 15 parsecs of the Sun. Their states of evolution, masses, ages and galactic orbits were determined: details of the selection process are fully discussed in Porto de Mello et al. (2006). In our selection process, we eliminate stars with masses outside the interval from 0.7 to 1.2 solar masses (taking into account the stellar metallicity): in this interval, the optimum tradeoff between stellar lifetime and the position of the CHZ (Franck et al. 2000) is found. Stars with too little mass (and luminosity) present less favorable conditions owing to their initial prolonged phase of very high chromospheric activity, with probable high impact on young planetary atmospheres due to the very close position of their CHZ. Tidally locked rotational periods of possible habitable planets are also a problem for these stars, again due to very close-in CHZs. More massive stars evolve too rapidly and do not allow enough time for the atmospheric oxygenation of habitable planets. We also eliminate multiple stars (for which a full dynamical analysis of the planetary stability within the CHZ will be presented elsewhere), excessively young stars (for which no atmospheric oxygenation is yet expected), stars with short-period giant planetary companions (potentially disruptive to the orbital stability of Earth-like planets inside the CHZ), and stars with low metallicities (for which the formation of Earth-like planets inside the CHZ is presumed to be much less likely). Among the total population of 487 stars within 15 parsecs of the Sun, 96 are Sun-like, but only 34 remain after all the criteria above are applied.

2. Discussion and Results

The focus of the present paper is on the analysis of the stellar galactic orbits, discussed by Porto de Mello et al. (2006) only superficially. The stellar orbits in the disk of the Galaxy, and the passages through the galactic spiral arms, is a rarely mentioned factor of stability which is drawing more interest as one regards how important it may be for long-term planetary climate stability, with a possible bearing on mass extinctions. The Sun lies very near the co-rotation radius (Lépine et al. 2001), where stars revolve around the Galaxy in the same period as the gravitational density wave perturbations sweep across the galactic disk, creating the spiral arms. In such a situation, the passages across the spiral arms, and potential encounters with star-forming regions (where supernova explosions generally occur) and giant molecular clouds, are supposedly minimized, unless other parameters come at play. Controversy still surrounds the proposition that time spent inside or around spiral arms might be dangerous to biospheres and conducive to mass extinctions. Clube & Napier (1982) suggested that molecular clouds might disturb the solar Oort comet cloud provoking heavy bombardment episodes and mass extinctions. Leitch & Vasisht (1998) also proposed a correlation between the times of great mass extinctions in the Earth with the passages of the Sun across the galactic spiral arms, in the last ~1 Gyr. Gies & Helsel (2005) and de la Fuente Marcos & de la Fuente Marcos (2004) suggested that a higher exposure to cosmic rays near spiral arms may trigger increased cloudiness in Earth's atmosphere and ice ages. Also, Gehrels et al. (2003) pointed out the threat to Earth's ozone layer posed by supernova explosions, whenever the Sun approaches spiral arms and star forming regions. Here we present the results of detailed calculations of the history of spiral arm passages in the last 500 million years, for the 13 best "biostar" candidates, those nearer than 10 parsecs and highlighted by Porto de Mello et al. (2006). The description of the galactic potential and positioning of the spiral arms is that of Dias & Lépine (2005). Each star's trajectory was integrated 3-dimensionally in a coordinate frame centered at the co-rotation radius, co-moving with the spiral wave pattern. We calculated the total time spent inside the spiral arms in the last 500 million years (Myr), when the spiral arm position could be traced with very good accuracy. The Sun's orbit is noteworthy for its exceptionally low eccentricity and proximity to the co-rotation radius, but it also has a low vertical velocity component (towards

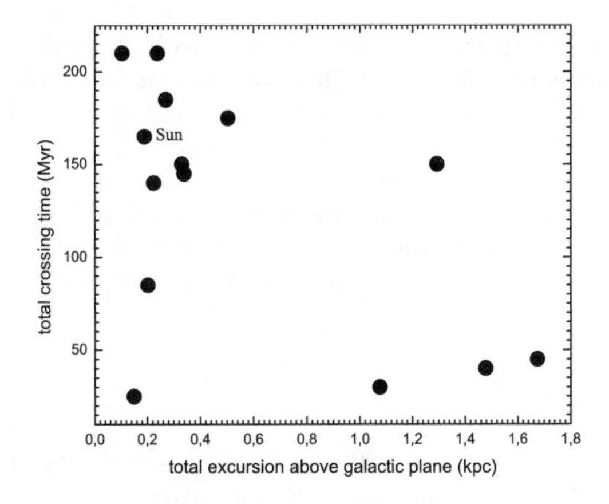

Figure 1. Total time spent by each star crossing spiral arms, in the last 500 million years, against the total excursion above the galactic plane, in kiloparsecs. A clear trend for stars which travel far above the galactic plane to spend a small fraction of time inside the spiral arms is seen.

the galactic poles). Thus the Sun rarely approaches the spiral arms, but the time spent crossing is long (Fig. 1). Other stars can have much more eccentric orbits but also large vertical components. These parameters are known to be linked to the stellar age: older stars have had more time to suffer gravitational perturbations, and their cumulative effect statistically increases both the orbital eccentricity and vertical velocity components. Thus, older stars generally spend long times below and above the galactic disk even when crossing the arms, since the galactic disk is only ~0.1 kiloparsec thick. Stars with very high eccentricities have spent as little time as 40 Myr years inside the arms, compared to 170 Myr for the Sun, in the last 500 Myr.

3. Conclusions

We found that, among the nearby stars, ~7% are optimally interesting targets for astrobiology, and only ~2% have properties truly close to those of the Sun, confirming, for a larger sample, the results of Porto de Mello et al. (2006). We also found, for a limited sample, that there is a large diversity of stellar orbits in the solar neighborhood, and the time fraction spent inside spiral arms can vary from 6% to 40%. The Sun, despite its proximity to the galactic co-rotation radius, has a low vertical velocity component and spends 30% of its lifetime crossing the spiral arms. There is a natural trend to consider that the Sun is a safe place for a biosphere, since mass extinctions (not necessarily linked to the spiral arms) have not been able to preclude the evolution of complex life. If, as many propose, the stellar galactic orbit is an important factor of long-term stability, we must conclude that most stars offer even safer abodes for life than the Sun. Within 10 parsecs, 60% of the astrobiologically interesting stars spend less time crossing spiral arms than the Sun, a fact mostly owed to high velocity components away from the galactic plane and more eccentric orbits. Our preliminary result suggests revision of the notion that proximity to the spiral arm co-rotation radius necessarily offers a safe place to develop a biosphere over long time scales. If the Sun, under a "pessimistic" approach (Ward & Brownlee 2000), is regarded as a safer than average place for the development of a biosphere, then we must conclude that spiral arm crossings pose no great danger to life.

4. Jargon

Parsec: 3.2616 light-years, or $\sim$3.09 x 10^{13} km; **Metallicity**: the stellar abundance of elements heavier than helium; **Supernova**: the extremely violent explosions that mark the deaths of very massive stars; **Oort cloud**: a spherical cloud of comets lying out to about a light-year from the Sun; **Molecular clouds**: interstellar regions where dust and gas densities are orders of magnitude greater than in the intervening medium; **Chromospheric activity**: stellar radiative flux, mainly in the ultraviolet and X-ray, induced by surface magnetic activity tied to rotation.

Acknowledgments. We acknowledge financial support from CNPq/Brazil (grant 476909/2006-6), FAPERJ (grant APQ1/26/170.687/2004) and partial travel support from the organizers.

References

Clube, S. V. M., Napier, W. M. 1982, Spiral arms, comets and terrestrial catastrophism. Q. J. R. Astron. Soc., 23, 45

de la Fuente Marcos, R., de la Fuente Marcos, C. 2004, On the correlation between the recent star formation rate in the Solar Neighbourhood and the glaciation period record on Earth. New Astronomy, 10, 53

Dias, W, S., Lépine, J. R. D. 2005, Direct Determination of the Spiral Pattern Rotation Speed of the Galaxy. ApJ, 629, 825

Franck, S., Block, A., von Bloh, W., Bounama, C., Schellnhuber, H. J., Svirezhev Y. 2000, Habitable zone for Earth-like planets in the solar system. Planetary and Space Science, 48, 1099

Gehrels, N., Laird, C. M., Jackman, C. H., Cannizzo, J. K., Mattson, B. J. 2003, Ozone Depletion from Nearby Supernovae. ApJ, 585, 1169

Gies, D R., Helsel, J. W. 2005, Ice Age Epochs and the Sun's Path through the Galaxy. ApJ, 626, 844

Leitch, E. M., Vasisht, G. 1998, Mass extincitions and the Suns encounters with spiral arms. New Astron., 3, 51

Lépine, J. R. D., Mishurov, Y. N., Dedikov, S. Y. 2001, A new model for the spiral structure of the Galaxy: superposition of 2- and 4-armed patterns. ApJ, 546, 234

Porto de Mello, G. F., del Peloso, E. F., Ghezzi, L. 2006, Astrobiologically Interesting Stars within 10 parsecs of the Sun. Astrobiology, 6, 308

Ward, P.D., Brownlee, D. 2000, Rare Earth. Copernicus Edition, Springer-Verlag, New York

Bioastronomy 2007: Molecules, Microbes, and Extraterrestrial Life
ASP Conference Series, Vol. 420, 2009
K. J. Meech, J. V. Keane, M. J. Mumma, J. L. Siefert, and D. J. Werthimer, eds.

Dynamical Stability of Habitable Planets in Astrobiologically Interesting Binary Stars

T. Michtchenko[1] and G. F. Porto de Mello[2]

[1]*Instituto Astronômico, Geofísico e de Ciências Atmosféricas, USP, Rua do Matão, 1226, Cidade Universitária, 05508-090, São Paulo, Brazil, e-mail: tatiana@astro.iag.usp.br*

[2]*Universidade Federal do Rio de Janeiro/Obs. do Valongo, Ladeira do Pedro Antonio 43, 20080-090, Rio de Janeiro, Brazil, e-mail: gustavo@ov.ufrj.br*

Abstract. Binary stars are universally thought as second rate sites for the location of habitable planets. It is still open to debate, in current planet-forming theory, whether planetary formation similar to that of single stars can proceed in multiple systems, and whether, once formed, these planets have a substantial probability of remaining in stable orbits inside the stellar continuously habitable zones (CHZs) for lengths of time compatible with the evolution of life. Here we consider binary stars having masses, ages and luminosities compatible with the long term permanence of habitable telluric planets inside their CHZ. We investigate the dynamical stability of the system composed by the binary and an Earth-like planet within the CHZ of the primary and secondary component: α Centauri was chosen as the main study case. We obtain the boundaries of the system's secular stability as a function of mass, semi-major axis and inclination. We conclude that stable planetary orbits do exist in the CHZs of these stars, but for those systems with periods similar to or less than $\sim$100 years, short-period perturbations lead to oscillations in the planetary semi-major axis up to $\sim$100 times larger than those suffered by Earth's orbit. This might be of consequence for the long term planetary climate stability and the periodical onset of ice ages.

1. Introduction

Porto de Mello et al. (2006) analyzed temperatures, metallicities, masses, luminosities and ages of the solar-type stars within 10 parsecs of the Sun, selecting non-binary stars with masses and ages in an optimal interval for the maintenance of telluric planets inside the continuously habitable zone (Franck et al. 2000) for the $\sim$3 Gyr (gigayears) necessary for a high atmospheric oxygenation. These stars are proposed as optimal targets for the future spaceborne interferometric probes seeking to identify directly extrasolar Earth-sized planets and measure the infrared spectroscopic biomarkers (des Marais 2002), thereby revealing the presence of photosynthetic life. Some of the stars considered by Porto de Mello et al. (2006) were rejected solely by being members of binary or multiple star systems. Controversy still surrounds the possibility of planetary habitability in multiple systems. Planetary stability around binary stars is possible in definite configurations, around the system's center of mass and around each of the components, given that the planets orbital radius is a factor of 5 or more larger than the binary separation (external case, planet around center of mass), or vice-versa (internal case, planet orbiting one of the components) (Quintana et al. 2002; Holman & Wiegert 1999; Pendleton & Black 1993). However, it is a well-founded expectation that the evolution of protoplanetary disks and coalescing planetesimal is probably different in a binary system as compared to an isolated star. We further note that the very few planetary systems loosely similar to our own, among the ~250 discovered so far, with giant planets orbiting at > 3 AU, in approximately circular orbits, were all found around single stars. Planetary systems in multiple stars are prone to dynamical instabilities which are absent from single stars. In the case of binaries with eccentric orbits, excitation of resonances may lead to strong instabilities and highly eccentric planetary orbits, endangering planetary climate stability. The probability of planetesimal condensation is constrained by the inclination between the accretion disk and the binary orbital plane: for large inclination angles the stability of planetary orbits is strongly reduced (Quintana et al. (2002)).

In this work, we consider the binary stars nearer than 10 pc of the Sun, with astrobiological interest according to Porto de Mello et al. (2006). They are α Centauri, η Cassiopeiae, HD 10360/10361 and HD 156274. These systems are all long period binaries with a Sun-like star as primary. We present the dynamical analysis of the shortest-period binary of this group, and therefore the most critical case: α Centauri. These results should apply to the other systems as well, as discussed below.

2. Dynamical Analysis

We analyze the motion of a hypothetical planet at the outermost edge of the CHZ of each of the component stars is studied, with the other star considered to be a perturber. The stable planetary-type orbits, where the planet orbits the center of mass of the stars instead of one of the stars, all lie well outside the CHZs for this system and need not be considered. In the α Centauri system, respectively, for the primary and secondary stars, the present inner habitability edges are 1.2 AU and 0.7 AU, and the outer edges are 1.7 AU and 1.1 AU. The solutions found for this system can certainly be applied to the other three systems, which have much longer periods and more distant secondaries. Details of the method are found in Michtchenko et al. (2002) and Michtchenko & Malhotra (2004).

In order to obtain the source of possible dynamical instabilities inside the CHZs, we applied both analytical and numerical approaches. The adopted planetary mass in the calculations is one Earth mass. The motion of the planet around one of the stars, with the second star as the perturber, is considered in a Jacobian coordinate system and the invariant plane is taken as the reference plane. Hamiltonian theory then describes the motion of the 3-body system (Brouwer & Clements 1961), as a function of the semi-major axis of the orbits, the position vector relative to the center of mass and the distance between the three bodies. To study the secular behavior of this system, we performed numerically the averaging of the Hamiltonian with respect to the mean anomalies of the planet and the perturber: a full description is given in Michtchenko & Malhotra (2004). The analysis of the topology of the phase space of the Hamiltonian system allows us to estimate the range of the eccentricity/inclination variations without time-expensive numerical integrations of the equations of motion. This analysis has shown that the planetary motion is stable, for eccentricities smaller than ~0.3, for inclinations as large as 35 degrees. For larger values of the inclination, the eccentricity/inclination variations are mutually enhanced and instability may occur.

In the construction of the numerical dynamical maps, a grid of initial conditions was defined on the plane of planetary eccentricity and inclination. The numerical integration time-span of each initial condition was one million years. The orbital paths of the planets obtained through direct numerical integrations were Fourier-transformed, and the information contained in the power spectra of the orbital elements was used in the construction of the dynamical maps (Michtchenko et al. 2002). The analysis of the structure of the dynamical maps on the representative planes revealed important features of the three-dimension secular dynamics in the region of the α Centauri system. The eccentricity-inclination plane is dominated by the regular secular motion of the system, but there are regions where the onset of chaos becomes possible, and planetary orbits would be highly unstable.

The high inclination domains at inclinations larger than 35 degrees show a complex dynamical behavior with several regimes of resonant motion. The dominating behavior is the eccentricity-inclination coupling, or Lidov-Kozai resonance, characterized by the coupled large variation of the eccentricity and inclination of the inner planet, originating very strong instabilities. The stable domains where small amplitude oscillations are located lies in the vicinity of the family of periodic solutions with inclinations less than 30 degrees. Due to the relatively large mutual stellar separation in the α Centauri, the mean-motion resonances of low order are absent in the CHZs of both components of the α Centauri system, and stable planetary orbits are possible in both stars.

It is expected, however, that the planetary climate stability in CHZs depends also on the amplitudes of the short-period perturbations. For instance, the amplitude of these terms in the variation of the Earth's semi-major axis is of the order of $\sim 10^{-4}$ AU. In the case of the α Centauri system, due to the large mass of the perturbers, the magnitude of the short periodic terms is of the order of $\sim 10^{-2}$ AU at the outer edge of the CHZ. This amplitude decreases rapidly with decreasing distance from the central star, that is, in the inner edge of the CHZ, but, even so, remain of the order of ~10 times as large as those undergone by the Earth's orbit. Note that, despite the relatively large amplitudes of these oscillations, the planetary orbits still remain inside the range of the CHZ. For the amplitude of these oscillations to approach values as small as those Earth's orbit, for the case of a planet orbiting the more massive primary star, and therefore subjected to less perturbation than a planet orbiting the secondary star, planetary distances must fall near ~0.8 AU, which lies well inside the inner edge of the CHZ of the primary, and thus habitable planets should not be possible there. In this sense, the α Centauri system does not allow planetary orbits as dynamically stable as the Earth's in what concerns the long term oscillations of semi-major axis and eccentricity. It is thought that these oscillations may be important factors in driving the cycle of global glaciations on Earth, the so-called Kroll-Milankovich cycles. Yet, our present understanding does not allow us to rule out these higher amplitudes for the short-period perturbations in the

α Centauri system's hypothetical terrestrial planets as necessarily critical to the maintenance of planetary climate stability.

Finally, for the other binary stars mentioned above, planetary orbits inside the CHZs of each component are certainly possible, and more stable than those in the α Centauri system.

3. Conclusions

We show that stable planetary orbits in the α Centauri system, as well in the binary stars Eta Cassiopeiae, HD 10360/10361 and HD 156274 are possible for a wide range of values of planetary eccentricities and semi-major axes, for both the primary and secondary components. For the α Centauri system, however, which has the shortest period of the binaries considered here, small amplitude oscillations of the planetary semi-major axis are much stronger than for Earth's orbit, and this fact may have important consequences for the long term climate stability. These factors are thought to drive the Milankovich glaciation cycles undergone by the Earth, and in this sense these stars offer less stable ecospheres than single stars like the Sun.

Acknowledgments. We acknowledge financial support from CNPq/Brazil (grant 476909/2006-6), FAPERJ (grant APQ1/26/170.687/2004), FAPESP and partial travel support from the organizers.

References

Brouwer, D., Clements, G. M. 1961, Methods of Celestial Mechanics. Academic Press Inc., New York

des Marais, D. J., Harwit, M. O., Jucks, K. W., *et al.* 2002, Remote Sensing of Planetary Properties and Biosignatures on Extrasolar Terrestrial Planets. Astrobiology, Volume 2, Issue 2, pp. 153-181

Franck, S., Block, A., von Bloh, W., Bounama, C., Schellnhuber, H. J., Svirezhev Y. 2000, Habitable zone for Earth-like planets in the solar system. Planetary and Space Science, 48, 1099

Holman, M. J., Wiegert, P. A. 1999, Long term stability of planets in binary systems. AJ, 117, 621

Michtchenko, T. A., Lazzaro, D., Ferraz-Mello, S., Roig, F. 2002, Origin of the Basaltic Asteroid 1459 Magnya: A Dynamical and Mineralogical Study of the Outer Main Belt. Icarus, 158, 343

Michtchenko, T. A., Malhotra, R. 2004, Secular dynamics of the three-body problem: application to the Nu Andromedae planetary system. Icarus, 168, 237

Pendleton, Y. J., Black, D. C. 1993, Further studies on criteria for the onset of dynamical instability in general three-body systems. AJ, 88, 1415

Porto de Mello, G. F., del Peloso, E. F., Ghezzi, L. 2006, Astrobiologically Interesting Stars within 10 parsecs of the Sun. Astrobiology, 6, 308

Quintana, E. V., Lissauer, J. J., Chambers, J. E., Duncan, M. J. 2002, Terrestrial planet formation in the Centauri system. ApJ, 576, 982

Bioastronomy 2007: Molecules, Microbes, and Extraterrestrial Life
ASP Conference Series, Vol. 420, 2009
K. J. Meech, J. V. Keane, M. J. Mumma, J. L. Siefert, and D. J. Werthimer, eds.

Orbital Stability of Earth-Type Planets in Binary Systems

J. Eberle, M. Cuntz, and Z. E. Musielak

Department of Physics, University of Texas at Arlington, Arlington, TX 76019-0059

Abstract. About half of all known stellar systems with Sun-like stars consist of two or more stars, significantly affecting the orbital stability of any planet in these systems. Here we study the onset of instability for an Earth-type planet that is part of a binary system. Our investigation makes use of previous analytical work allowing to describe the permissible region of planetary motion. This allows us to establish a criterion for the orbital stability of planets that may be useful in the context of future observational and theoretical studies.

1. Introduction

Observational evidence for the existence of planets in stellar binary (and higher order) systems has been given by Patience et al. (2002), Eggenberger et al. (2004), Eggenberger & Udry (2007), and others. Eggenberger & Udry presented data for more than thirty systems, mostly wide binaries, as well as several triple star systems, with separation distances as close as 20 AU (GJ 86). These observations are consistent with the finding that binary (and higher order) systems occur in high frequency in the local Galactic neighborhood (Duquennoy & Mayor 1991; Lada 2006; Raghavan et al. 2006; Bonavita & Desidera 2007). The fact that planets in binary systems are now considered to be relatively common is also implied by the recent detection of debris disks in various main-sequence stellar binary systems using the *Spitzer Space Telescope* (e.g., Trilling et al. 2007).

In the last few decades, significant progress has been made in the study of stability of planetary orbits in stellar binary systems. Most of these studies focused on S-type systems, where the planet is orbiting one of the stars with the second star to be considered a perturbator. Recently, David et al. (2003) investigated the orbital stability of an Earth-mass planet around a solar-mass star in the presence of a companion star and determined the planet's ejection time for systems with a variety of orbital eccentricities and semimajor axes.

In our previous work (Stuit 1995; Musielak et al. 2005), we studied the stability of both S-type and P-type orbits in stellar binary systems, and deduced orbital stability limits for planets. These limits were found to depend on the mass ratio between the stellar components. This topic was recently revisited by Cuntz et al. (2007) and Eberle et al. (2007), who used the concept of Jacobi's integral and Jacobi's constant (Szebehely 1967; Roy 2005) to deduce stringent criteria for the stability of planetary orbits in binary systems for the special case of the "coplanar circular restricted three-body problem." In this paper, we present case studies of planetary orbital stability for different stellar mass ratios and different initial planetary distances from its host star.

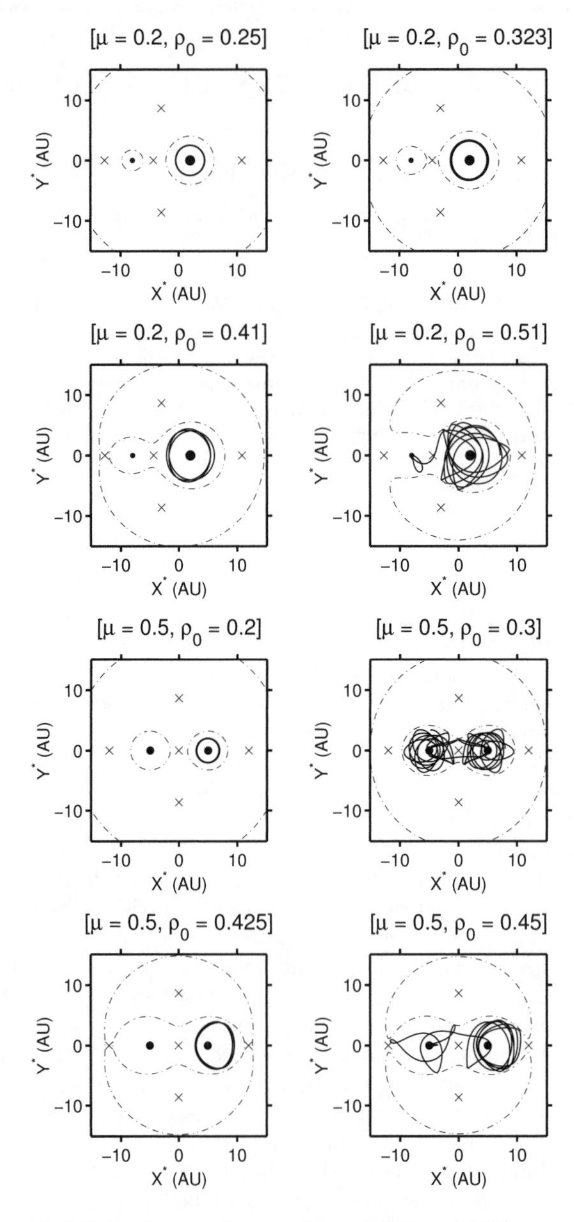

Figure 1. Model simulations for $\mu = 0.2$ and 0.5, and different values of ρ_0. Each panel shows the primary (*large dot*) and secondary (*small dot*) star, the planetary orbit (*solid line*), the five Lagrange points, and the "zero velocity contours" (*dash-dotted lines*). The Lagrange points are denoted as L2, L1, L3, respectively, from left to right along the line connecting the two stars, and as L4 (*top*) and L5 (*bottom*) apart from this line. For $\mu = 0.2$, the critical values $\rho_0^{(1)}$, $\rho_0^{(2)}$, and $\rho_0^{(3)}$ are given as 0.353, 0.420, and 0.692, respectively, and for $\mu = 0.5$, they are given as 0.251, 0.442, and 0.442, respectively.

2. Methods and Results

In the so-called coplanar circular restricted three-body problem the two stars are assumed to orbit each other in circles and their masses are much larger than that of the planet. In our case, it is assumed that the planetary mass is 1×10^{-6} of the mass of the star it orbits; also note that the planetary motion is constrained to the orbital plane of the two stars. In addition, it is assumed that the initial velocity of the planet is set for an initially circular orbit, and that it is in the same direction as the orbital velocity of its host star. This star shall be the more massive of the two stars. Furthermore, the starting position of the planet is to the right of its host star along the line joining the binary components (3 o'clock position). The mass ratio μ of the two stars is defined as $\mu = M_2/M$ with $M = M_1 + M_2$, where M_1 and M_2 are the masses of the primary and secondary star, respectively. Additionally, ρ_0 denotes the planet's relative initial distance $\rho_0 = R_0/D$ from its host star, with D as distance between the two stars and R_0 the initial planetary distance from the primary star.

In the following, we illustrate the transition from stability to instability by progressively increasing the value of ρ_0 for binary systems with a fixed mass ratio μ, given as $\mu = 0.2$ and 0.5, respectively. For both $\mu = 0.2$ and 0.5, we present the resulting planetary orbits for four different values of ρ_0 (see Fig. 1). Each panel shows the primary (larger dot) and secondary (smaller dot) star, the planetary orbit (solid line), as well as the zero velocity contours (dash-dotted lines). The upper four panels of Fig. 1 refer to the stellar mass ratio $\mu = 0.2$, whereas the lower four panels refer to $\mu = 0.5$,

Let us first focus on the case studies for $\mu = 0.2$ and $\rho_0 = 0.25$ and 0.323. Both values of ρ_0 are smaller than the critical value of $\rho_0^{(1)} = 0.353$, indicating that the planetary orbits are stable. The fact that we restricted the time of the simulation to 50 yrs is inconsequential owing to the fact that the orbital stability of the planet is guaranteed by the analytical properties of the system, namely $\rho_0 < \rho_0^{(1)}$; see Cuntz et al. (2007) and Eberle et al. (2007) for a more extended discussion. Also note that the zero velocity contour changes between the two panels due to the increase in ρ_0 by getting closer to the Lagrange point L1, although its topology remains unaltered. Moreover, we show two cases of unstable orbits by choosing $\rho_0 = 0.41$ and 0.51, respectively. Since both values exceed $\rho_0^{(1)}$, the zero velocity contour opens at L1, providing the possibility for the planet to be captured by the secondary star. For $\rho_0 = 0.51$, the zero velocity contour even opens at L2 because of $\rho_0 > \rho_0^{(2)}$. In case of $\rho_0 > \rho_0^{(3)}$ (not shown here), the contour would even open at L3, providing a further type of opportunity for the planet to escape from the binary system.

The results for $\rho_0 = 0.41$ show that the planetary orbit is unstable but still remains within the sphere of gravitational influence of the primary star. The situation is different for $\rho_0 = 0.51$ where the planet reaches the secondary star only after a few irregular orbits about the primary star have been completed. A more detailed analysis shows that the planet first encountered the secondary star after 39.6 yrs at a minimal distance of at most 0.1 AU. A second even closer encounter occurred after 41.2 yrs, when the simulation was stopped because the planet entered the Roche limit of the secondary star. The general behavior of the model with $\rho_0 = 0.51$ is due to the fact that ρ_0 exceeds both $\rho_0^{(1)}$ and $\rho_0^{(2)}$, which in principle allows the planet to escape from the binary system through the L2 point.

Our results for $\mu = 0.2$ demonstrate different cases of orbital stability and instability. Similar results are obtained for $\mu = 0.5$, albeit quantitative differences due to the different values of μ, $\rho_0^{(1)}$, $\rho_0^{(2)}$, and $\rho_0^{(3)}$. We find again that orbital stability is obtained if $\rho_0 < \rho_0^{(1)}$, whereas for larger values of ρ_0 instability is expected to emerge. Highly unstable cases are found for $\rho_0 = 0.3$ and 0.45.

3. Conclusions

For the special case of the "coplanar circular restricted three-body problem", we applied stringent mathematical criteria that allow to precisely determine whether a planetary orbit in a stellar binary system is stable or unstable. This is accomplished by comparing the planet's relative initial distance ρ_0 to the critical values $\rho_0^{(1)}$, $\rho_0^{(2)}$, and $\rho_0^{(3)}$, defined for a fixed stellar mass ratio μ. An adequate way of demonstrating this different type of behavior is the assessment of the topology of the zero velocity contour, given by the μ and ρ_0 values of the system. In this case, planetary orbital stability is obtained if the contour is completely closed around the primary star and planet. All numerical case studies show a behavior consistent with this theoretical prediction. Note however that for generalized binary systems, other methods are required to determine the long-term stability of planetary orbits (e.g., Holman & Wiegert 1999; David et al. 2003). Important applications of our work include contesting numerically deduced stability limits for cases where analytically deduced results exist.

Acknowledgments. This work has been supported by the Alexander von Humboldt Foundation (Z. E. M.).

References

Bonavita, M. & Desidera, S. 2007, A&A, 468, 721
Cuntz, M., Eberle, J., & Musielak, Z. E. 2007, ApJ, 669, L105
Danby, J. M. A. 1988, Fundamentals of Celestial Mechanics (Richmond: Willmann-Bell, Inc.)
David, E.-M., Quintana, E. V., Fatuzzo, M., & Adams, F. C. 2003, PASP, 115, 825
Duquennoy, A., & Mayor, M. 1991, A&A, 248, 485
Eberle, J., Cuntz, M., & Musielak, Z. E. 2007, A&A, submitted
Eggenberger, A., & Udry, S. 2007, in Planets in Binary Star Systems, ed. Haghighipour (New York: Springer), in press
Eggenberger, A., Udry, S., & Mayor, M. 2004, A&A, 417, 353
Holman, M. J., & Wiegert, P. A. 1999, AJ, 117, 621
Lada, C. J. 2006, ApJ, 640, L63
Musielak, Z. E., Cuntz, M., Marshall, E. A., & Stuit, T. D. 2005, A&A, 434, 355
Patience, J., et al. 2002, ApJ, 581, 654
Quintana, E. V., Lissauer, J. J., Chambers, J. E., & Duncan, M. J. 2002, ApJ, 576, 982
Raghavan, D., et al. 2006, ApJ, 646, 523
Roy, A. E. 2005, Orbital Motion (Bristol and Philadelphia: Institute of Physics Publ.)
Stuit, T. D. 1995, M.S. thesis, University of Alabama in Huntsville
Szebehely, V. 1967, Theory of Orbits (New York and London: Academic Press)
Trilling, D. E., et al. 2007, ApJ, 658, 1289

Bioastronomy 2007: Molecules, Microbes, and Extraterrestrial Life
ASP Conference Series, Vol. 420, 2009
K. J. Meech, J. V. Keane, M. J. Mumma, J. L. Siefert, and D. J. Werthimer, eds.

Characterizing Other Earths: Why and How

A. Quirrenbach

ZAH, Landessternwarte, Königstuhl 12, D-69117 Heidelberg, Germany

Abstract. The past few years have seen dramatic progress in studies of extrasolar planets, driven by technological advances that push detection limits and broaden the suite of tools at our disposal. Several techniques have been proposed for the detection of habitable planets around nearby Solar-type stars: transit photometry and spectroscopy, direct imaging, nulling interferometry and astrometry. They all require substantial further technology development and large investments in (mostly space-based) facilities. Therefore one has to ask the question how these different approaches compare to each other, what kind of planets each one can detect, and what information can be obtained from them. If one is interested in complicated issues like the potential habitability of a planet (or even the actual presence of life on it), a single technique can hardly provide conclusive answers. One needs to consider orbit, mass, age, surface and atmospheric properties of the planet; all of these are directly related to the host star. Therefore results from different techniques have to be combined intelligently to form a comprehensive picture, and to exclude misinterpretations of limited data sets. I will discuss how various observing techniques can help characterize other Earths, and how they can complement each other.

1. Introduction

The discovery and study of extrasolar planets has over the past ten years provided new insight in the formation and evolution of planetary systems. Technical progress will provide new tools to determine the properties of exoplanets, and thus lead to further refinements in our understanding of the processes that have also shaped our own Solar System. In the 2015-2025 time frame it will be possible to search for analogs of our Earth orbiting stars in the Solar neighborhood, to perform spectroscopic analyses of these Earth twins, and to search for the signature that life may imprint on the composition of their atmospheres. In this way, we are starting to address some of the oldest philosophical questions of mankind with modern scientific tools.

2. Astrometry

The principle of planet detection with astrometry is similar to that behind the Doppler technique: the presence of a planet is inferred from the motion of its parent star around the common center of gravity. In the case of astrometry one observes the two components of this motion in the plane of the sky; this gives sufficient information to solve for the orbital elements without the $\sin i$ ambiguity plaguing Doppler measurements. In addition, the astrometric method can be applied to all types of stars (independently of their spectral characteristics), it is less susceptible to noise induced by the stellar atmosphere, and it is more sensitive to planets with large orbital semi-major axes. From Kepler's Laws it follows immediately that the astrometric signal θ of a planet with mass m_p orbiting a star with mass m_* at a distance d in a circular orbit of radius a is

$$\begin{aligned} \theta &= \frac{m_p}{m_*}\frac{a}{d} = \left(\frac{G}{4\pi^2}\right)^{1/3} \frac{m_p}{m_*^{2/3}} \frac{P^{2/3}}{d} \\ &= 3\,\mu\mathrm{as} \cdot \frac{m_p}{M_\oplus} \cdot \left(\frac{m_*}{M_\odot}\right)^{-2/3} \left(\frac{P}{\mathrm{yr}}\right)^{2/3} \left(\frac{d}{\mathrm{pc}}\right)^{-1} . \end{aligned} \tag{1}$$

NASA's Space Interferometry Mission (SIM Planet Quest, Unwin et al. 2008) will push the precision of astrometric measurements far beyond the capabilities of any other project currently in existence or under development. SIM is essentially a single-baseline interferometer with 30 cm telescopes on a 9 m baseline. SIM is a pointed mission, i.e., targets can be observed whenever there is a scientific need (subject only to scheduling and Solar exclusion angle constraints), and the integration time can be matched to the desired signal-to-noise ratio. In its "narrow-angle" mode (i.e., for measurements of a target with respect to reference stars within a field of $\sim 1^\circ$), SIM will provide an accuracy of $\sim 1\,\mu$as for each measurement. This makes it possible to carry out a high-precision survey of ~ 200 nearby stars reaching down to $1\ldots3\,\mathrm{M}_\oplus$ (depending on stellar mass and distance), and a less sensitive survey of ~ 2000 stars establishing better statistics on massive planets in the Solar neighborhood. In addition, SIM will observe a sample of pre-main-sequence stars to investigate the epoch of planet formation.

3. Biomarkers

Life on Earth has a profound impact on the composition of the atmosphere, producing abundant oxygen, and methane in extreme thermodynamic disequilibrium. It is plausible to assume that any form of abundant life on an extrasolar planet might transform the planetary atmosphere in a similar way, and thus produce a signature that can be detected spectroscopically (Fig. 1, see also Des Marais et al. 2002).

The mid-infrared spectral region ($\sim 6\ldots20\,\mu$m) contains a number of important spectral features that can be used to diagnose the presence of H_2O, CO_2, and O_3, which serves as a proxy for O_2 (Angel et al. 1986, Léger et al. 1996). Methane also has a strong absorption band in this spectral region, but it would be extremely difficult to detect in an Earth-like atmosphere. In the past, however, before the rise of oxygen, biogenic CH_4 may have been more abundant in the Earth's atmosphere by a factor $100\ldots1000$ (Catling et al. 2001); at this level it would also be detectable and serve as an indicator of bacterial life.

There are of course also many molecular absorption bands in the visible and near-IR spectral ranges, for example the oxygen A band at 7600 Å (e.g. Owen 1980). Ozone has a broad band near 6000 Å, and a very strong band in the UV that is responsible for the short-wavelength cutoff in the transmission of our atmosphere. Water also has very strong absorption bands; the gaps between these bands define the astronomical J, H, K, and L bands in the near-IR. In

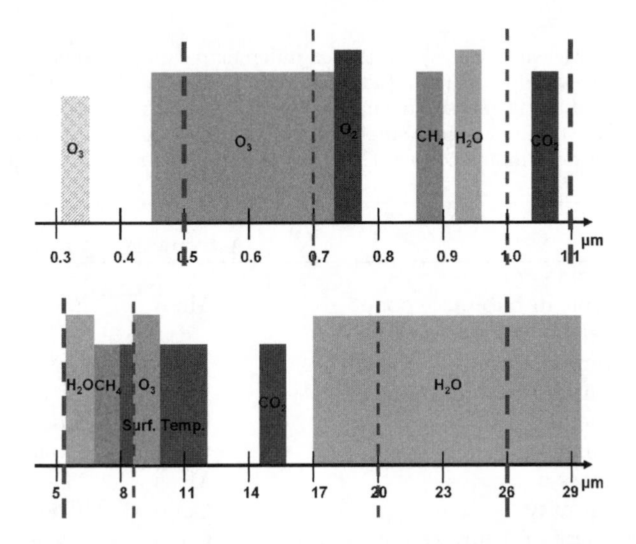

Figure 1. Important diagnostic bands in the visible (top) and mid-infrared (bottom) wavelength range. The dashed vertical lines indicate the wavelength coverage needed for a basic (water, oxygen) and extended analysis of the planetary atmosphere.

addition, the CH_4 and CO_2 bands in the $2\,\mu$m spectral region could serve as a diagnostic of elevated levels of these gases.

4. Coronography and Nulling Interferometry

Of particular interest are rocky planets that orbit their parent stars at a distance where the temperature is such that water can be liquid at the surface. Over the past few years, ESA and NASA have formulated rather similar goals for a mission — known as "Darwin" in Europe and as "Terrestrial Planet Finder" (TPF) in the US — aimed at performing detailed studies of Earth-like planets in these habitable zones of some 150 solar-type stars. Darwin/TPF therefore must have an angular resolution of ~ 35 mas, corresponding to 0.7 AU at 20 pc. The biggest challenge, however, is rejecting the light from the bright star, so that the much fainter planet can be observed.

At visible wavelengths, the contrast between the Earth and the Sun is a few times 10^{-10}. Building a telescope that can give such a dynamic range over a separation of ~ 35 mas is a daunting task. This problem can be solved with a space telescope equipped with an optimized coronograph and a deformable mirror for suppressing scattered light (Ford et al. 2004).

To achieve the required angular resolution in the mid-infrared an interferometer with baselines of at least $50\ldots100$ m is required. It appears that mounting each telescope and the beam combiner on a separate spacecraft will be much easier than building a single large structure, and therefore a "free-flyer" approach has been generally adopted. In an infrared interferometer the starlight can be suppressed by introducing an achromatic π phase shift in one arm of the

Table 1. Measurement synergies for different techniques aimed at characterizing terrestrial exoplanets (adapted from Beichman et al. 2007). "Meas" indicates a directly measured quantity; "Est" indicates a quantity that can be roughly estimated from a single mission; and "Coop" indicates a quantity that is best determined cooperatively using data from several missions.

	Astrometry	Visible	Mid-IR
Orbital Parameters			
Stable orbit in habitable zone	Meas	Meas	Meas
Characteristics for Habitability			
Planet temperature	Est	Est	Meas
Temp. variability due to eccentricity	Meas	Meas	Meas
Planet radius	Coop	Coop	Meas
Planet albedo	Coop	Coop	Coop
Planet mass	Meas	Est	Est
Surface gravity	Coop	Coop	Coop
Atmospheric and surface composition	Coop	Meas	Meas
Time variability of composition		Meas	Meas
Presence of water		Meas	Meas
Planetary System Characteristics			
Influence of other planets	Meas	Est	Est
Orbit coplanarity	Meas	Est	Est
Comets, asteroids, and zodiacal dust		Meas	Meas
Indicators of Life			
Atmospheric biomarkers		Meas	Meas
Red absorption edge of vegetation		Meas	

interferometer, so that there is destructive interference for the light from the star arriving on-axis (Bracewell 1978).

5. Synoptic Information for Planet Characterization

Whether information is gathered in the mid-IR or in the visible / near-IR, careful modeling of the planetary atmosphere is required to rule out an abiotic origin of O_2, CH_4, or any other presumed indicator of life (e.g. Schindler & Kasting 2000). One has to keep in mind that the biosphere, atmosphere, hydrosphere, and lithosphere interact in many complicated ways. Even reconstructing the processes that lead to the rise of oxygen in our own atmosphere remains a challenging task, in spite of the wealth of available data (Kasting 2001). It will certainly be much more difficult to interpret the glimpses of atmospheric chemistry that we may be able to obtain by remote sensing of extrasolar terrestrial planets. It is therefore important to develop complementary observing capabilities, which include dynamical mass measurements in addition to spectroscopy in multiple wavebands (see Table 1). As pointed out by Beichman et al. (2007), a synoptic approach is needed to resolve ambiguities in the information from one mission alone, and to enable robust modeling of planetary properties.

References

Angel, J.P.R., Cheng, A.Y.S., & Woolf, N.J. 1986. Nature 322, 341-343

Beichman, C.A., et al. 2007. In *Protostars and planets V.* Eds. Reipurth, B., Jewitt, D., & Keil, K., p. 915-928

Bracewell, R.N. 1978. Nature 274, 780-81

Catling, D.C., Zahnle, K.J., & McKay, C. 2001. Science 293, 839-843

Des Marais, D.J., et al. 2002. Astrobiology 2, 153-181

Ford, V.G., et al. 2004. In *Optical, infrared, and millimeter space telescopes.* Ed. Mather, J.C., SPIE Vol. 5487, pp. 1274-1283

Kasting, J.F. 2001. Science 293, 819-820

Léger, A., et al. 1996. Icarus 123, 249-255

Owen, T. 1980. In *Strategies for the search for life in the Universe.* Ed. Papagiannis, M.D., pp. 177-185

Schindler, T.L., & Kasting, J.F. 2000. Icarus 145, 262-271

Unwin, S., et al. 2008. Taking the Measure of the Universe: Precision Astrometry with SIM PlanetQuest. PASP 120, 38

Andreas Quirrenbach

Session XII

Biosignatures and Life Beyond Earth

Victoria Medows

Giovanna Tinetti

Bioastronomy 2007: Molecules, Microbes, and Extraterrestrial Life
ASP Conference Series, Vol. 420, 2009
K. J. Meech, J. V. Keane, M. J. Mumma, J. L. Siefert, and D. J. Werthimer, eds.

Search For Chiral Signatures in the Earthshine

M. F. Sterzik[1] and S. Bagnulo[2]

[1]*European Southern Observatory, Santiago, Vitacura, Chile*

[2]*Armagh Observatory, College Hill, Armagh BT61 9DG, Northern Ireland, UK*

Abstract. We describe an experiment to search for circular polarization in the spectrum of the Earthshine as induced by biotic material on the surface of the Earth. Organic material on Earth is abundant and its helical molecular structure is known to produce circular polarization of reflected diffuse light up to levels of one percent in the visible wavelength regime. We use the spectropolarimetric capability of FORS1 mounted at the Very Large Telescope in Paranal/Chile to detect such features. Our experiment is a benchmark required for future attempts to detect biotic material on other astronomical objects. We present the idea of our experiment, an estimate of the instrumental sensitivity required, and a short description of the observations. We expect to reach a final accuracy of 10^{-5} to measure Stokes V/I.

1. Motivation

From the ground, the globally integrated terrestrial spectrum can be measured by observing the light reflected by the daytime Earth onto the dark side of the lunar surface (Earthshine). Dependent on the relative phase angle of the Earth surface reflected from the Moon, and the Earths global cloud coverage, the enhanced reflectivity caused by ground vegetation ("Vegetation Red Edge," VRE) can be recovered at a wavelength around 700nm. Observations of this kind have been analysed by various groups, but the detectability of the VRE feature is difficult and not unambiguous (see, e.g., the review of Arnold (2007)).

An alternative avenue to detect possible biosignatures from remote sensing is to seek for key signatures of life through circular polarimetry (CP). It is well-known that optical activity can be induced by the interaction between light and chiral molecules of which living material is composed (Wolstencroft 2004). Homochirality, i.e. the exclusive use of L-amino acids and D-sugars in biological materials, causes a significant induction of circular polarization in the diffuse reflectance spectra of biotic material (Wolstencroft et al. 2004). Among many other biopolymers, photosynthetic pigments (e.g. chlorophyll) induce between 0.1% and 1% circular dichroism in its absorption bands (Houssier & Sauer 1970).

In a pioneering study Sparks et al. (2005) searched for chiral signatures on Mars stemming from living material. He obtained high-quality circular imaging polarimetry of the martian surface during opposition in 2003. Although the polarization noise levels were well below 0.1%, i.e. approaching the instrumental capabilities, no regions with significant circular polarization were found. Today, this null result may not seem too surprising, as recent results from the Mars-

Express mission indicate no evidence for liquid water anywhere on its surface. This, together with the early loss its atmosphere makes primitive surface-life on Mars highly unlikely (Bibring et al. 2005).

The Earth is the only location where we unambiguously know that organic material is abundant. We therefore searched for optical activity caused by biotic homochirality in the Earthshine. Our experiment is intended as a benchmark for further attempts to characterize potentially living material in a broader, astronomical, context.

2. Observations

Earthshine observations intrinsically yield disk-integrated Earth spectra, i.e. do not contain spatial information, and resemble spatially unresolved observations of the Earth. We obtained data in two observing epochs (end of November 2006, and middle of December 2006), which allow us to compare different phases of the Earthshine contributing surface areas.

FORS1, mounted at the Cassegrain focus of the Very Large Telescope on Cerro Paranal in Chile is an unique instrument at a 8m class telescope that offers both circular and linear spectropolarimetric observing modes through superachromatic $\lambda/4$ and $\lambda/2$ retarder plates mounted in front of a Wollaston prism. We refer to the FORS1 user manual for technical details. The scientific performance and sensitivity of this mode has already been demonstrated, and accuracies of 10^{-4} to measure Stokes V/I in high S/N (≈ 2000) spectra of magnetic stars have been obtained (Bagnulo et al. 2002). Note that the Earthshine has an approximate surface brightness of about $V \approx 12 - 13/''^2$, depending on the actual phase and illumination. High signal levels using the full dynamic range of the detector can therefore easily be obtained within short exposure times, important to repeat the polarimetric cycles and to increase the S/N level. As the lunar surface is spatially extended, and covers a significant fraction of the detector, an accuracy of 10^{-5} in high S/N re-binned and multi-stack spectra of the Earthshine can be obtained. We have chosen to record low-resolution spectra in the red wavelength band, covering approximately 600nm-1000nm, which includes the VRE signature around 700nm. Note that this set-up contrasts to the one used by Sparks et al., who used two discrete narrow-band interference imaging filters (centered at 378nm and 953nm) for their circular polarization detection experiment on Mars.

2.1. Sensitivity Requirement

The degree and spectral shape of the CP signal as expected in diffuse reflectance that arises from living material on Earth (mainly vegetation canopy) is not known. Laboratory measurements of a few species of plant leaves have been reported by Wolstencroft et al. (2004), and CP of about 1% varies slowly over the visible wavelength range. No better a priori knowledge of the spectral dependency of the CP signal expected from Earth's vegetation is available.

Two effects have to be considered when estimating the final CP signal strength expected in the Earthshine. On Earth, the surface filling factor of vegetation is not only dependent on the geometrical arrangement of oceans, ice, deserts and vegetated land-masses, but also on the global cloud coverage and its distribution at the time of observation. Cloud cover maps are available through the International Satellite Cloud Climatology Project, and can, in principle, be folded with the geometrical distribution of vegetated areas. Previous analysis show that on average about 60% of the Earths surface are affected by clouds. A cloud-free vegetation cover between 0% and up to 50% (in the maximum, typically 20%) can be expected for different observing phases (Montanes-Rodriguez et al. 2006). We have estimated the cloud-free vegetation area during the times of our observations, and estimate the effective contribution of vegetation to about 1% for our November observations, and to about 10% for December. In other words, the CP signal as expected from diffuse reflectance of a homogenous vegetation distribution is diluted by about 0.01 and 0.1 for the November resp. December observations. On the Moon, depolarization caused by scattering on the lunar surface dilutes the CP signal in the Earthshine further. From linear polarization measurements the depolarization strength of the lunar surface has been estimated to be around 10(Dolfuss 1957; DeBoo et al. 2005), but no more precise value can be attributed to this effect. We conclude that an accuracy of $\approx 10^{-4 \ldots -5}$ is required to determine Stokes V/I in order to positively infer the presence of a CP signal caused by vegetation in Earthshine observations.

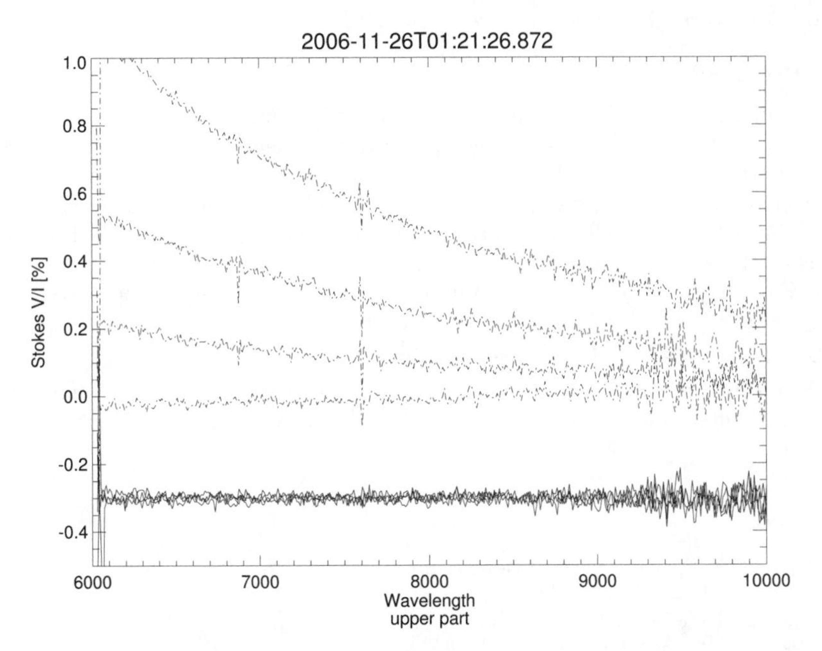

Figure 1. Spectrum of Stokes V/I as measured in the Earthshine on November 26, 2006. The dashed lines correspond to different regions on the detector. The highest degree of CP is measured in the blue, and at the edges of the field. The socalled "null polarization" spectrum is also given (full lines), shifted by -0.3%. The CP signal is dominated by cross-talk from linear polarization.

2.2. Data Signatures

For both epochs we observed both lunar sides (the bright and the dark side, corresponding to Moonshine and Earthshine). The long slit was positioned perpendicular to the lunar limb in the center of the detector, which allows to record the contributions of the sky background. All spectropolarimetric observations were performed in a standard fashion, i.e. by rotating the quarter-wave plate by 90deg in order to remove the instrumental polarization. We have also measured the linear polarization of the Earthshine, which is mainly caused by Rayleigh scattering in the Earths atmopshere. Actually, the dominant contribution to the measured Stokes V/I stems from cross-talk from linear polarization. Its spectral and spacial signatures has to be carefully analysed and removed. In Figure 1 we demonstrate this effect. Stokes V/I has been determined for four different regions across the detector. The effect of increasing cross-talk can be noted in the blue part of the spectrum, and with increasing distance from the detector center. In the same Figure (full lines) we plot the "null polarization" spectrum for each region (Donati et al. 1997). This quantity is particularly useful to estimate the noise level of the data. The r.m.s. of the null polarization is of the order of 10^{-4}, and spectral re-binning, and stacking of more polarization measurements indeed allow to reach the desired accuracy of 10^{-5}.

2.3. Prospect

The careful analysis of the instrumental crosstalk is required to remove the dominant contribution in V/I. Ancillary calibration measurements have therefore been performed. Additional analytic model descriptions will be employed to further characterize the instrumental cross-talk. The data obtained by us have the potential to detect CP signatures down to a level of $V/I \approx 10^{-5}$. Investigations of the strength, and spectral shape of CP in vegetation typical for the Earths integrated reflectance spectrum are largely absent, and deserve systematic follow-up. Intensive laboratory measurements of the CP of different plant and leaf species are therefore highly desirable.

Acknowledgments. The measurements have been obtained through the Directors Discretionary Time under program ID 278.C-5017 at the Very Large Telescope

References

Arnold, L., 2007, Space Science Review, in press, arXiv:0706.3798
Bagnulo, S., et al., 2002, A&A 389, 191.
Bibring et al., 2005, Science, 307, 1576.
DeBoo et al., 2005, App.Opt., 44, No.26, 5434.
Dolfuss, A., 1957, Supplements aux Annales dAstrophysique, 4, 3.
Donati, J.-F., et al. 1997, MNRAS 291, 658.
Houssier & Sauer, 1970, J. Am. Chem. Soc. 92, 779.
Montanes-Rodriguez, P., et al., 2006, ApJ 651, 544.
Sparks, W.B., Hough, J.H., Bergeron, L.E., 2005, Astrobiology 5, 737
Wolstencroft, R.D., Tranter, G.E., & D.D. Le Pevelen, 2004, in: Bioastronomy 2002, IAU Symposium 213, 149
Wolstencroft, R.D., 2004, in: Bioastronomy 2002, IAU Symposium 213, 154

Nick Woolf

Stephane Dumas

Bioastronomy 2007: Molecules, Microbes, and Extraterrestrial Life
ASP Conference Series, Vol. 420, 2009
K. J. Meech, J. V. Keane, M. J. Mumma, J. L. Siefert, and D. J. Werthimer, eds.

Detection of Endolithes Using Infrared Spectroscopy

S. Dumas, Y. Dutil, and G. Joncas

Dept. de physique, de génie physique et d'optique et Observatoire du mont Mégantic, Univsersité Laval, Québec, Canada, G1K 7P4

Abstract. On Earth, the Dry Valleys of Antarctica provide the closest martian-like environment for the study of extremophiles. Colonies of bacterias are protected from the freezing temperatures, the drought and UV light. They represent almost half of the biomass of those regions. Due to their resilience, endolithes are one possible model of martian biota.

We propose to use infrared spectroscopy to remotely detect those colonies even if there is no obvious sign of their presence. This remote sensing approach reduces the risk of contamination or damage to the samples.

1. Introduction

Space exploration is a difficult task and the search for life is no different. The equipment size a probe can bring severely limit the scope of the search. Every sample cannot be analyzed and selecting those that can be is not trivial. This project investigates the possibilities of remote detection using infrared spectroscopy in order to select those few samples to pick and analyze further.

2. Infrared Spectroscopy

The reason for using IR spectroscopy is mainly that almost all biological marker molecules (i.e. biomarkers) will show some spectral features in the near to far IR (Hand *et al.* 2005). Furthermore, the spectroscopy will not destroy the sample (e.g. there is no contact with the sample, no contamination). Previous techniques of detection tended to destroy, or damage, the endolithes or the environment in which they lived (e.g. by using electron-microscope, chemical and biological or analysis).

Biomarkers are an important source of information in the search for evidence of life in geological samples. Even if the organisms are dead or dormant, it is still possible to detect their presence, byproducts or even their chemical alterations of the environment.

In searching for extraterrestrial life, it is important to have a minimum of preconception about them in order to find it.

3. Endolithes

Endolithes are organisms that live inside rocks or in the pores between mineral grains. There are thousands of known species of endolithes, including members from Bacteria, Archaea, and Fungi. They represent near half the Earth's biomass and also present an ideal model of life for Mars (Ascaco 2002; Hand *et al.* 2005).

This study used two groups of samples. One from the Guelph region (west of Toronto, Ontario) and the other near Eureka on the Ellesmere Island in the Nunavut (Omelon 2006). Both regions have different geology and climate.

4. Methodology

Samples were scanned using a technique called diffuse scattering. The IR beam was directed toward the sample and a series of mirrors redirected the scattered beam to the IR sensor. The spectra were obtained using an IR Nicolet spectrometer.

The apparatus used to collect the scattered IR beam could not receive large rock samples. It was necessary to break them into smaller pieces. The same device was very sensible to the angle of incidence and reflectance. It was important to position both mirrors near the vertical above the sample. The adjustment of each mirrors, in order to optimize the reception of light, was very time consuming. Those adjustments were performed using an aluminum plate as the sample to maximize the received flux to the sensor. When the rock sample was placed in the light path, the total flux drops a lot but the signal-to-noise ratio was high enough.

Spectra in middle IR (132) and near IR (49) on 15 samples from the two groups were taken. The best results were from the Middle IR. Most of the spectra were taken at a resolution of 4 or 8 cm^{-1}. After analysis, it appears that a spectral resolution of 8 cm^{-1} is enough for our purpose. The middle IR spectra (from 4000 to 650 cm^{-1}) were then processed using Principal Component Analysis in order to classify them.

5. Principal Component Analysis

Principal Component Analysis (Marchi 2007) is a technique of Factorial Analysis (e.g. Multivariate Statistics). It is often used to find a new coordinate system in which the original data will be better aligned on some axes (e.g. principal components).

The technique can be summarized by the equation $A = UWV^T$ where A,U,W and V are matrices.

A is a $m \times n$ matrix containing the spectra on each row. The value of each absorption band is then regrouped in the columns which are the variables on which the PCA works. U, also a $m \times n$ matrix, contains the spectra in the new coordinate system. The matrix V, where VV^T is an identity matrix, contains the eigenvectors. W is a diagonal matrix containing the squared root of the eigenvalues and provides a clue about how many principal components (PC) can be used to describe the data. It is important to have a matrix A with more columns than rows (ie. $n > m$) else W becomes a singular matrix and the whole PCA fails.

To facilitate data manipulation and visualization for this paper, we have taken only the first three principal components. The result of the PCA process is illustrated in figure 1. The process can be extended to as many components

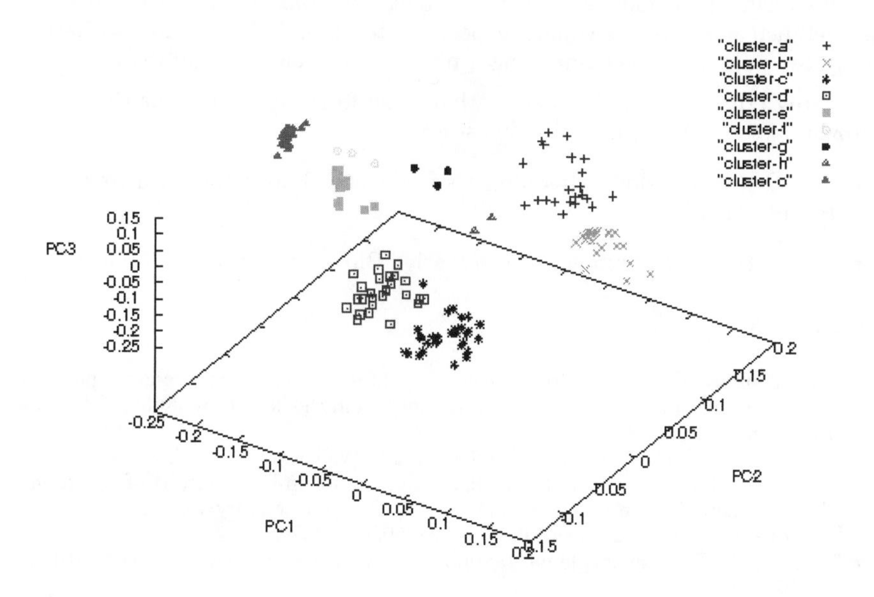

Figure 1. *Spectra plotted in the PCA space.*

as needed. Based on the eigenvalues from our data, the first five components are relevant, the others being buried in the noise.

We have identified several clusters. Cluster E groups the spectra containing features showing the presence of organic compounds. Cluster E is very close to another cluster of points calculated from organic spectra used as a reference (i.e. cluster O). Further, most of the spectra of cluster E have been taken directly from green regions visible on the samples.

6. Results and Conclusions

Our results show that it is possible to detect biological signatures using a spectrometer operating in the middle infrared range. However, it is not possible to highlight a particular region, or regions, in a spectrum to be used to identify biomarkers. The interdependence of the absorption bands related to the living are too complex to simply isolate a few.

The proposed technique calls for a more subtle approach by comparing witness spectra and an unknown spectrum by plotting them in the PCA's space. If the test spectrum, once projected into the PCA's space, is close to the reference group then the probability of it is containing biomarkers is high. Adding a non-organic spectra to the PCA space as references may improve the identification scheme.

This technique could be used to pinpoint potential life harboring rock for more detailed analysis. It could be possible to extend the method to better identify the unknown spectrum using a more precise reference database.

Acknowledgments. We wish to thank the following people for their help in providing the samples used for this study :

- Christopher Omelon, Microbial Geochemistry Laboratory, University of Toronto Geology
- Uta Matthes, Department of Integrative Biology, University of Guelph

References

Ascaco, C., 2002, New approach to the study of Antarctic lithobiontic microorganisms and their inorganic traces, and their application in the detection of life in Martian rocks, Int. Microbiol

Hand, K.P., Carlson, R.W., Sun,H., Anderson, M, Wadsworth, W., Levy, R., 2005, Utilizing active mid-infrared microspectrometry for in-situ analysis of cryptoendolithic microbial communities of Battleship Promontory, Dry Valleys, Antartica, Astrobiology and Planetary Missions, vol 5906, p.302

Marchi, S., 2007, Extrasolar planet taxonomy : a new statistical approach, astro-ph/0705.0910v1

Omelon, C., 2006, Environmental controls on microbial colonization of high Arctic cryptoendolithic habitats, Polar Biology, V.30

Bioastronomy 2007: Molecules, Microbes, and Extraterrestrial Life
ASP Conference Series, Vol. 420, 2009
K. J. Meech, J. V. Keane, M. J. Mumma, J. L. Siefert, and D. J. Werthimer, eds.

Interplanetary Transport of Microorganisms: Survival, Growth, and Adaptation

W. L. Nicholson,[1] P. Fajardo-Cavazos,[1] and A. C. Schuerger[2]

[1]*Departments of Microbiology & Cell Science, University of Florida, Space Life Sciences Laboratory, M6-1025/SLSL, Kennedy Space Center, FL 32899*
[2]*Plant Pathology, University of Florida, Space Life Sciences Laboratory, M6-1025/SLSL, Kennedy Space Center, FL 32899*

Abstract. The primary objective of our current work is to investigate the factors in the martian environment limiting survival and growth of terrestrial bacteria known to contaminate robotic spacecraft, primarily bacterial spores. In Mars simulations, solar UV is the most rapidly lethal factor to cells and spores, but is easily shielded. Results from experiments under simulated Mars conditions indicate that growth of terrestrial bacteria is further constrained by extreme low pressure, atmospheric composition, and limited availability of liquid water and organic nutrients.

1. Introduction

Interplanetary transport of microbes can be envisioned to occur either naturally as a consequence of impacts (lithopanspermia) or as a result of human spaceflight activities. In either case, the probability of interplanetary transfer of life by either natural or human means is a function of: (*i*) the initial population size (i.e., the bioload); survival of (*ii*) launch, (*iii*) transit through space, (*iv*) entry and deposition; and (*v*) survival and proliferation on the recipient planet. Modeling this process for testing lithopanspermia and for mitigation of forward and back contamination by human spaceflight calls for accurate simulation of all aspects of transfer, driven by realistic experimentally-derived survival and growth data (reviewed in Fajardo-Cavazos et al., 2007; Nicholson et al., 2000; 2005).

1.1. Lithopanspermia

Lithopanspermia theory postulates that microbes could be transported between planets within rocks ejected into space as a result of impacts. A number of lines of evidence lend support to the theory, including the existence on Earth of meteorites originating from Mars and the Moon and a well-understood physical mechanism for impact ejection. The physics of impacts predicts that around the impact site there exists a transient *spallation zone* where interaction of the direct and reflected shock wave can accelerate near-surface rocks to escape velocity with relatively little heating and shock damage. Spallation theory predicts that only rocks close to the surface are ejected; furthermore, the current collection of martian meteorites are of igneous origin. Various microbes have been shown to survive the physical stresses of ejection, space transit, hypervelocity atmospheric entry and deposition. Lithopanspermia theory postulates that the rate of exchange of life-bearing rocks among the inner terrestrial planets would have been much higher towards the end of the Late Heavy Bombardment period (>3.5 Ga), coincident with the time at which early life was presumably taking hold on Earth. During this early stage in planetary evolution, the environments of Earth and Mars were likely to have been much more similar, giving interplanetary passengers a better chance to survive and thrive in their new homes (see Section 1.4.).

1.2. Panspermia Resulting from Human Spaceflight

Within the past 50 years humans have left Earth and traveled to other planets and moons within the solar system, with robotic probes and manned exploration missions. Because microorganisms are ubiquitous on Earth, spacecraft leaving Earth to explore places such as Mars carry along populations of terrestrial microorganisms as accidental contaminants of spacecraft. The stresses placed upon microorganisms during transfer by human and robotic spaceflight are much gentler than those imposed by lithopanspermia, hence the probability of survival is higher. One of the major challenges facing mission planners and explorers alike is protection of pristine environments from "forward contamination" by terrestrial microorganisms, as well as potential "back contamination" of Earth environments by hypothetical extraterrestrial microbes (Rummel, 2001; 2002). However, despite the most rigorous planetary protection efforts, it is a virtual certainty that Earth microbes have, and will continue to, hitchhike from Earth to Mars (reviewed in Fajardo-Cavazos et al., 2007; Schuerger, 2004).

1.3. Bacterial Spores as Models for Interplanetary Transfer

Spores of the model organism *Bacillus subtilis* have long been used for panspermia research because of their notorious resistance to harsh environments including space (reviewed in Nicholson et al., 2000). But do bacterial spores actually find themselves in a position to be transported from Earth to Mars and survive the journey? Regarding lithopanspermia, we have demonstrated that *Bacillus* spp. spores are indeed found in the interiors of near-subsurface igneous rocks such as basalt and granite. The *Bacillus* species we recovered were closely related by 16S rDNA analyses to a number of *Bacillus* species previously found inhabiting extreme endolithic environments worldwide and contaminating spacecraft and their assembly facilities. Regarding human-directed panspermia, sampling studies have uncovered a number of spore-forming *Bacillus* spp. as dominant contaminants of ultra-clean Spacecraft Assembly Facilities and spacecraft destined for Mars. These spores exhibit a high degree of resistance to harsh environments prevailing in space and on Mars, suggesting that the ultra-clean spacecraft assembly facility environment actually selects for the hardiest spores (Fajardo-Cavazos et al., 2007; Schuerger, 2004). One of these isolates, B. *pumilus* SAFR-032, exhibited the highest degree of resistance (Kempf et al., 2005; Link et al., 2003) and is the first spacecraft-associated bacterium ever to have its genome sequenced. *B. subtilis* strain 168 and *B. pumilus* SAFR-032 have recently been chosen for an upcoming space-exposure experiment called PROTECT, to be mounted on the EXPOSE facility of ESA's Columbus module of the ISS, scheduled to launch on STS-122 on December 6, 2007. The purpose of the PROTECT experiment is to assay survival, viability, DNA photochemistry, and mutagenesis of spores exposed for 1–1.5 years to the space environment.

1.4. Present-day Mars Environment

In order for interplanetary transfer to be ecologically relevant, the transferred microbes must not only be able to survive, but to proliferate in their new environment. Environments considered "extreme" here on Earth are actually rather mild compared to the present-day surface of Mars. Mars' current cold conditions and low-pressure CO_2 atmosphere were likely initiated by faster interior cooling and the running down of its magnetic dynamo some 3.5 billion years ago. Without its protective magnetosphere in place, much of the martian atmospheric greenhouse was thought to have been blown away by the solar wind, resulting in the frozen desert we see on the martian surface today (Bennett et al., 2006). The major physical factors lethal or inhibitory to microbial growth and survival on Mars are: high solar UV and cosmic ray flux, low-pressure anoxic atmosphere, scarce or absent liquid water and organic nutrients, extreme cold, and potentially highly-oxidizing soils (Fajardo-Cavazos et al., 2007; Nicholson et al., 2005; Schuerger, 2004). Results from our recent experiments indicate that conditions on the surface of Mars pose formidable obstacles to the survival and growth of terrestrial microorganisms.

2. Results

2.1. Mars Simulations

In previous publications we have extensively described the construction and features of our Mars simulation facilities. For photographs and details, refer to Schuerger et al. (2003, 2006b).

2.2. UV Effects

The amount of UV penetrating to the surface of a planet is mainly a function of the distance of that planet from the sun and the composition of the planet's atmosphere (Nicholson et al., 2005). The atmosphere of Mars consists mainly of carbon dioxide ($\sim 95\%$) with extremely low levels of oxygen ($\sim 0.13\%$). CO_2 efficiently absorbs solar UV radiation shorter than ~ 190 nm. Thus the UV environment on the Martian surface is much richer in UVC and UVB than Earth's surface, including UV in the 260–280 nm range, which is absorbed efficiently by biological molecules such as DNA, RNA, and proteins. Thus, on Mars solar UV is rapidly lethal to microorganisms. Using simulated Mars environmental conditions obtained in a Mars Simulation Chamber (MSC), we demonstrated that sun-exposed surfaces received high enough levels of UV radiation to inactivate even the hardiest *Bacillus* spp. spores by up to 6 orders of magnitude within a few hours under simulated clear sky conditions (optical depth 0.5) (Schuerger et al., 2003; 2006b). Therefore, long-term microbial survival on Mars is likely possible only in locations shielded from solar UV.

2.3. Effects of Atmospheric Composition

The atmospheres of Earth and Mars are strikingly different in their composition. The martian atmosphere is predominantly CO_2 (95.3%), with traces of N_2 (2.7%), Ar (1.6%), O_2 (0.13%) and water vapor ($\sim$ 0.03%). We tested the ability of a number of spacecraft-contaminating *Bacillus* spp. spores to germinate and grow in Earth vs. a "martian" ($\sim$ 99% CO_2) atmosphere under conditions otherwise conducive to growth: temperature (30°C), pressure (1013 mbar), and abundant water and nutrients. All of the *Bacillus* spp. were able to grow in the high-CO_2 atmosphere, but growth was significantly slower and less luxurious (Schuerger and Nicholson, 2006). This is not be too surprising, since *Bacillus* spp. are generally aerobes or facultative anaerobes, preferring oxygen. However, facultative species are also able to grow without oxygen, using (*i*) fermentation of sugars or (*ii*) anaerobic respiration employing nitrate instead of oxygen as the terminal electron acceptor. We recently expanded on these observations by testing 31 sporulating and non-sporulating bacterial species, isolated from a variety of different environments including spacecraft and their assembly facilities, for their ability to grow in a simulated Mars atmosphere when other conditions were conducive to growth (Schuerger et al., 2006a). Fifteen of these species were completely unable to grow in $\sim$ 99% CO_2 at 1013 mbar. Growth was reduced to various degrees in the other 16 species, but a few bacteria tolerated the high-CO_2 atmosphere well, including *Enterococcus faecalis, Escherichia coli, Paenibacillus pabuli, Proteus mirabilis, Serratia liquifaciens, Staphylococcus aureus*, and *Terracoccus luteus* (Schuerger et al., 2006a).

2.4. Low Pressure (Hypobaric) Effects

The atmospheres of Earth and Mars differ dramatically in their pressure. The atmospheric pressure of Mars is only about 0.7% that of Earth, averaging $\sim$ 7 mbar and varying from a high of $\sim$ 10 mbar (Hellas Basin) to a low of $\sim$ 1 mbar (Olympus Mons). What is the effect of extreme hypobaria on the growth of microorganisms? We have exposed a large variety of bacteria to a range of pressures approaching Mars surface pressure, using either Earth or simulated Mars gas composition, with all other conditions conducive to growth. To date, without exception, we observed that bacterial growth ceased at pressures of 25 mbar or below, whether the microbes were incubated on the surface of semisolid agar medium or submerged in liquid culture, regardless of whether they are aerobes, anaerobes, or facultative microorganisms (Schuerger et al., 2006a; Schuerger and Nicholson, 2006). In addition, spore germination showed an even higher inhibitory threshold, and spores were unable to germinate at 35 mbar or below (Schuerger and Nicholson, 2006). The underlying molecular mechanism of this inhibition is currently unknown, but it does not appear to be lethal; when microbes were restored to Earth-normal pressure, germination and growth resumed (Nicholson and Schuerger, personal observations). Therefore, low pressure appears to be a formidable barrier for growth of terrestrial microorganisms on the martian surface.

2.5. Temperature Effects

The mean surface temperature of Mars is -61°C and ranges from nighttime lows of around -125°C to afternoon highs in Hellas Basin of perhaps $+20$°C or above. Prokaryotes themselves exhibit a wide range of temperatures at which they can grow, and microbes known as *psychrophiles* have been isolated from extremely cold Earth environments. Some of these organisms have been shown to grow in the laboratory at temperatures down to −10 to −12°C (Bakermans et al., 2003; Breezee et al., 2004), and retain some metabolic activity down to −15 to −20°C (Christner, 2002; Junge et al., 2006). However, the bacteria that have been isolated and cultivated from spacecraft and their assembly facilities to date are mostly mesophiles, with growth temperature minima ranging down to +10 to +15°C. When we cultivated seven spacecraft-contaminating *Bacillus* spp. on rich medium, Earth atmosphere, and various temperatures, we found that none was able to grow at or below +10°C; Furthermore, when this experiment was repeated using simulated Mars atmosphere, none was able to grow at +15°C or below (Schuerger and Nicholson, 2006). Therefore, most of the time, and on most of the locations on the surface of Mars, it is likely too cold to support the growth of common mesophilic spacecraft contaminants.

2.6. Combined Effects

In the experiments above, when physical parameters characteristic of the martian environment (atmospheric gas composition, pressure, and temperature) were altered and tested singly, it was found that terrestrial microbes were either slowed down significantly or entirely unable to grow as these parameters approached the values actually encountered on the martian surface. Furthermore, a microbe on the surface of Mars would encounter all these harsh physical parameters simultaneously. We have shown that CO_2-dominated atmosphere, low pressure, and low temperature actually combine to produce even greater inhibitory effects than does any one parameter applied by itself (Schuerger and Nicholson, 2006).

3. Discussion

Since the time of their formation, the environments of Earth and Mars have diverged significantly, to the extent that it may be difficult, if not impossible, for Earth-adapted microbes commonly found on spacecraft to survive or flourish if deposited directly into the Mars surface environment—the environmental shock may just be too great. However the possibility exists that certain Earth microbes might be able to live on Mars under special conditions. For example, mircoorganisms have been found living happily in hypersaline Antarctic lakes where temperatures can reach −14°C (Gilbert et al., 2004). These observations, coupled with the recent findings by the MER rovers of past saline environments at the martian surface (Squyres et al., 2004) strengthen the notion that martian microbes might inhabit cold salty subsurface aquifers. Our current experiments are aimed at determining if terrestrial microbes can survive and grow under accurate simulations of the conditions prevailing in these "special regions" of Mars (Beaty et al., 2006).

Acknowledgments. We thank Brent Christner for helpful discussions and José Robles for his excellent assistance with the manuscript. This work has been supported by grants from the NASA Exobiology Program (NNA04CI35A), the NASA Planetary Protection office (NAS2–00087, NNA05CS68G, and

NNA06CB58G), and the University of Florida / University of Central Florida Space Research Initiative (20020023/21988 and 00056802).

References

Bakermans, C., Tsapin, A.I., Souza-Egipsy, V., Gilichinsky, D.A., Nealson, K.H. 2003, Reproduction and metabolism at $-10°C$ of bacteria isolated from Siberian permafrost. Environ. Microbiol. 5, 321–326.

Beaty, D. et al. 2006. Findings of the Mars Special Regions Science Analysis Group. Astrobiology 6: 677–732.

Bennett, J.O., S. Shostak, and B. Jakosky. 2006. Life in the Universe, 2nd ed. Addison Wesley, 485pp.

Breezee, J., N. Cady, and J.T. Staley. 2004. Subfreezing growth of the sea ice bacterium "*Psychromonas ingrahamii*". Microb. Ecol. 47: 300–304.

Christner, B.C. 2002. Incorporation of DNA and protein precursors into macromolecules by bacteria at $-15°C$. Appl. Environ. Microbiol. 68: 6435–6438.

Fajardo-Cavazos, P., A.C. Schuerger, and W.L. Nicholson. 2007. Testing interplanetary transfer of bacteria by natural impact phenomena and human spaceflight activities. Acta Astronautica 60: 534–540.

Gilbert, J.A., P.J. Hill, C.E.R. Dodd, and J. Laybourn-Parry. 2004. Demonstration of antifreeze protein activity in Antarctic lake bacteria. Microbiology 150: 171–180.

Junge, K., H. Eicken, B.D. Swanson, and J.W. Deming. 2006. Bacterial incorporation of leucine into protein down to $-20°C$ with evidence for potential activity in subeutectic saline ice formations. Cryobiology 52:417-429.

Kempf, M. J.; Chen, F.; Kern, R.; Venkateswaran, K. 2005. Recurrent Isolation of Hydrogen Peroxide-Resistant Spores of Bacillus pumilus from a Spacecraft Assembly Facility. Astrobiology, Vol. 5, Issue 3, pp. 391-405.

Link, L., Sawyer, J., Venkateswaran, K., and Nicholson, W. 2003. Extreme spore UV resistance of Bacillus pumilus isolates obtained from an ultraclean spacecraft assembly facility. Microb. Ecol. 47, 159-163.

Nicholson, W.L., N. Munakata, G. Horneck, H.J. Melosh, and P. Setlow. 2000. Resistance of bacterial endospores to extreme terrestrial and extraterrestrial environments. Microbiol. Mol. Biol. Rev. 64: 548–572.

Nicholson, W.L., A.C. Schuerger, and P. Setlow. 2005. The solar UV environment and bacterial spore UV resistance: considerations for Earth-to-Mars transport by natural processes and human spaceflight. Mutat. Res. 571: 249–264.

Rummel, J.D. 2001. Planetary exploration in the time of astrobiology: protecting against biological contamination, Proc. Natl. Acad. Sci. USA 98: 2128–2131.

Rummel, J.D. 2002. Report on COSPAR/IAU Workshop on Planetary Protection, prepared by the COSPAR Planetary Protection Panel, Paris, France: COSPAR.

Schuerger, A.C. 2004. Microbial ecology of the surface exploration of Mars with human operated vehicles. pp. 363–386. in: C.S. Cockell (Ed.), Martian Expedition Planning, American Astronautical Society publication AAS 03–322, Univelt Publishers, Santa Barbara, CA, 2004.

Schuerger, A.C., B. Berry, and W.L. Nicholson. 2006a. Terrestrial bacteria typically recovered from Mars spacecraft do not appear able to grow under simulated martian conditions. Lunar and Planetary Science Conference 37: 1397.

Schuerger, A.C., R. L. Mancinelli, R. G. Kern, L. J. Rothschild, C. P. McKay. 2003. Survival of endospores of *Bacillus* subtilis on spacecraft surfaces under simulated martian environments: implications for the forward contamination of Mars. Icarus 165: 253–276.

Schuerger, A.C. and W.L. Nicholson. 2006. Interactive effects of hypobaria, low temperature, and CO_2 atmospheres inhibit the growth of mesophilic *Bacillus* spp. under simulated martian conditions. Icarus 185: 143–152.

Schuerger, A.C., J.T. Richards, D.A. Newcombe, and K. Venkateswaran. 2006b. Rapid inactivation of seven *Bacillus* spp. under simulated Mars UV irradiation. Icarus 181: 52–62.

Squyres, S.W. et al. 2004. In situ evidence for an ancient aqueous environment at Meridiani Planum, Mars. Science 306:1709–1714.

Natalia Gontareva

Carl Pilcher

Session XIII

SETI

Frank Drake

Guillermo Lemarchand

Bioastronomy 2007: Molecules, Microbes, and Extraterrestrial Life
ASP Conference Series, Vol. 420, 2009
K. J. Meech, J. V. Keane, M. J. Mumma, J. L. Siefert, and D. J. Werthimer, eds.

The Lifetime of Technological Civilizations and their Impact on the Search Strategies

G. A. Lemarchand

FCEN-CEA; Universidad de Buenos Aires; C.C. 8-Suc. 25; C1425FFJ Buenos Aires; Argentina. E-mail: lemar@correo.uba.ar

Abstract. Several very long-term biological and ecological studies have shown that different species on Earth emerge, develop, and become extinct with similar evolutionary patterns. Assuming the Principle of Mediocrity, we can expect that the distribution of different technological galactic species will behave with similar evolutionary patterns to the terrestrial ones. Based on the behavior of different centennial societal indicators we estimate the minimum lifetime of our terrestrial civilization. With this value, we infer a minimum threshold for the galactic density of technological civilizations. We apply these results to analyze the probability of a random detection by existing state-of-the-art SETI programs.

Introduction

The key factor for the success of the Search for Extra Terrestrial Intelligence (SETI) research program is our knowledge about the distribution of technological civilizations within the galaxy. A low distribution implies the "lack of success," while a high distribution implies a "possible detection" of technological extraterrestrial activities using our present radioastronomical and optical instrumentation.

By definition, a technological civilization has the technical capability to communicate or generate any kind of technological activities that can be intentional or unintentionally spread into the cosmos and detected from interstellar distances. Our species has reached the last stage when our first strong radio transmissions left the terrestrial ionosphere into outer space (~70 years ago).

In 1961, Frank Drake (Pearman 1963) proposed an equation to estimate the number of technological civilizations in our galaxy. Several calculations assigning different values to *Drake Equation's* factors showed that the number of technological galactic civilizations (N) is strongly dependent of the L factor, or the lifetime of a technological civilization in years (Kreifeldt 1973; Oliver 1975).

The *Principle of Mediocrity* proposes that our planetary system, life on Earth and our technological civilization are about "average" in the universe, and that life and intelligence will develop by the same rules of natural selection wherever the proper surroundings and the needed time are given (Hoerner, 1961). In other words, anything particular to us is probably average in comparison to others.

Several very long-term biological and ecological studies have shown that different species on Earth, emerge, develop and become extinct with similar evolutionary patterns (Charnov 1993; Gurney & Nisbet 1998). Human species may be not an exception. The *Homo sapiens* has broken the ecological "law" that establish that big, predatory animals are rare. Two crucial innovations in particular have enabled our species to alter the planet to suit ourselves and thus permit unparalleled expansion: speech (which implies instant transmission of an open-ended range of conscious thoughts) and agriculture (which causes the world to produce more human food than unaided nature would do). However, natural selection has not equipped us with long-term sense of self-preservation.

Based in our previous works (Lemarchand, 2000-2006), here, we will show that humans are facing a new type of macro-transition: the technological one. Lemarchand (2004) defined the *Technological Adolescent Age* (TAA) as the stage in which an intelligent species has the capability to become extinct due to: (1) self-destructive technological behavior (e.g. global war, terrorism, etc.); (2) environmental degradation of the home planet (e.g. global warming, overpopulation, increasing rates of species extinction, etc.); or simply by (3) the misdistribution of physical, educational and economical resources (difference between the degree of development among developed and developing societies). The last factor might cause the collapse of the civilization due to the tensions generated by the inequities among different fractions of the global society.

Here we present a semi-empirical approach to estimate the period of time that may last with this new macro-transition. Our humankind needs to pass from the TAA into a *Technological Mature Age* (TMA), in which we would learn how to live in harmony with the members of our species and the environment, and learn how to manage efficiently the increase of our power over nature at their different dimensions. If we considered all the steps that let the appearance of our technological civilization in the universe as "average," we could also assume that this TAA transition could be a typical evolutionary stage for all the hypothetical galactic civilizations.

A systematic study of the relevant indicators of the long-term behavior of our technological civilization is useful, not only to make an estimation of the possible value of L, but most importantly to identify which are the most variables that we should encourage to improve and change, in order to optimize the lifetime of our present terrestrial civilization. The comprehension of the evolution of these long-term societal patterns could help us to design different strategies to avoid our self-annihilation.

The TAA might be considered as a societal phase transition similar to the one that appears after the invention of the agriculture and cities (e.g. emergence of civilizations). In a broad sense, the TAA might be the bottleneck for the evolution of any technological civilization in the galaxy. Assuming the Principle of Mediocrity the TAA might be used as "a low threshold lifetime value" for similar galactic technological species.

In the following sections, we present some preliminary results that show the long-term evolution of several societal indicators and we applied those results to estimate the values of L, N, and the density of galactic civilizations $\delta(N)$.

The Lifetime of the Terrestrial Technological Civilization

In this section, we analyze the temporal evolution of several societal indicators to understand the time constants at which societies organized themselves in order to produce changes in a macro-behavior level.

Since the invention of mass destruction weapons (e.g. nuclear, chemical, biological, etc.) we have for the first time -in our evolutionary history- the possibility of becoming extinct as a terrestrial species. The large investment in military and defense applications of S&T research, development, and innovation activities, generated an exponential growth in the weapon's lethality (Fig.1). To show this effect we apply a *Theoretical Lethality Index* (TLI) to different types of weapons from 1780 to 2000. Colonel T.N. Dupuy (1979) originally proposed the TLI as the maximum number of casualties per hour that a weapon can generate, taking into account: rate of fire, number of targets, relative effectiveness, range effects, accuracy, reliability, etc. Our results show that the TLI rate increased in a factor of 60,000,000 during the last 100 years or 600,000 per year. This fact implies that we reached the technological capability to self-annihilate our civilization within a short period of time.

On the other hand, Sullivan et al (1978) showed that terrestrial radio transmissions increased their transmitting power at least in 4-5 orders of magnitude from 1945 to 1980, and showed how our terrestrial signals could be detect by nearby extraterrestrial technological civilizations. For example, the Ballistic Missile Early Warning System (BMEWS) radars sweep out a large fraction of full-sky with extraordinary powerful transmitters. With the exception of intermittent transmissions from planetary radars (e.g. Arecibo, DSN, etc.), the continuous BMEWS sky-surveys provide by far the most intense signals that leak from our planet to a large fraction of the galactic neighborhood.

It interesting to observe that our species reached the technological detectability level by a galactic neighbor civilization at the same time we reached the human self-annihilation technological capability.

Using data from 1800 to 1930, Richardson (1945) discovered that the distributions of wars over time follow a power-law. Levy (1983) introduced a definition of intensity of a war, I, as the ratio of battle deaths to the population at the time of the war. In our previous works (Lemarchand 2004, 2006) we have represented the distribution of the number of battles N_B against the intensity I. When we consider these distributions of deadly quarrels using normalized values of technologies, we found a correlation between the coefficient of lethality and the slope of the distribution that follows very well the behavior of a self-organized criticality (SOC) system. Our results show a very similar slope to the ones found by Clauset et al. (2007) in their analyses of the frequency of severe terrorist events worldwide between 1968 and 2006. Both studies cover the dynamics of the deaths generated by violence among humans from a few persons to tenths million people. In a frequency versus intensity log-log graph the slopes found in both studies are close to -1.4. In one way, this shows that the dynamics on inter-human violence is governed by the same type of processes observed in a great variety of other complex systems that have the SOC property (Jensen, 1998). A rough analysis shows that there is a small probability to have a violent event at which the complete human population will self-annihilate sometime, between the next decades to the next hundreds years.

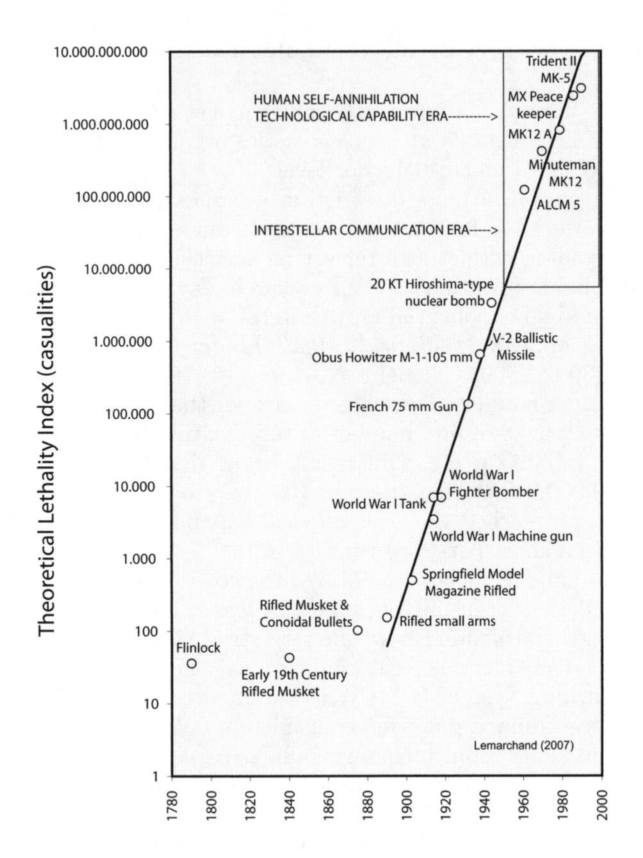

Figure 1. Historical evolution of weapon's lethality. Here we applied the *Theoretical Lethality Index* (Dupuy, 1979) to different weapons from 1780 to 2000. This index takes into account the physical characteristics of the weapons such as the rate of fire, number of targets, relative effectiveness, range effects, accuracy, reliability, etc. The vertical axis represents the theoretical maximum number of deaths that each weapon can generates per hour. The TLI increased in a factor of 60,000,000 during the last 100 years or at a rate of 600,000 per year.

Democracy is viewed as a mechanism of collective choice and a form of social organization that can be considered a superior substitute for other such mechanisms or forms of organization. In this way, democracy could be considered as a disembodied technology of government and social organization. As such, as well as any other embodied or disembodied technology, democracy is expected to grow, or diffuse, over time, amongst the world's population. To quantify this process, Lemarchand (2004) applied a logistic model to analyze, between 1800 and 2002, the diffusion of democracies as a fraction of the world population. The Pearson coefficient for the correlation between the real data and the model is $R^2 = 0.96$. The model predicts the annual growth rate of change for the global

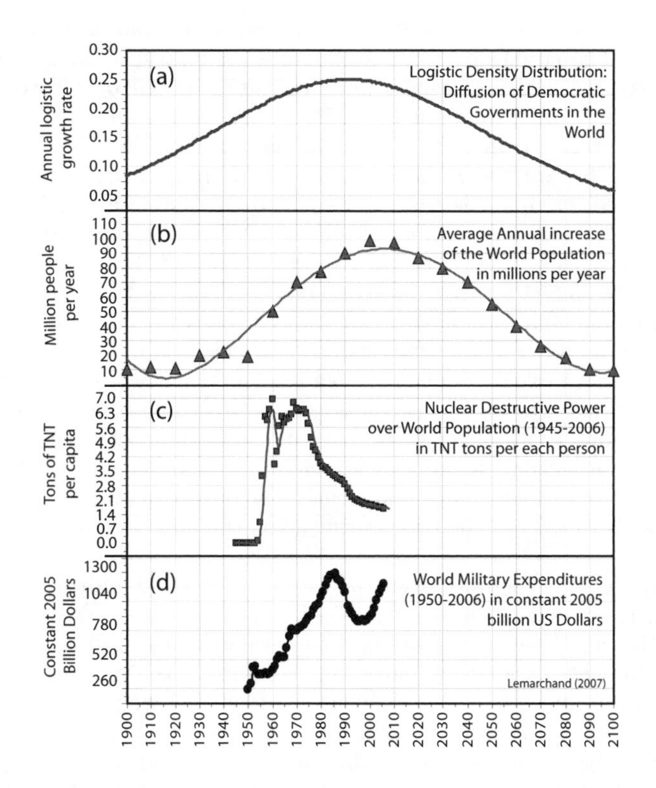

Figure 2. (a) Logistic density distribution (1900-2100) for the diffusion of democratic governments as a fraction of the world population (Lemarchand, 2004), using POLITY IV database from 1800 to 2002. We estimated the constants of the logistic growth model that fitted with $R^2 = 0.96$. (b) The world demographic transition according to UN estimations of the annual world population increases between 1900-2100. (c) Quotient between the distribution of destructive power accumulated in worldwide nuclear arsenals in megatons per year and the world population per year (1945-2006), expressed in tons of TNT per person. The present value is 1.7 tons of TNT per person. (d) Evolution of world military expenditures (1950-2006) in constant 2005 billion US dollars.

democratization human process. In Fig. 2 (a) we show this behavior between 1900 and 2100.

Kapitza (1996) proposed a mathematical model of the world population growth that showed a blow-up and self-similar regime. His model predicts the *Demographic Transition* or the well-established and observed change in the pattern of growth of all populations. This transition has been experienced by all developed countries; it began there at the end of 18th Century with the Industrial Revolution. At the present period, we are at the height of the transition on a global scale (see Fig. 2 (b) where we represent the UN estimations for the past and future increases in the world population). Following the mathematical model, the absolute rate of global population growth is expected to peak in the year 2007, but the relative growth rate reached its maximum value of 1.7 per year in 1989. The equations show that "for a planetary behavior" the human population started with its demographic transition by 1960 and would end by 2050. For the first time, since the invention of agriculture and cities, 12,000 ago, we are facing a phase-transition within the human reproductive regime.

In Fig. 2 (c) we show the distribution of the global destructive power deployed in nuclear arsenals per year divided by the total world population (1945-2006). The units are number of tons of TNT per person. Nowadays, there are almost 1.7 tons of TNT per each single human on Earth. By the end of 2006 there were 26,854 nuclear warheads deployed across USA, Russia, Britain, France and China. This amount is still many times over the minimum number needed to start a global climatic catastrophe (nuclear winter).

Due to the complexity, reliability and aging of early warning systems the rate of false alarms of a nuclear attack is relatively high. It is usually believe that during a major international crisis there would be insufficient time to distinguish false alarms from actual warning of an enemy attack. The available historical data suggests that a false alarm sufficiently severe to trigger a strategic nuclear attack would occur 50% of the time during a lengthy international crisis (Wallace et al. 1986).

Finally, Fig. 2 (d) shows the annual evolution of world military expenditures (1950-2006) expressed in constant 2005 billion US dollars. To scale this huge amount with something more familiar to us, we can take into account that the aggregated budget for space exploration (combining the annual budget of all national and international space agencies worldwide) represents less than 3.5% of the annual world military expenditures.

The long-term evolution of the different societal indicators analyzed here, show a transition that started after the WWII and that might end after the middle of XXI Century. In a first approach this period covers almost 200 years (see Fig. 2 (a) and (b)), with a highest peak within 1985-2015.

The population studies and their models show that we are facing a demographic transition similar to the one we already have 12,000 years ago with the invention of agriculture and cities. We are also facing a democratization transition worldwide that started 200 years ago and will last in 100 years. The studies about the distributions of wars over the last 500 years and the frequency levels of terrorist attacks show that the deadly violence among humans follows a self-organized criticality behavior. The SOC model predicts the possibility of having one event at which the whole world population will become extinct.

Our analysis shows the most dangerous combination of elements, which could start the end of our technological civilization within the next few decades. The following elements favored a human extinction event: *opportunity* (demographic and democratic transitions, Fig. 2(a) and (b)), *probability* (SOC human violence dynamics), *necessary technology* (Fig.1 and Fig. 2 (c)) and substantial *funding* (Fig. 2(d)).

If we want to avoid our species self-annihilation, we must change the rules of the inter-human interaction within the limits of this particular period, defined as TAA. In order to improve the lifetime of a technological civilization it is impossible to have superior science and technology and inferior morals. This combination is dynamically unstable and we can guarantee a self-destruction within the lifetimes of any advanced society (10^2 to 10^6 years?). At some point, in order to avoid their self-annihilation, all the intelligent species in the universe must have to produce an ethical breakthrough among the members of their societies in order to live in harmony between them and their planetary environment. Otherwise, the probability of global extinction might be very high and consequently their societal life expectancy very short.

The Impact of the Lifetimes of Civilization on the Search Strategies

Assuming the Drake Equation, the number of technological civilizations in the galaxy is represented by $N \sim (\Pi_f \times L)$. Here Π_f is the product among: the rate of galactic star formation; the fraction of stars forming planets; the number of planets per star with environments suitable for life; the fraction of suitable planets on which life develops; the fraction of life-bearing planets on which intelligence appears; and the fraction of intelligent cultures which are communicative in an interstellar sense. According to recent estimations, the most probable value could be close to $\Pi_f \simeq 1.5$. If we assume the so-called Principle of Mediocrity, we might also assume that the average lower boundary for a technological civilization lifetime with interstellar communication capabilities might be close to $L_{min} \sim 200$ years. Therefore, the lowest limit for the number of technological civilizations in our galaxy is $N_{min} \sim 300$. Assuming that all the technological civilizations N, reside in the galactic disk of radius R_g, thickness H_g the galactic density of civilizations will be

$$\delta(N) = \frac{N}{\pi {R^2}_g H_g} = \frac{300}{\pi 15^2\, 0.6} \times \text{ civilizations} \times \text{kpc}^{-3} \approx 0.7 \text{ civilizations} \times \text{kpc}^{-3}$$

The last result proposes a lowest threshold of one galactic civilization per each 1.42 kpc^3. A value $L_{min} \simeq 200$ yrs also implies that other civilizations should not have enough time to develop extremely powerful transmitters, so we implicitly are assuming a similar technological level to the terrestrial one. If these terrestrial-type civilizations are distributed uniformly but randomly, the probability of finding another galactic civilization in a sphere of radius R surrounding the Earth is:

$$P = 1 - e^{-\delta(N)\frac{4}{3}\pi R^3}$$

Extraterrestrial intelligence (ETI) should have similar transmitting capabilities to ours $EIRP_{Arecibo} \simeq 2 \times 10^{13}$ W. We can reasonable assume a typical equivalent isotropic radiated power of $EIRP_{ETI} \approx 10^{14}$ (we have the technology now to do so). Taking into account the present sensitivities of our SETI equipments, an Arecibo-type observation will be able to detect a 10^{14} W transmitting signal from a distance $R \simeq 1.4$ kpc. A fully developed (350 antennas) Allen Telescope Array (ATA) will be able to detect the same signal from a distance $R \simeq 0.35$ kpc. If we assume that *we are aiming our antenna to the correct place in the sky at the exact time at which the ETI signal is passing through Earth*, Arecibo's observations will have a probability $P \approx 0.8$ to detect it, while the chances of an fully developed ATA array will only be $P \approx 0.02$. It is clear that the last assumption is the trickiest one. We cannot expect an omnidirectional continuous time transmission strategy from ETIs. We need to look for new strategies to coordinate rational schemes for locations and timings between "transmissions" and "receptions" (Lemarchand, 1994).

Only those civilizations located at distances ≤ 0.02 kpc (≈ 70 ly) have the possibility to detect artificial radio signals coming from Earth, and consequently we are able to detect hypothetical "replies" only from stars that are at distances ≤ 0.01 kpc. We might also receive signals from those terrestrial-type civilizations, that might already detected Earth's size planets with instruments similar to *Kepler Mission* ones (search space $R \approx 1$ kpc) and eventually detected biosignatures in our planetary atmosphere with similar instruments to the ones we have under construction. These type of instruments already have the requirements to detect Eath-size planets where hypothetical nearby terrestrial-type galactic civilizations live ($L_{min} \approx 200$ yrs and $R \approx 1$ kpc). If our terrestrial civilization is an "average-type" galactic civilization, and if all the civilizations pass through the same TAA with $L_{min} \approx 200$ yrs, our analysis is in complete agreement with 47 years of SETI null results.

Acknowledgments. This research was supported by a Foundation for the Future (Seattle, USA) research grant. SETI Activities in Argentina are supported by The Planetary Society. My participation in this meeting was possible thanks to a generous travel grant from the organizers and The Planetary Society.

References

Charnov, E. L. 1993, *Life History Invariants*, Oxford Univ. Press, Oxford.

Clauset, A.; Young, M.; and Gleditsch, K.S. 2007, *J. Confl. Res.*, **51**: 58.

Gurney, W. S. C., & Nisbet, R. M. 1998, *Ecological Dynamics*, Oxford Univ.Press, Oxford.

Hoerner, S. von 1961, *Science*, **134**: 1839.

Jensen, H. J. 1998, *Self-Organized Criticality*, Cambridge. Univ. Press, Cambridge.

Kapitza, S. P. 1996, *Physics-Uspekhi*, **29**: 57.

Kreifeldt, J. G. 1973, *Icarus*, **14**: 419.

Lemarchand, G. A. 1994, *Ap&SS*, **214**: 209.

Lemarchand, G. A. 2000 in *When SETI Succeeds: The Impact of High-Information Contact*, ed. A. Tough, The Foundation for the Future, Bellevue, p.153.

Lemarchand, G. A. 2004, in *IAU Symposium 213: Bioastronomy 2002, Life Among the Stars*, eds. R.P. Norris and F. H. Stootman, A.S.P., San Francisco, p.460.

Lemarchand, G. A. 2006 in *The Future of Life and the Future of our Civilization*, ed. V. Burdyuzha, Springer, Dordrecht, p.457.

Levy, J. S. 1983, *War in the Modern Great Powers*, Univ. of Kentucky Press, Lexington.

Oliver, B. 1975, *Icarus*, **25**: 360.

Pearman, J. P. T. 1963, in *Interstellar Communication: The Search for Extraterrestrial Life*, ed. A. G. W. Cameron, W. A. Benjamin, New York, p. 287.

Richardson, L. F. 1945, *Nature*, **155**: 610.

Sullivan, W.T.; Brown, S. and Wetherill, C. 1978, *Science*, **199**: 377.

Wallace, M.D.; Crissey, B.L.; and Sennot, L.I. 1986, *J. Peace Res.*, **23**, 9.

Paul Shuch

Seth Shostak

Bioastronomy 2007: Molecules, Microbes, and Extraterrestrial Life
ASP Conference Series, Vol. 420, 2009
K. J. Meech, J. V. Keane, M. J. Mumma, J. L. Siefert, and D. J. Werthimer, eds.

New Strategies for SETI

S. Shostak

SETI Institute, 515 N. Whisman Rd., Mountain View, CA 94043

Abstract. It has been more than four decades since Frank Drake conducted the first, modern radio SETI experiment by searching solar-type star systems for persistent (minutes or hours), narrow-band signals (Drake 1960). Today, this approach remains a SETI strategy of choice (Tarter 1997). In this paper, we question the plausibility of this scheme by considering the problem from the point of view of the putative transmitting society. We deduce that very short, repetitive signal bursts might be a more likely signal sent from societies seeking to attract our attention.

1. Historic Strategies

SETI is primarily a search for electromagnetic (EM) signals. For largely historic reasons, serious SETI experiments were first performed at radio wavelengths; optical searches were attempted only several decades later.

Typically, SETI has looked for signals – or signal components – that maximize detectability for a given energy cost, and that approximately satisfy the transform relationship $\Delta\nu\Delta \mathrm{t} = 1$, where ν and t are respectively the frequency width and duration of the signal. In the radio regime, this has traditionally meant looking for very narrow, persistent carriers (or slowly pulsed carriers) of $\Delta\nu \sim 1$ Hz or less. In the optical, $\Delta \mathrm{t} \sim 10^{-9}$ second or shorter pulses have been sought, which would be visible over a correspondingly wide range of frequencies.

While such strategies simplify the detection problem, they hardly cover the gamut of possible signal types: either spectral regime could be used to send everything from extremely short pulses to spread-spectrum signals, and searching for these other emission types is something that SETI experiments of the future will presumably do. We will consider below a transmitting strategy that gives some insight into what might be the preferred signaling mode.

In terms of where on the sky to search, SETI experiments have historically fallen into two camps: (1) sky surveys and (2) targeted searches. We briefly consider each of these:

(1) Sky surveys make no assumptions about the location of extraterrestrial intelligence, but do so at the expense of reduced observing time – and therefore reduced sensitivity – for each patch of sky.

(2) Targeted searches are a straightforward derivative of the pioneering Project Ozma experiment, and sequentially examine specifically chosen locales. For most targeted searches, these locales were the positions of single F, G and K type stars (similar to the Sun). However, the candidate list has grown of late. Theoretical studies that indicate that planets could exist around many double

star systems (Quintana & Lissauer 2007), as well as the discovery of suggestive dust disks around double stars (Trilling et al. 2007) has obviated the usual SETI restriction to single star systems only. This approximately doubles the number of candidate stars.

In addition, other theoretical studies have shown that habitable planets could exist around the very common M-dwarf stars (Tarter et al. 2006), and several special objects, including black holes, supermassive stars, and certain types of binaries have also been proposed for SETI scrutiny.

In spite of a wide range of suggested targets, the approaches actually pursued by SETI practitioners are usually dictated by what is technically possible for us, as potential receivers, rather than what might make the most sense from the senders' point of view. In the following, we take the latter approach, and consider the possible implications for SETI.

2. Transmitting Strategies

In SETI's first few decades, it was assumed that a realistic, and possibly dominant signal type to be expected from extraterrestrial societies would be leakage radiation from broadcasts, radar, or other "domestic" activities.

However, scarcely more than a century after the invention of radio, much of the leakage radiation produced on our own planet seems destined to disappear. Large transmitters for television and FM radio, often operating with effective radiated powers measured in megawatts, will soon be supplanted by coaxial cable, fiber optic, and direct satellite broadcast. The drive toward greater efficiency will inevitably reduce leakage radiation, which is, after all, wasted energy.

It seems a reasonable hypothesis that detectable extraterrestrials will emit little leakage. These are, on statistical grounds, more technically advanced than we are, and their wasteful broadcast era will be long behind them. Consequently, if we detect their presence, it is because they are deliberately transmitting in our direction.

Given this assumption, what would prompt an extraterrestrial intelligence to target Earth? (Note that this is not a query as to their motivation for getting in touch with other intelligence, but merely to ask why they would send any signals to us in particular.) One possible answer is that they are aware of our existence, that Earth has distinguished itself with signs of intelligence. That seems unlikely. *Homo sapiens* has been broadcasting easily detected evidence of its presence only since the 1930s. Prior to that, there was little sign of human existence that could reach across interstellar space. Consequently, no star system farther than about 75 light-years will know we are here, and none more than half that distance has had time enough to send a signal our way – a signal our SETI programs might find.

Within 38 light-years of Earth are approximately 2,000 star systems. This is a small number. The most optimistic published estimates of the count of galactic worlds that are broadcasting detectable signals are of order 10^6 (Dick 1996). Even if these sanguine guesses are correct, only one in several times 10^5 star systems will host a technically active society, a number that is two orders of magnitude larger than the number that have heard from us and are close enough to have replied.

In short, it can be said that, unless the galaxy is far more fecund than even our optimistic estimates surmise, then no society is sending transmissions to Earth because of their knowledge of *Homo sapiens*. (Note that it can be said with only somewhat less confidence that no one in the galaxy knows of our existence, period. An exception to this would be if there were a probe near to Earth able to spot pre-radio human activity. Such a probe could have long ago sent information about us to its extraterrestrial parent.)

While intelligence was, until recently, cryptic, the presence of high levels of oxygen in Earth's atmosphere for more than 2 10^9 years means that any civilization that has visually detected our planet will know that biology is extant here.

Given these simple constraints on the knowledge that any extraterrestrials will have, there are only a few reasonable transmission strategies that would produce signals in our skies:

(1) An omnidirectional, or nearly omnidirectional broadcast throughout the Galaxy. While this is, of course, possible, it is costly in terms of the energy required – at least if the signal is to be detectable by contemporary SETI experiments. For example, a highly sensitive radio sky survey, SERENDIP III (Bowyer et al. 1997), had a sensitivity to narrow-band signals of about 3 10^{-25} watts/m^2. An omindirectional, continuous beacon at 25,000 light-years would need to operate at $\sim 10^{17}$ watts to produce a signal of this strength. Obviously, we can't state categorically what power levels are impractical for an advanced civilization, but we note that this number is comparable to the total solar insolation of Earth, and any transmitter bound to a planet would disrupt the climate if it consumed energy at this rate (Rebane 1993).

Because detectability scales only with the square root of the integration time, a shorter, intermittent (flashing) signal could produce a comparably detectable signal using a few factors of two less power in the radio. This assumes that each of the flashes lasts ~ 1 second (to minimize bandwidth spreading), and that the repeat interval is no more than a few tens of seconds, to be commensurate with current abilities to detect pulses. This is hardly a dramatic improvement.

A "beamed" beacon that either flashes or rotates, and that covers only the densest layer of the galactic plane (e.g., one or two thousand light-years thick at the Sun's location) could reduce power requirements between one and two orders of magnitude over that of the omnidirectional beacon. Nonetheless, this still requires more than the current rate of energy usage by the entire human populace and is therefore, one might hazard, non-negligible even for alien societies.

(2) A less wasteful transmitting strategy would be to narrowly target intended recipients. As noted, Earth will not be a target if the list is comprised only of worlds determined to have intelligence, since our presence is not yet known to putative senders. This situation could be used to augur against doing any SETI until our own signals have reached at least 10^5 star systems. But the assumption that extraterrestrials would target only worlds from which evidence of technology has been detected could founder if the typical light travel time (two-way) between a transmitting civilization and its intended recipients is comparable to the typical lifetime of technological societies. In that case, many transmissions that are sent as reactions to received signals would only reach (technologically) dead worlds.

Summarizing this line of argument: (a) light travel times between even nearby societies will be $\geq 10^3$ years if the number of such transmitting civilizations in the galaxy is $\leq 10^4$. This means that two-way communication could be tedious at best, and probably impractical. Consequently, transmissions would be "instructional" in the sense that they are designed to inform the recipients, but not to foster interaction. (b) Estimates of the technological lifetime of civilizations – while highly uncertain – have often suggested that this might be no more than a millennium or so (Shostak 2008). This is, of course, the same order of magnitude as the conservatively estimated light travel time between neighboring societies. (c) This equivalence in time scales could encourage some transmitting worlds to get in touch with recipients early in their technological development with a "ping" that would, for example, be a pointer to an information-rich source. Waiting for evidence of intelligence before transmitting would frequently result in a signal that arrives too late.

The only obvious strategy for targeting worlds with young technologies, a desideratum to ensure that the signal reaches them while they are still able to detect it, is to select recipients on the basis of long-lived biological signatures, such as the atmospheric oxygen noted above. Therefore, a viable approach for a transmitting society is to sequentially (and repeatedly) ping large numbers of bio-worlds (Shostak 2005). The shorter the ping, the shorter the repeat interval, and the greater the likelihood of detection. The ping itself would not convey much information beyond the critical fact of where on the sky it comes from. It is certainly true that if we were to find such a repeated ping, we would expend time and resources to study – with great sensitivity – that celestial direction. A low-power, omnidirectional source broadcasting the "message" would eventually be found, since the incentive and crucially important location information would both exist.

From the sender's point of view, a practical pinging project might use an electronically steerable array. If such an instrument had a maximum separation of elements of $\sim$5 10^8 wavelengths, then the transmitted beam would be $\leq$10 AU nearly everywhere in the Galaxy, or roughly the diameter of Jupiter's orbit. This should be adequate to encompass the conventional habitable zone of nearly all stars. At 21 cm wavelength, this is an array that is 10^5 km in size – not impossible for a space-based instrument. At 2 cm wavelength, it would fit within a single continent. The required power to produce a radio signal that we could detect today, is watts to megawatts (depending on the length of the ping), which is enormously less costly than the omnidirectional schemes described earlier.

While we have made these arguments on the basis of radio SETI, analogous logic also applies to optical experiments.

3. Conclusions

Today's SETI strategies are a strange marriage of short dwell times (1 - 2 sec for radio sky surveys, 100 – 200 sec for targeted searches, and a maximum of 600 seconds for the Lick Observatory optical SETI search) and long expectations. At least for the radio searches, if the signal isn't on the air for at least days (and years in the case of experiments that use off-line processing), then we can't find it with conviction. Indeed, it's often assumed that, since we generally look at targets only once in our radio studies, the transmitters are targeting us for at least thousands of years.

Since *Homo sapiens* has only started to fill nearby space with evidence of its presence, it is probable that no extraterrestrials know that our solar system has technology capable of receiving their signals. But continuous (or even pulsed) signals that are broadcast to large swaths of the Galaxy – the type of indiscriminate transmission that would reach us as a non-specific target – are energetically very costly.

In the face of this high cost, one might expect that extraterrestrial broadcasters might wait until they had received a signal from a world before directing a specific, targeted response. However, if the technological lifetime of the average society is only thousands of years or less, then a better approach is to ping a large number of solar systems in which worlds with biology have been detected, on the assumption that some fraction of these will have produced intelligence. This could be done in a rapid, and at very low-cost (in terms of energy) with tightly focussed, electronically steerable beams.

This efficient and seemingly efficacious transmitting strategy suggests that our SETI experiments might better eschew the search for continuous or nearly-continuous signals in favor of short, intermittent, and repeated pinging signals that are intended to be pointers to those who are trying to get in touch. The implications for SETI are to adapt our strategies and our equipment to better be able to find these types of signals.

References

Bowyer, S., Werthimer, D., Donnelly, C., Cobb, J., Ng, D. and Lampton, M. 1997, "Twenty years of SERENDIP, the Berkeley SETI effort: past results and future plans," *Astronomical and Biochemical Origins and the Search for Life in the Universe*, eds. C. Cosmovici, S. Bowyer, and D. Werthimer, Editrice Compositori (Bologna), 667

Dick, Steven J. 1996, *The Biological Universe: The Twentieth-Century Extraterrestrial Life Debate and the Limits of Science*, Cambridge University Press (Cambridge)

Drake, F. 1960, "How Can We Detect Radio Transmissions from Distant Planetary Systems," *Sky and Telescope* **39**, 140

Quintana, Elisa V., & and Lissauer, Jack J. 2007 "Terrestrial planet formation in binary star systems," arxiv:0706.3444, May 23

Rebane, K. K. 1993, "The search for extraterrestrial intelligence and ecological problems," in *The Third Decennial US-USSR Conference on SETI*, ed. S. Shostak, ASP Conference Series, **47**, Astronomical Society of the Pacific, (San Francisco), 219

Shostak, S. 2005, "Short-Pulse SETI", presented at 56th International Astronautical Congress, Oct 17-21, Fukuoka, Japan, IAC-05-A4.1.07

Shostak, S. 2008, "The lifetime of technological civilizations," *Cosmos and Culture: Cultural Evolution in a Cosmic Context*, eds. Steven J. Dick and Mark Lupisella, NASA History Series (Washington, DC)

Tarter, J. C. 1997, "Results from Project Phoenix: looking up from down under," *Astronomical and Biochemical Origins and the Search for Life in the Universe*, eds. C.B. Cosmovici, S. Bowyer, and D. Werthimer, Editrice Compositori (Bologna) 633

Tarter, J. C., Backus, P. R., Mancinelli, R. L., Aurnou, J. M., Backman, D. E., Basri, G. S., Boss, A. P., Clarke, A., Deming, D., Doyle, L. R., Feigelson, E. D., Freund, F., Grinspoon, D. H., Haberle, R. M., Hauck, S. A. II, Heath, M. J., Henry,

T. J., Hollingsworth, J. L., Joshi, M. M., Kilston, S., Liu, M. C., Meikle, E., Reid, I. N., Rothschild, L. J., Scalo, J. M., Segura, A., Tang, C. M., Tiedje, J. M., Turnbull, M. C., Walkowicz, L. M., Weber, A. L., and Young, R. E. 2006, arxiv.org/pdf/astro-ph/0609799

Trilling, D. E., Stansberry, J. A. Stapelfield, K. R., Rieke, G. H., Su, K. Y. L., Gray, R. O., Corbally, C. J., Bryden, G., Chen, C. H., Boden, C. H., and Beichman, C. A. 2007 "Debris disks in main sequence binary systems," arXiv:astro-ph/0612029v1

Steven Dick

Bioastronomy 2007: Molecules, Microbes, and Extraterrestrial Life
ASP Conference Series, Vol. 420, 2009
K. J. Meech, J. V. Keane, M. J. Mumma, J. L. Siefert, and D. J. Werthimer, eds.

Cultural Evolution and SETI

S. J. Dick
NASA HQ

Abstract. The Drake Equation for the number of radio communicative technological civilizations in the Galaxy encompasses three components of cosmic evolution: astronomical, biological and cultural. Of these three, cultural evolution totally dominates in terms of the rapidity of its effects. Yet, SETI scientists do not take cultural evolution into account, perhaps for understandable reasons, since cultural evolution is not well-understood even on Earth and is unpredictable in its outcome. But the one certainty for technical civilizations billions, millions, or even thousands of years older than ours is that they will have undergone cultural evolution. Cultural evolution potentially takes place in many directions, but this paper argues that its central driving force is the maintenance, improvement and perpetuation of knowledge and intelligence, and that to the extent intelligence can be improved, it will be improved. Applying this principle to life in the universe, extraterrestrials will have sought the best way to improve their intelligence. One possibility is that they may have long ago advanced beyond flesh-and-blood to artificial intelligence, constituting a postbiological universe. Although this subject has been broached, it has not been given the attention it is due from its foundation in cultural evolution. Nor has the idea of a postbiological universe been carried to its logical conclusion, including a careful analysis of the implications for SETI. SETI scientists, social scientists, and experts in AI should consider the strengths and weaknesses of this new paradigm.

1. The Importance of Cosmic Evolution

The evolution of the universe and its constituent parts from the Big Bang to the present has been confirmed only during the last 50 years. Cosmic evolution (Chaisson 2001, 2006), now known from spacecraft to encompass 13.7 billion years, has three components: astronomical, biological and cultural. If primarily astronomical, ending in planets, stars and galaxies, we live in a physical universe. If it typically proceeds to life, mind and intelligence, we live in what I term the "biological universe" (Dick 1996). And if intelligence is common, inevitably resulting in cultural evolution, I would argue that we may live in a "postbiological universe." A postbiological universe is one in which flesh-and-blood intelligence has been largely replaced by artificial intelligence. This possibility has been broached before (MacGowan & Ordway 1966; Davies 1995; Shostak 1998) but not seriously considered or taken to its logical conclusion.

I argue that if extraterrestrial intelligence exists, cultural evolution must be taken as seriously as other components of cosmic evolution. We know from Earth that the rate of cultural evolution totally dominates other forms of cosmic evolution. Ten thousand years ago terrestrial intelligence was not much different than it is today, still less astronomical objects. But cultural evolution has caused vast changes in human life since that time. If intelligent life exists in the universe, we must take into account the long time span during which that intelligence may have evolved culturally. In the fashion of Olaf Stapledon's Last and First Men and Star Maker, we must contemplate the possibilities of cultural evolution over thousands, millions, or billions of years. We cannot, of course, predict what might have happened, but we can pose plausible scenarios, scenarios that may have implications for our search. Here I explore one such scenario.

2. The Age of Extraterrestrial Intelligence and the Lifetime of Civilizations

Pioneers and practitioners of the Search for Extraterrestrial Intelligence (SETI) have often stated that extraterrestrial intelligence would be much older than terrestrial intelligence (Drake 1976, Shklovskii & Sagan 1966), and that therefore SETI programs stand to inherit the knowledge and wisdom of the universe. But they assume that ETI will just be some more advanced form of terrestrial intelligence and culture. This ignores what we observe on Earth:- a rapid pace of cultural evolution. How much time might extraterrestrials have had for such evolution ? Again, cosmic evolution holds the key to the possibilities by setting the parameters. Results from the Wilkinson Anisotropy Mapping Probe (WMAP) place the age of the universe at 13.7 billion years, with 1uncertainty, and confirm that the first stars formed at about 200 million years after the Big Bang (Bennett 2003; Seife 2003). The first stars were massive, but Sun-like stars and rocky planets may have formed about 12.5 billion years ago (Larson 2001). If Earth's history is any guide, it may have taken another 5 billion years for intelligence to evolve, so that the first intelligence in the universe could have evolved some 7.5 billion years ago, and 4-5 billion years ago in our own Milky Way Galaxy. These conclusions are essentially in line with those of other astronomers, who have estimated from different but reinforcing lines of evidence the age of extraterrestrial civilizations at several billion years (Kardashev 1997; Livio 1999; Norris 2000).

But this assumes that once intelligence arises it and the civilizations it may form will have indefinite longevity. This gives rise to the famous debates over L (the lifetime of a technological civilization) in the Drake Equation (Dick 1996). Beyond the single data point of Earth, values assigned to L depend largely on whether one is optimistic or pessimistic about the survival of civilizations. Taking radio communications (or computers) as the origin of a technological civilization, L is around 100 years on Earth; but it is also true that Homo sapiens has managed to nurture civilization in some form on our planet over the last five millennia. Nuclear war, mass extinctions, supernovae and gamma ray bursters are also problematic for the longevity of civilizations, but not necessarily fatally limiting (Chapman & Morrison 1994; Raup 1992; Scalo & Wheeler 2002). Counterbalancing the possibilities for destruction is the rapid pace of cultural evolution. Unlike biological evolution, L need only be thousands of years for cultural evolution to have drastic effects on civilization. In the end a value for L cannot be reached by rigorous deduction. But even the possibility of long lifetimes for civilizations should lead us to explore the likely evolution and nature of such civilizations and the intelligence which constitutes them.

3. Cultural Evolution

Compounding the problem of cultural evolution among extraterrestrials (and partially explaining why it has not been taken seriously before) is our lack of knowledge about the mechanism and direction of cultural evolution on Earth. Darwin is justly famous not for proposing biological evolution, but for arguing that the mechanism for biological evolution is natural selection. Unlike natural selection for biology, there is no consensus on mechanisms for cultural evolution, even though it takes place before our eyes. Recent Darwinian models for cultural evolution abound, but remain highly controversial, ranging from Dennett's general application of "Darwin's Dangerous Idea" to E. O. Wilson's sociobiology, Boyd and Richerson's detailed version of gene-culture coevolution, and Richard Dawkins' memes (Lalande 2002).

Lacking a robust theory of cultural evolution and its mechanisms, we are reduced to an extrapolation of current trends. There are many possibilities, but in sorting priorities I argue that all other possibilities are subservient to intelligence. I therefore adopt what I term the central principle of cultural evolution (Dick 2003), which I refer to as the Intelligence Principle: the maintenance, improvement and perpetuation of knowledge and intelligence is the central driving force of cultural evolution, and that to the extent intelligence can be improved, it will be improved. In Darwinian terms, knowledge has survival value, or selective advantage, as does intelligence at the species level, a fact that may someday be elucidated by an evolutionary theory of social behavior, whether group selection (Wilson 2002), selfish gene theory, evolutionary epistemology, or some other Darwinian model. The intelligence principle implies that, given the opportunity to increase intelligence (and thereby knowledge), whether through biotechnology, genetic engineering or artificial intelligence, any society would do so, or fail to do so at its peril.

There are many possibilities for improving intelligence, but perhaps the strongest Earth analog, admittedly taken from our parochial current point of view in space and time, is artificial intelligence (AI). Two decades ago Moravec (1988) predicted a postbiological Earth. Kurzweil (1999, 2005) has also argued and elaborated the AI claim, envisioning the takeover of biological intelligence, not by hostile machines but by willing humans who have their brains scanned, uploaded to a computer and live their lives as software running on machines. One can argue whether or not AI is possible (the weak vs strong AI debate), and progress in AI has admittedly been slower than many had hoped or predicted. But many have argued that progress has been accelerating in recent years, and we need to keep in mind the long time spans over which extraterrestrial intelligence may have evolved.

4. Implications for SETI

A postbiological universe has considerable implications for SETI. Environmental tolerance and availability of resources beyond the planetary realm means that SETI searches for postbiologicals need not be confined to planets around Sun-like stars, nor to planets at all. Postbiologicals could roam the Galaxy as von Neumann machines, Bracewell probes or smart microprobes. The Intelligence Principle tending toward the increase of knowledge and intelligence implies that postbiologicals would be most interested in civilizations equal to or more advanced than they, perhaps leaving us to intercept communications between postbiologicals rather than communications directly beamed toward us. This only makes the search more difficult. For similar reasons, postbiologicals might be more interested in receiving information than sending.

One may certainly debate these arguments, particularly strong AI, the definition of intelligence, the validity of the Intelligence Principle, and the nature and necessity of progress. SETI proponents, social scientists, cultural evolutionists and artificial intelligence experts should discuss the possibilities, and devise a plan of action that might supplement the current search based on an almost five-decade-old paradigm that sees extraterrestrials as humanoids with minds, motives and cultures similar to ours.

The postbiological universe may or may not currently be reality, and the postbiological Earth may or may not become reality. But cultural evolution as an integral component of cosmic evolution is unquestionably a reality. Whether or not extraterrestrials are postbiological, and whether or not our human successors become postbiological, one thing is certain: they will not be like us.

References

Bennett, C. L. et al., 2003. First year Wilkinson Microwave Anisotropy Probe (WMAP) observations: Preliminary maps and basic results, Ap.J Suppl, 148, 1-27.

Chapman, C.R. and D. Morrison, 1994. Impacts on the Earth by asteroids and comets: assessing the hazard, Nature, 367, 33-34.

Dick, S. J. 1996. The Biological Universe: The Twentieth Century Extraterrestrial Life Debate and the Limits of Science, Cambridge University Press, Cambridge.

Chaisson, E. 2001. Cosmic Evolution: The Rise of Complexity in Nature, Harvard University Press, Cambridge, Ma.

Chaisson, E. 2006. Epic of Evolution: Seven Stages of the Cosmos, Columbia University Press, New York.

Davies, P. Are We Alone? Philosophical Implications of the Discovery of Extraterrestrial Life, Basic Books, New York, 1995.

Dick. S. J. 2003. Cultural evolution, the postbiological universe and SETI, International Journal of Astrobiology, 2, 65-74

Drake, F. 1976. On hands and knees in search of Elysium, Technology Review, 78, 22-29

Kardashev, N. 1997. Cosmology and civilizations, Astrophys. Space Science, 252, 25-40.

Kurzweil, R. 1999. The Age of Spiritual Machines: When Computers Exceed Human Intelligence, Penguin Books, New York

Kurzweil, K. 2005. The Singularity is Near: When Humans Transcend Biology, Viking, New York

Lalande, K. and G. R. Brown, 2002. Sense & Nonsense: Evolutionary Perspectives on Human Behavior, Oxford University Press, Oxford.

Larson, R.B. and V. Bromm, 2001. The first stars in the universe, Scientific American, 285, 64-71

Livio, M. 1999. How rare are extraterrestrial civilizations and how did they emerge? Astrophysical Journal, 511, 429-431.

MacGowan, R. and F. I. Ordway, III, 1966. Intelligence in the Universe, Prentice-Hall, Englewood Cliffs, N.J.

Moravec, H. 1988. Mind Children: The Future of Robot and Human Intelligence, Harvard University Press, Cambridge, MA

Norris, R. 2000. How old is ET? In A. Tough (ed), When SETI Succeeds: The Impact of High-Information Contact, Foundation for the Future, Bellevue, WA, 103-105

Raup, D. 1992. Extinction: Bad Genes or Bad Luck, Norton, New York, 1992.

Scalo, J. and J. C. Wheeler, 2002. Astrophysical and astrobiological implications of gamma-ray burst properties, Astrophysical Journal, 566, 723-787.

Seife, C. 2003. MAP glimpses Universe's rambunctious childhood, Science, 299, 991, 993.

Shklovskii, J. and C. Sagan, 1966. Intelligent Life in the Universe, Holden-Day, San Francisco, CA.

Shostak, S. 1998. Sharing the Universe: Perspectives on Extraterrestrial Life, Berkeley Hills, Berkeley, CA.

Wilson, D.S. 2002. Darwin's Cathedral: Evolution, Religion and the Nature of Society, University of Chicago Press, Chicago.

Richard Carrigan

Bioastronomy 2007: Molecules, Microbes, and Extraterrestrial Life
ASP Conference Series, Vol. 420, 2009
K. J. Meech, J. V. Keane, M. J. Mumma, J. L. Siefert, and D. J. Werthimer, eds.

Progress on IRAS-based Whole-Sky Dyson Sphere Search

R. A. Carrigan, Jr.

Fermi National Accelerator Laboratory, Batavia, IL 60510, USA

Abstract. A Dyson Sphere (DS) is a hypothetical construct of a star purposely shrouded by a thick swarm of broken-up planetary material to better utilize all of the stellar energy. A clean DS identification would give a signature for intelligence at work elsewhere in the galaxy. A comprehensive search for Dyson Spheres has been carried out using the IRAS Low Resolution Spectrometer data base. Only a few candidates have been found and even these are not compelling. This search imposes a significant constraint on one class of astroengineering projects since the IRAS search had enough sensitivity to see an infrared object with the power output of the sun out to 400 pc. This reach encompasses a region that contains more than a million solar type stars.

1. Introduction

Dyson Spheres, shells surrounding stars to harness energy, were suggested by Dyson (1960). Advanced civilizations might undertake such an astroengineering project. Signatures for some grand technologies on this scale have been reviewed by Lermachnd (1995). Searching for signatures of these projects, Cosmic Archaeology or CA, represents a different approach to finding intelligence elsewhere in the Universe. Unlike SETI signals generated as beacons or for communication (Tarter 2000), the creation of a CA signature like a Dyson Sphere would not have required an active strategy on the part of the originating "civilization".

Dyson's suggestion posits that an advanced civilization might break up a star's planets into small planetoids or chip-sized fragments to form a loose shell that would collect all the light from the star. If the visible light was totally absorbed by the swarm a pure DS signature would be an infrared object with luminosity equivalent to the invisible star and a blackbody distribution with a temperature corresponding to the radius of the planetoid swarm. For the case of the Sun with the planetoids at the radius of the Earth the temperature would be approximately 300 K.

An earlier article (Carrigan 2004) reported a search for Dyson Spheres using IRAS filter data. It also reviewed the DS conjecture in more detail and discussed earlier searches (Jugaku & Nishimura 2002; Slysh 1985; Timofeev et al. 2000). The infrared satellite IRAS identified 250,000 infrared point sources and scanned 98% of the sky. IRAS and the IRAS database were discussed in the earlier article. In many ways the IRAS data base is ideal for a DS search because it covers the 8 to 100 micron infrared regime characteristic of a DS with both filters and a spectrograph and also covers the whole sky.

2. Dyson Sphere Surrogates and Aliases

A number of astronomical objects have infrared signatures that somewhat mimic a Dyson Sphere. Indeed, the birth and death phases of many stars are characteristically associated with heavy dust clouds around the star. Proto stars and planetary nebula congeal out of dust, typically in a time short on stellar scales while a dying star may blow off thick dust clouds in an extreme mass loss phase, again in a relatively short time. Since the associated times are short, the objects are rare. Some objects with resemblances to Dyson Spheres include stars with thick dust shells, regions of dust in a galaxy and very young stars that form in regions of dust, Miras stars, planetary nebula, as well as Asymptotic Giant Branch stars (AGB), and post AGB stars associated with the late phase of a star's history. More detail on these signatures is given in (Carrigan 2006).

3. The IRAS LRS Search

For a comprehensive DS search it is useful to have a whole sky survey. This rules out point and shoot infrared satellite instruments like the Spitzer Telescope[1]. In addition, the search device needs to be sensitive to temperatures ranging from roughly 100 to 600 K. Wavelengths associated with this temperature interval span 10 to 100 μm. This eliminates 2MASS as a search tool because it only reaches 2.17 μm (Skrutskie et al. 2006). Generally ground-based infrared telescopes don't work well because of the high sky background. Finally good angular resolution would be helpful to rule out associations with nearby stars. If an IRAS and a 2MASS source can be correlated, one can take advantage of the better 2MASS angular resolution and pointing accuracy. Overall, the database from the IRAS satellite (Beichman 1987) is the best existing resource available to address these requirements.

IRAS carried a Low Resolution Spectrometer (LRS) spanning 8 to 22 μm as well as the four filters centered at 12, 25, 60, and 100 μm. The LRS spectrometer sensitivity extended below 2 Jy in the 12 and 25 μm filters for the search reported below.

The Dyson Sphere search reported here used the 11224 source extended Calgary atlas for the IRAS Low Resolution Spectrometer (Kwok et al. 1997). The original search presented earlier (Carrigan 2004) used the 12, 25, and 60 μm filters to do a blackbody fit. The requirement that the IRAS quality factor for the 60 μm filter FQUAL(60)$>$ 1 was dropped for this new search to address the problem that low temperature DS sources would have small values in the 60 μm filter. Using the LRS data also overcomes the impact of cirrus and zodiacal light on the 60 μm IRAS filter as well as the associated impossibility of fitting a Planck distribution with just two filter points and still obtaining a measure of the statistical significance. From the 11224 source sample 10982 satisfy the requirement FQUAL(12), FQUAL(25)$>$ 1.

Dyson's original hypothesis envisioned spheres constructed with a radius to the star corresponding to the temperature range where water is a liquid. A broader picture might include a temperature span that incorporates the range of

[1]See http://ssc.spitzer.caltech.edu/documents/Spitzer_PocketGuide.pdf

temperatures over which some automata could operate. The initial temperature span for this search was set somewhat arbitrarily to be $100 < T1 < 600K$ where T1 is the color temperature (Carrigan 2004) based on a polynomial in the flux ratio f_{12}/f_{25}, f_{12} is the IRAS FLUX(12), f_{25} is the FLUX(25), and c_k are the fitting constants. The polynomial constants are c_0 = 98.0, c_1 = 433.1, c_2 = -386.0, c_3 = 240.2, c_4 = -47.0. Applying these temperature cuts leaves a sample of 6521 sources.

To limit the sample for individual source investigation, the LRS spectra for an earlier sample of 387 sources selected using the 12, 25, and 60 μm filters were examined and compared to the IRAS classification. The Calgary atlas assigns each LRS source to one of a number of groups on the basis of spectrographic features and background continuum shape. The spectra most similar to a Planck distribution fell in the C, F, H, P, and U Calgary groups. Carbon stars are classified as C. These are late type stars such as evolved cool giants with circumstellar shells or clouds of carbon dust material. Often they have an 11.3 μm SiC dust emission line. F objects typically have featureless spectra and are often oxygen or carbon rich stars with small amounts of dust. H cases have a red continuum often with a 9.7 μm absorption feature. Typically for the H cases the blackbody distribution has a temperature below 200 K. P cases can have a red continuum with a sharp rise at the blue end of the spectrum or either an 11.3 or 23 μm PAH emission feature. Sources in the U group have "unusual spectra showing a flat continuum" where the nature of the source is typically unknown. The other categories were not investigated for various reasons. For instance, A group stars had a 9.7 μm absorption feature. Initially the P group was not analyzed. Later the group was investigated further but eventually all of the P classification candidates were rejected. Selecting on C, F, H, P, and U resulted in a sample of 2240 sources.

The original search focused on "pure" Dyson Spheres where there would be no visible source. Sample selection was done by taking objects with the IRAS variable IDTYPE equal to 0. Strictly IDTYPE 0 corresponds to no catalog identification of the object. This requirement was relaxed for this new search to include cases where IDTYPE 2, an identified stellar object. After the IDTYPE selection 1230 objects remained. Many sources in the sample are associated with near-by objects that are either visible or 2MASS stars. Once more it is emphasized that this cut allows the possibility of both pure and partial Dyson Spheres.

A large fraction of the 1230 source sample was fitted using a weighted and an unweighted least squares. The weighted least squares consisted of the sum of the squares of the normalized Calgary corrected spectral values multiplied by the wavelength subtracted from the frequency form of the Planck distribution and weighted by the reciprocal of the Planck distribution. The original IRAS set contains some correction problems that appear when the short and the long wavelength spectrographic sets are compared. The Calgary-corrected values (Cohen et al. 1992) were chosen rather than the original IRAS set. The least squares was minimized by adjusting the temperature in the Planck distribution and the relative normalization. An unweighted least squares was also used. The least squares statistics were used to appraise but not reject sources.

Each source in the 1230 source sample was examined using the Strasbourg SIMBAD viewer[2] to see if there was a nearby optical source. The spectrum for the source was reviewed by plotting the IRAS filter values and the Calgary corrected spectroscopy data sets as well as the three filter values for the brightest, most proximate 2MASS source and also the fitted spectrum. Direct scanning of the spectra for non-Planck shapes, obvious spectral lines, and large data scatter, or for visual stars coupled with other discriminants eliminated about 80% of the 1230 object sample.

IRAS 17446-4048, classified as C by Calgary, illustrates a common source type in the sample. The LRS spectrum follows a Planck distribution with an added 11.3 μm SiO emission feature typical of carbon stars. The background below the emission line for these C cases is often close to a Planck distribution. For 17446-4048 the 2MASS points lie on the blackbody distribution. Several filter points are also available from the DIRBE experiment (Diffuse IR Background Explorer) on COBE. These have large errors and an angular resolution of only 0.7 degrees but do suggest a blackbody distribution that continues below the LRS wavelengths. This type of source was readily eliminated when the line was more than several sigma above the background.

After the initial scan about two thirds of the remaining 218 "possible" sources were eliminated because of identification of the source with a known object like a carbon star, often in conjunction with a nascent spectral line or a large deviation in the spectrum that made it difficult to identify the spectrum with a Planck distribution.

At this point sources in the H group were reexamined. IRAS 13129-6211 illustrates such a case. The distribution looks like it could be a Planck spectrum due to a Dyson Sphere but has another explanation. It may be near an HII region. There is also a nascent 11.5 μm absorption feature. The Planck temperature is 132 K. Typically H sources have low temperatures. In this case the source is near a visible star. About 2/3 of the H sources like 13129-6211 were in bright sky regions with high F[100] values (F[100] $>$ 100). Eventually all of the H sources were excluded because they could be explained by another mechanism. This left a sample of 39 sources. For that sample, regions with high F[100] values had least squares values that were high, typically due to dust, so that the blackbody fits were not good. This was not so true for the F[60] filter. A cut on bright sky in the F[100] filter eliminated 14 more sources so that 25 "interesting" sources remained.

The unweighted least squares was used at this point to cut out 11 sources because they had least squares values $>$ 0.25. Typically for this situation the correlation coefficient for the Planck fit was below 0.6-0.7, it was difficult to unfold Gaussian lines, and a polynomial was better than a Planck fit. Five more sources with least squares above 0.19 were cut because they had suspect features such as an association with a very bright visible or 2MASS star. This left only nine candidates.

IRAS 20369+5131 is the closest approximation to a Dyson Sphere spectrum in the nine source sample. MSX infrared filter data confirms the LRS distribution. (MSX is the Midcourse Space Experiment.) Several filter points

[2]The Strasbourg SIMBAD viewer is at http://simbad.u-strasbg.fr/sim-fid.pl

from DIRBE at small wavelengths lie well above the extrapolated blackbody distribution, MSX, and the closest 2MASS source. It may have been that the wide angular resolution of DIRBE overlapped a second source. The blackbody temperature is 376 K. While this looks like a pure Dyson Sphere SIMBAD classifies it as a carbon star. Typically this classification would be based on other spectra in the millimeter regime. However no such information was found in a literature search. The source is above the galactic plane and in a direction away from the galactic center. Using a bolometric formula developed by Slysh[6] one can estimate a bolometric distance of 42 pc for a DS with the luminosity of the sun.

Two other somewhat similar sources were identified; IRAS 19159+1556 and IRAS 16406-1406. Nether one was a particularly strong candidate. For the other six sources a polynomial fit to the distribution is typically as good as the Planck fit. For the future it may be useful to look at all nine sources both with radio telescopes and infrared resources such as the Spitzer GLIMPSE database.

None of the 1230 source sample was in lists of Dyson Sphere-like candidates identified by (Slysh 1985) or (Timofeev et al. 2000) in their IRAS DS searches. None of the nine sources were searched by the Phoenix SETI program. Two of the nine sources (IRAS 19159+1556, an "interesting" case and IRAS 20035+3242, a "challenged" source) were scanned in the SERENDIP SETI survey. No SETI signals were found at the level of 10^{-26} W/m^2 sampled over ~1 Hz windows in the frequency range from 424-436MHz.

4. Summary

This Dyson Sphere search has looked at many of the LRS sources with temperatures under 600K. The IRAS LRS reached down to filter fluxes of 1-2 Janskys in the F[12] and F[25] filters so that it could in principle have found sun-sized Dyson Spheres out to 400 pc. The results of this search suggest that at best there are only a few pure Dyson Sphere signatures with IDTYPE = 0, 2 out of the IRAS LRS sample in the $100 < T < 600$ K temperature region. With several possible exceptions all the sources identified in this search have some more conventional explanation other than as a Dyson Sphere candidate. Pure Dyson Spheres are not easy to find!

Acknowledgments. The author would like to thank K. Volk, J. Annis, and C. Beichman for guidance.

References

Beichman, C., Ann. Rev. Ast. 25, 521-63 (1987).

Carrigan, R., IAC-04-IAA-1.1.1.06, 55th International Astronautical Congress 2004 - Vancouver, Canada.

Carrigan, R., IAC IAC-06-A4.1.12 (2006).

Cohen, M., Walker, R., & Witteborn, F., AJ 104, 2030 (1992).

Dyson, F., Science 131, 1667 (1960).

Jugaku, J., & Nishimura, S., Bioastronomy 2002: Life Among the Stars, R. Norris and F. Stootman, eds, IAU Symposium, 213, 437 (2002)

Lemarchand, G., SETIQuest, Volume 1 Number 1, p. 3. On the web at http://www.coseti.org/lemarch1.htm.

Skrutskie, M. F., et al., AJ 131, 1163 (2006).
Slysh, V. I., The Search for Extraterrestrial Life: Recent Developments, M. D. Papagiannis (Editor), Reidel Pub. Co., Boston, Massachusetts, 1985, pg 315
Tarter, J., Ann. Rev. Astron and Astrophys. 48, 511 (2000).
Timofeev, M. Y., Kardashev, N. S., & Promyslov, V. G., Acta Astronautica Journal, 46, 655 (2000)
Kwok, S., Volk, K., and Bidelman, W. P., ApJS, 112, 557 (1997).

Douglas Vakoch

Bioastronomy 2007: Molecules, Microbes, and Extraterrestrial Life
ASP Conference Series, Vol. 420, 2009
K. J. Meech, J. V. Keane, M. J. Mumma, J. L. Siefert, and D. J. Werthimer, eds.

Anthropological Contributions to the Search for Extraterrestrial Intelligence

D. A. Vakoch

California Institute of Integral Studies and SETI Institute, 1453 Mission Street San Francisco, CA 94103

Abstract. Three recent annual conferences of the American Anthropological Association (AAA) have included symposia on the Search for Extraterrestrial Intelligence (SETI). This paper reviews these symposia, which dealt with themes associated with the overarching AAA conference themes for each year: in 2004, the SETI session addressed Anthropology, Archaeology, and Interstellar Communication: Science and the Knowledge of Distant Worlds; in 2005, it dealt with Historical Perspectives on Anthropology and SETI; and in 2006, the session examined Culture, Anthropology, and SETI. Among the topics considered in these symposia were analogues for contact with extraterrestrial intelligence (ETI), examining anthropologists' experience in the field encountering other cultures-past and present. Similarly, the methodologies of archaeologists provide analogies for making contact with temporally distant civilizations, based on reconstructions from fragmentary information. Case studies helped make such analogies concrete in the symposia. The challenges of comprehending intelligences with different mental worlds was explored through a study of the meetings of Neanderthals and Homo sapiens, for example, while the decryption of Mayan hieroglyphics provided lessons on understanding others of own species.

1. Science and the Knowledge of Distant Worlds

During the past two decades, anthropologists have been contributing to discussions about the possibility of life beyond Earth through annual CONTACT conferences (Riner, 2005), which bring together science fictions writers, anthropologists, and others from a variety of disciplines. Until recently, however, these deliberations have reached few scholars in the broader community of anthropologists and archaeologists. This paper provides an overview of some of the contributions that anthropologists have made to the Search for Extraterrestrial Intelligence (SETI) through a review of symposia held at three recent annual conferences of the American Anthropological Association (AAA).

The first AAA session dealing specifically with SETI was held during the 2004 annual meeting, held in Atlanta, Georgia, which had as its theme "Magic, Science, and Religion." Tapping into the theme through an examination of the nature of scientific knowledge, this SETI symposium, called "Anthropology, Archaeology, and Interstellar Communication: Science and the Knowledge of Distant Worlds," built upon the work of one of the symposium presenters, anthropologist Ben Finney, who spent a year in the 1980s doing field work with scientists leading NASA's SETI program. Finney's critique of that community's standard assumption-that science provides a universal language for constructing interstellar messages-was met with considerable resistance (Finney, 2004).

The 2004 AAA session continued that examination of the SETI community's basic assumptions about the nature of science and interstellar message design, showing how anthropology and archaeology can contribute to scientific practice by critiquing theoretical presuppositions of scientists and suggesting alternative perspectives. Several speakers (e.g., Campbell, 2004) drew upon insights gained through their participation in the workshop "Employing Experience with Decoding Ancient Human Languages to Encode and Decode Interstellar Messages," held in 2003 at the Fifth World Archaeological Congress (WAC-5). In the search for a universal language to overcome cultural differences between humans and extraterrestrials, many have emphasized knowledge that would be shared by human and extraterrestrial scientists (Vakoch, 2004). For example, the metal plaques borne on two Pioneer spacecraft, launched by NASA in the 1970s, indicate our location in the galaxy relative to prominent pulsars that slowly and systematically change frequency over time-locating the spacecrafts' launch in both time and space. And indeed, for an extraterrestrial civilization that values technical intelligence over social intelligence (Barkow, 2004), such a description might be the start of an ideal message. Nevertheless, although our scientific understanding of our place in the universe reveals much about the values of the culture that sent these probes, alternative notions of place are missed in messages that restrict themselves to scientific descriptions of our locality (O'Leary and Thomas, 2004).

Anthropology offers analogies to understanding the Other, which may help communicate broader notions of culture and increase the likelihood that messages will be intelligible (Denning, 2004). Simulations based on anthropological models of first contact between terrestrial cultures, for example, might help evaluate the adequacy of current protocols that guide responses to detecting a signal from extraterrestrials (Funaro, 2004). And yet, the delay of centuries or millennia between each exchange makes interstellar dialogue impossible, except as a dialogue across generations-at least as seen from the perspective of human interlocutors whose lifespans are measured in decades. In this scenario, initial misunderstandings could go uncorrected for a long time.

Archaeology also provides analogies for comprehending distant extraterrestrial civilizations through material artifacts. Yet such analogies have problems (Wason, 2004). Though spacecraft to other worlds can bear messages on metal plaques or recordings, more realistic attempts to communicate via radio waves or laser pulses provide no such tangible artifacts, but bear messages through inaudible signals that may be decodable only through scientific practices.

The attempt to communicate with other worlds has parallels with attempts to communicate across vast times with future generations of humans. Some of the same challenges of creating intelligible messages can be found in attempts to mark radioactive waste sites, in the hope of helping future generations avoid this danger (Drake, 2004). Indeed, recent intentional transmissions from Earth to other worlds may in part reflect a desire to preserve human culture, using other potentially more stable civilizations as a repository of knowledge that may not endure on Earth. The content of these messages may reveal much about contemporary human values (Harrison, 2004).

2. Historical Perspectives on Anthropology and SETI

In keeping with the theme of the AAA's 2005 annual meeting, "Bridging the Past into the Present," held in Washington, DC, that conference's SETI session was titled "Historical Perspectives on Anthropology and the Search for Extraterrestrial Intelligence (SETI)" and was later featured in Anthropology Today, a leading international anthropology journal (Dick 2006a; Forbess, Candea, and Olson 2006). As presenters in this symposium noted, when astronomers conducted the first SETI project in 1960, listening for radio signals from other stars, their efforts meshed well with then current functionalist accounts of cultural universals. Examining the possibility that human and extraterrestrial intelligence (ETI) would have enough in common to make contact across interstellar space, NASA and other space organizations engaged a handful of anthropologists through workshops and symposia. The 2005 AAA session built on the work of the prior year's symposium to explore the contributions of anthropology to SETI, and it identified several unresolved issues for future work. Several papers in this session reviewed the history of anthropologists' involvement in SETI, examining the impact of changing theoretical emphases within anthropology (e.g., Dick, 2005). The session's presenters included anthropologists who personally participated in these discussions during the past three decades (e.g., Urbanowicz, 2005). On longer timescales, the evolution of intelligence, culture, and communication were central themes of the session. The challenge of portraying diverse terrestrial cultures in interstellar messages as historically embedded was also addressed, drawing on the most extensive message sent into space thus far, which itself reflected cultures as seen in the 1970s (Vakoch and Tuli, 2005). The session also examined ways that functionalism of the 1940s through 1960s provided a framework consistent with early discussions of the possible prerequisites of cultures beyond Earth (Chick, 2005). The plausibility of finding cultural universals shared by humans and ETI was reexamined, given that these species may differ significantly in morphologies, dominant sensory modalities, and environments.

Building on the 2004 AAA session "Anthropology, Archaeology, and Interstellar Communication," the 2005 session also considered analogues for contact with ETI, examining anthropologists' experiences in the field encountering other cultures-past and present. The challenges of comprehending intelligences with different mental worlds were explored through a study of the meetings of Neanderthals and Home sapiens, hard as they may be to reconstruct (Wason, 2005). Using an alternative approach of role-playing, for over two decades annual CONTACT conferences have brought together anthropologists, space scientists, and scholars from other disciplines, with each conference culminating in a simulation of first contact between humans and otherworldly intelligence (Riner, 2005). The methodological challenges of finding appropriate analogues for contact were discussed, with special emphasis on the difficulty of anticipating interspecies encounters when there is no direct physical contact, but only the exchange of information across interstellar space, which is presupposed in SETI strategies.

The 2005 session also examined the potential value of anticipatory anthropology for the SETI community, in which anthropologists attempt to guide SETI scientists in preferred directions. The timing is apt for such increased involvement of anthropologists, as SETI programs expand exponentially in listening capability in the near future due to technological advances.

3. Culture, Anthropology, and SETI

In keeping with the AAA's 2006 conference theme of "Critical Intersections/Dan 4gerous Issues," in that year the SETI symposium emphasized the intersection of multiple disciplinary perspectives from the social sciences. This symposium, titled "Culture, Anthropology, and the Search for Extraterrestrial Intelligence (SETI)," was held in San Jose, California.

Presenters at that session argued that SETI is, on one hand, a straightforward experimental endeavor - astronomers searching our galaxy for signals of artificial origin. SETI is based on the assumption that we are the youngest technology in the Milky Way Galaxy now capable of interstellar information exchange, and thus we should listen rather than transmit (Tarter, 2006; Vakoch, 2006). In creating the possibility of contact with alien beings, however, SETI has also created a space for abundant speculation about the nature of societies on other worlds. Inevitably, such speculations draw upon inferences about the probability and essential characteristics of extraterrestrial intelligent beings and technology, but also upon prevalent ideas about how cultures and civilizations work here on Earth (Denning, 2006). But what, if anything, is universal about culture? Attempts to create unified theories of everything have long been popular in the physical sciences, but distinctly less so within contemporary anthropology. Nonetheless, past efforts have led to the development of interdisciplinary and comprehensive theoretical systems (such as Living Systems Theory) intended to yield universal principles (Harrison, 2006). And yet, nonnomological cultural practices, which cannot be anticipated before they appear, are arguably the real basis of the extreme behavioral plasticity underlying our special evolutionary-adaptive heritage (Whitaker, 2006). But if this is so, does this imply that intelligent species, wherever evolved, must display an (at least) equally various and unpredictable cultural repertoire? Matters become even more complex given the potentially long lifetimes of technological civilizations, extending to millions or billions of years, making it likely that flesh and blood intelligence will have evolved to artificial intelligence. Whatever the nature of extraterrestrial intelligence, communication with it requires reflection onwhat constitutes auniversal principle and what is relativeto a particular context (Bronson, 2006). For example, the perceptual systems of Homo sapiens constitute a species-specific user interface whose symbols are shaped by evolution not to match or approximate an external, objective, world but instead are shaped to radically simplify and reformat properties of that world in a manner that informs adaptive behavior. If evolution operates on life wherever it appears in the universe, then the perceptual systems of any alien intelligence also constitute a user interface adapted to its own niche (Hoffman, 2006). But how can we translate between distinct user interfaces adapted to distinct niches?

Mathematics has been considered a likely subject matter that technically advanced human and extraterrestrial intelligence would have in common. However, human mathematics is a byproduct of a particular kind of psychological evolution enhanced by a particular (and historically rare) kind of cultural evolution (Narens, 2006). Research in cognitive science suggests that mathematics is an artifact derived from the specific structure of human embodiment (Woolf, 2006). That life forms evolving on other worlds would have this same structure and hence develop the same mathematical system is dubious. It follows, therefore, that reliance on mathematics as a major means of communication with extraterrestrial intelligence may be misplaced. Anthropologists are in an excellent position to contribute to these varied issues by virtue of their knowledge of the general principles of cultural evolution, and the extent to which these principles are universal (DeVito, 2006; Dick, 2006b).

References

Barkow, J. H., 2004, Extraterrestrial Indigenous Knowledges. Paper presented in the Session "Anthropology, Archaeology, and Interstellar Communication: Science and the Knowledge of Distant Worlds," 103rd Annual Meeting of the American Anthropological Association, Atlanta, Georgia, December 16, 2004.

Bronson, M. C., 2006, Xenolinguistics 101: What SETI Can Learn from the Study of Human and Non-Human Communication. Paper presented in the Session "Culture, Anthropology, and the Search for Extraterrestrial Intelligence (SETI)," 105th Annual Meeting of the American Anthropological Association, San Jose, California, November 17, 2006.

Campbell, J. B., 2004, Human Cultural Diversity, the Diversity of Intelligence and the Assumed Universality of Radio Communication. Paper presented in the Session "Anthropology, Archaeology, and Interstellar Communication: Science and the Knowledge of Distant Worlds," 103rd Annual Meeting of the American Anthropological Association, Atlanta, Georgia, December 16, 2004.

Chick, G., 2005, Biocultural Prerequisites for the Development of Advanced Technology. Paper presented in the Session "Historical Perspectives on Anthropology and the Search for Extraterrestrial Intelligence (SETI)," 104th Annual Meeting of the American Anthropological Association, Washington, DC, December 3, 2005.

Denning, K., 2004, Archaeology, SETI, and the Inalienable. Paper presented in the Session "Anthropology, Archaeology, and Interstellar Communication: Science and the Knowledge of Distant Worlds," 103rd Annual Meeting of the American Anthropological Association, Atlanta, Georgia, December 16, 2004.

Denning, K., 2006, Theorizing Civilization through the Ultimate Others. Paper presented in the Session "Culture, Anthropology, and the Search for Extraterrestrial Intelligence (SETI)," 105th Annual Meeting of the American Anthropological Association, San Jose, California, November 17, 2006.

DeVito, C. L., 2006, The Role of Anthropology in Constructing Interstellar Messages. Paper presented in the Session "Culture, Anthropology, and the Search for Extraterrestrial Intelligence (SETI)," 105th Annual Meeting of the American Anthropological Association, San Jose, California, November 17, 2006.

Dick, S. J., 2005, The Role of Anthropology in SETI: An Historical View. Paper presented in the Session "Historical Perspectives on Anthropology and the Search for Extraterrestrial Intelligence (SETI)," 104th Annual Meeting of the American Anthropological Association, Washington, DC, December 3, 2005.

Dick, S. J., 2006a, Anthropology and the Search for Extraterrestrial Intelligence: An Historical View, Anthropology Today, 22, 3-7.

Dick, S. J., 2006b, Cultural Evolution and SETI. Paper presented in the Session "Culture, Anthropology, and the Search for Extraterrestrial Intelligence (SETI)," 105th Annual Meeting of the American Anthropological Association, San Jose, California, November 17, 2006.

Drake, F. D., 2004, Prevalence of Anthropomorphism in Our Messages to the "Others": Examples from the Voyager Record and the Radioactive Waste Markers Project. Paper presented in the Session "Anthropology, Archaeology, and Interstellar Communication: Science and the Knowledge of Distant Worlds," 103rd

Annual Meeting of the American Anthropological Association, Atlanta, Georgia, December 16, 2004.

Finney, B., 2004. An Anthropologist Communicating with SETI Scientists. Paper presented in the Session "Anthropology, Archaeology, and Interstellar Communication: Science and the Knowledge of Distant Worlds," 103rd Annual Meeting of the American Anthropological Association, Atlanta, Georgia, December 16, 2004.

Forbess, A., Candea, M., and Olson, V. A., 2006, Bringing the Past into the Present. Anthropology Today, 22, 23-26.

Funaro, J., 2004, Simulations for Designing Contact Protocols. Paper presented in the Session "Anthropology, Archaeology, and Interstellar Communication: Science and the Knowledge of Distant Worlds," 103rd Annual Meeting of the American Anthropological Association, Atlanta, Georgia, December 16, 2004.

Harrison, A. A., 2004, Speaking for Earth: Transmitting Cultural Values across Deep Space and Time. Paper presented in the Session "Anthropology, Archaeology, and Interstellar Communication: Science and the Knowledge of Distant Worlds," 103rd Annual Meeting of the American Anthropological Association, Atlanta, Georgia, December 16, 2004.

Harrison, A. A., 2006, An Application of Living Systems Theory to Extraterrestrial Cultures. Paper presented in the Session "Culture, Anthropology, and the Search for Extraterrestrial Intelligence (SETI)," 105th Annual Meeting of the American Anthropological Association, San Jose, California, November 17, 2006.

Hoffman, D., 2006, Niche, User Interface, and Interstellar Communication. Paper presented in the Session "Culture, Anthropology, and the Search for Extraterrestrial Intelligence (SETI)," 105th Annual Meeting of the American Anthropological Association, San Jose, California, November 17, 2006.

Narens, L., 2006, Mathematical Universals for Extraterrestrial Communication. Paper presented in the Session "Culture, Anthropology, and the Search for Extraterrestrial Intelligence (SETI)," 105th Annual Meeting of the American Anthropological Association, San Jose, California, November 17, 2006.

O'Leary, B., and Thomas, J., 2004, 'There's No Place Like Home': The Human Sense of Place in Interstellar Messages. Paper presented in the Session "Anthropology, Archaeology, and Interstellar Communication: Science and the Knowledge of Distant Worlds," 103rd Annual Meeting of the American Anthropological Association, Atlanta, Georgia, December 16, 2004.

Riner, R., 2005, Afterthoughts from CONTACT as Forethoughts to Contact with ETI. Paper presented in the Session "Historical Perspectives on Anthropology and the Search for Extraterrestrial Intelligence (SETI)," 104th Annual Meeting of the American Anthropological Association, Washington, DC, December 3, 2005.

Tarter, J. C., 2006, SETI: '42' May Be the Answer, or at Least the Beginning of an Answer. Urbanowicz, C. F., 2005, On Intelligence: SETI and Terrestrial Intelligence. Paper presented in the Session "Historical Perspectives on Anthropology and the Search for Extraterrestrial Intelligence (SETI)," 104th Annual Meeting of the American Anthropological Association, Washington, DC, December 3, 2005.

Vakoch, D. A., 2004, In Our Image: Science and Culture in Interstellar Messages. Paper presented in the Session "Anthropology, Archaeology, and Interstellar Communication: Science and the Knowledge of Distant Worlds," 103rd Annual Meeting of the American Anthropological Association, Atlanta, Georgia, December 16, 2004.

Vakoch, D. A., 2006, Evolution, Culture, and Extraterrestrial Altruism: The Importance of Active SETI in a Selfish Universe. Paper presented in the Session "Culture, Anthropology, and the Search for Extraterrestrial Intelligence (SETI),"

105th Annual Meeting of the American Anthropological Association, San Jose, California, November 17, 2006.

Vakoch, D. A., and Tuli, K., 2005, Portraying Indian Cultures in the Voyager Recordings: Tradition, History, and Meaning in Interstellar Messages. Paper presented in the Session "Historical Perspectives on Anthropology and the Search for Extraterrestrial Intelligence (SETI)," 104th Annual Meeting of the American Anthropological Association, Washington, DC, December 3, 2005.

Wason, P. K., 2004, Inferring Intelligence: Prehistoric and Interstellar. Paper presented in the Session "Anthropology, Archaeology, and Interstellar Communication: Science and the Knowledge of Distant Worlds," 103rd Annual Meeting of the American Anthropological Association, Atlanta, Georgia, December 16, 2004.

Wason, P. K., 2005, Encountering Alternative Intelligences: Cognitive Archaeology and SETI. Paper presented in the Session "Historical Perspectives on Anthropology and the Search for Extraterrestrial Intelligence (SETI)," 104th Annual Meeting of the American Anthropological Association, Washington, DC, December 3, 2005.

Whitaker, M. P., 2006, What Would Be Alien?: Some Cultural Anthropological Reflections on Not Being Human. Paper presented in the Session "Culture, Anthropology, and the Search for Extraterrestrial Intelligence (SETI)," 105th Annual Meeting of the American Anthropological Association, San Jose, California, November 17, 2006.

Woolf, S. L., 2006, Mathematics as a Human Artifact. Paper presented in the Session "Culture, Anthropology, and the Search for Extraterrestrial Intelligence (SETI)," 105th Annual Meeting of the American Anthropological Association, San Jose, California, November 17, 2006.

Paolo Musso

Bioastronomy 2007: Molecules, Microbes, and Extraterrestrial Life
ASP Conference Series, Vol. 420, 2009
K. J. Meech, J. V. Keane, M. J. Mumma, J. L. Siefert, and D. J. Werthimer, eds.

Von Neumann's Probes And Their Implications For SETI

P. Musso

University of Insubria, Department of Sciences of Communication, email:musso.pl@libero.it

Abstract. Despite a very widespread opinion, the possibility for an advanced civilization of building automatic probes to colonize the Galaxy would have negative implications for SETI. But it seems that this possibility should be excluded.

1. Possibilities on Colonization of the Galaxy

The Von Neumann's Probes hypothesis, i.e. the idea that the colonization of the Galaxy may be done through automatic self-replicating mechanical probes, instead of by intelligent living beings, has sometimes been considered as a possible way to enter in touch with ETs, complementary to the traditional SETI based on radio waves. On the contrary, such a hypothesis, if correct, would be fatal for SETI. Indeed, applying the well known Fermi Paradox to VNPs it follows that, if they can be constructed by an advanced civilization, and such civilizations do exist in the Galaxy, then VNPs should already be here. In fact, it is possible that, for various reasons, *some* civilizations decide not to use VNPs, or to program them excluding any contact with other civilizations, forever or for a while, until they reach a certain degree of technological development. But it is *very* unlikely that *all* the civilizations existing in the Galaxy do the same, if we assume, according to the Copernican Principle, that our place in the Universe is not special, but in the average. So, it follows that, if ETs exist and VNPs are possible, then at least *some* civilizations should have sent them, and thus we should already be in touch with one or more kinds of VNPs. Since it is not the case, then, according to the Fermi Paradox, either ETs do not exist, at least into our Galaxy (and therefore SETI can never succeed), or VNPs are impossible. Despite a very widespread opinion, I think that the second answer is the most probable one. VNPs, indeed, would be *very* advanced machines, probably the most advanced ones we can imagine. Now, any advanced technology requires very sophisticated instruments to find the exotic materials it is based on, and very complex machines to be built up. In turn, all those instruments and machines are produced by very similar complex processes, and so on. At the end of the story, we have to recognize that even to produce an ordinary personal computer (or any other similar gear) it is necessary nothing less than *our whole technological civilization.* In the same way, to reproduce themselves VNPs (which would be *much* more complex than a PC) would have to reproduce in advance the whole civilization (*much* more complex than ours) which created them, or, at least, a substantial part of it. It is very hard to understand how VNPs could do this in a time short enough to begin to reproduce themselves before going all out of order. The only conceivable solution to this problem seems to be sending a *very* huge amount of VNPs, with an even huger amount of instruments and machines. But this would make the enterprise much more difficult than a mission with a crew of living beings. Also living beings, in fact, would have the problem of reproducing their whole civilization on the new planet they have landed on, before being ready to prepare another mission. But, at least, they would have not to wait for this in order to reproduce themselves! In other words, very likely the most efficient VNPs are intelligent living beings themselves. If so, these are bad news, since we can no more look at VNPs as a possibility to get in touch with ETs. But there are also good news. Indeed, if VNPs are impossible, then: 1) the absence of VNPs on Earth does not imply that ETs do not exist; 2) an invasion of hostile alien VNPs as a consequence of active SETI can be excluded; 3) more generally, this makes much more unlikely the possibility of a physical contact with an alien civilization as a consequence of active SETI, and, therefore, it makes active SETI much more likely not to be dangerous for us.

Eric Korpela

Bioastronomy 2007: Molecules, Microbes, and Extraterrestrial Life
ASP Conference Series, Vol. 420, 2009
K. J. Meech, J. V. Keane, M. J. Mumma, J. L. Siefert, and D. J. Werthimer, eds.

SETI with Help from Five Million Volunteers: The Berkeley SETI Efforts

E. J. Korpela,[1] D. P. Anderson,[1] R. Bankay,[1] J. Cobb,[1] G. Foster,[1] A. Howard,[1] M. Lebofsky,[1] G. Marcy,[1] A. Parsons,[1] A. Siemion,[1] J. Von Korff,[1] D. Werthimer,[1] and K. A. Douglas[2]

[1]*University of California, Berkeley, CA, USA, 94720-7450*

[2]*School of Physics, University of Exeter, Exeter, United Kingdom*

Abstract. We summarize radio and optical SETI programs based at the University of California, Berkeley.

The ongoing SERENDIP V sky survey searches for radio signals at the 300 meter Arecibo Observatory. The currently installed configuration supports 128 million channels over a 200 MHz bandwidth with 1.6 Hz spectral resolution. Frequency stepping allows the spectrometer to cover the full 300MHz band of the Arecibo L-band receivers. The final configuration will allow data from all 14 receivers in the Arecibo L-band Focal Array to be monitored simultaneously with over 1.8 billion simultaneous channels.

SETI@home uses desktop computers volunteers to analyze over 100 TB of at taken at Arecibo. Over 5 million volunteers have run SETI@home during its 10 year history. The SETI@home sky survey is 10 times more sensitive than SERENDIP V but it covers only a 2.5 MHz band, centered on 1420 MHz. SETI@home searches a much wider parameter space, including 14 octaves of signal bandwidth and 15 octaves of pulse period with Doppler drift corrections from -100 Hz/s to +100 Hz/s.

The ASTROPULSE project is the first SETI search for μs time scale pulses in the radio spectrum. Because short pulses are dispersed by the interstellar medium, and amount of dispersion is unknown, ASTROPULSE must search through 30,000 possible dispersions. Substantial computing power is required to conduct this search, so the project will use volunteers and their personal computers to carry out the computation (using distributed computing similar to SETI@home).

The SEVENDIP optical pulse search looks for ns time scale pulses at visible wavelengths. It utilizes an automated 30 inch telescope, three ultra fast photo multiplier tubes and a coincidence detector. The target list includes F,G,K and M stars, globular cluster and galaxies.

Introduction

At the University of California, Berkeley, we are conducting five SETI searches that are roughly orthogonal to each other in search space. These five searches are summarized in Table 1.

The SERENDIP V sky survey covers a relatively broad range of radio frequencies, but not as sensitively as SETI@home. The SETI@home sky survey is more sensitive and examines a much wider variety of signal types than SERENDIP, but only covers a narrow band centered on the 21 cm Hydrogen

Program Name	Timescale	Wavelength
SERENDIP	1 s	radio
SETI@home	1 ms – 10 s	radio
ASTROPULSE	μs	radio
SEVENDIP	ns	optical

Table 1. SETI programs at the University of California, Berkeley

line (a "magic frequency"). The ASTROPULSE program is the first search for μs time scale radio pulses. The SEVENDIP optical pulse search is sensitive to low duty cycle ultra-short pulses (eg: pulsed lasers). The optical continuous search is sensitive to narrow band long duty cycle signals (eg: continuous visible lasers).

We describe each of these programs below.

Optical SETI

There is no clear wavelength choice for SETI. Microwave, IR and visible wavelengths all have advantages and disadvantages, depending on what factors another civilization chooses to optimize (power, size, bandwidth, and/or beam size). Although optical photons require more energy to generate than radio photons, optical beam sizes are typically much smaller, and directed interstellar communication links can be more efficient (Lampton 2000; Townes 1961, 1997).

SEVENDIP (Search for Extraterrestrial Visible Emissions from Nearby Developed Populations)

The SEVENDIP program at Berkeley searches for nanosecond time scale pulses, perhaps transmitted by a powerful pulsed laser operated by a distant civilization. The target list includes mostly nearby F,G,K and M stars, plus a few globular clusters and galaxies. The pulse search utilizes Berkeley's 0.8 meter automated telescope at Leuschner observatory and specialized instrumentation to detect short pulses. A similar instrument has been developed at Harvard University (Howard et al. 1999).

The SEVENDIP instrument uses beam splitters to feed light from the telescope onto three high speed photomultiplier tubes (Wright et al. 2001). These tubes have a rise time of 0.7 ns and are sensitive to 300 - 700 nm wavelengths. The three signals are fed to high speed amplifiers, fast discriminators, and a coincidence detector. Three detectors are needed to reject "false alarms," which can be caused by radioactive decay and scintillation in the PMT glass, cosmic rays, and ion feedback. These false alarms can happen often in a single PMT, but almost never occur in three PMT's simultaneously.

The Leuschner pulse search has examined several thousand stars so far, each star for one minute or more. The experiment's sensitivity is 1.5×10^{-17} W/m^2 for a 1 ns pulse, which corresponds to 1.5×10^{-28} W/m^2 average power if the pulse duty cycle is one nanosecond every 100 seconds.

The SERENDIP V Arecibo Sky Survey

The SERENDIP SETI program began 25 years ago; it has gone through four generations of instrumentation and has observed on 14 radio telescopes. During these twenty five years, SERENDIP's sensitivity has improved by a factor of ten thousand and the number of channels has increased from one hundred to more than one hundred million (Werthimer et al. 1997; Bowyer et al. 1997).

The latest SERENDIP sky survey, SERENDIP V, began in earnest in 2009. Observations are ongoing. The survey utilizes the National Astronomy and Ionospheric Center's 305 meter radio telescope in Arecibo, Puerto Rico. The survey thoroughly covers 25% of the sky (declinations from +3 to +33 degrees) and has moderate coverage from -2 to +3 and +33 to +38 degrees. Each of the 10 million beams will be observed an average of three times during the five year survey. Multiple observations are needed because sources may scintillate (Cordes, Lazio & Sagan 1997) or have short duty cycles, and many of our robust detection algorithms require multiple detections.

The sky survey utilizes a real time 128 million channel FFT spectrum analyzers to search for narrow band radio signals in a 300 MHz band centered at the 21 cm Hydrogen line (1420 MHz). The currently installed system consists of one such instrument. We are working towards a final configuration consisting of 14 of these instruments which will allow simultaneous analysis of data from all of the 14 receivers of the Arecibo L-band Focal Array (ALFA) The system has a 0.6 second integration time, 1.6 Hz resolution, and a sensitivity of 10^{-24} W/m^2.

SERENDIP V conducts observations continuously whenever the ALFA array is in uses and simultaneously with ongoing astronomy programs. SERENDIP data analysis is described by Cobb et al. (2000). Information on signals whose power exceeds 16 times the mean noise power are logged along with baseline power, telescope coordinates, time and frequency. This data is transmitted to Berkeley in real time; then, radio frequency interference (RFI) rejection algorithms are applied to the data, off-line, at UC Berkeley. After the RFI is rejected, computers search for candidate signals. SERENDIP's candidate detection algorithms are sensitive to several types of signals, which, individually or combined, may trigger an event to be noted for further study. These algorithms test for beam pattern matching, linear drift rates, regularly spaced pulses, multiple frequencies (particularly those periodic in frequency), and coincidence with nearby stars, globular clusters, or extra-solar planetary systems. Every few months, the entire data base is scanned for multiple detections – "signals" that are detected again when the telescope revisits the same sky coordinates. We test how well these multiple detections fit a barycentric reference frame. We also apply another test that allows much higher frequency separation, which is necessary if transmitters are not corrected for their planet's rotation and revolution. Data are simultaneously sent to Cornell University for analysis using other techniques.

Potential candidates are scored and ranked by the probability of noise causing that particular detection. In cases where multiple detections have been made, a joint probability is assessed. These joint probabilities are used for comparing candidates against each other and generating a prioritized candidate list for re-observation.

The SETI@home Sky Survey

SETI@home data comes from the same piggyback receiver that SERENDIP uses at the Arecibo radio telescope. Whereas SERENDIP analyzes this data primarily using a special-purpose spectrum analyzer and supercomputer located at the telescope, SETI@home records the data, and then distributes the data through the internet to hundreds of thousands of personal computers. This approach provides a tremendous amount of computing power but limits the amount of data that can be handled. Hence SETI@home covers a relatively narrow frequency range (2.5 MHz) but searches for a wider range of signal types, and with improved sensitivity (Anderson et al. 2000; Sullivan et al. 1997).

SETI@home was launched on May 17, 1999. In its 10 years of operation, it has attracted over 5 million participants. Together the participants have contributed over 2×10^{27} floating point operations making SETI@home the largest computation ever performed SETI@home is also one of the largest supercomputers on our planet, currently averaging 3.5 PFLOP actual performance. Users are located in 226 countries, and about 50% of the users are from outside the U.S.

Although SETI@home has 1/80 the frequency coverage of SERENDIP V, its sensitivity is roughly ten times better. The SETI@home search also covers a much richer variety of signal bandwidths, drift rates, and time scales than SERENDIP V or any other SETI program to date.

Primary data analysis, done using distributed computing, computes power spectra and searches for "candidate" signals such as spikes, gaussians, and pulses. Secondary analysis, done on the project's own computers, rejects RFI and searches for repeated events within the database of candidate signals.

SETI@home covers a 2.5 MHz bandwidth centered at the 1420 MHz Hydrogen line from each of the 14 ALFA receivers (7 beams $\times$ 2 polarizations). The 2.5 MHz band is recorded continuously onto SATA disks tapes with one bit complex sampling. Disks are mailed to UC Berkeley for analysis.

SETI@home data disks from the Arecibo telescope are divided into small "work units" as follows: the 2.5 MHz bandwidth data is first divided into 256 sub-bands; each work unit consists of 107 seconds of data from a given 9,765 Hz sub-band. Work units are then sent over the Internet to the client programs for the primary data analysis.

Because an extraterrestrial civilization's signal has unknown bandwidth and time scale, the client software searches for signals at 15 octave spaced bandwidths ranging from 0.075 Hz to 1220 Hz, and time scales from 0.8 ms to 13.4 seconds. The rest frame of the transmitter is also unknown (it may be on a planet that is rotating and revolving), so extraterrestrial signals are likely to be drifting in frequency with respect to the observatory's topocentric reference frame. Because the reference frame is unknown, the client software examines about 1200 different Doppler acceleration frames of rest (dubbed "chirp rates"), ranging from -100 Hz/sec to +100 Hz/sec.

At each chirp rate, peak searching is done by computing non-overlapping FFTs and their resulting power spectra. FFT lengths range from 8 to 131,072 in 15 octave steps. Peaks greater than 24 times the mean power are recorded and sent back to the SETI@home server for further analysis.

Besides searching for peaks in the multi-spectral-resolution data, SETI@home also searches for signals that match the telescope's Gaussian beam pattern. Gaussian beam fitting is computed at every frequency and every chirp rate at spectral resolutions ranging from 0.6 to 1220 Hz (temporal resolutions from 0.8 ms to 1.7 seconds). The beam fitting algorithm attempts to fit a Gaussian curve at each time and frequency in the multi-resolution spectral data. Gaussian fits whose power exceeds the mean noise power by a factor of 3.2 and whose reduced χ^2 of the gaussian fit is less than 1.42 are reported to the SETI@home servers. More details of the SETI@home analysis can be found in Korpela et al. (2001)

SETI@home also searches for pulsed signals using a modified Fast Folding Algorithm (Staelin 1969) and an algorithm which searches for three regularly-spaced pulses.

To determine signals of interest the data for each sky position which has recently received new potential signals is examined by our Near-Time Persistency Checker (NTPCkr). This program scores candidates based upon the probability that the assemblage of potentials signals seed could be due to random fluctuations in the noise background. This score included likelyhood of multiple detections in any reference frame, those that repeat in the barycentric frame, that match the antenna beam pattern, or detections coincident with newly detected planets, nearby stars (from the Hipparcos catalog) or galaxies. We generate a ranked list of our best candidates for reobservation.

Most of the signals found by the client programs turn out to be terrestrial based radio frequency interference (RFI). We employ a substantial number of algorithms to reject the several types of RFI (Cobb et al. 2000) from the best signals. Once RFI rejection has been performed on a candidate group, it is re-scored by the NTPCkr.

The SETI@home screen saver program is available for mac, windows and many versions of unix. Participants can download the client software at: `http://setiathome.berkeley.edu`.

SETI@home has clearly shown the viability of volunteer based distributed computing for other scientific problems. To this end we have developed an infrastructure dubbed BOINC (Berkeley Open Infrastructure for Network Computing). which can be used for other applications. See K. Douglas et al. this volume.

ASTROPULSE

Radio SETI searches to date have concentrated on narrowband signals as opposed to a wideband signals such as a pulses. The ASTROPULSE project is the first SETI search for μs radio pulses. ASTROPULSE detects pulse widths ranging from 1 μs to 1 ms. Such pulses might come from extraterrestrial civilizations, evaporating black holes, gamma ray bursters, certain supernovae, or pulsars. The ASTROPULSE program mines the SETI@home data archive for serendipitous detections of such events.

One of the unique features of this search is that it is the first pulse search to use coherent de-dispersion in a "blind" fashion - we have no previous knowledge of a specific dispersion measure (DM) to examine. The reason this search has never been attempted before is due to the enormous computing power required.

The computing problem is eminently parallel in nature. Similar to SETI@home, ASTROPULSE uses volunteers and their personal computers to carry out the computation. ASTROPULSE uses a general purpose distributed computing system we have developed (BOINC).

Thus far, ASTROPULSE has looked at about a year of SETI@home data, resulting in 82.2 million potential detections. We are currently working on RFI rejection techniques to clean the dataset of RFI, primarily due to aviation radars. More details of ASTROPULSE and other searches for pulsed signals that we are performing can be found in Von Korff et al. (this volume)

Acknowledgments. This work has been supported by the Planetary Society, the Seti Institute, the University of California, Sun Microsystems and donations from individuals around the planet. Key hardware was donated by Network Appliance, Xilinx, Fujifilm Computers, Toshiba, Quantum, Hewlett Packard, and Intel Corp. We receive excellent support from the staff of the Arecibo Observatory, a facility of the NSF, managed by Cornell University. SETI@home and Astropulse have been supported in part by NSF Grants AST-0808175 and AST-0307956, and NASA Grant NNX09AN69

References

Anderson, D., Werthimer, D., Cobb, J., Korpela, E., and Lebofsky, M. 2000. "SETI@home: Internet Distributed Computing for SETI," in *Bioastronomy 99: A New Era in Bioastronomy*, ed. G. A. Lemarchand and K. J. Meech, p. 511

Bowyer, S., Werthimer, D., Donnelly, C., Cobb,J., Ng, D., and Lampton, M. 1997. "Twenty Years of SERENDIP, the Berkeley SETI Effort: Past Resultsand Future Plans," in *Astronomical and Biochemical Origins and the Search for Life in the Universe*, ed. C. B. Cosmovici, S. Bowyer, and D. Werthimer, p. 667

Cobb, J., Lebofsky, M., Werthimer, D., and Bowyer, S. 2000. "SERENDIP IV: Data Acquisition, Reduction and Analysis," in *Bioastronomy 99: A New Era in Bioastronomy*, ed. G. A. Lemarchand and K. J. Meech, p. 485

Cordes, J., Lazio, T., and Sagan, C. 1997. "Scintillation-Induced Intermittency in SETI," *the Astrophysical Journal*, 487, 782

Howard, A., Horowitz, P., Coldwell, C., Latham, D., Papaliolios, C., Stefanik, R., Wolff, J., and Zajac,J. 2000. "Optical SETI at Harvard-Smithsonian" in *Bioastronomy 99: A New Era in Bioastronomy*, p. 545

Korpela, E., Werthimer, D., Anderson, D., Cobb, J. & Lebofsky, M. 2001. "SETI@home–Massively distributed computing for SETI," *Computing in Science and Engineering*, **v3n1**, 78

Lampton, M. 2000. "Optical SETI: The Next Search Frontier," in *Bioastronomy 99: A New Era in Bioastronomy*, ed. G. A. Lemarchand and K. J. Meech, p. 565

Montebugnoli, S., Monari, J., Orfei A, et al. 2000. "Setitalia - SETI in Italy," in *Bioastronomy 99: A New Era in Bioastronomy*, ed. G. A. Lemarchand and K. J. Meech, p. 501

Reines, A., and Marcy, G. 2002. "Optical SETI: A Spectroscopic Search for Laser Emission from Nearby Stars" *Publications of the Astronomical Society of the Pacific* April

Staelin, D. H. 1969. in *Proc. IEEE*, 57, 724

Stootman, F., DeHorta, A., Wellington, K., and Oliver, C. 2000. "The Southern Serendip Project," in *Bioastronomy 99: A New Era in Bioastronomy*, ed. G. A. Lemarchand and K. J. Meech, p. 491

Sullivan, W., Werthimer, D., Bowyer, S., Cobb, J., Gedye, D., and Anderson, D. 1997. "A New Major SETI Project Based on Project SERENDIP Data and 50,000 Personal Computers," in *Astronomical and Biochemical Origins and the Search for Life in the Universe*, ed. C. B. Cosmovici, S. Bowyer, and D. Werthimer, p. 729

Townes, C. 1961. "Interstellar and Interplanetary Communication by Optical Masers," in *Nature* 190, 205

Townes, C. 1997 "Optical and Infrared SETI," in *Astronomical and Biochemical Origins and the Search for Life in the Universe*, ed. C. B. Cosmovici, S. Bowyer, and D. Werthimer, p. 585

Werthimer, D., Bowyer, S., Ng, D., Donnelly, C., Cobb, J., Lampton, M., and Airieau, S. 1997. "The Berkeley SETI Program: SERENDIP IV Instrumentation," in *Astronomical and Biochemical Origins and the Search for Life in the Universe*, ed. C. B. Cosmovici, S. Bowyer, and D. Werthimer, p. 683

Wright, S., Drake, F., Stone, R., Treffers, R., and Werthimer, D. 2001 "An Improved Optical SETI Detector," SPIE proceedings on Optical SETI, ed. S. Kingsley.

Dan Werthimer

Bioastronomy 2007: Molecules, Microbes, and Extraterrestrial Life
ASP Conference Series, Vol. 420, 2009
K. J. Meech, J. V. Keane, M. J. Mumma, J. L. Siefert, and D. J. Werthimer, eds.

Sustainability: A Tedious Path to Galactic Colonization

Y. Dutil and S. Dumas

Dept. de physique, de génie physique et d'optique et Observatoire du mont Mégantic, Université Laval, Québec, Canada, G1K 7P4

Abstract. Civilization cannot sustain an exponential growth for long time even when neglecting numerous laws of physics! In this paper, we examine what are fundamental obstacles to long term survival of a civilization and its possibility to colonize the Galaxy. Using the solar system as a reference, resources available for sustained growth are analyzed. Using this information, we will explore the probability of discovering a civilization at its different stage of energy evolution as estimating some possible value of L, the typical life time of an extra-terrestrial civilization.

1. Introduction

Under the general assumption in SETI, extra-terrestrial civilization must be older than ours. However, little discussion has been made on how they should reach such a long lifetime.

2. What Physicists Have to Say?

In 2010, our civilization will consume 17 TW of power (EIA 2006). We might wonder what the physical limits to our power consumption are and what rate of growth is sustainable over time. We argue than the first growth crisis will be provoked by climate change caused directly by energy production. We arbitrarily fix this limit at 1 W/m^2 or 127 TW. This is slightly lower than the climate change produced by the increase of green house gases (GHG) on Earth (1.6 W/m^2, IPCC2007). With a 2% growth rate, we will reach this limit within a century. This first step is highly critical since to grow further a civilization must have complete control over all biophysical parameters of its ecosystem.

The following steps can be described by the classical Kardashev civilization types Kardashev (1964). Keeping the growth rate at 2%/yr, we would control all the light falling on Earth in 466 years (type I); in 1,533 years, we would control all the power output of our Sun (type II) and in 2,729 years, we would be able to control all the power of the Milky Way (level III)! Nevertheless, energetic considerations restrict the typical growth rate over a million years at a few ppm per year, which goes down to a few ppm over a billion years. In consequence, a civilization must stops growing very soon in its history (after a few hundred years at most).

These physical limitations leave few possibilities for sustainable civilization. They could be photosynthesis limited ($\sim$ 10 TW), climatically limited (127 TW) or solar flux limited (174,000 TW). It should be pointed out than the only fuel that can power a civilization over a billion years is deuterium.

3. What Economists Have to Say?

The simplest economic model for growth has been proposed by Malthus in 1798. It simply states that population growth will be exponential while resources growth must be arithmetic which lead to a reduce wealth unless the population growth is stopped. Since then various economists have studied the population behavior and resources stocks for the case with renewable resources. These models are of the Ricardo-Malthus type Ricardo (1817); Malthus (1798) and are a subtype of Lotka-Voltera predator-prey models Lotka (1925); Volterra (1926). These models produce three types of solutions: extinction, oscillation around a fixed point and stable steady states.

Such models have been successfully applied to population collapse of the Easter Island Brander & Taylor (1998). Generalizations of this model Reuveny & Decker (2000) indicate than the *only escape of Malthusian trap is through institutions restricting utilization of resources.* This restriction itself is very difficult to implement effectively.

Sustainability is even harder to achieve when non renewable resources are modeled. Some authors argue that technology will compensate for the natural capital loss Solow (1997); Stiglitz (1997) but others consider this as impossible Georgescu-Roegen (1971); Daly (1997). Using a completely different approach (system dynamic analysis), Meadows et al (1972) came to the same conclusion.

Limitation of the action of the technology to insure sustainability has already being pointed out by the English economist Jeavons in 1865 Jevons (1865). Any technological amelioration leading to an improved efficiency will increase the affordability of this technology, which will then increase resources consumption. A modern formulation, known as the *Khazzoom-Brookes Postulate* Brookes (1990); Khazzoom (1980), argues that energy saving innovations can end up causing even more energy to be used as the money saved is spent on other goods and services which themselves require energy in their production.

4. What Anthropologists Have to Say?

Anthropologists have uncovered a rare example of strong sustainability for human civilization on a small pacific island: Tikopia. Against unfavorable odds, Tikopian have managed to survive on this isolated ecosystem for three millennia. Archaeological records show a first phase of sharp decline in forest areas, increased erosion, depletion of fish stocks and extinction of bird species, closely paralleling indigenous population growth. However, a striking finding by archaeologists is that the phase of environmental degradation was followed by a progressive historical change in society's resource-management practices.

Tikopian took effective measures between 1000 and 1800 AD to stabilize their population at approximately 1,281 to 1,323 people. They accomplished their goal by infanticide, abortion, and decreeing that only first-born sons could have children. In addition, the inhabitants shifted from *slash and burn* practices to sustainable agriculture. Doing so, they have replaced the island natural ecosystem by an artificial one that mimics the structure and interrelationship found in natural ecologies. Finally, they eliminated pigs, despite the value Polynesians placed on them, because they damaged gardens and ate food than human could consume Firth (1983); Kirsch (2000).

Amazingly, Tikopia sits on the cyclone belt, so every year its inhabitants deal with cyclones, every five or ten years, bringing heavy winds. Not is only Tikopia's society sustainable but it is also very resilient.

5. What Ecologists Have to Say?

Ecosystems are by definition sustainable. Selective pressures are hypothesized to drive evolution in one of two stereotyped directions: r- or K-selection MacArthur *et al.* (1967); Pianka (1970). These terms, r and K, are derived from standard ecological algebra, as illustrated in the simple Verhulst equation of population dynamics:

$$\frac{dN}{dt} = rN\left(1 - \frac{N}{K}\right)$$

where r is the growth rate of the population N, and K is the carrying capacity of its local environmental setting. Typically, r-selected species produce many offspring, which are, comparatively, less likely to survive to adulthood. Whereas K-selected species invest more heavily in the nurture of fewer offspring, which have a better chance of surviving to adulthood.

In unstable or unpredictable environments r-selection predominates, where the ability to reproduce quickly is crucial, and there is little advantage in adaptations that permit successful competition with other organisms, because the environment is likely to change again. In stable or predictable environments K-selection predominates, as the ability to compete successfully for limited resources is crucial. Populations of K-selected organisms typically are very constant and close to the maximum that the environment can bear. It should be pointed out than in natural ecosystem biodiversity tend to increase both the stability and the productivity of ecosystem Johnson *et al.* (1996); McGrady-Steed *et al.* (1997).

It is likely that any sustainable civilizations would follow a similar evolution trajectory. Therefore, alien civilizations are likely to be extremely complex, very efficient with a very low rate of growth.

6. Conclusion

A general conclusion is than strong sustainability can only be achieved by the implementation of strict control on the exploitation of resources. This control must be effectively achieved on a society which is likely to be extremely complex. The same society must also be very resilient to any large scale perturbation. These levels of social technologies are largely beyond our capabilities.

Since civilization collapse by resources exhaustion can happen on a very short time scale and social organization to avoid this is very difficult to put in place, it is likely that most civilization are short lived (few hundred years) and therefore unlikely to engage in galactic colonization. The lucky ones that are sustainable will have a zero growth rate, which would largely reduce the pressure for galactic colonization.

References

Brander, J.A., Taylor, M.S., 1998, The simple economics of Easter Island: a Ricardo-Malthus model of renewable resource use, Am. Econ. Rev. 88, 119-138

Brookes, Len, 1990, Energy Efficiency and Economic Fallacies, Energy Policy, Vol. 18, No. 2, pp 199-201

Daly, H. E., 1997, Georgescu-Roegen versus Solow/Stiglitz, Ecological Economics, 22, 261-266

Firth, R ., 1983, We, the Tikopia. Stanford University Press, Palo Alto, CA

Georgescu-Roegen, N., 1971 The Entropy Law and the Economic Process, Cambridge MA USA: Harvard Univ. Press.

Jevons, W. S., 1865, The Coal Question: An Inquiry Concerning the Progress of the Nation, and the Probable Exhaustion of Our Coal-Mines Published: London: Macmillan and Co.

Johnson, K. H., Vogt,K. A., Clark,H. J., Schmitz, O. J., Vogt,D. J., 1996, Biodiversity and the productivity and stability of ecosystems, Trends in ecology and evolution, vol.11,no9,pp.372-377

Kardashev, N. S., 1964, Transmission of information by extraterrestrial civilizations, Sov. Astron. 8, 217-220

Khazzoom, J. D., 1980, Economic Implications of Mandated Efficiency Standards for Household Appliances, Energy Journal, Vol.1, No.4, pp21-39

Kirsch, P., 2000, On the road of the winds: an archaeological history of the Pacific Islands before European contact, University of California Press, Berkeley, CA

Lotka, A. J., 1925, Elements of physical biology. Baltimore: Williams and Wilkins Co.

MacArthur, R. and Wilson, E. O., 1967, The Theory of Island Biogeography, Princeton University Press, ISBN 0-691-08836-5M

Malthus, T., 1798, An Essay on the Principle of Population, Penguin, New York

McGrady-Steed, J., Harris, P. M. and Morin, P. J., 1997, Biodiversity regulates ecosystem predictability, Nature, vol 390, 162-165

Meadows, D. H., Meadows, D. L., Randers, J. and Behrens III, W.W., 1972, The limits to growth: a report for The Club of Rome's project on the predicament of mankind, New York : Universe Books, 205 p.

Pianka, E. R., 1970, On r and K selection. American Naturalist 104, 592-597.

Reuveny, R., and C. S. Decker, 2000, Easter Island: Historical Anecdote or Warning for the Future? Ecological Economics 35, no. 2: 271-87.

Ricardo. D., 1817, Principles of Political Economy and Taxation

Solow, R. M., 1997, Georgescu-Roegen versus Solow/Stiglitz, Ecological Economics, 22, 267-268

Stiglitz, J. E. 1997 Georgescu-Roegen versus Solow/Stiglitz. Ecological Economics, 22: 269-270

Verhulst, P. F., 1838, Notice sur la loi que la population poursuit dans son accroissement, Corresp. Math. Phys. 10, 113-121.

Volterra, V., 1926, Fluctuations in the abundance of a species considered mathematically, Nature 118, 558-560

Bioastronomy 2007: Molecules, Microbes, and Extraterrestrial Life
ASP Conference Series, Vol. 420, 2009
K. J. Meech, J. V. Keane, M. J. Mumma, J. L. Siefert, and D. J. Werthimer, eds.

Spin-Off Successes of SETI Research at Berkeley

K. A. Douglas,[1] D. P. Anderson,[2] R. Bankay,[2] H. Chen,[2] J. Cobb,[2] E. J. Korpela,[2] M. Lebofsky,[2] A. Parsons,[2] J. Von Korff,[2] and D. Werthimer[2]

[1]*School of Physics, University of Exeter, Exeter, United Kingdom*

[2]*Space Sciences Laboratory, University of California Berkeley, Berkeley CA, USA 94720*

Abstract. Our group contributes to the Search for Extra-Terrestrial Intelligence (SETI) by developing and using world-class signal processing computers to analyze data collected on the Arecibo telescope. Although no patterned signal of extra-terrestrial origin has yet been detected, and the immediate prospects for making such a detection are highly uncertain, the SETI@home project has nonetheless proven the value of pursuing such research through its impact on the fields of distributed computing, real-time signal processing, and radio astronomy. The SETI@home project has spun off the Center for Astronomy Signal Processing and Electronics Research (CASPER) and the Berkeley Open Infrastructure for Networked Computing (BOINC), both of which are responsible for catalyzing a smorgasbord of new research in scientific disciplines in countries around the world. Futhermore, the data collected and archived for the SETI@home project is proving valuable in data-mining experiments for mapping neutral galatic hydrogen and for detecting black-hole evaporation.

1. The SETI@home Project at UC Berkeley

SETI@home is a distributed computing project harnessing the power from millions of volunteer computers around the world (Anderson 2002). Data collected at the Arecibo radio telescope via commensal observations are filtered and calibrated using real-time signal processing hardware, and selectable channels are recorded to disk. These disks are shipped to UC Berkeley, where the data are distributed over the Internet in the form of "work units" to volunteers who use spare cycles on their computer to search for patterns in the recorded noise. Processed work units are returned to UC Berkeley and stored in databases for further statistical analysis in the search for signals indicative of extra-terrestrial intelligence.

While the direct product of this research has been a series of null results, the technology developed for this project is widely applicable, and has spun off several derivative projects based around the real-time signal processing architecture used at the telescope (and other radio telescopes worldwide), the distributed computing architecture used for off-line data processing, and the expansive archives of survey data collected for analysis. We illustrate the foundations of this research in the SETI@home project, and present several applications of the technology SETI@home has developed and generalized for widespread use.

2. The Center for Astronomy Signal Processing and Electronics Research

In 2002, the SETI@home project submitted a proposal to the National Science Foundation to build a multipurpose signal processing board based on Field Programmable Gate Array (FPGA) processors that would be used to build a new SETI spectrometer at Arecibo, but would be flexible and reconfigurable for general radio astronomy use. The SERENDIP V processing board funded by this successful grant proposal was conscripted by more than 12 projects in applications ranging from pulsar timing surveys (Demorest et al. 2004) to interferometric correlation (Chippendale et al. 2005) to galactic hydrogen surveys (Heiles et al. 2004). The success of this project highlighted a need in the radio astronomy community for generic, commodity solutions for high-performance signal processing systems. Our expertise in correlator and spectrometer development and our experience in designing the SERENDIP V spectrometer let us to establish an offshoot group: the Center for Astronomy Signal Processing and Electronics Research (CASPER).

The modular FPGA-based hardware and open-source signal processing libraries developed by CASPER (Parsons et al. 2006) are now being used for interferometry, beam-forming, spectroscopy, VLBI and pulsar timing at Arecibo, the Allen Telescope Array, GMRT (India), MeerKAT (South Africa), the Deep Space Network (NASA), Square Kilometer Prototype (Bologna), MIT/Haystack Observatory, Harvard/Smithson-ian SMA, NRAO (Green Bank and the VLA), and the Precision Array for Probing the Epoch of Reionization (Australia). CASPER is currently developing a next-generation spectrometer for planetary missions that will orbit Mars, Venus and Earth: NASA's MARVEL, VESPER and CAMEO sub-millimetre spectrometers.

3. The Berkeley Open Infrastructure for Network Computing

As more volunteers joined the SETI@home distributed computing project, the Berkeley Open Infrastructure for Network Computing (BOINC) was developed in order to make effective use of the increases in computing power and the diverse computational resources made available by SETI@home users. BOINC provides a non-commercial middleware system for volunteer computing, originally developed to support SETI@-home, but intended to be useful for other applications in areas as diverse as mathematics, medicine, molecular biology, climatology, and astrophysics. The intent of BOINC is to make it possible for researchers to tap into the enormous processing power of personal computers around the world (Anderson 2003). Today, BOINC users together comprise the world's largest supercomputer.

The BOINC project is a Berkeley-led and NSF-funded program which manages the top-level software common to over 50 distributed computing projects. BOINC projects include Einstein@home, a search for gravitational waves; protein folding with Rosetta@home; global climate modelling with Climateprediction.net; as well as HIV, cancer, and malaria drug research. Users are able to allocate fractions of their personal computing resources to any number of these

Table 1: The 10 most popular volunteer computing projects using the BOINC infrastructure as of September 1, 2009

Projects	Users	Added Today	Hosts	Added Today	Countries	Total FLOP	PFLOP/s Today
SETI@Home	1,010,446	+362	2,418,560	+871	234	4.7×10^{27}	8.0
MilkyWay@home	31,857	+119	69,363	+222	159	5.7×10^{25}	5.9
World Community Grid	255,427	+217	777,876	+1,383	216	1.4×10^{26}	3.3
Einstein@Home	236,724	+82	982,947	+976	211	1.2×10^{26}	2.0
AQUA@home	6,738	+20	13,052	+32	114	1.2×10^{25}	0.5
GPUGRID	5,584	+16	9,102	+29	93	3.0×10^{25}	1.5
Climate Prediction	201,702	+114	390,945	+206	210	7.9×10^{25}	0.8
Rosetta@Home	260,677	+164	778,085	+431	220	6.7×10^{25}	0.9
PrimeGrid	26,420	+26	76,204	+72	158	1.2×10^{25}	0.3
ABC@home	22,978	+23	63,054	+42	149	1.7×10^{25}	0.2

projects, providing a valuable service to projects that might otherwise have been unable to afford access to the computing power necessary for their research.

4. Data Mining Experiments with SETI@home Data

While the main goal of the SETI@home project is to detect signals of extraterrestrial and intelligent origin, the data collected have been used for other scientific projects, including the mapping of neutral galactic hydrogen (SETHI), and searching for short-timescale bursting phenomena (AstroPulse).

The 21-cm spectral line of neutral hydrogen (H I) is one of the best tracers of the Galaxy's interstellar medium. The SETI@home spectrometer has been (non-uniformly) sampling the H I sky visible from Arecibo since 1999. Millions of spectra have been combined to create three-dimensional spectral emission "datacubes" covering 7200 square degrees of the sky (Douglas and Korpela 2009).

AstroPulse is a distributed computing project fashioned after SETI@home. It will search for pulses of emission from pulsars, black holes, and other exotic objects that have been dispersed by the interstellar medium. More details are given in another articles in these proceedings (Von Korff et al. 2009).

In July 2006 a new SETI@home data recorder was installed at Arecibo to perform commensal observations with *all* observations done with the ALFA receiver. This multibeam system provides unprecedented sensitivity to weak celestial (and man-made) emission and greatly increases our data rate. These advances ensure that SETI@home will continue to lead the search for intelligent life beyond Earth for years to come.

Acknowledgments. SETI@home runs largely on the donations of its many volunteers. Without this support, the spin-off advances described in these proceedings would not have been possible. SETI@home and Astropulse are also supported by NSF Grant AST-0808175 and NASA Grant NNX09AN69G. The SETHI project was funded by NSF Grant AST-0307956. The GALFA 21-cm line surveys are funded by NSF Grant AST-0709347.

References

Anderson, D. P., J. Cobb, E. Korpela, M. Lebofsky, D. Werthimer, "SETI@home: An Experiment in Public-Resource Computing," in Communications of the ACM, Nov. 2002, Vol. 45 No. 11, pp. 56-61.

Anderson, D. P., "Public Computing: Reconnecting People to Science," presented at the Conference on Shared Knowledge and the Web, Residencia de Estudiantes, Madrid, Spain, Nov. 17-19 2003.

Chippendale A. P., R. Subrahmanyan, R. D. Ekers, "The Cosmological Reionization Experiment," presented at *New Techniques and Results in Low Frequency Astronomy*, Hobart, Dec. 2005.

Demorest, P., R. Ramachandran, D. Backer, R. Ferdman, I. Stairs, and D. Nice, "Precision Pulsar Timing and Gravity Waves: Recent Advances in Instrumentation," in *Bulletin of the American Astronomical Society*, Dec. 2004, pp. 1598–+.

Douglas, K. A., and Korpela, E. J. 2009, in preparation

Heiles, C., J. Goldston, J. Mock, A. Parsons, S. Stanimirovic, and D. Werthimer, "GALFA Hardware and Calibration Techniques," in *Bulletin of the American Astronomical Society*, Dec. 2004, pp. 1476–+.

Parsons, A., D. Backer, C. Chang, D. Chapman, H. Chen, P. Crescini, C. de Jesus, C. Dick, P. Droz, D. MacMahon, K. Meder, J. Mock, V. Nagpal, B. Nikolic, A. Parsa, B. Richards, A. Siemion, J. Wawrzynek, D. Werthimer, and M. Wright, "PetaOp/Second FPGA Signal Processing for SETI and Radio Astronomy," in *Asilomar Conference on Signals and Systems, Pacific Grove, CA*, Nov. 2006, pp. 2031–2035.

Von Korff, J. et al. 2009, this volume

Bioastronomy 2007: Molecules, Microbes, and Extraterrestrial Life
ASP Conference Series, Vol. 420, 2009
K. J. Meech, J. V. Keane, M. J. Mumma, J. L. Siefert, and D. J. Werthimer, eds.

Astropulse and Fly's Eye: SETI Searches for Transient Radio Signals Using Distributed Computing

J. Von Korff,[1,2] A. Siemion,[3,4] E. Korpela,[1] D. Werthimer,[1,3] P. McMahon,[3,5] J. Cobb,[1] M. Lebofsky,[1] D. Anderson,[1] B. Bankay,[1] G. Bower,[4] G. Foster,[4] J. van Leeuwen,[4] W. Mallard,[3] and M. Wagner[3]

[1]*University of California, Berkeley - Space Sciences Lab, Berkeley, CA 94720*

[2]*vonkorff@ssl.berkeley.edu*

[3]*University of California, Berkeley - Berkeley Wireless Research Center, Berkeley, CA 94720*

[4]*University of California, Berkeley - Department of Astronomy, Berkeley, CA 94720*

[5]*Stanford University - Department of Computer Science, Stanford, CA, 94305*

[6]*University of California, Berkeley - Department of Physics, Berkeley, CA 94720*

Abstract. Berkeley conducts 7 SETI programs at IR, visible and radio wavelengths. Here we review two of the newest efforts, Astropulse and Fly's Eye.

A variety of possible sources of microsecond to millisecond radio pulses have been suggested in the last several decades, among them such exotic events as evaporating primordial black holes, hyper-flares from neutron stars, emissions from cosmic strings or perhaps extraterrestrial civilizations, but to-date few searches have been conducted capable of detecting them.

We are carrying out two searches in hopes of finding and characterizing these μs to ms time scale dispersed radio pulses. These two observing programs are orthogonal in search space; the Allen Telescope Array's (ATA) "Fly's Eye" experiment observes a 100 square degree field by pointing each 6m ATA antenna in a different direction; by contrast, the Astropulse sky survey at Arecibo is extremely sensitive but has 1/3,000 of the instantaneous sky coverage. Astropulse's multibeam data is transferred via the internet to the computers of millions of volunteers.

The Fly's Eye was successfully installed at the ATA in December of 2007, and to-date approximately 450 hours of observation has been performed. We have detected three pulsars (B0329+54, B0355+54, B0950+08) and six giant pulses from the Crab pulsar in our diagnostic pointing data. We have not yet detected any other convincing bursts of astronomical origin in our survey data.

1. Astropulse Telescope and Hardware

Arecibo Observatory scans approximately $\frac{1}{3}$ of the sky, between declinations of -1.33 and 38.03 degrees. This means that Astropulse cannot see the galactic center (around -29 dec) but can see many known pulsars, including the Crab.

The ALFA receiver has 7 dual-polarization beams on the sky, each with a $3.5'$ beamwidth. The central beam has a gain of 11 K Jy^{-1}, and the other beams have 8.5 K Jy^{-1} The system temperature is 30 K.

We collect and store the 14 signals, saving a copy of each file at NERSC, the National Energy Research Scientific Computing Center. In all, we have taken 100 TB of data so far from ALFA multibeam. Volunteers' PCs will then process the data and send the results back to Berkeley, see Section 2.3..

2. Astropulse Algorithm and its implementation

2.1. Dedispersion

Between a radio pulse's source (i.e. ET, pulsar, or neutron star) and our detector, a radio pulse must travel through the Interstellar Medium (ISM), causing dispersion (Wilson et al. 2009) so that the high frequency components travel slightly faster.

We have to reconstruct the original pulse, bringing together the component frequencies and reassembling them so that the signal has a short duration again.

We have two choices for our methodology: coherent dedispersion and incoherent dedispersion. Astropulse uses coherent dedispersion, whereas other radio surveys use incoherent dedispersion. Incoherent dedispersion is much more computationally efficient, and for longer timescales it's almost as good as coherent dedispersion. However, Astropulse would be unable to examine the 0.4 μs timescale without coherent dedispersion.

Incoherent dedispersion means that the signal is divided up into sub-bands, and the power vs. time of each sub-band is recorded. The method is called "incoherent" for this reason – the phase information about individual frequencies is lost; only the total power of each subband is retained. Then, the sub-bands are realigned at all possible dispersion measures, in an effort to find one DM at which the components align to produce a large power.

However, this project is limited in two ways. First, the goal of recording power vs. time makes sense only on a timescale greater than $\frac{1}{d\nu}$, where $d\nu$ is the width of each sub-band. This is because of time-frequency uncertainty. Second, if $\Delta\tau$ is the time over which the pulse is dispersed, then in each sub-band the pulse is dispersed by $\Delta\tau \cdot \frac{d\nu}{\Delta\nu}$. So the method cannot localize the pulse better than this. Combining these two limits, we find that the minimum timescale for incoherent dedispersion happens when

$$dt = 12.8\mu\, s(\frac{DM}{56.791})^{0.5}(\frac{\nu_0}{1.42\text{GHz}})^{-1.5} \qquad (1)$$

Where we have calibrated to the Crab's DM, and ν_0 is the center frequency. For the Crab pulsar, this is a limit of 12.8 μs. For a more distant source, the limit might be as much as 50 μs. But coherent dedispersion deals with amplitude rather than power, and attains arbitrary resolution.

We perform coherent dedispersion using a deconvolution, which can be accomplished using FFTs in time $O(N \log N)$.

2.2. Algorithm Logic

Astropulse loops through the data at several nested levels, starting at a dm of 896 samples (49.5 pc cm^{-3}):

1. Large and small dm chunk: blocks of 128 or 16 dms at a time.
2. Data chunks of size $32768 = 2^{15}$ samples. Compute FFT of the data for use in convolution.
3. All dms within a small dm chunk, and sign of the dm (positive and negative.)

   We consider negative dms for a few reasons: First, extraterrestrial intelligences might communicate using signals dispersed with negative dms. And second, signals detected at both positive and negative dm might be a sign of RFI.
4. Combine samples at scales (or "co-adds") of 2^0 to 2^9 samples.
5. Samples within the data chunk

   If the power in one sample (or rather 2^ℓ samples) is above a certain threshold, then Astropulse reports a pulse.

The Fast Folding Algorithm The Fast Folding Algorithm (FFA) is described in Staelin (1969). In our case, it operates on a time series of bins, not samples. Each bin is a sum of either 16 or 128 samples, depending on whether it was created during the large or small dm chunk loop. The FFA runs a nested loop, similar to the single pulse algorithm, looping over frequencies (the highest level), then subfrequencies, coadds, and bins (the lowest level).

2.3. BOINC

Astropulse runs on the BOINC platform, which stands for "Berkeley Open Infrastructure for Network Computing." BOINC is a set of programs that organizes volunteers' home computers to perform scientific calculations. In any BOINC project, a researcher has a computing problem that can run in parallel, that is, on several machines at once. Perhaps the problem involves searching a physical space (for Astropulse, the sky), and performing the same computation on each point in that space (for Astropulse, dedispersion.) The first BOINC project, SETI@home, searched exclusively for narrowband transmissions across the sky, which could be a signal from extraterrestrial intelligence. Or, the space could instead be a parameter space, for instance a space of potential climate models (climateprediction.net) or protein configurations (Rosetta@home).

The researcher for a BOINC project need not be affiliated with UC Berkeley, or with the BOINC development team at Berkeley (although we happen to be so affiliated.) BOINC is open source, and can be downloaded, compiled, and operated by anyone with sufficient technical skills; many projects currently exist outside Berkeley.

Likewise, volunteers need no particular technical knowledge. They just visit the BOINC web page and download the “client” programs. Astropulse has access to around 500,000 volunteers, each of whose machines might have 2 GFLOPs of processing power, and be on 1/3 of the time, for a total of 300 TFLOPs – approaching the speed of the world’s fastest supercomputer.

3. Fly’s Eye

The Allen Telescope Array has several advantages over other telescopes worldwide for performing transient searches, particularly when the search is for bright pulses. The ATA has 42 independently-steerable dishes, each 6m in diameter. The beam size for individual ATA dishes is considerably larger than that for most other telescopes, such as VLA, NRAO Green Bank, Parkes, Arecibo, Westerbork and Effelsberg. This means that the ATA can instantaneously observe a far larger portion of the sky than is possible with other telescopes. Conversely, when using the ATA dishes independently, the sensitivity of the ATA is far lower than that of other telescopes.

The Fly’s Eye instrument was purpose built to search for bright radio pulses of millisecond duration at the ATA. The instrument consists of 44 independent spectrometers using 11 CASPER IBOBs. Each spectrometer processes a bandwidth of 210MHz, and produces a 128-channel power spectrum at a rate of 1600Hz (i.e. 1600 spectra are outputted by each spectrometer per second). Therefore each spectrum represents time domain data of length 1/1600Hz=0.000625s=0.625ms, and hence pulses as short as 0.625ms can be resolved[1].

We have to-date performed roughly 400 hours of observing with the Fly’s Eye.

3.1. Fly’s Eye Offline Processing

The analysis required for the Fly’s Eye experiment is, in principle, fairly simple – we wish to search over a wide range of dispersion measures to find large individual pulses. Specifically our processing requires that all the data be dedispersed with dispersion measures ranging from 50 cm^{-3} pc to 2000 cm^{-3} pc. At each dispersion measure the data needs to be searched for ‘bright’ pulses.

The processing chain is in practice significantly more complicated than this description suggests. Processing is performed on compute clusters, with input data formatted, divided and assigned to worker nodes for processing. In the worker node flow, the data is equalized, RFI rejection is performed, and finally a pulse search is performed through the range of dispersion measures. The results are written to a database where they can be subsequently queried. The key feature of the results is a table that lists, in order of decreasing significance, the pulses that were found and the dispersion measures they were located at.

---

[1]Pulses of duration <0.625ms can also be detected provided that they are sufficiently bright, but their length cannot be determined with a precision greater than the single spectrum length.

Average power equalization is performed on the frequency spectrum equalized values $P_i'(t)$. We compute the average power over all frequency channels for a single integration (time sample t). The power average is defined as $\overline{P'(t)} = \frac{1}{N}\sum_{i=0}^{N-1} P_i'(t)$. N is the number of channels (for Fly's Eye this is always 128). The motivation for why it is possible to normalize the power is that we expect pulses to be dispersed over many time samples, so this procedure should not remove extraterrestrial pulses.

Our strategy for mitigating constant narrowband RFI is simply to identify the channels that are affected, and to exclude them from further processing. This channel rejection is typically performed manually by looking at a set of spectra and identifying obviously infected channels, which are then automatically excluded in subsequent processing runs.

Intermittent RFI is often quite difficult to automatically distinguish from genuine astronomical pulses, and we followed a conservative approach to try to ensure that we do not accidentally excise dispersed pulses. Our statistic for intermittent RFI is the variance of a single channel over a 10 minute data chunk, $\sigma_i^2 = \left(\frac{1}{T_0}\sum_{t=0}^{T_0-1}(P_i''(t))^2\right) - \left(\frac{1}{T_0}\sum_{t=0}^{T_0-1}(P_i''(t))\right)^2$. Curve-fitting determines a σ_i^2 outside which it is likely that channel i contains time-varying RFI. Future reprocessing will likely use a more robust method, such as that based on a kurtosis estimator Nita et al. (2007).

Our final RFI mitigation technique is manual – in our results it is easy to see high-σ hits that are a result of RFI: these hits appear as simultaneous detections at many dispersion measures.

3.2. Detection of Giant Pulses from the Crab Nebula

A suitable test of transient detection capability is to observe the Crab pulsar and attempt to detect giant pulses from it. We were able to detect several such pulses, which appeared only at the expected dispersion measure. Wideband RFI appeared across a wide range of dispersion measures.

Acknowledgments. Astropulse has been supported in part by NSF Grants AST-0808175 and AST-0307956, and NASA Grant NNX09AN69G. We are grateful for the support of the NAIC Arecibo Observatory, a facility of the NSF, managed by Cornell University. We would also like to thank the Allen Telescope Array a facility managed by the University of California and the SETI Institute.

References

Nita G. M., Gary D. E., Liu Z., 2007, PASP 119, 805
Staelin D. H., 1969, Proc. IEEE 57, 724
Wilson T. L., Rohlfs K., Hüttemeister S., 2009, Tools of Radio Astronomy, Springer

Claudio Maccone

Bioastronomy 2007: Molecules, Microbes, and Extraterrestrial Life
ASP Conference Series, Vol. 420, 2009
K. J. Meech, J. V. Keane, M. J. Mumma, J. L. Siefert, and D. J. Werthimer, eds.

Spectral Line Measurements in Exceptionally Low SNR Achieved by Virtue of the KLT (Karhunen-Loève Transform)

C. Maccone,[1] S. Pluchino,[2] and F. Schillirò[3]

[1]*Member of the International Academy of Astronautics*

[2]*Visiting Research Fellow, IRA-INAF Radiotelescopes at Medicina, Bologna, Italy*

[3]*Technology Researcher, IRA-INAF Radiotelescope at Noto, Siracusa, Italy*

Abstract. A little-known tool for spectral line measurements is the KLT (Karhunen-Loève Transform). This mathematical algorithm is superior to the classical FFT in that: 1) The KLT can filter signals out of the background noise over both wide and narrow bands. On the contrary, the FFT rigorously applies to narrow-band signals only. 2) The KLT can be applied to random functions that are non-stationary in time, i.e. whose autocorrelation is a function of the two independent variables t_1 and t_2 separately. Again, this is a sheer advantage of the KLT over the FFT, since the FFT rigorously applies to stationary processes only, i.e. when the autocorrelation is a function of the absolute value of the difference of t_1 and t_2. 3) The KLT can detect signals embedded in noise to unbelievably small values of the Signal-to-Noise Ratio (SNR), like 10^{-3} or so. This particular feature of the KLT is described in detail in this paper.

1. Introducing the KLT

The Karhunen-Loève Transform (KLT) is possibly the most advanced mathematical algorithm available in 2007 to achieve both noise filtering and data compression in all fields of science. The KLT can work with signals of any kind: both narrow-band and wide-band, both stationary and non-stationary, and all embedded in lots of noise, i.e. with a very low signal-to-noise ratios (SNRs). A concise description is at the site http://en.wikipedia.org/wiki/Karhunen-Loève_theorem and a more profound mathematical description is found in the first author's books (Maccone 1994 & 2009). One starts with:

$$X(t) = \sum_{n=1}^{\infty} Z_n \varphi_n(t), \tag{1}$$

for $0 \le t \le T$ with $E\{X(t)\} \equiv 0$.

The infinite series in Equation 1 is called the KL expansion of the stochastic process $X(t)$ over the finite time $0 \le t \le T$. In radio astronomy $X(t)$ is "whatever gets inside the radio telescope", i.e. the background noise plus a possible very weak signal buried in the noise. The KL expansion (1) separates the behavior in time (expressed by the functions $\varphi_n(t)$) from the behavior in probability, expressed by the random variables Z_n. It can be proven that the functions $\varphi_n(t)$ are othonormal over $0 \leq t \leq T$ and they are the eigenfunctions of the autocorrelation of $X(t)$, i.e. the solutions to the "KLT integral equation"

$$\int_0^T E\{X(t_1)X(t_2)\}\varphi_n(t_2)\,dt_2 = \lambda_n\,\varphi_n(t_1). \tag{2}$$

Here the λ_n are corresponding eigenvalues and they turn out to always be a decreasing sequence of positive numbers that actually are the variances of the corresponding random variables Z_n. Moreover

$$E\{Z_m Z_n\} = \lambda_n\,\delta_{mn}. \tag{3}$$

This means that Z_n are uncorrelated to each other, and, for Gaussian $X(t)$, are statistically independent. In practice, one gets the input process $X(t)$ over $0 \leq t \leq T$, computes its autocorrelation $E\{X(t_1)X(t_2)\}$, solves the integral equation (2) and so gets both the eigenvalues λ_n and the eigenfuntions $\varphi_n(t)$. Finally, the random variables Z_n are obtained by projecting the known $X(t)$ over the "eigenvectors" $\varphi_n(t)$ as

$$Z_n = \int_0^T X(t)\,\varphi_n(t)dt. \tag{4}$$

2. Continuous Time vs. Discrete Time in the KLT

The KL expansion in continuous time, t, is what we have described so far. This may be more "palatable" to theoretical physicists and mathematicians inasmuch as it may be related to other branches of physics, or of science in general, in which the time obviously must be continuous. For instance, the first author spent fifteen years of his life (1980-1994) to investigate mathematically the connection between the KLT and Special Relativity. But let us go back to the time variable t in the KL expansion (1). If this variable is discrete, rather than continuous, then the picture changes completely. In fact, the integral equation (2) now becomes a system of simultaneous algebraic equations of the first degree, that can always be solved! The difficulty here is that this system of linear equations is huge (hundreds or thousands of elements are the rule for autocorrelation matrices in SETI and in other applications, as image processing and the like). So, the KLT may be practically impossible to find numerically because even the most powerful computers may get stuck, unless we resort to simplifying tricks of some kind. This is just what the authors did in 2007 for the SETI-Italia program (Schillirò et al. 2007).

3. Bordered Autocorrelation Method (BAM) for the KLT of STATIONARY PROCESSES

We now confine ourselves to **stationary** $X(t)$ over a discrete set of instants $t = 0, ..., N$. In this case, the autocorrelation of $X(t)$ becomes the Toeplitz matrix:

$$Toeplitz = \begin{bmatrix} R_{xx}(0) & R_{xx}(1) & R_{xx}(2) & ... & ... & R_{xx}(N) \\ R_{xx}(1) & R_{xx}(0) & R_{xx}(1) & ... & ... & R_{xx}(N-1) \\ R_{xx}(2) & R_{xx}(1) & R_{xx}(0) & ... & ... & R_{xx}(N-2) \\ ... & ... & ... & R_{xx}(0) & ... & ... \\ R_{xx}(N) & R_{xx}(N-1) & ... & ... & R_{xx}(1) & R_{xx}(0) \end{bmatrix}$$

We may choose N at will but, clearly, the higher N, the more accurate the KLT of $X(t)$. On the other hand, the final instant T in the KLT (1) can be chosen at will and now is $T = N$. So, we can regard $T = N$ as a sort of 'new time variable' and even take derivatives with respect to it! We'll do so in a moment! But let us now go back to the Toeplitz Array. If we let N vary as a new free variable, that amounts to bordering it, i.e. adding one (last) column and one (last) row to the previous correlation T. This means to solve again the system of linear algebraic equations of the KLT for $N + 1$, rather than for N. So, for each different value of N, we get, for instance, a new value of the first eigenvalue λ_1 now regarded as a function of N, i.e. $\lambda_1(N)$. But now the surprise comes! This $\lambda_1(N)$ turns out to be proportional to N. And such a function $\lambda_1(N)$, of course has a derivative, $\frac{\delta\lambda_1(N)}{\delta N}$ that can be computed numerically as a new function of N. And this derivative turns out to be a constant with respect to N. This fact paves the way to a new set of applications of the KLT to Astrobiology as well as to other fields of science!

4. Reconstructing a STATIONARY $X(t)$ by virtue of its FIRST eigenvalue ONLY!

The first eigenvalue $\lambda_1(N)$ also is the (by far) largest, i.e. the most important one, since it is the variance of the most largely-scattered random variable Z_1. In other words, even if you study the first eigenvalue of the KLT only, you actually study the bulk of your data! And $\lambda_1(N)$ is called "dominant". Well, numeric simulations performed by authors Schillirò et al.(2007) lead to the results shown in 4 plots in Figure 1. The first plot is the ordinary Fourier spectrum of a pure tone at 300 Hz buried in noise with SNR=0.5 (this is about the lowest SNR beyond which the FFT starts failing to denoise a signal). The second plot shows the first KLT dominant eigenvalue $\lambda_1(N)$ over 1200 time samples. Clearly, this $\lambda_1(N)$ is proportional to N.

So, its derivative, $\frac{\delta\lambda_1(N)}{\delta N}$ is a constant with respect to N. But then we may take the Fourier transform of such a constant and clearly we get a Dirac delta function, i.e. a peak just at 300 Hz. In other words, we have KLT-reconstructed the original tone by virtue of our BAM. The fourth plot below is such a BAM-reconstructed peak, and it is of course identical to the third plot, that is the ordinary FFT of first KLT eigenfuction (obtained by solving the full and long system of N algebraic first-degree equations).

Let us now do the same again, but with an incredibly low SNR of 0.005. Results are presented in Figure 2. Poor Fourier here is in a mess! No classical FFT spectrum can be identified at all! But for the KLT, no problem! The second plot clearly shows that $\lambda_1(N) \propto N$. The third plot (KLT FAST way) is the NEAT KLT spectrum of the 300 Hz tone obtained by the FFT of $\frac{\delta\lambda_1(N)}{\delta N}$

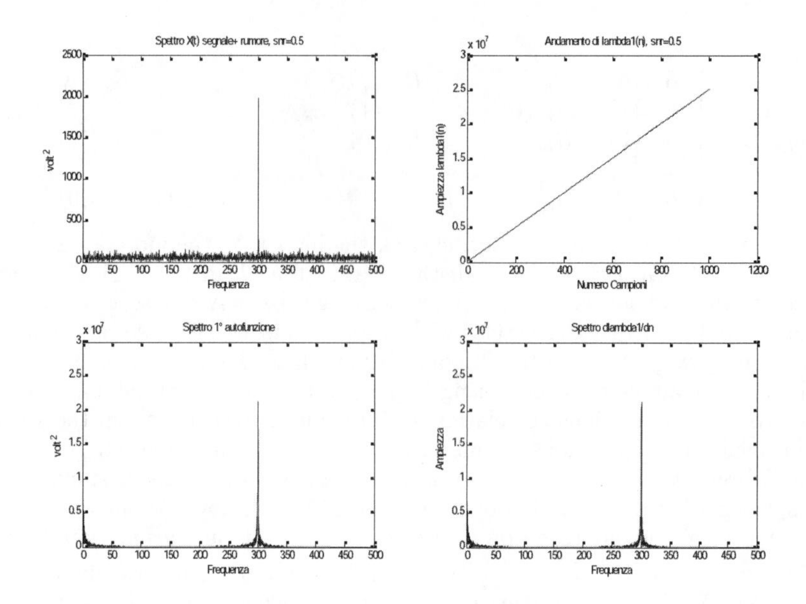

Figure 1. Numeric simulations of the first eigenvalue of the KLT.

and this is just the same as the third plot (KLT SLOW way). This proves the superior behavior of the KLT.

5. Conclusions

Let us summarize the results mathematically described in this paper. When the stochastic process $X(t)$ is stationary (i.e. it has both mean value and variance constant in time), then there are two alternative ways to compute the first KLT dominant eigenfunction (that is the roughest approximation to the full KLT expansion, that may be "enough" for practical applications!): 1)Either you compute the first eigenvalue and get the first eigenfunction or 2)(BAM) You compute the derivative of the first eigenvalue with respect to $T = N$ and then Fourier-transform it to get the first eigenfunction. In practical numerical simulations of the KLT it may be less time-consuming to choose option 2) rather than option 1). In either case, the KLT of a given stationary process can retrieve a sinusoidal carrier out of the noise for values of the signal-to-noise ratio (SNR) that are three orders of magnitude lower than those that the FFT can still filter out. In other words, while the FFT (at best) can filter out signals buried in a noise that as a SNR of about 1 or so, the KLT can, say, filter out signals that have a SNR of, say, 0.001 or so. This is the superior achievement of the KLT with respect to the FFT.

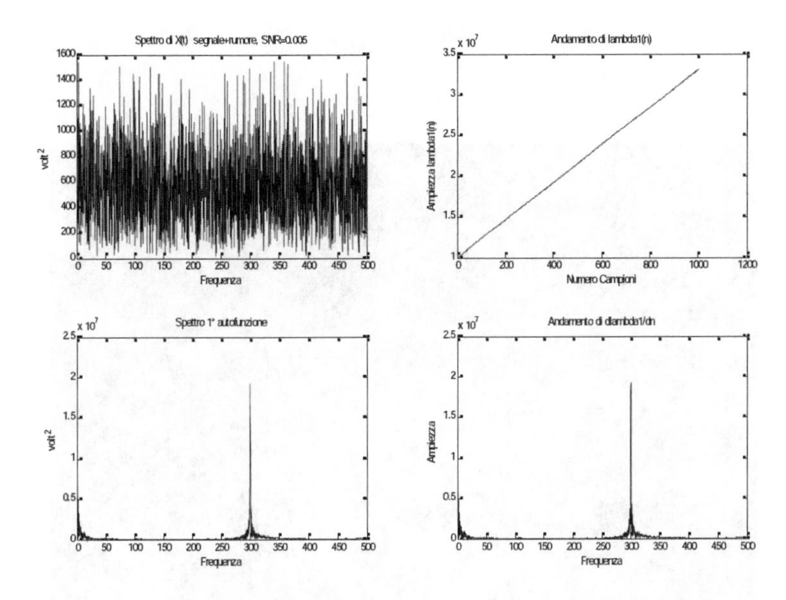

Figure 2. Numeric simulations of the first eigenvalue of the KLT with low SNR.

References

Maccone, C. 1994, "Telecommunications, KLT and Relativity Volume 1", a book published by IPI press, Colorado Springs, Colorado, USA, ISBN # 1-880930-04-8. This book embodies the results of some thirty research chapters published by the author about the KLT in the fifteen years span 1980-1994 in peer-reviewed journals.

Maccone, C. 2009, "Deep Space Flight and Communications - Exploiting the Sun as a Gravitational Lens", a 400-pages treatise about FOCAL space mission that embodies and updates all previously published material about FOCAL. ISBN 978-3-540-72942-6 Springer Berlin Heidelberg New York. Library of Congress Control Number: 2007939976. ©Praxis Publishing Ltd, Chichester, UK, 2009

Schillirò, F., Pluchino, S., Maccone, C., Montebugnoli, S. 2007, Istituto Nazionale di Astrofisica (INAF) Istituto di Radioastronomia (IRA) Rapporto Tecnico - "La KL Transform: considerazioni generali sulle metodologie di analisi ed impiego nel campo della Radioastronomia", Technical Report (available in Italian only).

Jill Tarter

Session XIV

Education and Public Outreach

Mary Kadooka

Pamela Harman

Bioastronomy 2007: Molecules, Microbes, and Extraterrestrial Life
ASP Conference Series, Vol. 420, 2009
K. J. Meech, J. V. Keane, M. J. Mumma, J. L. Siefert, and D. J. Werthimer, eds.

Astrobiology as a Context for High School Science: Teachers' Professional Development

P. K. Harman and E. K. DeVore

Education and Public Outreach, SETI Institute.
Email: pharman@seti.org

Abstract. Astrobiology is an interdisciplinary science investigating life in the universe. Where did life originate? What is its future? Are we alone in the universe? These major questions drive astrobiology research, are investigated through understanding evolution in the broadest sense, and are stimulating questions for the classroom. *Voyages Through Time* is a tested and effective astrobiology curriculum, emphasizing inquiry, examination of evidence, and the nature of science. The SETI Institute's NASA Astrobiology Institute (NAI) Lead Team Education and Public Outreach (EPO) effort developed Astrobiology Summer Science Experience for Teachers (ASSET) as a six day science teacher professional development institute capitalizing upon the intriguing science and the exemplar curriculum.

1. Introduction

Research based standards for science teacher professional development are described in detail in the National Science Education Standards (NSES). The standards can be summarized as learning about science content, learning about science pedagogy, the lifelong learning process, and coherency of professional development. The SETI Institute's NAI EPO effort developed Astrobiology Summer Science Experience for Teachers (ASSET) in alignment with these standards. Astrobiology science talks, practice with *Voyages Through Time*^{TM} curriculum, an inquiry immersion experience, curriculum implementation planning, leadership skill development, and outreach planning are woven through the six-day workshop. The achievement of goals is measured by daily workshop formative evaluation and summative evaluation at the conclusion.

ASSET leverages the NAI EPO investment, reaching teachers, students and the general public. Astrobiology reaches the classroom via *Voyages Through Time* (VTT). VTT was developed by a SETI Institute partnership with NASA Ames Research Center, the California Academy of Sciences, and San Francisco State University with major funding from the National Science Foundation (Grant# 9730693), NASA, Hewlett Packard Company, Foundation for Microbiology, SETI Institute, and Educate America. The curriculum is aligned with NSES Content Standards and AAAS Benchmarks for Science Literacy. VTT was successfully field-tested in 28 states and more than 80 high school classrooms across the U.S.A. As the participants – a blend of individuals and teams from thirty states – complete at least two outreach activities in the next two years, they will contribute to the public understanding of astrobiology. The

SETI Institute supports the ASSET network outreach efforts with outreach or workshop materials and science speakers.

2. Design

Design of ASSET began with a SETI Institute Education and Outreach staff retreat, focused upon developing staff understandings of the learning mechanisms for both adults and students, and raising staff awareness of the NSES science teacher professional development standards. A common language was developed, and ASSET goals were defined to

1. increase astrobiology science content knowledge,
2. increase understanding of inquiry as a basis performing science and as an instructional approach,
3. increase familiarity with the VTT Scope and Sequence, and
4. increase understanding of the elements and strategies of professional development.

The goals were introduced at the beginning of the summer institute, summarized as astrobiology science content, VTT pedagogy and practice, inquiry as the basis of science and as an instructional approach, and essential elements of professional development design. The week long schedule-at-a-glance was color-coded to indicate the alignment of goals with activities, and it promoted participant awareness of the goals. Objectives were defined for each goal, based upon the needs of the participants. Participants' needs assessment was performed via a survey of undergraduate and graduate study, teaching experience, leadership experience, and classroom practices regarding the teaching of evolution that was a part of the application for ASSET. Metacognitive reflection concluding each day was connected to the NSES Inquiry Spectrum and the NAI Roadmap to emphasize the relationship between the day's activities and the goals of the summer institute experience.

Professional development strands in the schedule aligned with learning goals after participants' needs were assessed, and reflected best practices in science teaching. Each day's sessions – astrobiologists' specific content talks, classroom connections made for each VTT module, inquiry immersion experiences, NSES professional development awareness sessions, implementation planning and outreach planning – had a defined objective that was evaluated and deemed successful. Participants made poster presentations, and led VTT classroom activities to demonstrate their personal classroom expertise.

3. Recruitment and Selection

Teachers were recruited from the across the U.S. via the SETI Institute Education Electronic Newsletter, national, state, and local science teacher associations, state and county offices of education and the NAI newsletter and teams. The ASSET application, available on-line at the SETI Institute EPO page, required applicants to describe how evolution is taught in their classroom and any challenges they experience with the topic, to list their personally desired outcome from attending ASSET, the school administration's commitment to implement at least one curriculum module, and their proposed outreach plan. Teachers were encouraged, though not required to apply as teams. The applications were scored using a rubric, giving priority to those from schools with large populations of underserved and underrepresented students.

4. Evaluation

Immediate and daily evaluation generated minor daily schedule changes. Summative evaluations have generated design alterations over the course of four summer programs, 2004 – 2007. Daily evaluation forms listed the objective for each session with a Likert scale, (*http://en.wikipedia.org/wiki/Psychometrics*) a type of psychometirc response scale often used in questionnaires (*http://en.wikipedia.-org/wiki/Questionnaire*). The participant's responses were tabulated and analyzed. In addition, participants were invited to comment upon each activity in the program as a part of the daily evaluation. Program modifications were made to ensure high rates of objective achievement. Outreach and implementation plans and pre- and post-concept maps were collected with the final evaluation forms. Significant gains in astrobiology science content understanding was reflected in the differences between the pre- and post-concept scores.

5. Findings

To date, 80 teachers and ten NAI EPO professionals from a total of 30 states, a Department of Defense teacher and a Chilean PhD candidate on Chilean national scholarship, have attended ASSET. All ASSET teachers have implemented at least one *Voyages Through Time* module. Curriculum implementation has been accomplished to varying degrees ranging from complete adoption as a 9th grade course to using a single module in a discipline-specific course. To date, 67% of the 2004 – 2006 ASSET alumnae have completed the outreach commitment. Teachers and NAI EPO professionals who attended were enthusiastic about both the curriculum and the ASSET institute format. Each participant committed to conducting at least two outreach activities in the following 2 years. Activities have included:

- Mentoring colleagues
- Presenting at science teacher conferences – including National Science Teachers Association, state and local science teacher associations, National Association of Biology Teachers, and American Association of Physics Teachers
- Facilitating online courses for students and teachers
- Astrobiology Day
- Mother - daughter science day

- Local workshops through science academies and museums.

NAI EPO leads recommended further involvement by other EPO professionals in the NAI teams, and used the ASSET format as a professional development model at NAI institutions. Outreach and classroom resources were distributed via CD. The SETI Institute maintains an e-mail listserve that has provides each annual cohort a means of maintaining contact and to share science news and classroom experiences. SETI Institute provides print and digital media for outreach events as needed to ASSET alumnae.

Acknowledgments. ASSET is funded by the NASA Astrobiology Institute, Educate America, SETI Institute and Lockheed Martin.

References

American Association for the Advancement of Science 1993, Benchmarks for Science Literacy, (New York: Oxford University Press, Inc)

National Research Council 2000, How People Learn: Brain, Mind, Experience and School, ed. J. D. Bransford et al (Washington DC: National Academy Press)

National Research Council 2005, How Students Learn: Science in the Classroom, ed. M. S. Donovan & J. D. Bransford (Washington DC: National Academy Press)

National Research Council 2000, Inquiry and the National Science Education Standards, (Washington DC: National Academy Press)

National Research Council 1996, National Science Education Standards, (Washington DC: National Academy Press)

Bioastronomy 2007: Molecules, Microbes, and Extraterrestrial Life
ASP Conference Series, Vol. 420, 2009
K. J. Meech, J. V. Keane, M. J. Mumma, J. L. Siefert, and D. J. Werthimer, eds.

Astrobiology for Knowledge Transfer

C. A. Price

AAVSO/Tufts University

Abstract. For over a century educators have been debating the issue of knowledge transfer, whether concepts learned in one domain can transfer to another. Research in the last few decades has found that successful transfer is remarkably rare in education, especially in science and mathematics. Astrobiology as a domain may offer an opportunity for successful knowledge transfer due to its inherent multidisciplinary nature.

1. A New Definition: Transfer as Preparation For Future Learning (PFL)

Transfer research has long been mired in the debate over general transfer. Early experiments routinely failed to successfully detect general transfer (Lave 1988; Singley & Anderson 1989). In the 1980's, the situated cognition theorists offered an alternative approach based on the belief that what is learned cannot be separated from how and where it is learned (Brown, Collins, & Duguid 1989; Clancey 1997; Lave 1988). Accordingly, prior experiments were flawed in that they were conducted in unrealistic laboratory environments and thus were really only tests of sequestered problem solving (SPS). Whenever possible, studies need to consider prior knowledge, the environment, and the social context of the experiment. When solving a problem, students need access to the same resources which they would have in real life.

This literature led to a discussion of the concept of transfer itself, including the proposals of alternative terminology (Bransford & Schwartz 1999; Carraher, Nemirovsky, & Schliemann 1995; Hammer et al. 2004; Lobato 2003). Most of these alternatives regard transfer as an educational process or activity to evaluate (i.e., how can we measure transfer?) as opposed to an event to detect (i.e., did transfer happen or not?). One of the most influential papers on the subject is by Bransford and Schwartz (1999), who suggest the term Preparation For Future Learning (PFL) instead of transfer. The new terminology reflects the core goal of transfer: to supply the learner with a skill and/or conceptual knowledge to solve future, unforeseen problems in the real world.

2. Transfer and Conceptual Change In Science Education

Transfer in science education is usually discussed within the domain of conceptual change. diSessa (2006) defines conceptual change as building new ideas in the context of old ones, a process similar to many of the transfer definitions. He clarifies: The "conceptual" part of the conceptual change label must be treated less literally. Various theories locate the difficulty in such entities as "beliefs," "theories," or "ontologies," in addition to concepts. Thus, the definition of concept needs to be flexible.

Three early theories of conceptual change emerged in the 1980's (diSessa 2006). The first assumes that prior beliefs get replaced when new beliefs are seen as incommensurable with the old beliefs by the student. This leaves little room for transfer as one concept is replaced with another. The second early theory is based on the theory-theory developmental model. One of the most famous misconception studies (McClosky 1983a, 1983b) uses theory-theory belief in the scientific nature of the mind to explain how learners develop their own coherent and complex concepts and theories which they must deal with in the learning process. Finally, there is the rational view. This pragmatic view hold that learners need rational reasons to replace existing ideas and was championed in a seminal paper by Posner et al. (1982) and later refined to include motivational and other situational factors by Strike and Posner (1992).

In a review of conceptual change research, Duit (2003) believes the solution to the problems of science education lie in multi-framework approaches. He advocates for the merging of affecting and cognitive domains to overcome misconceptions in science education. He says, "It is fruitful to merge ideas of conceptual change and theories on the significance of affective factors." Doing do, he argues, would address the need to stimulate the learner interests. In particular, he feels, "Multi-perspective frameworks have to be employed in order to adequately address the complexity of the teaching and learning processes."

3. Metacognition, Transfer and Science Education

The National Research Council Committee on the Development of the Science of Learning called for the use of various meta-cognitive tools, such as self-monitoring and reciprocal teaching, to facilitate transfer of science knowledge (Bransford, Brown and Cocking 1999). Metacognition is also suggested as one of the four mechanisms of transfer by Sternberg and Frensch (1993). Their Set mechanism considers the state of mind ("set") of the learner, who needs to be actively attempting to transfer while learning. Georghiades (2000) proposes that metacognition can foster transfer and durability in science education. Furthermore, Georghiades and Paraskevas (2004) warn of a theory-practice gap emerging in the field and suggest, "...the use and training of metacognition in natural contexts."

Goldstone and Son (2005) studied interactions with computer simulations of complex adaptive systems. They found the best transfer occurring when the simulation began in a concrete phase and ended in an idealized phase. This led to three pedagogical recommendations to foster transfer in simulations: vary the appearance of materials, distinguish the simulation in the lesson, and fade the concreteness of the simulation over time.

Van Gog, Paas, and Merrinboer (2004) studied the literature regarding the use of worked problems/examples in science education and proposed that they would be more effective if accompanied by process-explanations. That is, the students need to know the why as well as the how. Their interpretation of Cognitive Load Theory and the earlier studies on worked-examples illustrate that learners can apply a worked-example to a similar example that immediately follows. For durability and the ability to choose when to use the new knowledge, learning, "...must involve both knowledge of its domain and of its teleology."

4. A Sample Astrobiological Mechanism for Transfer

Possible astrobiological applications of Perkins three mechanisms for knowledge transfer (Perkins et al. 1998):

1. Abstraction: Transfer can occur by applying similar deep principles to different subjects. An astrobiological example could be applying the concept that life can form in extreme conditions deep the in ocean to the possibility of life forming in extreme, deep water conditions on other worlds, such as Europa.

2. Transfer by affordances: Another transfer mechanism relies on similar action opportunities ("affordances"). If the learner can identify similar results and opportunities from the different knowledge concepts then transfer of knowledge may occur. For example, if a learner discovers in a chemistry class that water freezes at 0 degrees Celsius, while thinking of astrobiology they may realize that it would be difficult for water-based life to form on planets colder than 0 degrees Celsius, which leads to understanding of the Habitable Zone.

3. High road and low road transfer: This are two related, yet distinct mechanisms. A "low road" transfer relies on fairly direct relationships between two concepts and results in a semi-automatic transfer. For example, learning about the components of the Earth atmosphere in an Earth Science class could help a learner identify what components to look for in the atmosphere of other planets that could support life. Low road transfers are often reflexive. Alternatively, a "high road" transfer requires conscious, applied mental effort in comparing abstractions and is usually not reflexive. It can support the far transfer of knowledge from one area to another. For example, knowledge of fiber optic cable communication technology and techniques could be applied to optical search for extra-terrestrial intelligence (Optical SETI) light analysis algorithms.

Acknowledgments. The author would like to thank the Astrobiology Institute at the University of Hawaii for a travel grant to present this poster at the Bioastronomy 2007 meeting.

References

Barnett, S. M., & Ceci, S. 2002. When and Where Do We Apply What We Learn? A Taxonomy for Far Transfer. Pyschological Bulletin. 128, 4, 612-637

Bransford, J. D., Brown, A. L., & Cocking, R. R. 1999. How people learn: Brain, mind, experience, and school. Washington, DC: National Academy Press.

Bransford, J. D., & Schwartz, D. L. 1999. Rethinking transfer: A simple proposal with multiple implications. Review of Research in Education. 24, 61-100.

Carey, S. 1986. In Ontogeny and Phylogeny of Human Development, S. Strauss (Ed.). Norwood, NJ: Ablex.

Caraher, D., & Scliemann, A. D. 2002. The Transfer Dilemma. Journal of the Learning Sciences. 11, 1, 1-24.

Clancey, W. J. 1997. Introduction: What Is Situated Cognition? In Clancey, W. J. (ed.), Situated Cognition. New York, NY: Cambridge University Press.

diSessa, A. A. 2006. A History of Conceptual Change Research. R. Sawyer (Ed.). The Cambridge Handbook of the Learning Sciences. Cambridge. Cambridge University Press

Duit, R., & Treagust, D. F. 2003. Conceptual change: a powerful framework for improving science teaching and learning. International Journal of Science Education. 25, 6, 71-688

Georghiades, P. 2000. Beyond conceptual change learning in science education: focusing on transfer, durability and metacognition. Educational Research, 42, 2, 119-139.

Georghiades, P., & Paraskevas, A. 2004. From the general to the situated: three decades of metacognition. International Journal of Science Education. 26, 3, 365-383.

Goldstone, R. L. & Son, J. Y. 2005. The transfer of scientific principles using concrete and idealized simulations. The Journal of the Learning Sciences. 14, 1, 69-110.

Lave, J. 1988. Cognition in practice. Boston, MA: Cambridge.

Lobato, J. 2003. How Design Experiments Can Inform a Rethinking of Transfer and ViceVersa. Educational Researcher. 32, 1, 17.

Perkins, D. N., & Salomon, G. 1998. Are cognitive skills context-bound? Educational Researcher. 18, 1, 16-25

Posner, G. J., Strike, K. A., Hewson, P. W., & Gertzog, W. A. 1982. Accommodation of a scientific conception: Toward a theory of conceptual change. Science Education. 66, 2, 211-227.

Singley, M. K., & Anderson, J. R. 1989. The Transfer of Cognitive Skill. Cambridge, Massachusetts: Harvard University Press.

Strike, K. A. & Posner, G. J. 1992. A revisionist theory of conceptual change. In R. A. Duschl & R. Hamilton (eds.), Philosophy of science, cognitive psychology, and educational theory and practice. Albany, NY: SUNY Press.

Van Gog, T., Paas, F., & Merrinboer, J. J. G. 2004. Process-Oriented Worked Examples: Improving Transfer Performance Through Enhanced Understanding. Instructional Science. 32, 83-98.

Bioastronomy 2007: Molecules, Microbes, and Extraterrestrial Life
ASP Conference Series, Vol. 420, 2009
K. J. Meech, J. V. Keane, M. J. Mumma, J. L. Siefert, and D. J. Werthimer, eds.
%documentclass[11pt,twoside]article

Development, Evaluation, and Dissemination of an Astrobiology Curriculum for Secondary Students: Establishing a Successful Model for Increasing the Use of Scientific Data by Underrepresented Students.

L. Arino de la Rubia,[1] J. Butler,[1] T.Gary,[1] S. Stockman,[2] M. Mumma,[3] S. Pfiffner,[4] K. Davis,[5] and J. Edmonds[6]

[1]*Institute for Understanding Biological Systems, Tennessee State University, 3500 John A. Merritt Blvd, RASP Building, Nashville, TN 37209-1561*

[2]*NASA Science Mission Directorate, NASA Headquarters, Mail Suite 3C26, Washington, DC 20546-0001*

[3]*Goddard Center for Astrobiology, Solar System Exploration Division, Mailstop 690.3, NASA Goddard Space Flight Center, Greenbelt, MD 20771*

[4]*Center for Environmental Biotechnology, The University of Tennessee, 676 Dabney Hall, MS 1605, Knoxville, TN 37996*

[5]*Spectra Tech, Inc., 132 Jefferson Court, Oak Ridge, TN 37830*

[6]*Carnegie Institution of Washington, 1530 P Street NW, Washington DC 20005*

Abstract. The Minority Institution Astrobiology Collaborative began working with the NASA Goddard Center for Astrobiology in 2003 to develop curriculum materials for high school chemistry and Earth science classes based on astrobiology concepts. The Astrobiology in Secondary Classrooms modules are being developed to emphasize interdisciplinary connections in astronomy, biology, chemistry, geoscience, physics, mathematics, and ethics through hands-on activities that address national educational standards. Since this time, more NASA Astrobiology Institute Teams have joined this education and public outreach (EPO)effort. Field-testing of the Astrobiology in Secondary Classrooms materials began in 2007 in five US locations, each with populations that are underrepresented in the career fields of science, technology, engineering, and mathematics.

1. Introduction

The intent of the Astrobiology in Secondary Classrooms (ASC) project is to establish a successful model for creating the scientists of tomorrow by bringing powerful technology tools and current scientific data into an interdisciplinary curriculum focused on reaching all students. Goals for students participating in the ASC curriculum in their classrooms include:

- An understanding of the research pursuits and findings of key astrobiology researchers
- An appreciation for scientific research and the current knowledge base available in astrobiology
- A high degree of scientific and technological literacy
- A desire to continue their studies in STEM areas, particularly in areas pertaining to astrobiology

The ASC curriculum tackles many of the current problems in science education by addressing curriculum issues as well as minimizing classroom limitations that affect science instruction, particularly in classrooms containing high numbers of students underrepresented in science careers. Many science curricula, including textbooks, lack connections among different academic disciplines and do not provide students with a coherent framework for both science literacy and content knowledge. The ASC modules are being developed using research-based teaching strategies designed to diminish achievement gaps and increase the participation of underrepresented groups in science, technology, engineering, and mathematics (STEM).The ASC project began in 2003 with a team of university faculty from minority serving institutions and teachers selected by members of the Minority Institution Astrobiology Collaborative (MIAC). Working with scientists at the Goddard Center for Astrobiology, the team developed the ASC curriculum framework. Now, through this network of minority-serving institutions, the ASC staff seeks to enable middle and high school teachers across the United States to include astrobiology-related activities in their classrooms. Major partners during the field-testing phase of the materials are sites designated as NASA Science, Engineering, Mathematics and Aerospace Academies (SEMAA). Partnerships with SEMAA programs and other minority serving locations allow for a focus on diversity when field-testing and developing the ASC curriculum in both formal and informal educational settings. There are currently field-testing sites in five different locations where more than 80 percent of the students are members of the Native American, African American, or Hispanic American communities.

2. Discussion

Research supports the use of astrobiology as a framework for increasing science literacy (Astrobiology Design Project Team, 2002; Carrapiço, et al. 2001; Rodrigues & Carrapiço, 2005; Slater, 2006; Staley, 2003; Tang, 2005) because of its interdisciplinary nature. Furthermore, partnerships between curriculum developers, teachers, professional scientists and NASA researchers will provide the "real-world" contexts that are recognized as a vital part of science literacy and increasing student interest and understanding of STEM areas.The pedagogical side of the ASC curriculum has been grounded in three evidenced-based practices shown to increase achievement among all students and specifically among ethnically diverse students:

1. The *Five Standards for Effective Pedagogy* developed by the Center for Research on Education Diversity and Excellence (CREDE) provide a framework for culturally relevant instruction (Tharp, et al., 2003). The ASC Curriculum incorporates these principles in each of the modules in recognition of the importance of cultural awareness and the dynamics of learning in diverse settings (Lee & Luykx, 2006; Aikenhead, 2001; Lynch, et al., 2005).

2. The ASC Curriculum includes differentiated instruction that provides teachers with strategies for scaffolding that is a necessary part of effective teaching with varying levels of prior knowledge and understanding.

3. Civic Capacity is a principle whereby engaging students in community and through meaningful partnerships, academic enrichment can have significant and long-lasting effects. In their work with the NSF funded VISIT Teacher Enhancement Project, Hunter and Xie (2001) detailed the barriers for teachers accessing and using the vast amounts of data on the Internet. The ASC project will partner curriculum developers and teachers with astrobiology researchers to develop scientific data sets that are user-friendly in the high school classroom as well as provide much needed materials and laboratory supplies in order to overcome these barriers.

3. Program Evaluation

Evaluation of the ASC curriculum is in progress and includes iterative evaluation and implementation to modify modules. In the 2006 ASC teacher workshop held at Tennessee State University, 14 teachers were given pre-surveys, post-evaluations, and were asked to submit recommendations and participate in exit interviews. The surveys and evaluations used a Likert scale of 1 (poor) to 4 (excellent). Teachers scored the following items highly: workshop content, presentations by scientists, encouraging use of hands-on activities, and their perceived increase in knowledge about astrobiology. In fact, when rating the scientific content embedded in the ASC modules, 13 of the 14 teachers rated the activities as excellent (Table 1).

Table 1. Mean ratings for teacher response questionnaire statements by question type (N=14)

Statement	Mean
Use of technology: spectroscope and spectrometer	3.69
Biology and extremophile activities	3.92
Chemistry activities as they relate to astrobiology	3.92
Physics and astronomy activities including field trip to Dyer Observatory	3.92
Earth science activities analyzing rock samples collected on the geology fieldtrip	3.62

ASC modules were field tested in the 2006 high school SEMAA session at TSU. Students were asked to rank the activities they recommended as part of future SEMAA programs. Sixty percent (60%) indicated they would like to participate in astrobiology activities; Forty-seven percent (47%) indicated interest in activities concerning research in astronomy and the search for life in the universe. The 2007 SEMAA High School Summer Camp focused on the geoscience module of the ASC curriculum. Students completed a pre- and post-session survey using a Likert scale of 1 (poor) to 4 (excellent). Participants were asked about their interest levels in Earth science, engineering, space science, and general science as well as how much they felt they knew about technology uses in the geosciences, the importance of geoscience, and science/engineering careers. A paired sample *t*-test was used to calculate the significance of the mean differences in student responses from time 1 to time 2 in a single sample. These students reported a statistically significant increase in interest in space science, earth science, and engineering (Table 2).

Table 2. Mean ratings for student interest questionnaire statements by question type, paired t-test values, and probability (p) values

Question	Mean	Std. Dev.	*N*Pairs	*t*-Value	*p*
Space Science					
Pre-survey	2.46	0.877	13	-2.94	0.012*
Post-survey	3.00	0.913			
Earth Science					
Pre-survey	2.54	1.05	13	-2.89	0.014*
Post-survey	3.15	0.80			
Engineering					
Pre-survey	3.08	0.669	13	-2.57	0.026*
Post-survey	3.58	0.515			

*Significance at the $p < .05$ level

Activities at the 2007 summer camp involved using hand-held reflectance spectrometers to analyze rocks, learning about resonant frequencies, and simulated missions to Mars while looking for signs of life. A paired sample *t*-test was used to calculate the mean differences in student responses from time 1 to time 2 in a single sample. These students reported a statistically significant increase in understanding the area of technology uses in science and technology uses in geoscience (Table 3).

4. Summary

The ASC modules will provide a web-based interdisciplinary curriculum in astrobiology that is free and easily accessible by the public. The curriculum is designed to supplement existing state curricula by providing a framework that draws together all areas of science through engaging activities, providing teachers with activities that meet both state and national standards along with encouraging science literacy. Accomplishing this goal will involve modification of modules based on feedback from teachers during professional development and implementation with students in formal and informal educational settings. Research during the field-testing phase of the project will assess the impact of these crosscutting activities on student performance and attitudes about science along with student interest in STEM careers. More information about

Table 3. Mean ratings for student knowledge questionnaire statements by question type, paired t-test values, and probability (p) values

Question	Mean	Std. Dev.	*N* Pairs	*t*-Value	*p*
Technology In Science					
Pre-survey	3.00	0.95	13	-2.80	0.017*
Post-survey	3.42	0.67			
Technology In Geoscience					
Pre-survey	2.38	1.04	13	-2.25	0.044*
Post-survey	3.08	0.49			
Careers in the Geosciences					
Pre-survey	1.85	0.80	13	-2.67	0.020*
Post-survey	2.69	0.947			

*Significance at the $p < .05$ level

the ASC project is available at www.astroclassroom.org, the SEMAA program website can be found at www.semaa.net, and the MIAC website is located at www.miacnetwork.org.

Acknowledgments. Funding has been provided by the Goddard Center for Astrobiology NAI Team, The Indiana-Princeton-Tennessee Astrobiology Initiative, the NASA Astrobiology Institute Minority Institution Research Support (NAI-MIRS) Program, NASA Astrobiology Institute Education and Public Outreach, Carnegie Institution of Washington, Tennessee Space Grant Consortium, the National Science Foundation's Diversity in Geoscience Education Program, and the CREST Program.

References

Aikenhead, G. S. 2001, Integrating Western and aboriginal sciences: Cross-cultural science teaching. Research in Science Education. 31, 3, 337-355

Astrobiology Design Project Team, 2002, Search for Extraterrestrial Life: A Multi-Disciplinary Perspective. Publication No. COSPAR - 02-A-00039. Paris, France: 53rd International Astronautical Congress

Carrapiço, F., Lourenco, A., Fernandes, L., & Rodrigues, T. (2001). A Journey to the Origins: The Astrobiology Paradigm in Education. *Proceedings of SPIE*, 4495, 295-300.

Hunter, B., & Xie, Y. (2001). Data Tools for Real-World Learning. *Learning & Leading with Technology*, 28(7), 18-24.

Lee, O., & Luykx, A. (2006). *Science Education and Student Diversity.* New York: Cambridge University Press.

Lynch, S., Kuipers, J., Pyke, C., & Szesze, M. (2005). Examining the Effects of a Highly Rated Science Curriculum Unit on Diverse Students: Results from a Planning Grant. *Journal of Research in Science Teaching*, 42, 912-946.

Rodrigues, T., & Carrapiço, F. (2005). Teaching Astrobiology. A Scientific and a Cultural Imperative. *Proceedings of SPIE*, 5906.

Slater, T. F. (2006). Capturing Student Interest in Astrobiology Through Dilemmas and Paradoxes. *Journal of College Science Teaching*, 35, 42-45.

Staley, J. T. (2003). Astrobiology, the transcendent science: the promise of astrobiology as an integrative approach for science and engineering education and research. *Current Opinion in Biotechnology*, 14, 347-354.

Tang, B. L. (2005). Astrobiological Themes for Integrative Undergraduate General Science Education. *Astronomy Education Review*, 4, 110-114.

Tharp, R. G., Doherty, R. W., Echevarria, J., Estrada, P., Goldberg, C., Hilberg, R. S., & Saunders, W. M. (2003). *Research Evidence: Five Standards for Effective Pedagogy and Student Outcomes* (Rep. No. G1). Santa Cruz, CA: Center for Research on Education, Diversity, & Excellence.

Edna DeVore

Bioastronomy 2007: Molecules, Microbes, and Extraterrestrial Life
ASP Conference Series, Vol. 420, 2009
K. J. Meech, J. V. Keane, M. J. Mumma, J. L. Siefert, and D. J. Werthimer, eds.

The NASA Astrobiology Institute - Minority Institution Research Support Program: Strengthening the Astrobiology Community

T. Gary,[1] J. Butler,[1] L. Arino de la Rubia,[1] E. L. Myles,[1] K. Bradford,[2] M. Kirven-Brooks,[3] R. M. Ceballos,[4] L. Taylor,[5] B. Bell,[6] and G. Coulter,[7]

[1]*Institute for Understanding Biological Systems, Tennessee State University, 3500 John Merritt Blvd. Nashville, TN 37075*

[2]*Office of the Director, NASA Ames Research Center, Moffett Field, CA 94035*

[3]*NASA Astrobiology Institute, NASA Ames Research Center Mail Code N247-6, Moffett Field, CA 94035*

[4]*Native American Research Laboratory, Division of Biological Sciences, The University of Montana, 32 Campus Drive, HS505, Missoula, MT 59812*

[5]*Solar Astrophysics Laboratory, Lockheed Martin Space Systems Company, Bldg. 252, 3251 Hanover Street, Palo Alto Ca. 94304*

[6]*Minority University-SPace Interdisciplinary Network (MU-SPIN), NASA Goddard Space Flight Center Mail Code 606-3 Greenbelt, MD 20771*

[7]*Challenger Learning Center of Colorado 10215 Lexington Drive, Ste 110, Colorado Springs, CO 80920*

Abstract. This paper describes the history, purpose and successes of the NASA Astrobiology Institute Minority Institution Research Support Program (NAI-MIRS). This program is designed to provide support and training in astrobiology to a new generation of researchers from Minority Serving Institutions. The NAI-MIRS program provides sabbaticals, follow-up support, and travel opportunities for faculty and students from minority institutions. The purpose of this initiative is to increase the attendance and participation of underrepresented scientists in astrobiology research laboratories, at professional conferences, and as NAI Team members. As Minority Serving Institutions (MSIs) graduate a higher percentage of students of color entering graduate schools in science and engineering than their majority counterparts, support to MSIs from the NAI-MIRS program will encourage the growth of astrobiology-related programs at these institutions identifying talented researchers and providing an avenue to foster astrobiology research, increases awareness of astrobiology within minority communities. Achievements in astrobiology by the Minority Serving Institutions include the first direct detection of an extrasolar planet and a MSI graduate, LaTasha Taylor, featured in the journal Science as one of the first minority students to enter the NSF funded Astrobiology IGERT program. To date, the NAI-MIRS program has involved faculty members from the three major MSIs: Tribal Colleges and Universities, Historically Black Colleges and Universities, and Hispanic Serving Institutions and partnered with the Minority Institute Astrobiology Collaborative (MIAC).

1. Introduction: Astrobiology's Strength is Diversity

Astrobiology is the study of the origin, distribution and destiny of life in the universe. The goal of this field is focused on addressing three fundamental questions in science: 1) How did life begin and evolve, 2) Does life exist elsewhere in the universe and 3) What is the future of life on Earth and beyond ? To approach solutions to these questions, the field of astrobiology emphasizes interdisciplinary connections including involvement of multidisciplinary scientific fields, international partners, collaborations, and the inclusion of scientists of all ages, genders and ethnicities. This paper will highlight the NASA Astrobiology Institute Minority Institution Research Support (NAI-MIRS) program, one of the NASA funded projects commissioned to increase the involvement of the minority community in astrobiology.

2. NAI-MIRS Program Focuses on the Next Generation of Astrobiology

The NASA NAI-MIRS program focuses on faculty and students attending Minority Serving Institutions (MSIs). These institutions are divided into three main types: Historically Black Colleges and Universities (HBCUs), Hispanic Serving Institutions (HSIs), and Tribal Colleges and Universities (TCUs). There are more than 340 MSIs recognized by the US Dept. of Education, and combined, these institutions enroll more than two million Hispanic, African American or Native American students (Table 1).

Table 1. Underrepresented communities reached by MSIs† in the United States

Community	**Minority Serving Institution**	**Number of Institutions**	**Students of Color**
African American	Historically Black Colleges and Universities	103	370,000
Hispanic	Hispanic Serving Institutions	210	1,674,000
Native American	Tribal Colleges and Universities	33	17,000
Total Students		346 MSIs	$\sim 2,000,000$

†: Data taken from the White House Initiative on Minority Serving Institutions, Hispanic Association of Colleges and Universities, and American Indian College Fund (Humphreys 2005; Dept. of Education 2007).

The Higher Education Act of 1965 defines an HBCU as "...any historically black college or university that was established prior to 1964, whose principal mission was, and is, the education of black Americans." According to the President's Board of Advisors on Historically Black Colleges and Universities (2005), HBCUs were formed to eliminate the adverse residue from slavery, plus a century of legally sanctioned discrimination, against United States citizens of African descent. There are 103 HBCUs operating in the United States, most are located in the south. Although they constitute only 3 percent of America's institutions of higher education, HBCUs account for over 40 percent of bachelor's degrees earned by blacks in the physical, mathematical, biological, and agricultural sciences (Humphreys 2005). The nation's 33 TCUs, located in 12 states, had a total enrollment of about 17,000 full- and part-time students in 2005 (Dept. of Education, 2007). Hispanic-serving institutions are post-secondary institutions with a population of Hispanic students greater than 25 percent.

Research findings presented in the National Science Foundation publication on minorities in the areas of science, technology, engineering, and mathematics (STEM) state that the combined percent of Hispanic, African American and Native American students attending graduate school increased 40 percent over a ten year period between 1993 and 2003, while the percent of white non-Hispanic students decreased approximately 15 percent (NSF 2006).

Along with the demographic changes expected in the United States over the next few decades, the future workforce, including the next generation of Astrobiologists, will include significant numbers of underrepresented students. One of the best sources of this future workforce can be found at Minority Serving Institutes. The culturally supportive environments found at many of the MSIs are one reason that these schools graduate a higher percent of underrepresented students than most majority institutes. For example, HBCU's graduate 40 percent of all African Americans receiving bachelor's degrees in STEM areas (NSF 2002). African Americans who graduate from HBCUs in the sciences are more likely to go to graduate school and complete their doctoral degrees than African Americans graduating with a bachelor degree from other institutions (NSF 2000).

3. History and Purpose of MIRS

The NASA Astrobiology Institute (NAI) is a virtual organization that began in 1998 and represents a partnership between NASA and competitively selected NAI Teams (Blumberg 2003). The purpose of the NAI is to promote, conduct, and lead integrated multidisciplinary research in the field of astrobiology. As part of its mission, NAI initiated the Minority Institution Research Support (MIRS) program in 2002. This innovative program was created in part due to the passion and efforts of Karen Bradford and the leadership of Dr. Baruch Blumberg (Bradford and Wilmoth 2002; Goolish *et al.* 2005) in hopes of increasing the participation of minority faculty and students.

In 2005, the NAI partnered with the Minority Institute Astrobiology Collaborative (MIAC), a Tennessee State University (TSU) member, to manage and expand the NAI-MIRS program. MIAC is also a virtual organization (modeled after NAI Central) focused on expanding the astrobiology educational and research infrastructure within the minority community. The directors, Todd Gary (TSU) and Benita Bell (NASA Goddard) are the co-directors for MIAC with a mutual commitment to increasing interest and opportunities for growth among underrepresented populations (www.miacnetwork.org).

The NAI-MIRS program provides support for astrobiology research sabbaticals, follow-up support for laboratory or curriculum development, and travel for faculty and their students from minority institutions to attend astrobiology science conferences and participate in astrobiology research laboratories. The expected outcome is to increase diversity at all levels of the NAI by generating new collegial interactions, enlisting additional academic institutions in the NAI's participant base, and providing minority institutions and students with lasting professional relationships.

TSU is a Historically Black College or University (HBCU) located in Nashville, Tennessee with active astrobiology research and educational programs. Greg Henry, a TSU astronomer, made the first direct detection of an extrasolar planet, and has more than 200 refereed publications (Henry *et al.* 2000). Two TSU scientists, Todd Gary and Lewis Myles, have participated in the NAI-MIRS program (Goolish *et al.* 2005). Both of these scientists have received funding to support astrobiology from NSF, NASA, and TSU. Judy Butler coordinates astrobiology education and public outreach at TSU and is involved in developing the Astrobiology for Secondary Classrooms (ASC) curriculum, as part of the NASA Goddard Center for Astrobiology (NAI Team). She coordinated the publication of 10,000 copies of a 16-page astrobiology newspaper for middle school students, in partnership with the local Newspapers In Education program. Seven TSU students involved in astrobiology research at TSU now attend graduate school. One of these students, LaTasha Taylor, graduated from the University of Washington with a master's degree, through the NSF funded Astrobiology Interdisciplinary Graduate Education and Research Training (IGERT) program.

4. Successes of the NAI-MIRS Programs

The NAI-MIRS program is in its second year of management by MIAC, the Minority Institute Astrobiology Collaborative (MIAC), represented at Tennessee State University (TSU). This leadership by MIAC has led to greater visibility and collaboration within minority communities. Collaborators include AIHEC (the American Indian Higher Education Consortium), NoBCChe (National Organization of Black Chemists and Chemical Engineers), UNCF:SP (United Negro College Fund: Special Programs) and NASA Minority University SPace Interdisciplinary Network (Mu-SPIN). The NAI-MIRS website, www.nai-mirs.org, has received more than 17,000 hits since the launch in 2006.

The NAI-MIRS Program has successfully involved researchers from two Hispanic Serving Institutions (HSI), four Historically-Black Colleges and Universities (HBCU), and one Tribal College. Below are a few examples of NAI-MIRS awardees from around the country.

4.1. Impact of the NAI-MIRS Program in Minority Scientific Communities

Todd Gary from Tennessee State University, an HBCU, collaborated with Jim Lake from the NAI UCLA Team in the use of computational methods for determining evolutionary relationships and horizontal gene transfer. This NAI-MIRS experience was instrumental in engaging African American students in Astrobiology and the development of Astrobiology research and funding at TSU, The NASA Goddard's Astrobiology in the Secondary Classroom and the creation of a national organization focused on a Minority Institute Astrobiology Collaborative (www.miac.org).

Michael Ceballos, a NAI-MIRS sabbatical recipient in 2006, and the former Director of the Salish Kootenai College Molecular Biology and Biochemistry Research Laboratory, conducted research in viral ecology and evolution with Ken Stedman of Portland State University, co-chair of the NAI Virus Focus Group. This NAI-MIRS experience has been instrumental in engaging Native American students in Astrobiology, supporting research in the first basic research laboratory at a Tribal College and the first Native American Research Laboratory for basic sciences at any university in the nation.

Ceballos is currently a research assistant professor at the University of Montana where he has developed new facilities collectively called the Native American Research Laboratories that are dedicated to providing "hands-on" research opportunities for Native undergraduates and graduate students. Since many Native students come to the university from high schools in economically-challenged areas or from tribal colleges where access to state-of-the-art instrumentation and faculty research expertise is often limited, they are at a disadvantage when applying for graduate school or graduate research positions in university laboratories (NSF 2000). The Native American Research Laboratories (NARL) at the University of Montana is dedicated to addressing this educational disparity, so that Native students can be more competitive in applying for graduate positions. As part of the start-up of these labs, Ceballos was awarded a National Science Foundation Research Initiation Grant and a NASA EPSCOR award focused on topics in astrobiology. The amount of these two awards, totaling $1,000,000, represents a significant return on investment made during the term of Ceballos' NAI-MIRS sabbatical program. NASA Ames scientist Jonathan Trent, and NAI Virus Focus Group co-chair, Kenneth Stedman, of Portland State University, are serving as scientific advisor and collaborator, respectively, on the NARL research projects. In the first two years of operation NARL has served more than 55 students, two-thirds of whom are Native American representing more than 20 tribes from across the nation. These include undergraduates, graduates, and even high school summer interns all of whom have been actively engaged in astrobiology research as part of these projects.

In an NAI Internet-featured article titled "Astrobiologist Builds Native American Research Laboratory", Ceballos said the following about Native American scientific research. "It's time that we indigenous people cease being only the subjects of research and become part of the community doing the research. We often forget that indigenous people of the Americas were at the pinnacle of global scientific discovery and engineering achievement when Europe was in the

Dark Ages. Science is a part of our Native heritage that needs to be revived" (NAI 2007).

Abel Mendez, a NAI-MIRS sabbatical recipient in 2007 is a faculty member at the University of Puerto Rico at Arecibo, a HSI. Mendez worked with Chris McKay, of the NAI Ames Research Center Team, developing a field instrument designed to model microbial growth in the dynamic temperatures of natural environments. Mendez's work focuses on defining quantitative ways to measure planetary habitability, and engages Hispanic students in Astrobiology. Mendez presented a poster at AbSciCon 2006 on planetary habitability as well as at Bioastronomy 2007.

Don Walter is an astronomer who holds the rank of Professor of Physics at South Carolina State University, a HBCU located in Orangeburg, South Carolina. Walter completed a NAI-MIRS sabbatical with Mike Mumma and Michael DiSanti at The NASA's Goddard Center for Astrobiology (GCA) in 2007. Walter's summer work included optical studies of emission lines from comets and is expected to lead to the development of a program of sustainable cometary research at SCSU in partnership with GCA.

E. Lewis Myles, a NAI-MIRS sabbatical recipient in 2005 from Tennessee State University, a HBCU, worked with Jack Farmer, from the NAI alumni team at Arizona State University. As a result, he has established an astrobiology laboratory at TSU and incorporated astrobiology into the undergraduate curriculum.

Lee Anne Martinez, an Associate Professor of Biology at Colorado State University - Pueblo, a Hispanic Serving Institution, was a NAI-MIRS sabbatical recipient in 2007. Martinez began a genomic analysis of open water diatoms in the laboratory of Jim Lake, of the NAI UCLA team, to explore horizontal transfer of operational genes that may lead to the incorporation of endosymbionts by diatoms. Lee Anne's background includes nitrogen-fixation in diatom mats and this work will support current research at Colorado State-Pueblo.

Katherine Bancroft was the representative for a Tribal College Travel Award to the 2006 Astrobiology Science Conference in Washington, DC. Bancroft also attended the 2006 Astrobiology in Secondary Classrooms (ASC) workshop sponsored by the NASA Goddard Center for Astrobiology. Bancroft is currently coordinating the field-testing of the ASC curriculum and materials in an enrichment program sponsored by the three Paiute Shoshone reservations, Lone Pine, Big Pine, and Bishop in Owens Valley, California.

4.2. Impact on Students from MSI

The NAI-MIRS program provides funding for students of the sabbatical recipients to travel to astrobiology conferences. As a result, several of these students became involved in astrobiology research at the University of Washington, the Jet Propulsion Laboratory, the Woods Hole Marine Biological Laboratory, and the Centro de Astrobiología in Madrid, Spain and presented their research at astrobiology conferences (Taylor, L. *et al.* 2004; Taylor, T. *et al.* 2004). Seven TSU students who were involved in astrobiology research at TSU since the NAI-MIRS program began, have gone on to attend graduate school in STEM areas. One of these TSU students, LaTasha Taylor, graduated with a bachelor's degree in Aeronautics and a minor in Biology from Tennessee State University. Taylor participated in the NASA Reduced Gravity Student Flight Program, experiencing zero gravity onboard the NASA KC-135 aircraft. She went on to obtain her Master's Degree from the University of Washington College of Engineering. Currently she is a Space Systems Engineer in the Solar and Astrophysics Laboratory at Lockheed Martin's Advanced Technology Center in Palo Alto. She has been featured in the journal Science (Mervis 2006), is currently on the board of directors for the National Society of Black Engineers, was featured as a role model in a NASA education podcast (Nunley 2007) and served as Executive Officer and Crew Engineer for Crew 53 of the Mars Desert Research Station in the Utah desert. In this capacity, she conducted human factors engineering and biology research while assisting with the operations of the Mars-analogue simulation facility for a two-week tour of duty.

5. The Website and The Application Process

Information about the NAI-MIRS program is online at www.nai-mirs.org. This website also contains applications for the NAI-MIRS Sabbatical program. Two awards are given each year to faculty and/or researchers from minority serving institutions to engage in a sabbatical focused on research in astrobiology. To compete for these awards, the applicant is expected to propose a research project that will fit into the Astrobiology Roadmap (Des Marais *et al.* 2003), be mentored by an established researcher in the field,and demonstrate support from their home institution. The research project must have the potential for sustainability after the completion of the sabbatical. Each award will provide a stipend of $1,500 per week (up to 10 weeks); up to $5,000 for housing and travel; and $5,000 for supplies to support the research.

Two competitive awards, of approximately $5,000 each, will be available to scientists who have completed the NAI-MIRS research sabbatical program. These follow-up support awards will be available to continue that research and relationship established between the scientist at the minority institution and the scientist at the host institution. Follow-up support will be awarded based on demonstration of efforts toward building a research or academic infrastructure in astrobiology at the minority institution.

6. How Can You Be Involved ?

If You Are...

A faculty member at a Minority Institution? You can apply for a travel award or a summer sabbatical.

An astrobiologist and a member of an NAI team? You can be a mentor, working with a new colleague for up to a year, benefiting from fresh ideas and different perspectives.

An astrobiologist? You can recommend a colleague to take part in these programs.

Information on ways to become involved and participate is on the NAI-MIRS website: www.naimirs.org

7. Acknowledgements

Funding for the NAI-MIRS program has been provided by the NASA Astrobiology Institute. Additional support for travel has been provided by the Tennessee State University NASA University Research Center, the NASA Broker facilitator program at South Carolina State University and the NASA Goddard Minority University-Space Interdisciplinary Network(MU-SPIN) program.

References

Blumberg, B. S. 2003, The NASA Astrobiology Institute: Early History and Organization. Astrobiology 3(3), 463-470

Bradford, D.C. and Wilmoth, K. 2002, Collaborating with minority institutions: NAI minority institution faculty sabbatical. Astrobiology, 2(4), 471

Bell, B., Bowman, A., Butler, J, Gary, T., Myles, L., Laval, B., Rivera, M., Lowe, L., Jejelow, F., Perry, B. and Walter, D. 2004, MIAC, the first "virtual collaboration" of minority institutions focused on astrobiology. International Journal of Astrobiology 1(suppl. 2004), 91

Department of Education, Office of Postsecondary Education, White House Initiative on Tribal Colleges and Universities, Report to the President: 2004-05 Annual Performance Report on the Results of Federal Agency Actions to Assist Tribal Colleges and Universities and Recommendations to Strengthen Implementation of Executive Order 13270, President's Board of

Des Marais, D. J., Allamandola, L. J., Benner, S. A., Boss, A. P., Deamer, D., Falkowski, P. G., Farmer, J. D., Hedges, S. B., Jakosky, B. M., Knoll, A. H., Liskowsky, D. R., Meadows, V. S., Meyer, M. A., Pilcher, C. B., Nealson, K. H., Spormann, A. M., Trent, J. D., Turner, W. W., Woolf, N. J., and Yorke, H. W. 2003, The Astrobiology Roadmap. Astrobiology 3(2), 219-235

Goolish, E., Abe, S., Fletcher, J. and Bradford, K. (2005). Fellowship and Funding Opportunities available through the NASA Astrobiology Institute. Astrobiology 5(2):180

Henry, G.W., Marcy, G. W., Butler, R. P. and Vogt, S., S. 2000, A Transiting "51 Peg-like" Planet. The Astrophysics Journal 529, L41-44

Humphreys, J. 2005, Economic Impact of the Nation's Historically Black Colleges and Universities (NCES 2007– 178). U.S. Department of Education, National Center for Education Statistics. Washington, DC: U.S. Government Printing Office

Mervis, J. Graduate Training: Diversity remains elusive for NSF flagship program. 2006, Science 313, 1454

NASA Astrobiology Institute. 2007, Astrobiologist builds Native American Research Laboratory. NASA Astrobiology Institute Feature Stories. (http://nai.arc.nasa.gov)

Nunley, D. 2007, NASA Student Opportunity Educational Podcasts: Episode 34: LaTasha Taylor. http://www.nasa.gov/multimedia/podcasting/education/ED_NSO_Episode_34.-html

Taylor, L., Prieto-Ballesteros, O., Gómez-Elvira, J., Fernández-Remolar, D., Gómez, F., Parro, V., Amils, R., Myles, L., and Gary, T. 2004, Using Fluorescence Biological Analysis (FBA) to Examine the Subsurface of Europa for Signs of Life. International Journal of Astrobiology (suppl. 2004), 54

Taylor, T., Henry, G., Gary, T., Myles, L., & Burks, G. 2004, Solar Brightness Variability and the Evolution of Life on Earth. International Journal of Astrobiology (suppl. 2004), 46

Ed Prather

Erika Offerdahl

Session XV

Astrobiology Glossary

Bioastronomy 2007: Molecules, Microbes, and Extraterrestrial Life
ASP Conference Series, Vol. 420, 2009
K. J. Meech, J. V. Keane, M. J. Mumma, J. L. Siefert, and D. J. Werthimer, eds.

An Evolving Astrobiology Glossary

K. J. Meech[1] and W. W. Dolci[2]

[1]*Institute for Astronomy, 2680 Woodlawn Drive, Honolulu, HI 96822*

[2]*NASA Astrobiology Institute, NASA Ames Research Center, MS 247-6, Moffett Field, CA 94035*

Abstract. One of the resources that evolved from the Bioastronomy 2007 meeting was an online interdisciplinary glossary of terms that might not be universally familiar to researchers in all sub-disciplines feeding into astrobiology. In order to facilitate comprehension of the presentations during the meeting, a database driven web tool for online glossary definitions was developed and participants were invited to contribute prior to the meeting. The glossary was downloaded and included in the conference registration materials for use at the meeting. The glossary web tool is has now been delivered to the NASA Astrobiology Institute so that it can continue to grow as an evolving resource for the astrobiology community.

1. Introduction

Interdisciplinary research does not come about simply by facilitating occasions for scientists of various disciplines to come together at meetings, or work in close proximity. Interdisciplinarity is achieved when the total of the research experience is greater than the sum of its parts, when new research insights evolve because of questions that are driven by new perspectives. Interdisciplinary research foci often attack broad, paradigm-changing questions that can only be answered with the combined approaches from a number of disciplines. However, most scientists are trained in a specific discipline, which often narrows as one advances in a career. It is difficult to be both expert in a field and possess great breadth.

In order to facilitate a new style of interdisciplinary research, scientists need to be exposed to interdisciplinary approaches early in their careers, both through graduate training programs, such as the many astrobiology summer and winter schools that have become available (http://astrobiology.nasa.gov/nai/careers/), and through interdisciplinary workshops and meetings. As an outgrowth of discussions between graduate students and scientists attending the NASA Astrobiology Institute (NAI) general meeting in 2001, NAI supported the development of an *Astrobiology Primer* (Mix et al., ed. 2006) as a concise reference tool for astrobiology. While not intended to be a textbook, it was developed to be a resource for students newly entering the interdisciplinary field of astrobiology. At nearly the same time, an astrobiology glossary of about 1000 terms was published (Gargaud et al. 2006) as a companion to a special astrobiology issue of *Earth, Moon and Planets.* It is intended that this also grow into an evolving document that will be more like an astrobiology encyclopedia.

In a pre-conference effort to encourage the speakers at the Bioastronomy 2007 meeting to present their research in a manner accessible to the participants who were drawn from many disciplines, we developed a data-base driven on-line system for entry of discipline-specific terms and their definitions. Each presenter was asked to go through their presentations and select all the terms that were specific to their sub-discipline, but might not be familiar to other disciplines and to define them online. Many meeting attendees participated in this activity, often having several definitions for the same terms, but from different disciplines. The glossary below is the product of this activity. After each term and its definition, the discipline for which the term is defined is indicated in parentheses.

The glossary is available on the NASA Astrobiology Program website at: http://astrobiology.nasa.gov/glossary/. New terms may be submitted by filling out a form on the glossary website. In the spirit of Web 2.0, evolution of the glossary into a wiki or similar interactive community effort is being considered. An online interactive glossary could be used to promote interdisciplinary understanding at future astrobiology meetings, as a download for laptops or smart phones for example, or for communicating terms in real time during talks via social media such as Facebook or Twitter. Glossary entries could also be linked to stories on the Astrobiology Program website so that the reader can have easy access to definitions of unknown terms.

Acknowledgments. Many attendees of the Bioastronomy 2007 meeting contributed definitions to the glossary. In particular, we would like to thank MingHui Chen, who set up the database-driven interactive web page, and the Univ. of Hawai'i NAI postdocs for their help in proof reading: L. Chizmadia, N. Haghighipour, J. Keane, J. Kleyna, G. Phillip, J. Pittichová, K. Stockstill, and W. Zheng. This material is based upon work supported by the National Aeronautics and Space Administration through the NASA Astrobiology Institute under Cooperative Agreement No. NNA04CC08A issued through the Science Mission Directorate.

References

Mix, L. J., Armsron, J. C, Mandell, A. M., Mosier, A. C, Raymond, J., Raymond, S. N., Steward, F. J., von Braun, K., & Zhaxybayeva, O. 2006, *Astrobiology*, 6, 735

Gargaud, M., Claeys, Ph., Martin, H. 2006, *Earth, Moon & Planets*, 98, 319

- **16S rRNA 70S** – Ribosome is composed of 2 subunits: 50S and 30S. 30S subunit is composed of 21 ribosomal proteins and 16S ribosomal RNA. Evolution of 16S rRNA is extremely conserved throughout all life domains due to the crucial function in the production of proteins. 16S is the most used phylogenetic marker. (*Molecular Biology*)

- **accretion** – Accumulation of dust and gas into larger bodies such as stars, planets and moons, or as discs around existing body. An accretion disc (or accretion disk) is a structure formed by diffuse material in orbital motion around a central body. Instabilities within the disc redistribute angular momentum, causing material in the disc to spiral inward towards the central body. (*Astronomy*)

- **acetaldehyde** – CH_3CHO. (*Organic Chemistry*)

- **acetamide** – CH_3CONH_2. (*Organic Chemistry*)

- **acetic acid** – CH_3COOH. (*Organic Chemistry*)

- **acetonitrile** – CH_3CN, cyanomethane, methyl cyanide. (*Organic Chemistry*)

- **acetyltransferase** – An enzyme that catalyzes the transfer of acetyl (the acetic acid radical CH_3CO) groups from one compound to another. (*Biology*)

- **acidithiobacillus** – A genus of proteobacteria. The members of this genus used to belong to Thiobacillus, before they were reclassified in the year 2000. (*Biology*)

- **active SETI** – Deliberate transmission of messages from Earth, for the purpose of initiating communication with Extra-Terrestrial Intelligence. Also known as METI (Messaging to Extra-Terrestrial Intelligence). (*SETI*)

- **adenine** – One of the four nuclear bases of ADN and ARN, and a component of ATP. This molecule of global formula $H_5C_5N_5$ is a puric base (derivative of purine) and is made of two heterocycles of 5 and 6-atoms. It plays an important role in the living world as component of nucleotides. (*Biochemistry*)

- **aeolian** – Caused by, or carried by the wind. Also spelled eolian. (*Geology*)

- **aerogel** – A low-density solid-state material derived from silicone gel in which the liquid component of the gel has been replaced with gas. The result is an extremely low density solid with several remarkable properties, most notably its effectiveness as an insulator. It is nicknamed frozen smoke, solid smoke or blue smoke due to its semi-transparent nature and the way light scatters in the material. Aerogel was used as the high speed capture medium for comet dust by the Genesis and Stardust spacecraft. (*Chemistry, Space Exploration*)

- **aerosol** – This term technically refers to airborne solid particles (also called dust or particulate matter) or liquid droplets. (*Chemistry, Physics*)

- **Afrho** – Standard planetary science abbreviation ($Af\rho$) used in cometary studies. The value is a proxy for the dust production of a comet. (*Astronomy*)

- **albedo** – Albedo (often geometric albedo) is defined in various ways, but is fundamentally the ratio of the intensity of light diffusely reflected from a surface to the intensity of the incident light (usually sunlight). Fresh snow has an albedo of about 0.9, while powdered elemental carbon has albedo less than 0.05. (*Planetary Science*)

- **aliphatic hydrocarbons** – The simplest organic compounds, containing straight chains of carbon and hydrogen. (*Organic chemistry*)

- **alkaliphilic** – Organisms that grow optimally or very well at pH values above 9, often between 10 and 12, but cannot grow or grow only slowly at the near-neutral pH value of 6.5. (*Biology*)

- **Allen array** – A joint project of the SETI Institute and the University of California at Berkeley, the facility is a radio interferometer array of 6m antennas that operate at cm-wavelengths. Eventually the array will have 350 dishes and conduct targeted and survey SETI observations, as well as radio astronomy programs. (*SETI*)

- **Amadori rearrangement** – One of the steps in the Maillard reaction (see Maillard reaction). An organic reaction describing an acid or base catalyzed reaction important in carbohydrate chemistry. (*Organic Chemistry*)

- **amino acids** – A general class of organic molecules with a tetravalent carbon atom bonded to a hydrogen atom, an amino group (NH_2), a carboxyl group (COOH), and an organic group (R) giving a general formula $CHR(NH_2)COOH$; slight variations on this general formula are known. (*Organic Chemistry*)

- **amorphization** – The process of turning into an amorphous phase. Amorphous refers to a lack of order in the arrangement of the atoms. (*Physics, Geology*)

- **amorphous ice** – Phase of water ice in which there is no long-range order of the positions of the atoms (as opposed to crystalline ice; see definition of mineral). Amorphous ice forms when water vapor condenses on a surface maintained below $\sim$120K. (*Astrochemistry*)

- **amphibole** – An important group of rock-forming, hydrated (containing an OH group) silicate minerals that form prism or needle-like crystals (see definition of silicate). They have the inosilicate (double chains) structure, composed of double chain SiO_4 tetrahedra, linked at the vertices. They are generally generally dark-colored, and contain ions of iron, magnesium, calcium and aluminum in varying amounts. (*Geology*)

- **amphiphile** – A term describing a chemical compound possessing both hydrophilic and hydrophobic properties. Such a compound is called amphiphilic or amphipathic. This forms the basis for a number of areas of research in chemistry and biochemistry, notably that of lipid polymorphism. (*Chemistry*)

- **anaerobic chemoautotroph** – An organism that obtains energy from the metabolism of inorganic chemicals from the environment without the use of oxygen. Autotrophs synthesize organic chemicals necessary for growth from CO_2. (*Microbiology*)

- **anneal** – Change occuring in the structure of a material due to the application of heat and/or pressure. In this context, it refers to low temperature amorphous ices when the ice is heated, allowing for an internal re-structuring of the ice matrix. As the ice restructures, pore space can decrease and other trapped molecules can get released in the process, resulting in one possible cause of volatile escape in icy bodies at low temperatures. (*Astrochemistry, Physics*)

- **anoxic** – without oxygen. (*Biology, Chemistry*)

- **antiamoebin** – A type of polypeptide that is capable of forming ion channels in phospholipid membranes. It is produced by fungi of the species Emericellopsis and has antibiotic properties against the organisms responsible for amoebic dysentry. (*Biology*)

- **aphelion** – Term defining a specific point in an orbit. The prefix ap refers to the farthest point from the focus of an elliptical orbit, and helion (from helios) refers to the sun. Aphelion is the farthest point in an orbit about the sun. (*Astronomy*)

- **apparent brightness** – The brightness of an astronomical object, such as a star or a comet, as viewed or measured from the Earth. (*Astronomy*)

- **arc second** – An angular measure. Each degree of angle contains 3600 arc seconds, or 60 arc minutes. Each arc minute of angle contains 60 arc seconds. (*Astronomy*)

- **Archean** – Used to describe rocks from the Archeozoic time period. This geologic time is the earlier part of the pre-Cambrian time corresponding to 4.6 to 2.5 billion years before the common era; the later being the Proterozoic. (*Geology*)

- **Arecibo message** – A deliberate transmision from Earth on 16 November 1974, beamed toward the M-13 cluster at a frequency of 2380 MHz during ceremonies to dedicate the refurbished Arecibo Observatory. The Message consisted of 1679 bits, arranged in 73 rows $\times$ 23 columns (those both being prime numbers). It graphically described a binary number sequence, hydrocarbon chemistry, DNA, the human figure (including a height scale), our solar system, and the Arecibo dish transmitting the signal. (*SETI*)

- **Arkose** – A sandstone rich in feldspar, the family of K, Na, Ca aluminosilicate minerals that are abundant in igneous and metamorphic rocks. (*Geology*)

- **aromatic hydrocarbon** – A planar ring of six carbon atoms (usually described as having alternating single and double bonds) having bound hydrogen atoms. The simplest such molecule is benzene, which has a chemical formula C_6H_6. Other atoms and functional groups can replace hydrogens to make more complex molecules. Multiple fused aromatic hydrocarbons are called polycyclic aromatic hydrocarbons, or PAHs. (*Astrochemistry, Organic Chemistry*)

- **asteroid** – Small ($<$ 1000 km) rocky bodies orbiting the Sun. Many of them orbit between Mars and Jupiter (the main asteroid belt) and represent bodies left over from the era of planet formation. (*Astronomy*)

- **AU** – Astronomical Unit, which is the average distance from the Earth to the Sun (1.495×10^8 km). (*Astronomy*)
- **autotrophy** – Synthesizing organic molecules from inorganic ones. (*Microbiology*)
- **Bacillus subtilis** – A gram-positive soil bacterium that forms hardy dormant spores. Because the spores are highly resistant to a wide variety of physical insults, B. subtilis is widely used as a model organism for astrobiological and space flight studies. (*Biology*)
- **bacteriophage** – A virus that infects bacteria. (*Biology*)
- **banded iron formation** – (BIF) A rock that consists of alternating bands of iron-rich minerals, generally hematite (Fe_2O_3), and chert or fine-grained quartz (SiO_2). Often found in Precambrian sedimentary rocks. (*Geology*)
- **barycenter** – Center of mass in an orbital system. (*Astronomy*)
- **basal metazoans** – The groups comprising the root of the animal tree of life, namely Poriferans, Cnidarians, Ctenophores, and Placozoans. (*Biology*)
- **basalt** – A dark, fine-grained, extrusive (volcanic) igneous rock with a low silica content (45% to 52% SiO_2), but rich in iron, magnesium and calcium. It is composed largely of plagioclase feldspar and pyroxene, but may also be rich in olivine. Generally occurs in lava flows, but also as dikes. Basalt makes up most of the ocean floor and is the most abundant volcanic rock in the Earth's crust. (*Geology*)
- **baseline** – The vector between any two telescopes of an interferometer. It has both an azimuthal direction relative to the compass, and a length. (*Astronomy*)
- **basement** – Basement rocks refer to the oldest geologic layers in the stratigraphic column. These are usually metamorphic and can be also referred to as crystalline basement rocks. (*Geology*)
- **bathymetry** – Refers to the measurement of the depth of a body of water. *Oceanography*)
- **bioma** – Life zone with similar characteristics. (*Biology*)
- **biomonomers** – Molecules that are the buiding blocks of biopolymers: proteic amino acids for proteins; nucleotides for RNA or DNA. (*Biochemistry*)
- **biopolymers** – Biopolymers are a class of polymers produced by living organisms, also known as biological polymers. DNA, RNA, and proteins are all examples of bipolymers. (*Biology*)
- **biosensor** – A device for the detection of an analyte (chemical constituent) that combines a biological component with a physicochemical detector component. (*Biology*)

- **biosignature** – A signature in the environment (on Earth, elsewhere in the solar system, or in extrasolar planetary systems) that has been produced by life. (*Astrobiology*)

- **biosilification** – The process in which biosignatures are created by the interaction of organic materials (especially saccharides) with silicates. (*Microbiology, Geology*)

- **black holes** – When the density of matter becomes great enough the gravitational field becomes so strong that not even light can escape. The event horizon is the boundary between where light can and cannot escape. (*Astronomy*)

- **black smoker** – Also called hydrothermal vent. Structure observed on the ocean floor generally associated with mid ocean ridges. There, hot hydrothermal fluid, rich in base metal sulphides, enters in contact with cold oceanic water. Polymetallic sulphides and calcium sulphate precipitate progressively building a columnar chimney around the vent. (*Geology*)

- **bolide** – A bright meteor that explodes or strikes the earth – a fireball. (*Astronomy, Planetary Geology*)

- **brown dwarf** – A substellar object that lacks sufficient mass to ignite hydrogen fusion. (*Astronomy*)

- **carbonaceous chondrite** – A class of meteorites that have elemental abundance ratios of certain rock-forming elements that are nearly identical to the ratios observed in the sun. Carbonaceous chondrites include the carbon-rich meteorite classes CI, CM, and CR, and the chondrule-rich classes CV, CO, and CK, and the minor groups CB and CH. (*Cosmochemistry*)

- **carbonate** – A group of minerals composed primarily of a divalent cation (*e.g.*, Ca, Na, Mg, Fe., etc.) and CO_3. Carbonates can be formed through chemical precipitation (supersaturation due to temperature decrease), biochemical processes (*e.g.*, shells, coral), by metamorphic processes, igneous crystallization and by the evaporation of water. The two most abundant carbonate minerals in today's oceans are calcite and aragonite, two different arrangements of $CaCO_3$. Dolomite $(Ca,Mg)CO_3$ is another common carbonate mineral in ancient sediments. (*Geology*)

- **carcinogenesis** – The molecular process by which normal cells are transformed into cancer cells. (*Biology*)

- **cellular automaton, (CA)** – A computational model that is discrete in both space and time. Space is divided into boxes called cells, and their values are only calculated at discrete time instants (you could consider this as dividing space-time into boxes). CA uses a set of fixed rules to transition from one state (value of the cell) to another. (*Computer science*)

- **centaur** – A small solar system body object whose orbit around the Sun crosses the orbit(s) of one or more major planets. These objects are thought to have been ejected from the Kuiper Belt and their planet-crossing orbits are temporary (~10,000 years or less) because of dynamical interactions with the planets. Some Centaurs exhibit cometary activity. (*Planetary Science*)

- **chemotrophy** – Metabolic system using a chemical energy source, rather than light-energy. (*Microbiology*)

- **chert** – Chemical sedimentary rock, most commonly deposited on the seafloor by chemical precipitation from the ocean water, composed of very fine crystals of quartz or silica (SiO_2). (*Geology*)

- **chiral** – Used to describe an object that is non-superimposable on its mirror image, such as the human hand. In terms of chemistry, these objects are usually molecules. The two mirror images of a chiral molecule are referred to as enantiomers, the left and right hands. For organic molecules such as amino acids, chirality is a powerful tool that can be used to discriminate between abiotic (non-biological) and biotic (biological) origins. (*Chemistry, Organic Chemistry*)

- **CHON particle** – Cometary dust particle rich in Carbon, Oxygen, Nitrogen, Hydrogen. CHON particles were identified in comet Halley from in situ mass spectrometry during the 1986 Giotto mission. (*Astronomy, Astrobiology*)

- **chromatic complementary adaptation** – CCA. Adaptation of living organisms to the properties of light through the biosynthesis of pigments (optically active biomolecules) and development of coloration. CCA is essential for energy harvesting from the parent star in accordance with it's spectral characteristics (*i.e.* temperature). (*Biology*)

- **chromospheric** – Pertaining to the chromosphere, a complex warm gaseous layer that lies above the visible surface of the Sun. (*Astronomy*)

- **circularly polarized light** – Light with the electric field component (vector) tracing out a circle as the light propagates from the source to the observer. One proposal for the origin of chirality (handedness) of biomolecules, such as amino acids, is the influence of circularly-polarized light on pre-biotic material. (*Astronomy, Chemistry*)

- **clathrate** – A chemical substance consisting of a lattice of one type of molecule trapping and containing a second type of molecule. (The word comes from the Latin clathratus meaning furnished with a lattice). For example, a clathrate hydrate involves a special type of gas hydrate that consists of water molecules enclosing a trapped gas. A clathrate therefore is a material which is a weak composite, with molecules of suitable size captured in spaces which are left by the other compounds. They are also called host-guest complexes, inclusion compounds, and adducts (chiefly in the case of urea and thiourea). They used to be called molecular compounds. Clatrates are stable at high pressures and low temperatures. Scientists believe that compounds on the sea bed have trapped large amounts of methane in similar configurations, which is important for the energy industry. For planetary science clathrate destabilization may be important for volatiles and their impact on geological features and environments seen elsewhre in the solar system. (*Chemistry, Organic Chemistry*)

- **clay mineral** – Clay minerals are hydrous phyllosilicates, sometimes with variable amounts of iron, magnesium, aluminum, alkali metals (*e.g.* Na, K, etc.), alkaline earths (*e.g.* Mg, Ca, etc.) and other cations. Clays have structures similar to the micas and therefore form flat hexagonal sheets. Clay minerals are common weathering products (including weathering of feldspar, pyroxene and olivine) and low temperature hydrothermal alteration products. Clay minerals are very common in fine grained sedimentary rocks such as shale, mudstone and siltstone, in fine grained metamorphic slate and phyllite, and in meteorites which have experienced hydrothermal alteration or aqueous alteration. (*Geology*)

- **clonal** – Population generated upon replication of a single molecule or a single individual. A clonal population in virology designates an ensemble of progeny viruses that arise from replication of a single virus during a limited number of replication rounds. The standard method to isolate clonal populations of viruses is the development of lytic plaques. (*Biology*)

- **clone** – (1) A cell, group of cells, or organism that is produced asexually from and is genetically identical to a single ancestor. The cells of an individual plant or animal, except for gametes and some cells of the immune system, are clones because they all descend from a single fertilized cell and are genetically identical. A clone may be produced by fission, in the case of single-celled organisms, by budding, as in the hydra, or in the laboratory by putting the nucleus of a diploid cell into an egg that has had its nucleus removed. Clones of other cells and some plants and animals can also be produced in a laboratory. (2) A copy of a sequence of DNA, as from a gene, that is produced by genetic engineering. The clone is then transplanted into the nucleus of a cell from which genetic material has been removed. (Source: American Heritage Science Dictionary) (*Biology*)

- **CM meteorites** – The most abundant type of carbonaceous chondrite meteorite. Possesses essentially a solar composition, aside from the more-volatile elements (*e.g.* H, He, O, etc.); contains organic and hydrated compounds and is thought to represent relatively primitive solar-system material. (*Cosmochemistry*)

- **codon** – The genetic code is the set of rules by which information encoded in genetic material (DNA or RNA sequences) is translated into proteins (amino acid sequences) by living cells. Specifically, the code defines a mapping between tri-nucleotide sequences called codons and amino acids; every triplet of nucleotides in a nucleic acid sequence specifies a single amino acid. (*Biology*)

- **collective intelligence** – Evolved stage of advanced civilizations where the dominant paradigm to resolve conflicts is through collaboration and non-violence. Our planet might start to witness an Earth Community, where all earthlings have covered their basic needs. (*Natural Philosophy*)

- **color** – A measure of the relative fluxes or reflectivities at two wavelengths or wavelength intervals. In a magnitude system it is the difference between two magnitudes. (*Astronomy*)

- **column density** – The number of atoms of a certain element as seen along a line-of-sight projected to a unit area at the observer (often in units of atoms/cm^2). (*Astronomy*)

- **coma** – The nebulous luminescent cloud of volatile elements, molecules, and dust that has escaped from a cometary nucleus. The coma may extend 10^5-10^6 km in diameter, and constitutes the major portion of the visible "head" of a comet as seen from Earth. (*Astronomy*)

- **comet** – Small (~few km-sized) icy solar system body thought to be a little-altered remnant of the formation of the solar system, likely formed in the vicinity of the giant planets. (*Astronomy*)

- **cometary tail** – Most visible part of an active (outgassing) comet. There are two main tails: the blueish, straight ion tail (luminous from fluorescence of gas), and the white-yellow (luminous from reflected sunlight) curved, dust tail. These tails may reach up to a fraction of an astronomical unit in some cases. A neutral sodium tail has also been observed. (*Astronomy*)

- **commensal** – Literally means eating from the same plate. This refers to a symbiotic relationship between two organisms in which benefit to one organism does not affect the other. (*Biology*)

- **commensal** – Used to mean multiple-concurrent observations that can be conducted without detriment to one another. (*Astronomy*)

- **confusion limit** – Spectral line data at radio wavelengths where the data is no longer noise limited. Spectral lines cover the the full spectral range, such that no baseline noise is present. Can occur in dense molecule clouds. (*Astronomy*)

- **continuously habitable zone (CHZ)** – A region of space around a star where a planet can sustain liquid water at its surface, for at least part of its local year, considered over a long enough period for sophisticated life to emerge. (*Astronomy*)

- **CRISM** – Compact Reconnaissance Imaging Spectrometer for Mars. This was launched on the Mars Reconnaissance Orbiter in August 2005 and is currently mapping Mars. It is a visible-infrared hyperspectral mapper in the wavelength range from 0.36-3.92 μm. (*Remote Sensing*)

- **cryogenic volcanism** – The eruption of volatiles (usually water) on the surface of a planet or moon which then cools and freezes into a new layer of crust, similar to the eruption of silicate magma. (*Planetary Geology*)

- **cryostat** – A device used for maintaining the low temperature of an object. (*Physics, Chemistry*)

- **crystalline ice** – The phase of water ice exhibiting a structure with the long-range order of a mineral. Crystalline ice is typically found in its hexagonal form, but sometimes exists in the metastable cubic form. Upon warming, cubic ice will transform irreversibly to hexagonal ice. (*Chemistry, Astronomy*)

- **cyanhydric acid** – HCN, hydrogen cyanide. In prebiotic chemistry, may have been the starting material for purine synthesis. (*Organic Chemistry*)

- **cyanoacetylene** – HC_3N, cyanoethyne. (*Organic Chemistry*)

- **cyanobacteria** – From the Greek word kyan, meaning blue. Also known as Cyanophyta. Cyanophyta is a phylum (or division) of Bacteria that obtain their energy through photosynthesis. (*Biology*)

- **cyanogen** – C_2N_2. (*Organic Chemistry*)

- **damped Lyman-α system (DLA)** – Intervening absorbing gas between us and a background source of light such as a QSO (galaxies with bright nuclei due to powerful central black holes). DLAs are defined as those intervening sources with the greatest neutral hydrogen content, often presumed to be due to a galaxy. (*Astronomy*)

- **Darwin** – A European Space Agency project that will be a flotilla of free-flying spacecraft forming a space infrared interferometer devoted to the search of extrasolar planets and the study of their atmosphere composition. Molecules of CO_2, O_3 and H_2O have IR bands in the 6-20μm range; their simultaneous detection is under study as signature of (biological) photosynthetic processes. NASA's TPF-C and TPF-I projects (Terrestrial planet finder - coronagraph and -interferometer) have similar goals. (*Astronomy, Astrobiology*)

- **daughter species** – Produced by photodissociation (or other type of chemical splitting) of a parent molecule. For example, in the coma of a comet: OH from H_2O. The parent(s) of some commonly observed radicals (C_2, C_3, CN) are uncertain. (*Astronomy, Physics*)

- **debris disk** – A disk of dust particles around a star that is formed through collision of planetesimals (see planetesimal for its definition). (*Astronomy*)

- **derivatization** – A technique used in chemistry which transforms a chemical compound into a product that is more easily detected or analyzed. OPA/NAC derivatization is often used to label amino acids with a chiral fluorescent tag, increasing the detection sensitivity for the compound and enabling the separation of amino acid enantiomers using standard liquid chromatography techniques. (*Chemistry, Organic Chemistry*)

- **devolatilized** – Used to describe materials that have lost their volatile components (*e.g.* CO_2, H_2O, etc.). (*Geology*)

- **diagenesis** – The process by which sediment undergoes chemical and physical changes during its lithification (conversion to rock). Compaction, leaching, cementation, and recrystallization are all forms of diagenesis. Erosion and metamorphism are not. Oil, gas, and coal form through the diagenesis of organic sedimentary matter. (*Geology*)

- **diffuse interstellar bands** – Absorption profiles commonly observed towards reddened stars in the Milky Way. Current work points to organic molecules such as carbon chains and polycyclic aromatic hydrocarbons as the likely sources of the diffuse interstellar bands (DIBs). The hundreds of observed DIBs to date fall within the visible part of the electromagnetic spectrum. (*Astronomy*)

- **dihydroxyacetone** – $HOCH_2COCH_2OH$. (*Organic Chemistry*)

- **diurnal** – Relating to or occurring in a 24-hour period; daily. (*Astronomy, Geology*)

- **DNA**) – Deoxyribonucleic acid. This is a nucleic acid that contains the genetic information for all living organisms. DNA is composed of two long polymers of nucleotides, connected with sugars and phosphate groups. (*Biology*)

- **dosimeter** – A device that is used to measure an individual's exposure to a hazardous environment. A radiation dosimeter, for example, is a device that measures the cumulative dose of radiation. (*Physics*)

- **Drake Equation** – This equation (rarely also called the Green Bank equation or the Sagan equation) is a famous result in the speculative fields of exobiology, astrosociobiology and the search for extraterrestrial intelligence and identifies specific factors thought to play a role in the development of such civilizations. This equation was devised by Dr. Frank Drake (now Emeritus Professor of Astronomy and Astrophysics at the University of California, Santa Cruz) in the 1960s (first proposed in 1961) in an attempt to estimate the number of extraterrestrial civilizations in our galaxy with which we might come in contact. The main purpose of the equation is to allow scientists to quantify the uncertainty of the factors which determine the number of extraterrestrial civilizations. Although there is no unique solution to this equation, it is a generally accepted tool used by the scientific community to examine these factors. (*SETI*)

- **dwarf planet** – At its General Assembly in 2006, the International Astronomical Union which has the responsibility and authority for approving names of astronomical objects, officially determined that there are 8 planets in the solar system. A Dwarf planet is an object orbiting the sun that is not large enough (massive enough) to have its self-gravity pull itself into a round (or near-spherical) shape. (*Astronomy, Planetary Astronomy*)

- **dust** – Interstellar, interplanetary, cometary dust. Small solid particles (0.1 μ to 1 mm), generally made of silicates, metal ions and/or carbonaceous matter. (*Astronomy, Cosmochemistry*)

- **EA** – Ethylamine is a chemical compound with the formula $CH_3CH_2NH_2$. It has a strong ammonia-like odor. It is miscible with virtually all solvents and is considered to be a weak base, as is typical for amines. Ethylamine is widely used in chemical industry and organic synthesis. It is also used in cigarettes. (*Organic chemistry*)
- **EACA** – Epsilon amino-n-caproic acid or 6-aminohexanoic acid. Found in Nylon-6. (*Organic Chemistry*)
- **earthshine** – Light and energy radiated from Earth into space, usually used to refer to the illumination of the dark side (opposide the Sun) of the Moon. (*Astronomy*)
- **eccentricity** – Parameter (e) characterizing the shape (*i.e.* roundness) of an orbit. The eccentricity is equal to 0 for a circle, equal to 1 for a parabola, higher than 1 for a hyperbola and between 0 and 1 for an ellipse. (*Astronomy*)
- **ecologite** – A metamorphic rock formed under high pressure, dominated by garnets and olivine, with a composition similar to basalt. (*Geology*)
- **enantiomeric excess** – EE. The absolute difference between the mole fraction of the enantiomers present in a chiral compound. In the case of a mixture of two enantiomers whose respective concentrations are D and L with D greater than L, *i.e.* the enantiomeric excess in percent is $100 \times$ (D - L)/(D + L). (*Chemistry*)
- **enantiomers** – Two isomeric molecules that are non-superimposable mirror-images of each other. They have identical and physical properties except for their ability to rotate plane polarized light. (*Chemistry*)
- **endergonic** – Absorbing energy, in the form of work. (*Chemistry, Physics*)
- **endogenic** – Of or pertaining to a process originating on the Earth or other solar system body. Volcanism, glaciation, and tectonics are examples of endogenic processes. (*Geology, Planetary Astronomy*)
- **energy transduction** – Refers to any cellular mechanism used to transform a source of external energy (radiation, chemical) in energy useful for the cell (respiration, photosynthesis, fermentation). (*Microbiology*)
- **entropy** – Denoted by S, this is a measure of the extent of disorder. It can also be described as a quantitative measure of the amount of thermal energy not available to do work. (*Physics*)
- **enzyme** – Any of numerous proteins produced in living cells that accelerate or catalyze the metabolic processes of an organism. (*Biology, Biochemistry*)
- **EPO** – Education/Public Outreach Science. (*Education*)
- **equivalent width** – The width of a box reaching up to the continuum that has the same area as the observed spectral line. (*Astronomy*)

- **eukaryote** – A cell with a nucleus, an organism whose cells contain complex structures that are enclosed with a membrane. (*Biology*)
- **ethylene glycol** – CH_2OHCH_2OH. The simplest dialcohol. (*Organic Chemistry*)
- **evaporite** – A mineral that precipitates during the evaporation of salty water. Sulfates and carbonates are examples of evaporite minerals. (*Geology*)
- **evpatoria messages** – An ongoing series of powerful microwave transmissions from Earth, starting in 1999, directed toward specific sun-like stars, beamed from the Evpatoria radar telescope in the Ukraine. (*SETI*)
- **exergonic** – Releasing of energy, in the form of work. (*Chemistry*)
- **exogenic** – Planetary process orginating externally. Examples include comet or meteoroid impacts, radiation damage, tidal forces. (*Geology, Planetary Astronomy*)
- **exoplanet** – Synonym of extra-solar planet (*Astronomy*)
- **extinction** – Removal of astronomical radiation from the path due to absorption or scattering. (*Astronomy*)
- **extinction** – The end of a group of organisms or an organism. (*Biology, Ecology*)
- **extrasolar planet** – A planet that orbits a star other than our Sun. Since the first confirmed detection was announced in 1995, nearly 400 extrasolar planets have been detected in orbit around distant stars in our galaxy. Also extra-solar planet. (*Astronomy*)
- **extremophiles** – An organism, usually unicelular, that thrives in physically or chemically extreme conditions for growth and reproduction. Those conditions exceed optimal conditions for mesophilic organisms. (*Biology*)
- **feldspar** – A group of abundant rock-forming minerals of the general formula $XAl(Al,Si)_3O_8$ where X can be K, Na, Ca, Ba, Rb, Sr or Fe. Feldspars constitute 60% of the Earth's crust and occur in all types of rocks (igneous, sedimentary, metamorphic). Two common types of felspars are plagioclase feldspar (X=Na or Ca) and orthoclase (X=K). (*Geology*)
- **ferric iron** – Iron with an oxidation number of +3 (*i.e.* Fe^{3+}). (*Chemistry*)
- **ferrous iron** – Commonly known as Iron(II) oxide. It consists of the chemical element iron in the oxidation state of 2 bonded to oxygen (*i.e.* $Fe^{2+}O$). (*Astrobiology*)
- **FFT** – Fast Fourier Transformation. A mathematical algorithm to quickly compute a fourier transform. (*Astronomy, Physics*)

- **field of view** – The instantaneous region of the sky that is accessible to an observing instrument. (*Astronomy*)

- **fluorescence** – Atoms, molecules and solids that are excited to higher energy levels, can decay to lower energy levels by emitting electromagnetic radiation at specific wavelengths. (*Physics*)

- **flux** – The net energy flow across an element of area per second per wavelength interval. (*Astronomy, Physics*)

- **formaldehyde** – H_2CO, methanal. The simplest aldehyde. Polymerizes easily (*e.g.* POM). In prebiotic chemistry, may have been the starting point of ose formation (formose reaction). (*Organic Chemistry*)

- **formamide** – NH_2CHO. (*Organic Chemistry*)

- **formic acid** – HCOOH. (*Organic Chemistry*)

- **FTIR** – Fourier transform infrared spectroscopy. A measurement technique for IR spectroscopy, where EM radiation that has passed through a sample is passed through an interferometer and a fourier transform is used to mathematically produce the spectrum. (*Physics*)

- **full width half maximum** – FWHM. A measure, in either frequency or angular size, of the location where the measured intensity has been reduced to half its peak value. (*Astronomy, Physics*)

- **genome** – Hereditary information of an organism. It is constituted by a nucleotide sequence that can be DNA (in cellular genomes and in DNA viruses) or RNA (in RNA viruses). (*Biology*)

- **geophysiology** – The dynamic processes and feedbacks of the biosphere, life and its interacting environment. Geophysiology is analogous to the physiology of living organisms, but it need not imply that the biosphere is a superorganism as in some versions of Gaia theory. (*Geology*)

- **geothermal** – Thermal energy that comes from within the Earth. (*Geology*)

- **giant planet** – A general term indicating a planet similar to Jupiter in mass and size. (*Astronomy*)

- **Gibb's energy** – Amount of energy available from chemical reactions; $dG = dG_0 + RT \ln(Q$. dG = Gibbs energy of reaction, dG_0 = standard state Gibbs energy of reaction, R = universal gas constant, T = absolute temperature, Q = activity quotient of reaction. (*Chemistry, Physics*)

- **gigahertz** – GHz. A frequency of 1 billion hertz or 1 billion cycles per second. A radio signal with a frequency of 1 GHz has a wavelength of 30 cm. (*Astronomy, Physics*)

- **glucose** – A sugar ($C_6H_{12}O_6$) occurring widely in most plant and animal tissue. It is the principal circulating sugar in the blood and the major energy source of the body. (*Biology, Biochemistry*)

- **glutamic acid** – A non-essential amino acid, $C_5H_9NO_4$, occurring widely in plant and animal tissue and proteins, and having monosodium glutamate as a salt. Plays a central role in amino acid metabolism, acting as precursor and also acts as amino group donor for other amino acids. It is also a neurotransmitter, important in the nitrogen metabolism of plants; used in monosodium glutamate to enhance the flavor of meats. (*Biochemistry*)

- **glycine** – NH_2CH_2COOH. The simplest amino acid and the only one which is not chiral. One of the 20 proteic amino acids. (*Organic Chemistry, Biochemistry*)

- **glycoaldehyde** – CH_2OHCHO. The smallest molecule that contains both a hydroxyl group and an aldehyde group. (*Organic Chemistry*)

- **greenstone** – Metamorphosed basalts. Greenstone belts are common in the Archean. (*Geology*)

- **H-α emission** – Emission line of Hydrogen (n=3 to 2) at 653nm (red). This emission line is thought to trace accretion and circumstellar-material in young stars. (*Astronomy*)

- **Hadean** – Geological time period from the formation of the Earth 4.5 billion years ago to the beginning of the Archean at 3.8 billion years ago. (*Geology*)

- **Hapke theory** – Sattering model devised by Bruce Hapke, incorporating the optical properties of the particles making up a diffusely reflecting particular surface. The Hapke theory has been widely tested and is used by many investigators for the calculation of synthetic spectra of Solar System bodies by the incorporation of the complex optical refractive indices of the component(s) of the scattering surface. (*Physics, Planetary Astronomy*)

- **heliosphere** – The immense magnetic bubble carved out in space by the solar wind that defines the extent of the Sun's influence. The point at which the solar wind slows down as it interacts with the interstellar medium is called the termination shock. NASA Spacecraft Voyager 1 reached this boundary in 2003-2004 and is now passing through the heliosphere. (*Astronomy*)

- **hematite** – Also spelled haematite. The mineral form of Iron(III) oxide, (Fe_2O_3), one of several iron oxides. (*Geology*)

- **hertz** – Abbreviated as Hz. A unit of frequency equal to 1 cycle per second. The alternating current delivered by wall sockets in the US oscillates with a frequency of 60 Hz (50 Hz). (*Physics*)

- **heterocycles** – Cyclic compounds in which carbon is substituted by other elements, typically oxygen, nitrogen, or sulfur. (*Chemistry, Organic Chemistry*)

- **heterogeneous reaction** – The reaction between reactants that exist in different phases. For example, a common type of reaction studied in atmospheric science is that between a gas species and a solid substrate (ice crystal or aerosol particle). (*Chemistry*)

- **homochirality** – Term used to refer to a group of molecules that possess the same sense of chirality or handedness. Life on Earth is based on homochirality. For example, active amino acids that are incorporated into proteins and enzymes are in the L-form, while most biologically relevant sugars are in the D-form. (*Chemistry*)

- **hot-Jupiter** – A gas giant (Jupiter-like) planet in a very close orbit to its host star. The distances of hot-Jupiters to their parent stars are about 100 times smaller than the distance of Earth to the Sun. (*Astronomy*)

- **Huronian** – Glaciation(s) occurring at about 2.4-2.2 billion years ago. (*Geology*)

- **hydrogen bonding** – A weak bond in which the electron is shared with the hydrogen atom. (*Chemistry, Physics*)

- **hydrogeological** – Refering to the study of the water in the subsurface. (*Geology*)

- **hydrolysis** – A chemical reaction in which a chemical compound reacts with water, usually resulting in the formation of aqueous ionic compounds and/or hydrated molecules. (*Chemistry, Geochemistry*)

- **hydrophilic** – Molecules that can bond with water via hydrogen bonds. "Water-loving". (*Biology, Organic Chemistry*)

- **hydrophobic** – Physical propery of molecules that are repelled from water. (*Biology, Organic Chemistry*)

- **hydrothermal vent** – See Black smoker. (*Geology*)

- **IETI** – Invitation to ETI: a web-based SETI project for initiating dialog with possible extraterrestrial intelligent autonomous probes surveying our civilization. http://ieti.org. (*SETI*)

- **igneous** – Rocks formed by the crystallization (freezing) of magma (molten rock). (*Geology*)

- **imidazole** – An organic crystalline base, $C_3H_4N_2$, that is an inhibitor of histamine. (*Biochemistry*)

- **interferometer** – A larger telescope that is created by combining the signals from a number of smaller telescopes. The sensitivity of an interferometer is determined by the total collecting area of all the individual telescopes. The spatial resolution of an interferometer is determined by the longest baseline between any two telescopes. (*Astronomy*)

- **ion channels** – Pore-forming proteins that help to establish and control the small voltage gradient across the plasma membrane of all living cells by allowing the flow of ions down their electrochemical gradient. They are present in the membranes that surround all biological cells. (*Biology*)

- **IR spectroscopy** – Abreviation for Infra-red spectroscopy. Identification of substances by looking at how infrared light interacts with matter as a function of wavelength. This technique is very powerful in determining the presence of various minerals and compounds, including functional groups in organic compounds. (*Chemistry, Physics, Remote Sensing*)

- **IRAF** – Standard software for the analysis of astronomical data, primarily images and spectra. The acronym stands for Image Analysis and Reduction Facility. Developed at the National Optical Astronomy Observatories in Tucson starting in the 1980s. (*Astronomy*)

- **ISM** – Interstellar medium. The matter, *i.e.* gas (mostly H and He) and dust that exists in the space between stars and galaxies. (*Astronomy*)

- **ISO** – Infrared Space Observatory. Space telescope sensitive for infrared light designed and operated by the European Space Agency (ESA), in co-operation with ISAS/JAXA and NASA. Observed in the 2.5-240μm wavelength range from 1995 to 1997. (*Astronomy*)

- **isochronal ages** – An isochron is a line on a plot that shows the relation between a nonradiogenic isotope divided by the daughter isotope versus the parent isotope divided by the daughter isotope. For example, a line on a $^{232}Th/^{204}Pb$ versus $^{208}Pb/^{204}Pb$ plot or on a $^{40}K/^{36}Ar$ versus $^{40}Ar/^{36}Ar$ plot. A straight line on a plot of this kind is called an isochron and shows that the data points came from minerals which formed at the same time. (*Geochemistry, Cosmochemistry*)

- **isocyanhydric acid** – HNC. Isomer of cyanhydric acid HCN. (*Organic Chemistry*)

- **isomer** – One of two or more forms a chemical compound which have the same number and type of each atom but a different arrangement of atoms, but possessing different properties. There are structural isomers, geometric isomers, optical isomers, and stereoisomers. (*Chemistry, Biochemistry*)

- **isovaline** – A non-biological amino acid which has the structural formula $C(CH_3)(CH_2CH_3)(NH_2)COOH$. Isovaline has the highest abundance of any amino acid yet identified in CM meteorites Murchison and Murray. (*Organic Chemistry, Cosmochemistry*)

- **JFC** – Jupiter family comet. A dynamical classification for comets with orbital periods less than 20 years whose orbital dynamics are strongly influenced by gravitational interactions with the planet Jupiter. (*Planetary Astronomy*)

- **K/T boundary** – Few centimeter-thick sedimentary layer located at the Cretaceous(K)- Cenozoic(Tertiary,T) boundary. Its Iridium-enrichment is interpreted as having come from an exogenous source, a giant meteoritic impact 65 million years ago (the K/T event). (*Geology*)

- **keplerian speed** – The orbital speed of a body under the influence of gravity, usually of a substantially more massive body. For keplerian motion replace motion for speed in the above definition. (*Astronomy*)

- **kerogen** – Terrestrial kerogen is a dark, complex, macromolecular organic substance produced by geologic processing of biologic materials. Insoluble organic material in some classes of meteorites has a structural similarity to kerogen, but with no biological connotation. (*Organic Chemistry, Cosmochemistry*)

- **Kuiper belt** – A region of the Solar System beyond the planets extending from the approximately the orbit of Neptune (at 30 AU) to approximately 55 AU from the Sun. It is similar to the asteroid belt, although it is far larger; 20 times as wide and 20-200 times as massive. KBO orbits are of low inclination to the plane of the planets' orbits, and the Kuiper Belt where they occur is regarded as the source region of the short-period comets. It consists of small remnants from the formation of the Solar System and several dwarf planets (including Pluto). But while the asteroid belt is composed primarily of rock and metal, the Kuiper belt is composed largely of ices, such as methane, ammonia, and water. (*Planetary Astronomy*)

- **langmuir oscillation** – Three-dimensional macroscopic oscillations of dissociated H^+ protons on the net of hydrogen bonds. An important phenomenon in non-magnetized plasmas, with applications in astronomy (*e.g.* for solar radio emission). (*Astronomy, Chemistry, Physics*)

- **LC/FD-TOF-MS** – Liquid chromatography with UV fluorescence detection and time of flight mass spectrometry. This is a powerful analytical technique that can be used to separate a complex mixture of organic compounds (for example in carbonaceous meteorites) and identify each compound uniquely by the time it takes for the molecule to elute from the column, the exact mass of the compound, and UV fluorescence. (*Chemistry*)

- **limb** – Appendage of an animal such as an arm or a leg. (*Physiology*)

- **limb** – Large branch of a plant. (*Botany*)

- **limb** – The visible edge of a disk, such as the limb of the Sun or a planet, sometimes used as an adjective (*e.g.*, limb darkening). (*Astronomy*)

- **lithification** – A process where, under pressure, sediments compact and become rock. (*Geology*

- **lithopanspermia** – (Litho, rock; panspermia, literally = seeds everywhere). Originally proposed in the 19th century by Richter, Lord Kelvin, and von Helmholtz. In its current form, the theory that endolithic organisms can be transferred between planets as passengers inside impact ejecta. A theory of Hoyle and Wickramasinghe that suggests that seeds, spores and microbes exist in space and that life on Earth may have originated through panspermia. Lithopanspermia hypothesizes that microbes are incorporated into meteorites which then impact on planets. (*Astrobiology*)

- **lithotrophy** – Metabolic system that requires inorganic compounds as sources of energy. (*Geomicrobiology*)

- **long-period comet** – Comets with orbital periods > 200 years. Dynamically, these have probably evolved from more distant orbits in the Oort cloud. Short-period comets classified as Halley-family comets probably originated as long-period comets. (*Planetary Astronomy*)

- **luminosities** – An intrinsic measurement of the amount of energy a stellar body radiates per unit time. (*Astronomy*)

- **Lyman-α** – The n=2 to n=1 Hydrogen transmission at 121.6 nm. Lyman-alpha is measured in absorption in damped Lyman-alpha (DLA) galaxies to measure the total neutral hydrogen content (see Damped Lyman-Alpha Systems). (*Astronomy*)

- **Methylamine** – MA. The chemical compound with a formula of CH_3NH_2. It is a derivative of ammonia, wherein one H atom is replaced by a methyl group. MA is used both as a solvent and as a building block for the synthesis of other organic compounds. (*Organic Chemistry*)

- **magnetite** – A black, strongly magnetic opaque mineral of the spinel group and having the chemical formula Fe_3O_4 (containing both Fe^{2+} and Fe^{3+}). Magnetite is very common and widely distributed accessory mineral in rocks of all kinds, especially volcanic and metamorphic rocks. (*Geology*)

- **magnetospheric emission** – Relating to emissions from an astronomical object, usually a planet, that are powered by its magnetic field. Typically these emissions are generated in the radio part of the electromagnetic spectrum. In the solar system, Jupiter, Saturn, Uranus, Neptune, and Earth all have strong magnetic fields. Interactions between their magnetic fields and particles streaming outward from the Sun generate these magnetospheric emissions. By analogy, extrasolar planets have been hypothesized to be detectable from their magnetospheric emissions. (*Astronomy*)

- **magnitude** – A measure of the brightness of a star on a logarithmic scale. The original scheme, devised a couple thousand years ago, gave the brightest stars in the sky the lable of first magnitude or first importance, the next brightest second magnitude. This was quantized later, as the eye is a logarithmic detector, to a system where magnitude $= 2.5 \log_{10}$ (flux). (*Astronomy*)

- **Maillard reaction** – The reaction between amino acids and sugars. It yields a complex mixture of materials, including heterocyclic compounds and insoluble organic materials, similar to those that are found on the meteorites. (*Organic Chemistry*)

- **maser** – A device that produces coherent electromagnetic waves through amplification due to stimulated emission. Historically the term came from the acronym "microwave amplification by stimulated emission of radiation", although modern masers emit over a broad portion of the electromagnetic spectrum. Also applicable to astronomical sources where water molecules emit 22 GHz radiation due to a population inversion. (*Astronomy, Physics*)

- **mass independent fractionation** – A process in a three isotope system (*e.g.* ^{16}O, ^{17}O, ^{18}O) by which the isotopes are fractionated differently than what is expected from mass differences. For example, in the oxygen system there are three isotopes with masses 16, 17 and 18. Normal mass-dependent fractionation results in the reaction rate of 17 half-way between that of ^{16}O and $^{18}O16$. This results in a slope one-half line on a three isotope plot (*e.g.* terrestrial fraction line). However, mass-independent fractionation results in a trend different from the one-half slope. A common example of mass-independent fractionation is the production of ozone (O_3) from oxygen gas (O_2). It is unknown why this should be the case. (*Cosmochemistry, Geochemistry*)

- **megahertz** – Abbreviated MHz. A unit of frequency of 1 million hertz, or 1 million cycles per second. A radio signal with a frequency of 1 MHz has a wavelength of 30,000 cm = 3 km. (*Astronomy, Physics*)

- **meridian** – A longitudinal reference line that traverses the earth in a north-south direction. (*Geography*)

- **Meridiani planum** – A plain located 2 degrees south of Mars' equator, the landing site for the second of NASA's two Mars Exploration Rovers, named Opportunity. (*Planetary Astronomy*)

- **mesophile** – An organism whose optimal growth is at moderate temperatures, between 15-40°. (*Biology*)

- **metagenome** – Metagenomics, also called environmental genomics or community genomics, involves the use of molecular biology and bioinformatics to examine DNA extracted from an environmental sample, bypassing the need to cultivate (or even know) the organisms from which the DNA originates. The metagenome is the total genomic content of all organisms in a given sample. The global metagenome is the combined DNA content of all types of life on earth, the Bioastronomy2007 metagenome the combined DNA content of all meeting attendees etc. (*Biology*)

- **metallicities** – The fractional abundance of heavy elements (all elements in the periodic table except H and He) relative to H. (*Astronomy*)

- **metamorphic** – Metamorphic rocks are sedimentary or igenous rocks that have undergone mineralogical changes as a result of being subjected to high temperatures, high pressures, or both. (*Geology*)

- **metazoa** – Multicellular animals that make up the major portion of the animal kingdom; cells are organized in layers or groups as specialized tissues or organ systems. (*Biology*)

- **metazoan phylogeny** – Evolutionary history of the metazoa; branching patterns describing evolutionary relationships between the metazoa originating at the last common metazoan ancestor. (*Biology*)

- **meteoric** – Groundwater which originates in the atmosphere and reaches the zone of saturation by infiltration and percolation. (*Hydrology*)

- **meteorite** – Extraterrestrial object, fragment of an asteroid, of a planet (like Mars) or of the Moon that falls on the Earth surface. (*Planetary Astronomy*)

- **methyl formate** – $HCOOCH_3$. (*Organic Chemistry*)

- **methyl hydantoins** – Hydantoin is a heterocyclic (*i.e.* ringed) organic compound, also known as glycolylurea. (*Organic Chemistry*)

- **METI** – Messaging to Extra-Terrestrial Intelligence. Also known as Active SETI. (*SETI*)

- **mid ocean ridge flanks** – (MOR) Oceanic crust is formed at mid-ocean ridges which are places where magma upwells from the mantle to cool and create new crust. The flanks are the parts of the mid-ocean ridge where the elevation decreases to the ocean plateau, much like the flanks of a volcano which decrease down to meet the surrounding land. (*Geology*)

- **mineral** – A naturally-occurring, inorganic substance formed through geological processes having a definite chemistry and definite structure. (*Geology*)

- **Mojave desert** – An arid region of southeastern California and portions of Nevada, Arizona and Utah, that occupies more than 25,000 square miles. Situated between the Great Basin Desert to the north and the Sonoran desert to the south (mainly between 34 and 38 degrees N latitudes), the Mojave desert is a rainshadow desert and is defined by a combination of latitude, elevation, geology, and indicator plants. (*Geography*)

- **molecular clouds** – Collections of gas and dust found in interstellar space. Masses can range from one to a million solar masses. Densities can range from 1000 to a ten million particles per cubic centimeter. Chemical composition is mostly molecular hydrogen, helium and small amounts of simple inorganic and organic molecules. (*Astronomy*)

- **molecular flexibility** – The capacity to adopt different structural configurations as a function of interactions with the environment. (*BioChemistry, Chemistry*)

- **monophyletic clade** – A clade is a group of organisms consisting of a single common ancestor and all the descendants of that ancestor. Any such group is also considered a monophyletic group of organisms, and can be modeled by a cladogram, a diagram of the organisms in the form of a tree. (*Biology*)

- **montane** – Biogeographic term refering to the highland area located below the tree-line. (*Biology, Ecology*)

- **MS** – Mass spectrometer. Analytical device to determine the elemental composition of a sample. (*Chemistry, Geochemistry*)

- **Murchison** – A CM carbonaceous chondrite which fell in 1969 and was collected soon thereafter. Murchison has abundant organic compounds, including carboxylic acids, amino acids, and sugars. (*Cosmochemistry*)

- **mutagens** – Physical or chemical agents that increase the mutation frequency above the natural level of an organism. (*Biology*)

- **neoproterozoic** – Geologic time period between 1 billion and 0.54 billion years ago, the last phase of the Proterozoic. (*Geology*)

- **nitrogen fixation** – A biological process of conversion (by certain soil microorganisms, such as rhizobia) of atmospheric nitrogen into compounds that plants and other organisms can assimilate. (*Biology*)

- **NMR** – Nuclear Magnetic Resonance spectroscopy is an analytical technique that determines the chemical environment of specific elemental nuclei. It is most powerful in identifying H-, C-, P-, and F-bearing molecules. (*Chemistry*)

- **Noachian** – An epoch (named after Noachis Terra) related to the formation of Mars, between 3.8 and 3.5 billion years ago. (*Planetary Astronomy*)

- **non-resonant (hot-band) fluorescence** – An infrared radiation process in which (1) a molecule in its ground vibrational state (E0) is excited by solar infrared radiation into a higher vibrational state (E1), and (2) transits to an intermediate state (E2, $E0 < E2 < E1$) via emission of an infrared photon. H_2O "hot-band" fluorescence provides a means to study the water molecules in the comae of comets from ground-based observatories. (*Astronomy, Physics*)

- **non-rhizospheric** – Soil without presence of roots. (*Biology*)

- **nuclear spin** – A quantum-mechanical characteristic, representing the total angular momentum of atomic nucleus. Molecules containing two or more identical nuclei (*e.g.* H_2O, NH_3, CH_4) display nuclear spin modifications (also referred as isomers) grouped according to their total nuclear spin (I). The H_2O molecule is organized into two isomers depending on whether the nuclear spins of its H atoms are parallel (ortho-H_2O, $I = 1$) or antiparallel (para-H_2O, $I = 0$) (see also ortho-para ratio; nuclear spin temperature). (*Physics*)

- **nuclear spin temperature** – The temperature that would correspond to a given ortho-para ratio if in local thermodynamic equilibrium. Ortho-states are increasingly favored at higher spin temperatures and a statistical equilibrium ratio of 3/1 is reached at ~50 K (see also nuclear spin; ortho-para ratio). (*Physics*)

- **nucleotide** – Molecule made by condensation of a base (purine or pyrimidine), a ose and a phosphate group linked to the ose. Nucleotides are ribonucleotides when the ose is ribose. They are deoxyribonucleotides when the ose is deoxyribose. DNA is a polydeoxyribonucleotide while RNAs are polyribonucleotides. The symbol of a nucleotide is determined by the base (A for adenine, C for cytosine, G for guanine, T for thymine, U for uracil). (*Biochemistry*)

- **nucleus** – Solid part of a comet, of typical diameter 1-100 km. Made of ices (H_2O, CO, CO_2, and other minor volatile organic compounds, such as CH_3OH) and dust particles (silicates and organics). Also called a dirty snowball, according to F. Whipple. (*Astronomy*)

- **nucleus** – The control center of a cell, containing most of the genetic material. (*Biology*)

- **oligomerization** – A chemical process that converts monomers (small molecules) to a finite degree of polymerization (*i.e.* incomplete polymerization). (*Chemistry, Biochemistry*)

- **olivine** – A green to brown mineral, which consists of a solid-solution series from forsterite (Mg_2SiO_4) to fayalite (Fe_2SiO_4). Olivine is a common rock-forming mineral of low-silica igneous rocks, including basalt. It crystallizes early from a magma and weathers relatively easily at the surface of the Earth. (*Geology*)

- **OMEGA** – Acronym for Observatoire pour la Mine'ralogie, l'Eau, les Glaces, et l'Activite. OMEGA is a visible and infrared mineralogical mapping spectrometer that provides compositional information regarding the surface of Mars in 0.5 to 5.2μm region of the spectrum at a spatial resolution of 100 meters per pixel. (*Remote Sensing*)

- **Oort cloud** – Spherical halo of km-sized icy bodies surrounding the Sun. This extends out to approximately 200,000 times the distance from the Earth to the Sun and contains as many as a trillion bodies. (*Astronomy*)

- **ophiolite** – Sections of the oceanic crust and the subjacent upper mantle that have been emplaced within continental crustal rocks. (*Geology*)

- **ortho-para ratio** – OPR. The abundance ratio between ortho- and para-nuclear spin modifications (see also nuclear spin, nuclear spin temperature). (*Physics*)

- **outgassing** – Ejection of gasses that were frozen, absorbed or trapped in materials, such as comets. (*Astronomy*)

- **oxidative** – A process involving a reaction with oxygen. Any process by which the proportion of the electronegative constituent in a compound increase. (*Chemistry*)

- **oxide** – Minerals that are composed primarily of oxygen and one or more metals (*e.g.*, Ca, Na, Mg, Fe, Ti, etc.). Magnetite, hematite (iron oxide/rust), and ilmenite are oxide minerals. (*Geology*)

- **P-cygni profile** – A specific shape of a stellar emission line, from the prototype bright star P-Cygni. The blue side of the line shows strong absorption but the red side of the line is in strong emission. This comes from stellar winds or outflowing circumstellar material. (*Astronomy*)

- **paleoclimate** – A climate or condition (pressure, temperature) that existed in geologic past; paleo meaning old or ancient. (*Geology*)

- **panspermia** – Literally, seeds everywhere. The theory that life can originate anywhere in the universe given the proper conditions, and can be propagated through space from one location to another. (*Astrobiology*)

- **parent molecule** – Native molecule present as ice in the comet nucleus, released directly from the nucleus (*e.g.*, H_2O, CO, H_2CO, HCN, NH_3, CH_4, C_2H_2, C_2H_6). (*Astronomy, Physics*)

- **pc** – Abreviation of parsec, which is itself an abbreviation of parallax of one second of arc. A unit of length (1 pc $= 3.08 \times 10^{16}$ m), one parsec is defined to be the distance from the Earth to a star that has a parallax of 1 arcsecond. (*Astronomy*)

- **peptide bond** – A chemical bond between two molecules (typically amino acids) in which the amine group of one molecule reacts with the carboxyl group of another molecule forming a CO-NH bond. (*Organic Chemistry*)

- **per mil** – Per thousand, often abbreviated with a symbol similiar to a percentage sign with an extra o after the slash (o/oo). (*Geochemistry, Cosmochemistry*)

- **perihelion** – The closest approach distance to the Sun for a given orbit. (See aphelion). (*Astronomy*)

- **permafrost** – Soil at or below the freezing point of water (0°C or 32°F) for two or more years. (*Geology*)

- **permineralized** – Refers to fossils preserved by the infilling of empty spaces of an organism with minerals (commonly silica, calcite, or pyrite). The original organic and/or inorganic structures of such fossils are not replaced, but rather are surrounded by and infused with minerals. A more common–though now out of favor–term for permineralization is petrification. The types of organisms generally preserved by permineralization include microorganisms, plants, and fungi. (*Paleontology, Paleobiology*)

- **pH** – The symbol "pH" refers to a scale of acidity for liquids. The thermodynamic definition is pH = - log(a), where "log" is a base-10 logarithm and a represents the thermodynamic activity of hydrogen ion (actually hydronium, H_3O^+). For most work done under conventional laboratory conditions, the definition reduces to pH = -log[H^+] where [H^+] is the hydrogen ion (actually hydronium, H_3O^+) concentration in units of moles per liter. Although the latter definition is the simpler of the two, for some environments relevant to bioastronomy the former is needed for accuracy. Acidic solutions are taken as having pH < 7, alkaline (basic) ones as having pH > 7, and neutral ones as having pH $= 7$. (*Chemistry*)

- **phased array** – A type of interferometer in which the signals from all the individual telescopes are added together in such a way as to make them point at one direction in the sky, producing the small field of view of a single large telescope with a diameter equal to the longest baseline between any two telescopes of the interferometer. (*Astronomy*)

- **phospholipid** – A type of fat where one fatty acid has been replaced by a phosphate group. Phospholipids have a hydrophilic head and hydrophobic tails, and tend to arrange themselves into bi-layered structures. (*Organic Chemistry*)

- **phosphite** – $HPO_3{}^{2-}$, the anion of phosphorous acid, H_3PO_3, with phosphorus in a reduced oxidation state (3+) relative to orthophosphate. (*Chemistry*)

- **photometric** – A night is defined as photometric if there are no clouds (causing extinction of light) and the absolute brightness of celestial objects may be measured and their brightnesses calibrated. (See extinction) (*Astronomy*)

- **photometry** – Measurement of apparent brightness of an astronomical object (typically in the form of magnitude or flux). (*Astronomy*)

- **photomultiplier tube** – A photomultipier tube is a device that multiplies and converts photons to an electrical signal. (*Astronomy, Physics*)

- **phyllosilicate** – A type of structural silicate with typically parallel sheets of silicate tetrahedra (Si_2O_5). Common examples include clay minerals, serpentine and micas. (*Geology*)

- **phyllosilicate texture** – Phyllosilicate minerals form as sheets of Si and O atoms, resulting in flat, platy mineral form. A phyllosilicate texture may indicate that the rock has been recrystallized under high pressure and temperature (metamorphism). (*Geology*)

- **phylogenetic** – Relating to or based on evolutionary development or history. (*Biology*)

- **phylogeny** – (1) The evolutionary development and history of a species or higher taxonomic grouping of organisms. Also called phylogenesis. (2) The evolutionary development of an organ or other part of an organism. (*Biology*)

- **phylum** – A primary division of a kingdom, as of the animal kingdom, ranking next above a class in size. (*Biology*)

- **planet** – A celestial body that (a) is in orbit around the Sun, (b) has sufficient mass for its self-gravity to overcome rigid body forces so that it assumes a hydrostatic equilibrium (nearly round) shape, and (c) has cleared the neighborhood around its orbit (official definition from the International Astronomical Union). (*Astronomy, Planetary Astronomy*)

- **planetary nebula** – The often shapely result of the death throes of a Sun-like star. The star swells to an enormous size and ejects its outer atmosphere to form the nebula (*e.g.* Cateye Nebula, Hourglass Nebula) which consists of an expanding shell of glowing gas. The process generally leaves a relatively faint and compact object called a white dwarf at the center of the nebular structure. The planetary nebulae have nothing to do with planets. (*Astronomy*)

- **planetesimals** – Rocky objects with sizes of a few kilometers. Planetesimals are formed through sticking and growth of dust particles in the original gaseous disk around a star. (*Astronomy*)

- **polychromatic** – Consisting of many colors. In astronomy it refers to the ability to observe a wide range of frequencies at one time. (*Astronomy*)

- **polycyclic aromatic hydrocarbon** – Abbreviated PAH. These are relatively chemically stable compounds composed of numerous fused aromatic carbon rings (6-membered planar rings - see aromatic hydrocarbons). Such compounds are often found in organics-containing meteorites and comprise the bulk of terrestrial kerogens. (*Organic Chemistry, Astrochemistry*)

- **polymerization** – A chemical reaction in which two or more small molecules combine to form larger
molecules that contain repeating structural units of the original molecules. (*Chemistry, Organic Chemistry*)

- **polymorph** – A mineral that is identical to another in chemical composition but has a different crystal structure. An example is hexagonal and cubic water ice. (*Geology*)

- **POM** – Polyoxymethylene, a polymer of formaldehyde (H_2CO). (*Chemistry*)

- **prebiotic** – Of, relating to, or being chemical or environmental precursors of the origin of life (Prebiotic molecules). Existing or occurring before the origin of life. (*Biology, Chemistry*)

- **precession** – Change in the direction of the axis of a rotating object. (*Astronomy*)

- **prokaryote** – Cells that lack a membrane-bound nucleus; by contrast, in eukaryotes the chromosomal DNA is enclosed in a membrane to form a nucleus. (*Biology*)

- **prompt emission** – Radiation process in which (1) a radical is produced by photodissociation of a parent molecule (*e.g.*, OH from H_2O) in excited state, and (2) the dissociation product emits promptly (within ~10 milliseconds) transiting to a lower energy level. When observed in comets, infrared prompt emission from OH can be used as an indirect means to study the parent (H_2O) molecule. (*Astronomy, Physics*)

- **proplyd** – A rotating protoplanetary disk of dense dust and gas surrounding a young newly formed star. Since the very first protoplanetary disks were seen in the Orion Nebula (a very massive molecular cloud), in the literature proplyds refer just to this particular molecular cloud. For Astrobiology purposes, we can think that proplyds are the place were planets will be formed. (*Astronomy*)

- **proteins** – Any of a large class of complex organic chemical compounds that are essential for life. Proteins play a central role in biological processes and form the basis of living tissues. They consist of long chains of amino acids connected by peptide bonds and have distinct and varied three-dimensional structures. Enzymes, antibodies, and hemoglobin are examples of proteins. (*Biology*)

- **proteomics** – A new and evolving field of science that seeks to specify all the proteins produced by a cell in all types of situations and environments and to understand how they function. Because proteins are the product of information coded for in DNA, proteomics is closely allied to the study of the genome. (*Genetics*)

- **proterozoic** – Geologic time period from 2.5 to 0.54 billion years ago, the most recent subdivision of the Precambrian. (*Geology*)

- **proto-intelligence** – Current state of humankind where the dominant paradigm is violence. Proto-intelligence is the evolutionary stage that precedes the Collective Intelligence in advanced civilizations. (*Sociology*)

- **protoplanet** – A planet in the process of forming from material in a protoplanetary disk. Moon- to Mars-sized objects formed through collision and sticking of planetesimals. (*Astronomy*)

- **protosolar nebula** – Rotating disk of gas, dust and ice, from which the solar system is originated. (*Astronomy*)

- **protostar** – An object that forms by contraction out of the gas of a giant molecular cloud in the interstellar medium. The protostellar phase is an early stage in the process of star formation. (*Astronomy*)

- **PSN** – Protosolar nebula astronomy. (*Astronomy*)

- **purine bases** – A purine is a heterocyclic aromatic organic compound, consisting of a pyrimidine ring fused to an imidazole ring. Purines make up one of the two groups of nitrogenous bases. Pyrimidines make up the other group. These bases make up a crucial part of both deoxyribonucleotides and ribonucleotides, and the basis for the universal genetic code. (*Biology, Organic Chemistry*)

- **pyrimidine** – A heterocyclic aromatic organic compound similar to benzene and pyridine, containing two nitrogen atoms at positions 1 and 3 of the six-member ring. (*Biology, Organic Chemistry*)

- **pyrolysis-gas chromatography-mass spectrometry** – Commonly abbreviated py-gc-ms. An analytical technique used to measure the composition of solid (generally insoluble) organic materials by thermally breaking (pyrolizing) them into smaller molecular fragments. Such fragments are then separated by use of a gas chromatographic (GC) column and analyzed by use of a mass spectrometer (MS). (*Organic Chemistry*)

- **pyroxene** – An important group of rock-forming silicate minerals found in many igneous and metamorphic rocks. They share a common structure comprised of single chains of silica tetrahedra and they crystalize in the monoclinic and orthorhombic system. They usually consist of a diavalent cation (*e.g.* Fe, Mg, Ca, etc.) with a SiO_3. (*Geology*)

- **quartz** – The second most common mineral in the Earth's continental crust. It is made up of a lattice of silica (SiO_2) tetrahedra. (*Geology*)

- **racemic** – Of or relating to a chemical compound that contains equal quantities of dextrorotatory (D) and levorotatory (L) forms and therefore does not rotate the plane of incident polarized light. (*Chemistry, Biochemistry*)

- **racemization** – Process in which a particular enantiomer is converted into its mirror-image counterpart. (*Chemistry*)

- **radiation, EM** – Electromagnetic radiation, *e.g.*, light and other electromagnetic waves. (*Astronomy, Physics*)

- **radiation** – Ionizing radiation, high-energy atomic particles and gamma and x rays that can damage issue. (*Physics, Biology*)

- **radiation** – The rapid evolution of new families and species. (*Biology*)

- **raman spectroscopy** – A type of spectroscopy that measures frequency-shifted photons that have scattered off of a material (organic or inorganic) having been excited by laser light. The incident photons interact with the molecular bonds of the material (the bonds absorb energy from or impart energy to these photons, thus changing their frequency). The magnitudes of the frequency shifts are diagnostic of the types of bonds or molecular structures present. (*Chemistry, Cosmochemistry*)

- **rayleigh fractionation** – A process of crystallization which assumes that as soon as the mineral has crystallized into a solid, it is immediately removed from the melt liquid. Crystal removal can occur by several methods, the most common is gravitational settling to the bottom of the magma chamber. Also known as perfect fractional crystallization. (*Geology*)

- **rayleigh scattering** – The scattering of electromagnetic radiation by particles whose sizes are much smaller than the wavelength of the radiation. (*Astronomy, Physics*)

- **red dwarf** – The most common type of star. Characteristically, it is a small, cool, very faint, main sequence star whose surface temperature is under about 4,000 Kelvins. (*Astronomy*)

- **reddening** – E(B-V). The excess red light seen towards objects obscured by dust. Blue light is preferentially absorbed and scattered by dust relative to red light. (*Astronomy*)

- **redox** – Oxidation-reduction reactions. Chemical reactions that require electron transfer. (*Chemistry, Physics*)

- **redshift** – Denoted by the letter z, redshift quantifies the expansion rate of the Universe. Due to the expansion, objects are receding from Earth. The further away an object is the faster it is receding and, therefore, has a larger redshift. Due to the finite speed of light, the larger the redshift the further back in time we are viewing the object. (*Astronomy*)

- **reducing** – A condition in which electrons are available to be added to an ion or molecule, reducing the positive charge. For example, the reduction of Fe^{3+} to Fe^{2+} by the addition of an electron. This process is the opposite of oxidation by which electrons are removed, resulting in the increase of the positive charge of the ion or molecule. (*Chemistry, Geochemistry*)

- **reflectance spectrum** – The spectral energy distribution of sunlight diffusely reflected from a planetary surface or atmosphere across a region of the spectrum. (*Planetary Astronomy*)

- **reflectivity** – General term given to the response of a solid surface to incident radiation as a function of wavelength – *e.g.* the fraction of light reflected versus wavelength. Grey objects reflect all wavelengths of light equally, whereas red objects reflect more light at longer wavelengths than at short wavelengths. (*Astronomy, Physics*)

- **refractory** – (1) High temperature element or material, first to condense from gas, last to evaporate from solid. (2) Substance which remains solid in all temperature conditions available in a particular body of the solar system (*e.g.* dust particles in a comet). (*Cosmochemistry*)

- **resilient** – The capacity of a system, potentially exposed to energetic stress to adapt, by resisting or changing in order to reach and maintain an acceptable level of functioning and structure. (*Biochemistry, Chemistry*)

- **resonance** – A dynamical state occurring when the ratio of the periods of two planets are approximately equal to the ratio of two integers. For instance, Saturn and Jupiter are said to be in a near 5:2 resonance since their orbital periods are approximately 30 and 12 years, respectively. (*Astronomy*)

- **respiration** – Extracting energy by using chemio-osmotic processes to oxidize reduced substrates. (*Microbiology*)

- **retrosynthesis** – A method for designing the organic chemical synthesis. It starts from the target molecule to the starting materials, namely in the opposite direction from the traditional organic synthesis design method. The retrosynthesis fragments the target molecule into the subtargets, which are then fragmented further, all the way to the starting materials. (*Organic Chemistry*)

- **rhizosphere** – The soil zone that surrounds and is influenced by the roots of plants. (*Biology*)

- **rifampicin-resistance** – Rifampicin is a semisynthetic antibiotic that interferes with the synthesis of RNA and is used to treat bacterial and viral diseases. Rifampicin-resistance allows bacteria to grow despite the presence of the antibiotic. (*Biology*)

- **Rio scale** – An analytical tool for assessing the veracity and significance of candidate SETI signals received on Earth. (*SETI*)

- **short-period comet** – Dynamical classification for comets which have periods less than 200 years. Jupiter-family comets likely originated from the vicinity of the Kuiper belt and their orbits became influenced by the planet Jupiter. Halley-family short period comets probably originated in the Oort cloud. (*Planetary Astronomy*)

- **RNA polymerase** – (RPASE). A polymerase that catalyzes the synthesis of a complementary strand of RNA from a DNA template, or, in some viruses, from an RNA template. (*Biology*)

- **rock** – A rock is a naturally-occurring aggregate of one or more minerals. Rocks are classified as igneous, sedimentary, or metamorphic. (*Geology*)

- **Rosetta** – European Space Agency (ESA) mission to explore comet 67P/ Churyumov-Gerasimenko. Launched in 2004, it will reach the comet in 2014. An orbiter will follow the comet during one year, and a lander (Philae) will perform in situ analyses on the comet nucleus surface. (*Astronomy*)

- **rotational spectrum** – When excited above their lowest energy state, rotating molecules emit radiation. This radiation is related to the mass and structure of the molecule giving a characteristic pattern of spectral lines. (*Physics, Chemistry*)

- **San Marino Scale** – An analytical tool for assessing the potential impact of Active SETI signals transmitted from planet Earth. (*SETI*)

- **scalemic** – Any non-racemic chiral substance. (*Chemistry, Biochemistry*)

- **scattering model** – Model calculated with radiative transfer theory to describe quantitatively the diffuse reflectance of (usually) sunlight from a planetary surface or atmosphere. Scattering models are used to calculate synthetic spectra by incorporating the spectral distribution of the illuminating light and the optical properties of the reflecting surface. (*Planetary Science*)

- **schmidt telescope** – Wide angle photographic telescope, using a spherical mirror with a correcting lens to increase the field of view. (*Astronomy*)
- **sedimentary** – Rocks composed of cemented (lithified) sediments derived from pre-existing rocks, or they may be directly precipitated from solution (*e.g.*, limestone). (*Geology*)
- **semimajor axis** – Average distance from the Sun for a given orbit. (*Astronomy*)
- **sensitivity** – An observational measure of how faint a signal can be detected. Improved sensitivity can generally be obtained by using a larger telescope, observing for longer times, or employing filters in frequency or time that exclude unwanted noise. (*Astronomy, Physics*)
- **serine** – One of the 20 amino acids that is a common constituent of many proteins. (*Organic Chemistry, Biology*)
- **serpentine** – A group of common rock-forming hydrous magnesium iron phyllosilicate minerals ($(Mg,Fe)_3Si_2O_5(OH)_4$). It may contain minor amounts of other elements including Al, Cr, Mn, Co and Ni. In mineralogy and gemology, serpentine may refer to any of 20 varieties belonging to the serpentine group. (*Geology*)
- **SETI** – An acronym denoting the Search for Extraterrestrial Intelligence. These are generally efforts to eavesdrop on radio or light signals sent by other intelligence in the cosmos. (*Astronomy, Astrobiology*)
- **sidereal** – time to make one complete orbit around the sun, relative to the background stars. (*Astronomy*)
- **siderophors** – A molecular receptor that binds and transports iron. (*Biochemistry*)
- **SIDP** – Stratospheric Interplanetary Dust Particle. (*Astronomy*)
- **silicate** – A class of minerals consisting primarily of silicon and oxygen and one or more metals (Ca, Na, Fe, Mg, etc.). Common silicate mineral groups include feldspars, pyroxenes, olivines, clay minerals, and quartz. (*Geology*)
- **SMD** – Space Mission Directorate NASA. (*Astronomy*)
- **snowline** – A region around the Sun or other star beyond which water freezes and cannot exist in liquid form. (*Astronomy*)
- **snowline** – The point at which ice and snow cover the ground year round. (*Geology*)
- **solar wind plasma** – Flow of particles consisting primarily of high energy electrons and protons that are ejected from the upper atmosphere of the Sun. (*Astronomy*)

- **solid state C-13 NMR** – This method is used to study the chemical composition of the solid organic materials, such as the insoluble organic materials on meteorites or from the Maillard reaction. (*Chemistry, Organic Chemistry*)

- **space weathering** – The interaction of the sun's plasma (solar wind) with the surfaces of airless bodies (*e.g.* asteroids, the Moon) that results in the transformation of the original minerals into glassy material with interstitial metal blebs (Fe-Ni). This process makes the reflectance spectra more flat and redder than it would be so that the asteroid no longer resembles its daughter meteorites. (*Astronomy, Cosmochemistry*)

- **spatial resolution** – A measure of the smallest separation between points that can be discriminated. Better spatial resolution permits an astronomer to study finer details on the sky and can be achieved by using a single telescope with a larger diameter, or an interferometer with the individual telescopes separated over a larger distance. (*Astronomy*)

- **spectral imaging correlator** – A piece of instrumentation used with an interferometer that multiplies together the signals from every pair of telescopes to form a radio-image over a field of view that is determined by the size of the individual telescopes and with a spatial resolution determined by the longest baseline between any pair of antennas. Every pixel of the radio-image will have a number of frequency channels that is determined by the complexity of the electronics in the correlator, but is typically on the order of 1000. (*Astronomy, Physics*)

- **spectral line survey** – Obtaining radio astronomical data over a wide frequency range from many individual observations. (*Astronomy*)

- **spectrometer** – A device for measuring the intensity of radiation absorbed, reflected, or emitted by a material as a function of wavelength. (*Physics, Remote Sensing*)

- **spectropolarimetry** – The wavelength dependence of polarization - spectroscopy combined with polarimetry. (*Physics, Astronomy*)

- **spectroscopic orbit** – The orbital solution for a star derived from the radial velocities along the line of sight to the star. The radial velocities are determined from Doppler shifts in the spectrum of the star. Normally the inclination of the orbit to the line of sight is unknown, unless there is an eclipse (transit) or astrometric information. Without the orbital inclination, there is an ambiguity for a spectroscopic orbit with low velocity amplitude: the companion may have very little mass, or the orbit may be viewed nearly face on. (*Astronomy*)

- **stable equilibrium** – A condition in which all acting influences are canceled by others, resulting in a stable, balanced, or unchanging system. (*Chemistry, Physics*)

- **starshade** – A 16 petal occulter developed by University of Colorado's Dr. Webster Cash. The starshade is used like a coronagraph in space, along with a space based telescope for the mission New Worlds Observer. This mission (which is still in proposal form) will diffract light in such a way that it deconstructively interferes with itself, enabling small terrestrial planets to be imaged and their atmospheres analyzed. This mission has the potential to detect plant life on an exoplanet! This mission will also discover and image a plethora of new exoplanets, providing scientist with valuable knowledge on the formation of planets and solar systems. (*Astronomy*)

- **stellar companion** – A memeber of a dual-star (binary) system. (*Astronomy*)
- **Strecker** – Refers to a method of synthesizing racemic mixtures of alpha-amino acids, as reported by Adolph Strecker (1822-1871). Starting materials are an aldehyde or ketone and ammonium cyanide (or NH_3 and either HCN or a cyanide salt). (*Chemistry*)
- **Strecker degration of amino acids** – One of the steps in the Maillard reaction (see Maillard reaction). (*Organic Chemistry*)
- **stress** – Internal distribution of force per unit area that balances and reacts to external loads applied to a body. (*Geology, Physics*)
- **sublimation** – Process of a solid phase transforming directly to a gas phase. (*Chemistry, Physics*)
- **sulfate** – A class of evaporite minerals composed primarily of sulfur and oxygen (SO_4), plus one or more metals (*e.g.*, Ca, Na, Mg, Fe, etc.). Epsomite, gypsum, and anhydrite are common sulfate minerals. (*Geology*)
- **super-earth** – A terrestrial world greater than 1 Earth mass but no larger than about 5 Earth masses. The physical characterization of such worlds will be a challenge for Astrobiology. (*Astronomy*)
- **supernova** – A stupendous stellar explosion – the death throes of a star much more massive than our sun that spews heavier elements needed for life out into the galaxy and that sends out shock waves that can trigger the birth of new stars. All elements heavier than Fe are made in supernova explosions. (*Astronomy*)
- **synodic** – The adjective synodic refers to the time needed for an object to return to the same position in the sky, as seen from the Earth. An object's synodic period is not the same as its sidereal period. For example, Jupiter returns to the same position in the Earth's sky in about 12 months (synodic period), but requires about 12 years (sidereal period) to orbit the Sun once and return to the same position relative to the Sun and other stars of the Milky Way. (*Astronomy*)
- **tectonic** – Relating to, causing, or resulting from structural deformation of the earth's crust. (*Geology*)

- **teraflop** – The acronym "flop" stands for a "floating point operation" of a computer, so a teraflop is a trillion (million, million) flops. One application in astrobiology is in the comparison of computer performance for data processing, such as in SETI programs. (*Mathematics*)

- **terrestrial** – Resembling the Earth, such as the rocky objects Mercury, Venus, Earth, Mars, and Moon, which are often referred to as terrestrial bodies. (*Planetary Astronomy*)

- **terrestrial** – Referring to the Earth, as in the phrase terrestrial life versus alien life. (*Astrobiology*)

- **terrestrial** – Associated with the land area of the Earth, as in the distinction between terrestrial and marine life. (*Biology, Geology*)

- **TES** – Thermal Emission Spectrometer aboard the Mars Global Surveyor spacecraft, which launched in November 1996 and arrived at Mars in September 1997. The TES interferometer measures within the thermal infrared, from 6-50μm ($\sim$1650-200 cm^{-1}) with 5 and 10 cm^{-1} spectral sampling. (*Remote Sensing*)

- **THEMIS** – Thermal Emission Imaging System, aboard the Mars Odyssey spacecraft which launched in April 2001 and arrived at Mars in October 2001. THEMIS combines a 5-wavelength visual system with a 9-wavelength infrared imaging system. (*Remote Sensing*)

- **theory** – (1) Conjecture, hypothesis, or speculation that is not factual in nature. (2) Body of knowledge or practice, as in music theory or number theory. (3) Broadly-based and well-supported scientific explanation; similar to a scientific law but usually less mathematical and of broader application, as in the theory of plate tectonics, theory of relativity, or theory of evolution.

- **thermal intertia** – A measure of how much heat the upper 10-15 cm of a surface can store during the day and re-radiate at night. It is primarily dependent upon the physical properties of the surface (*e.g.*, particle size, degree of induration, rock abundance). (*Remote Sensing*)

- **thioformaldehyde** – H_2CS. (*Organic Chemistry*)

- **tholin** – Complex organic solid produced by energy deposition in cosmically abundant materials. Titan tholin was first produced by corona discharge in a gaseous mixture of methane and nitrogen to simulate photochemical processes (UV photolysis and/or charged particle irradiation) in Titan's atmosphere. Titan tholin and other tholins consist of amorphous and unstructured carbon, carbon nitrides, aromatic structures with varying degrees of order, aliphatic structures (small side chains and bridging units), and a wide array of other molecular structures. (*Astrobiology, Organic Chemistry, Planetary Astronomy*)

- **tidal heating** – The heating of the interior of one planetary body caused by stresses induced from the gravitational pull of another. (*Astronomy*)

- **tidal locking** – Tidal locking makes one side of an astronomical body always face another, for example, one side of the Earth's Moon always faces the Earth. A tidally locked body takes just as long to rotate around its own axis as it does to revolve around its partner. This synchronous rotation causes one hemisphere constantly to face the partner body. Usually, only the satellite becomes tidally locked around the larger planet, but if the difference in mass between the two bodies and their physical separation is small, both may become tidally locked to the other, as is the case between Pluto and Charon. (*Planetary Astronomy, Physics*)

- **transcribed genes** – A gene is a section of DNA which contains the "instructions" on how to make a protein. Proteins, which include enzymes involved in metabolism and structural elements such as actin and myocin in muscle, do most of the work in a cell. When a protein is needed by the cell the appropriate gene is activated and its coding sequence is copied in a process called transcription, producing an RNA copy of the gene's information. This RNA can then direct the synthesis of the protein in a process called translation. A transcribed gene is one which has been activated to have an RNA copy made of it so its protein product can be made and used. (*Biology*)

- **transit photometry** – A technique for detecting and studying extrasolar planets. Measurements are made of the brightness of a star with suspected planets as the latter pass in front of (transit) the star. The measurements are compared to those taken when the star and planets are in different relative positions. (*Astronomy*)

- **transit spectroscopy** – A technique designed to study the atmospheres of extrasolar planets. Spectral measurements are made of a star with suspected planets as the latter pass in front of (transit) the star. The measurements are compared to those taken when the star and planets are in different relative positions. (*Astronomy, Physics*)

- **transiting planet** – If the orbit of the planet around its host star is viewed edge on, then the planet may transit between the observer and the star, producing a dip in the light signal. Transiting planets are important because they establish the inclination of the orbit to the observer's line of sight, which allows a determination of the actual mass of the panet. The amount of light blocked gives the size of the planet. Together the mass and radius give the density and thus information about the nature of the planet. (*Astronomy, Physics*)

- **transmembrane** – Passing or occurring across a membrane. (*Biology*)

- **trimethylsilyl** – Three methyl groups (*i.e.* carbon bonded with three hydrogens) bonded to a silicon atom ($-Si(CH_3)_3$). (*Organic Chemistry*)

- **true polar wander** – Reorientation of the rotation axis of a planet relative to the surface geography due to mass redistribution. To be distinguished from apparent polar wander, which is due to plate tectonics. (*Astronomy, Planetary Science*)

- **tryptophan** – An amino acid essential in human nutrition. It is one of the 20 amino acids encoded by the genetic code (as codon UGG). (*Biochemistry*)

- **UBVRI** – One of many standard photometric systems in astronomy. Typically defined by the bandpasses of filters, the U band is in the blue-ultraviolet part of the spectrum, the B-band is blue, V-band is visual (wavelength centered near 0.55 μm), R is centered toward the red end of the optical portion of the spectrum, and I goes into the near infra-red. (*Astronomy, Physics*)

- **unstable equilibrium** – An equilibrium state of a system in which any departure of the system from equilibrium gives rise to forces or tendencies moving the system further away from equilibrium. (*Chemistry, Physics*)

- **urea** – An organic compound with the structural formula H_2N-C(=O)-NH_2 and empirical formula CH_4N_2O. (*Organic Chemistry*)

- **urease** – An enzyme that catalyzes the hydrolysis of urea into carbon dioxide and ammonia. (*Biochemistry*)

- **UTTR** – Utah Test and Training Range. Military area in northern Utah and the landing site for the Stardust spacecraft. (*Geography*)

- **UV** – Ultra-violet electromagnetic radiation, whose wavelength is shorter than visible light and longer than x-ray wavelengths in the range between 10 nm to 300nm. *Astronomy Physics*)

- **UVB** – Electromagnetic radiation closely related to visible light, but with a shorter wavelength, usually taken as 280-320 nm. UVB is emitted by the Sun by mostly absorbed by the atmospheric ozone layer before it reaches the ground. UVB is strongly absorbed by DNA and by proteins, damaging them. The ozone layer can be damaged by anthropogenic compounds and indirectly by astrophysical ionizing radiation sources. (*Physics, Photobiology*)

- **variable stars** – Stars with brightness fluctuations due to, expansion and contraction, rotation, or being eclipsed by a nearby object. One class of interest to astrobiologists are the T Tauri stars, thought to be young objects. Their variations are due, at least in part, to dusty orbiting circumstellar material. (*Astronomy*)

- **virus** – A non-living microscopic infectious agent that is capable of reproducing only inside of a host cell. (*Astrobiology, Biology*)

- **volatile** – Molecule or atom that sublimates at relatively low temperature (*i.e.* cometary ices). *Cosmochemistry, Planetary Astronomy*)

- **voltammetric** – The measurement of the potential energy per unit charge. (*Chemistry, Physics*)

- **voltammetry** – Electrochemical technique applying a voltage potential to a working electrode and measuring current proportional to concentration of the analyte present. (*Chemistry, Physics*)

- **white dwarf** – The final stage of evolution for stars of low and medium mass (~0.5-9 solar masses). A white dwarf has a very high density since its mass can equal that of the Sun while its volume is comparable to that of the Earth. Type Ia supernovae are thought to arise from interactions involving a white dwarf that gravitationally attracts material from a nearby giant star. (*Astronomy*)

- **XANES** – X-ray Absorption Near Edge Structure. A type of absorption spectroscopy. XANES spectra indicate the absorption peaks due to the photoabsorption cross section in the X-ray Absorption Spectra. XANES was used to determine the nature of the chemical bonding of C, N, and O of organic material embedded in comet grains returned from the Stardust mission. (*Chemistry, Physics*)

- **zircon** – A mineral belonging to the group of nesosilicates. Its chemical name is zirconium silicate and its corresponding chemical formula is $ZrSiO_4$. Uranium is often a trace element in zircon, making these minerals value for the dating of the formation of the igneous rock from which they formed. (*Geology*)

- **zodiacal light** – The Sun's light reflected off of Solar System dust. The dust is primarily in the same plane as the planets and the result of asteroid-asteroid collisions and comet erosion. (*Astronomy*)

- **zwitterion** – A molecule that, while neutral overall, contains separate atoms with formal positive and negative charge. The zwitterionic character of molecules influences their chemical reactions. Amino acids are probably the most-important zwitterions for astrobiology. (*Chemistry*)

Author Index

General Index

ASTRONOMICAL SOCIETY OF THE PACIFIC

The Astronomical Society of the Pacific (ASP) is an international, nonprofit, scientific and educational organization. Some 120 years ago, on a chilly February evening in San Francisco, astronomers from Lick Observatory and members of the Pacific Coast Amateur Photographic Association-fresh from viewing the New Year's Day total solar eclipse of 1889 a little to the north of the city-met to share pictures and experiences. Edward Holden, Lick's first director, complemented the amateurs on their service to science and proposed to continue the good fellowship through the founding of a Society "to advance the Science of Astronomy, and to diffuse information concerning it." The Astronomical Society of the Pacific (ASP) was born.

The ASP's purpose is to increase the understanding and appreciation of astronomy by engaging scientists, educators, enthusiasts and the public to advance science and science literacy. The ASP has become the largest general astronomy society in the world, with members from over 70 nations.

The ASP's professional astronomer members are a key components of our Society's membership. It is their desire to share the rich rewards of their work with the public which permits the ASP to act as a bridge, explaining the mysteries of the universe. For these members, the ASP publishes the Publications of the Astronomical Society of the Pacific (PASP), a well-respected monthly scientific journal. In 1988, Dr. Harold Macnamara, the PASP editor at the time, founded the ASP Conference Series at Brigham Young University. The ASP Conference Series shares recent developments in astronomy and astrophysics with the professional astronomy community.